STUDY GUIDE

WARREN BURGGREN, EDITOR
University of North Texas

CONTRIBUTORS:

BRIAN BAGATTO
University of Akron

JAY BREWSTER
Pepperdine University

LAUREL HESTER
University of South Carolina

BIOLOGICAL SCIENCE

FOURTH EDITION

SCOTT FREEMAN

Benjamin Cummings

Boston Columbus Indianapolis New York San Francisco Upper Saddle River
Amsterdam Cape Town Dubai London Madrid Milan Munich Paris Montréal Toronto
Delhi Mexico City São Paulo Sydney Hong Kong Seoul Singapore Taipei Tokyo

Editor-in-Chief: Beth Wilbur
Acquisitions Editor: Becky Ruden
Executive Director of Development: Deborah Gale
Editorial Project Manager: Kim Wimpsett
Project Editor: Brady Golden
Executive Marketing Manager: Lauren Harp
Managing Editor: Mike Early
Senior Production Project Manager: Shannon Tozier
Production Service: S4Carlisle Publishing Services
Illustrations: Imagineering Media Services
Text Design: S4Carlisle Publishing Services
Cover Design: Riezebos Holzbaur Design Group
Cover Production: Seventeenth Street Studios
Manufacturing Buyer: Michael Penne
Text and Cover Printer: Edwards Brothers Malloy
Cover Image: Black Swan—*Cygnus atratus*, © Eric Isselée/Fotolia

ISBN 10: 0-321-56168-6; ISBN 13: 978-0-321-56168-8
ISBN 10: 0-321-87698-9; ISBN: 978-0-321-87698-0 (Books a la Carte edition)

Benjamin Cummings
is an imprint of

www.pearsonhighered.com

1 2 3 4 5 6 7 8 9 10—EBM—14 13 12
Manufactured in the United States of America.

Contents

Preface

Introductory textbooks are written to provide a foundation for successful learning. In addition to the textbook, students also make use of supplementary learning materials associated with their text, typically including CDs, websites, and study guides. In our experience, a study guide can be a wonderful asset to students, helping them assess their progress and move to the all-important stage of thinking critically about the material they are learning. Unfortunately, it is also our experience that study guides are sometimes created as a hurried afterthought—simply a repackaging of the text material along with a series of multiple-choice questions to determine if biological facts and principles have been learned.

We wanted to do something different—and something far better. So, while Scott Freeman was writing the first edition of *Biological Science*, Brian Bagatto, Jay Brewster, Laurel Hester, and I were discussing with the then Project Manager, Karen Horton, how to produce a more interactive, effective, and enjoyable study guide that would be an important component of the *Biological Science* support package. As the first edition of the study guide emerged, we realized that it would be helpful to do more than just review the material found in the textbook. We also wanted to put together guidelines for how to study biology, how to write in biology, and how to understand why biological experiments underpin everything the student learns. Consequently, Part I of the book, unique among study guides, emerged as a "survival guide" for your biology course. Part II subsequently delved into the actual material presented in *Biological Science*, but in a way that challenges students to think critically and analytically about the material they are reading in the text and hearing in their lectures.

This fourth edition holds true to our original formula. Each chapter has a similar format to aid your learning, beginning with an outline of "Key Concepts" that summarizes the concepts discussed in the chapter. Woven through this outline are boxes entitled "Check Your Understanding," essentially self-tests where you can write down brief responses to questions to give you a sense of your own understanding of the material as you progress through the chapter. A section entitled "Integrating Your Knowledge" ties together concepts throughout the book, followed by a "Mastering Challenging Topics" section that walks you through the most difficult topics and concepts. Finally, a comprehensive section called "Assessing What You've Learned" tests your understanding of the material and helps you prepare for exams. Thus, each chapter provides a formula for successful learning, knowledge, and retention.

With this fourth edition of the Study Guide, we feel that we have fine-tuned the concept of the Study Guide as an important tool for mastering a course in biological sciences. Together with the fourth edition of Scott Freeman's *Biological Science* and the other support material available to the student and instructor, we hope this Study Guide will help instill our own love of biology, and biological experimentation, into future generations of biology students.

Our special thanks go of course to our Project Managers, Kim Wimpsett and Brady Golden, without whose enthusiastic support this project would not have been possible. We also thank the team of editors and other specialists at Benjamin Cummings. Kirsten Balayti copyedited this Study Guide, bringing to bear her language skills and attention to detail. The team at S4Carlisle Publishing Services also expertly formatted the pages.

Finally, we are extremely grateful to all of those important people in our lives—family, friends, and students—who put up with our temporary neglect while we focused on the creation of this fourth edition of the Study Guide.

Warren Burggren
Denton, Texas
February 2010

PART 1

Tools for Learning

Chapter 1

Experimentation and Research in the Biological Sciences

Experimentation is the process in which, under controlled conditions, we alter the course of an activity, process, or event to test a hypothesis and gauge the outcome of that alteration. Experimentation differs from observation in that the former involves active steps taken by an experimenter to change an event or process, and in doing so to understand its nature. Experimenting seems to be as innate to humans as breathing or eating. We have always tinkered with the world around us. In the early days of *Homo sapiens*, or humans, the tinkering may have had immediate, practical purposes—does the wood from the tree with the heart-shaped leaves burn hotter and longer in a cooking fire than the wood from trees with needles? Does an arrowhead with carefully sculpted edges bring down a deer more reliably than a simple triangular arrowhead? Finding answers to these questions by trying different approaches and learning from the outcomes was often a matter of life or death, so it is hardly a wonder that the tendency for "experimentation" appears to have been naturally selected for in human genes.

Modern humans carry out experimentation in the highly organized format we call *scientific research*. Some of the experimentation and research we carry out continues to be "life or death," such as the ongoing search for an HIV vaccine, or better screening procedures for various types of cancers. Such research is called *applied research* because of the hope that there may be an immediate application of the results of such experiments. Much of the current experimentation in biomedical research is considered to be applied research. However, a great deal of experimentation and research is considered to be *basic research* (sometimes called *curiosity-driven research*) that provides basic scientific knowledge and a foundation for applied research.

Descriptions of both basic and applied research form a central, integrated theme in Scott Freeman's textbook,

Biological Science. Rather than just asking you to memorize "facts," *Biological Science* offers important insights into the generation of the fascinating information you will be learning in this course, and it should leave you with a new appreciation for how scientific information is gathered.

In Part 1 of your study guide, we begin in **Chapter 1** by considering the process of experimentation and research that leads to the creation of new data in an attempt to answer questions posed by scientists. Then, in **Chapter 2** of Part 1, we will consider how data are tabulated and graphed for the most efficient communication of the patterns they portray.

A. KEY CONCEPTS OF EXPERIMENTATION

Setting up and carrying out successful experiments depends on careful organization and thought. Let's discuss a few important concepts of experimentation:

- Hypothesis testing
- Independent and dependent variables
- Experimental controls
- Precision and accuracy
- Consistent treatment of experimental populations

1.1 Hypothesis Testing

In **Chapter 1** of *Biological Science*, Scott Freeman discusses the creation of a hypothesis. His example stems from a question about why giraffes have long necks (see pp. 9–10 of your text). A hypothesis is a statement that is proposed to explain a certain set of events, which can be tested and proved or disproved. You might alternatively think of a hypothesis as a theory waiting to be put to the test. *Biological Science* poses the question, "Why do giraffes have long necks?" One hypothesis to explain this observation is: "The long necks of giraffes enable them to reach food that is beyond the reach of other mammals." In other words, when a scientist poses a hypothesis as a statement, he or she is initially "taking a position" when trying to answer the question. The scientist may have no idea what the answer to the question may be, but still suggests a possible explanation in the form of a hypothetical, positive statement. This isn't arrogance or wishful thinking—it's just the process of testing hypotheses.

Once a hypothesis is put forth, the next step is to design an experiment to test that hypothesis. The experimental design should be such that the experiment's results will either prove the hypothesis to be valid—the results support the proposed explanation—or the results will disprove, or falsify, the hypothesis. Proving a hypothesis is often a matter of gathering so much evidence in support of a hypothesis that any reasonable person would believe it to be proven. In reality, falsifying a hypothesis is often easier. The philosopher Karl Popper (1902–1994) explained the concept of falsifying hypotheses with an example about ravens. If researchers propose the hypothesis, "All ravens are black," then they might set out to collect as many black ravens as possible. After collecting hundreds of black ravens, they may believe the hypothesis to be true (that is, to be "proved"). However, the researchers can't be sure that there isn't a white raven out there somewhere! In contrast, if they set out to find a *white* raven and succeed, then the hypothesis that "all ravens are black" is completely and efficiently falsified. The researchers might then suggest the new hypothesis that "All ravens are black except for a very, very small population of white ravens."

Falsifying a hypothesis leads to the creation of a new and better hypothesis that takes new results and observations into account. When a hypothesis has undergone numerous modifications and can no longer be falsified, then scientists may permit themselves to refer to the hypothetical statement as a fact. When a fact or body of facts has stood the test of time, it may become a law.

✓ **CHECK YOUR UNDERSTANDING**

In *Biological Science*, Scott Freeman describes many interesting experiments performed by biologists. Turn to page 218 in Chapter 11 (How Do Genes Work?). Here, Freeman describes an experiment in the early 1940s by Geadle and Tatum, who investigated the source of hereditary material. Can you figure out what the original hypothesis proposed by the authors might have been? Try doing the same for the experiments by Hardy and Weinberg on allele frequencies that Freeman describes on page 436 of Chapter 25 in your textbook.

1.2 Independent and Dependent Variables

All experiments, and the tables and graphs that present the information from these experiments, possess both *independent* and *dependent* variables. It is crucial that you understand the difference between them. The independent variable is typically the variable that is controlled, or fixed, by the scientist. For example, consider a plant biologist who is interested in the adaptations of a certain species of cactus to the dry conditions of the Mojave Desert. (You should pay particular attention to this example, because we will use it repeatedly in this and the next few chapters.) The biologist wonders whether the amount of water available to cactus seedlings affects their growth rate. To test this, she might take cactus seeds from native plants and grow cactus seedlings in a greenhouse (**Figure 1.1**).

FIGURE 1.1   In this experiment carried out by a plant biologist, five groups of cactus seedlings are grown in a greenhouse and provided with different amounts of water each week.

Here are the experimental conditions: All the seedlings are exposed to a constant 18 hours of light. The amount of water that each pot receives is controlled as part of the experiment. Different pots of cactus seedlings are given different amounts of water, and the amount of growth stimulated by each watering regime is carefully measured and recorded each week. The amount of water given to the seedlings is the independent variable, because the amount of water the seedlings receive is fixed and independent of any physiological response the seedlings might have to the water. Their growth in response to watering is the dependent variable because the amount they grow more or less *depends* on the water they receive during the experiment.

The design of an experiment can be simple, as in this case involving cactus seedlings, or it can be very complex (for example, simultaneously varying water, nutrients, light, and genetic stock of the cactus and tracking any effects on seedling growth—an experiment with four independent variables). Nevertheless, in every experiment, you should be able to identify a dependent variable (the variable that unpredictably changes) and one or more independent variables (controlled as part of the experimental design). Once you have sorted out the meaning of dependent and independent variables, it will be easier to understand how tables and graphs present their information.

Find a few descriptions of experiments in *Biological Science*, and see if you can identify the independent and dependent variables in each.

1.3 Experimental Controls

Every student taking this course has carried out an experiment—both formally and informally. Your formal experience might include experiments in other college or high school science courses. Informally, think of the experiments you carry out every day. For example, have you ever tried to study less and still get the same grade, or tried to study more to get a better grade? Have you ever fiddled with a recipe to see if you can improve it? Have you ever checked to see whether you get better gasoline mileage with one brand of gas over another? These examples of real-life "tinkering" are actually experiments producing data.

While we all may be experimenters in our own right, what is often lacking in our informal experiments is a built-in control. A control is the portion of the experiment in which all possible variables are held constant, contrasting the portion of the experiment in which the variables are altered to observe the resulting effects. The presence of a control ensures that any effects observed in the experimental portion of the experiment can be attributed to the altered variable. For example,

you can change a recipe for cookies by adding more chocolate chips, but you don't know if the cookies taste better due to the extra chips or because you are especially hungry when you taste them. What this experiment needs is a *control*, which would involve baking one batch of cookies using the normal recipe side by side with another batch using more chocolate chips and then tasting the two batches at the same time.

Many otherwise good experiments can suffer from a lack of adequate controls. Controls help a scientist draw valid conclusions from the observed results of an experiment, which is why trained scientists are scrupulously careful to create and include controls with every experiment. For example, a team of researchers trying to test a new drug they are developing will test the drug on one experimental population of animals or patients while simultaneously giving a second, equivalent population a *placebo*, or inactive drug treatment, as the control. Any effects subsequently observed in the experimental population can then be attributed to the drug being tested.

To help develop the concept of controls a little further, let's return to our example about cactus seedlings. The average amount of water that falls in the seedlings' natural habitat during the period when cactus seeds might sprout is about 5–10 millimeters per week (mm/week). This quantity translates into about 20 milliliters (ml) of water per week, allowing the experimenter to give some of the seedlings the same amount of water they would normally receive in the wild. These plants constitute the control population for this experiment.

The experimental populations are those that receive more or less water than is given to the controls. (By the way, the term *population* is frequently used to describe a group of organisms, usually of the same species.) By setting up the experiment in this way, the plant biologist can be sure to take any extraneous or unexpected factors into account. For example, perhaps the amount of nutrients or light that all of the plants receive in the greenhouse is different than the amount they would normally receive in the wild. The biologist is still able to test the single effect of different amounts of watering by comparing all data to the control population. Returning to our example of changing a cookie recipe, this is similar to baking two batches of cookies—one the control batch and one the experimental batch.

Even if you are ravenous when you taste the two different populations of cookies, you still can make a comparison between the two recipes.

All of the experiments that Scott Freeman mentions in *Biological Science* involve careful use of controls, enabling them to survive scientific scrutiny to become landmark studies.

1.4 Precision and Accuracy

We often use the terms *precision* and *accuracy* fairly interchangeably; yet in the context of scientific experiments, they mean two quite different things. Let's consider a vegetable scale at the grocery store. Is it accurate, precise, or both? Let's say you place a bunch of broccoli on the scale and observe that it weighs 2.0 pounds. You then weigh the same bunch of broccoli five more times, and it weighs 2.0 pounds each time. In this case, we can say that the scale is "precise" because it repeatedly gives the same weight for the same object. But let's also assume that the broccoli *actually* weighs 1.7 pounds. In this case, the scale—while precise because it repeatedly gives the same weight of 2.0 pounds—is not accurate. It would be accurate only if it indicated that the broccoli weighed 1.7 pounds, and it would be

accurate *and* precise only if it repeatedly gave a weight of 1.7 pounds (**Figure 1.2**).

Scientists carrying out experiments must be both precise and accurate with their measurements. For this reason, experiments involving measurement with instruments (and most do) also involve the process of calibration. This means that an instrument is adjusted until it reliably produces the accurate measurement of a known standard. For example, rather than simply trusting the reading on a thermometer, a scientist will compare the temperature it reads against a known temperature (or a previously calibrated thermometer). Simple mercury thermometers can't be adjusted, but electronic ones can be adjusted to read the correct temperature. Ultimately, the accuracy of data generated depends on how carefully the scientist has calibrated the measurement devices used to collect data. The data on cactus seedling growth, for example, will be useful only if the ruler used for measuring plant height is accurate, the beaker used for measuring water is accurate, and the measurements and water amounts can be carried out precisely.

1.5 Consistent Treatment of Experimental Populations

When performing experiments, it is important to ensure that all experiments are carried out in exactly the same way every time. Otherwise, results may be due to

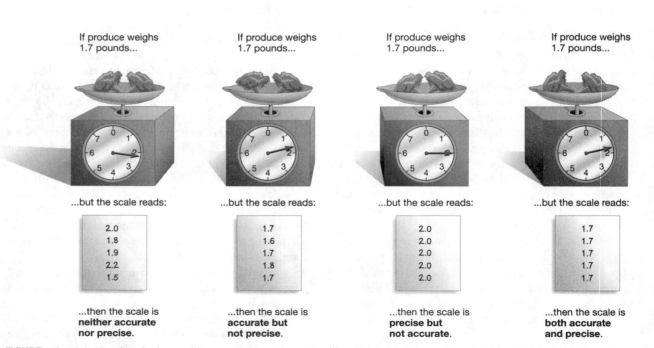

FIGURE 1.2 A measuring device such as a grocery scale can be (1) neither precise nor accurate, (2) accurate but not precise, (3) precise but not accurate, or (4) both precise and accurate.

experimental error and may not accurately reflect the results of the experiment. For example, in the case of the cactus seedlings, if some plants are measured in the morning and some in the late afternoon, wilting in the late afternoon sun might cause some plants to measure smaller than they actually are. If some plants are watered on Mondays and some on Tuesdays, some measurements taken Monday afternoon would be made on recently watered, non-wilted plants, while some plants measured will be dehydrated and wilted. Any height differences will be due to the day on which the measurements were taken rather than to how much water they received weekly.

✔ **CHECK YOUR UNDERSTANDING**

Let's imagine that you are carrying out an experiment to see if studying for an extra night each week will improve your grade point average. For the first two weeks, you study on Tuesday night in the quiet library. But in the following two weeks, you study on Friday night in your dorm room with the sounds of loud music and parties all around you. When you tally the scores from all of your quizzes, exams, and assignments for the month, you find that your GPA remains unchanged. Is there something wrong with your experimental design, or does studying just not matter?

B. SETTING UP AN EXPERIMENT

Scientists may spend weeks, months, or even years planning an experiment, and they may face limitations due to lack of manpower, funding, or equipment. Some experiments are relatively simple to set up and require little maintenance. Our plant biologist's experiment with cactus seedlings, for example, is fairly straightforward, requiring only that the biologist obtain the seedlings, plant them, water them regularly according to the experimental design, and record their growth. In contrast, research into the effects of a new drug on the heart rate of animals and humans may require the combined efforts of numerous graduate students, postdoctoral fellows, research scientists, and clinical staff, as well as large amounts of expensive equipment—and may take several years to complete. Yet, despite great differences in effort and expense to carry out research experiments, the processes of generating a hypothesis, testing the hypothesis by performing experiments, and analyzing the data are common to all good scientific experimentation.

✔ **CHECK YOUR UNDERSTANDING**

Try setting up your own hypothetical experiment using this scenario: Big Coal Company needs to build roads through some wilderness areas in its search for fresh coal fields. These roads will cross some streams that are important migratory routes for spawning salmon as they swim upstream. Thus, Big Coal needs to install culverts under the new roads as they are built to allow the water to flow under them. The project's chief engineer wants to install culverts with a 24-inch diameter, which will produce a water velocity of about 25 centimeters per second (cm/sec). The biological consultant hired by Big Coal fears that the fish in the stream won't be able to swim upstream against such a strong current and thinks they should use more expensive culverts with a 36-inch diameter, producing a water velocity of only 10 cm/sec. Can you design an experiment that would resolve this conflict? Construct a formal hypothesis. Think of the materials and supplies you would need, assuming that all costs will be covered by Big Coal Company. What would be the experimental procedure? What type of data would you collect? How would you present it to the engineer to make your case?

C. HOW MUCH DATA SHOULD BE COLLECTED?

You are probably starting to realize that biological research can be an elaborate and time-consuming enterprise. Poorly planned experiments can result in either an insufficient amount of data for accurate analysis and robust conclusions or the collection of more data than necessary, which is a waste of both time and money.

All scientific data—especially biological data—have inherent variation. The most important sources of this variation are:

- **Variation from inherent differences**—natural variation in the samples being examined. For example, some cactus seedlings will grow more rapidly than others simply because of their genetic disposition.

- **Variation from measurement error**—every scientific measurement has some inherent error. Different people measuring one seedling's height will produce different values, and the same person measuring one seedling's height at different times will produce different values. (Remember our earlier discussion of precision and accuracy.)

Let's return to the experiment with cactus seedlings. How many seedlings do you think the plant biologist should use? Should she set up just five flower pots with one seed each, giving each pot a different amount of water? This would be the easiest design, as well as simplest experiment to care for. Should she set up 20 pots? How about 100 or 500 pots? What criteria should she use to decide the size of her experiment?

Biologists typically spend a lot of time thinking about exactly how many replications—repetitions of an experiment—to carry out. Scientists want to ensure that the measurements they make are as accurate as possible, and one way to ensure accuracy is to take the average of several replicates. Of course, some experiments are relatively easy to replicate—for example, repeatedly measuring the height of a cactus seedling in a flower pot or growing several cactus seedlings rather than one. Consider, however, measuring the compression strength of the bite of a great white shark—and trying to get replicate measurements of the *same* great white shark, as well as several others in the same size range!

A variety of factors affect the design of an experiment with replicate measurements. Assuming that a researcher can manage the practical limitations of time and resources, the most important factor to consider when designing an experiment is the statistical reliability of the data produced. Statistics is a vibrant subbranch of mathematics, and to discuss the numerous statistical methods used to test the validity of biological research would be beyond the scope of this study guide. However, two simple and useful conventions to learn are the mean and the standard deviation. The mean is the average of all data points collected, and the standard deviation of the mean describes the variation in the data about the calculated mean value in statistical terms.[1] A large variation about the mean (that is, a large standard deviation) indicates the possibility that the data is inconclusive.

To ensure that statistically sound conclusions can be made about our example experiment, the plant biologist might decide to use one seedling in each of five flower pots at every watering amount, for a total of 25 flower pots (**Figure 1.1**).

This experimental protocol, or plan for the experiment, will allow the biologist to calculate for the five flower pots mean values and standard deviations that represent a single watering amount. Without going into statistical detail on the methods used for calculation, taking five accurate measurements (with replicate measurements taken on each seedling) often proves sufficient to determine statistical plausibility.

Carry out a simple experiment to illustrate measurement error. Using a ruler, tape measure, or even the tiles on a floor, measure the length of your normal walking stride. Now estimate the length of a corridor, the width of a small field, or some other distance by counting the number of paces you take as you walk from one end to the other. Do this five times. Ask five of your friends, relatives, or roommates each to estimate the length in question using the same method. Compare the results of your measurements. What would be the most *precise* way to measure this distance? What would be the most *accurate* way to measure this distance? If you owned a land surveying company and your reputation depended on accurate measurements, would you have a team of surveyors carry out the measurements, or would you ask one surveyor to repeat his or her measurements more than once?

In our example experiment (but not in an experiment involving bite strength of the great white shark!), it is a relatively small investment of time and money to compensate for error by growing additional seedlings. Thus, the biologist may elect to grow 10 seedlings at each watering level to ensure against the unexpected death of a few seedlings or other possible disasters, such as a dropped flower pot.

D. ANALYZING AND INTERPRETING DATA

Analyzing data is a crucial step in the experimentation process. It's not enough to simply produce and record data; a scientist must then attempt to make sense of the gathered information to decide whether to accept or reject the hypothesis. You have already undergone similar

---

[1]Biologists use several conventions to describe variation—standard deviation, standard error, confidence intervals, and so on. Calculating each of these can be complex, so a rule of thumb is useful. Generally, if the standard deviations about two means substantially overlap each other, then the two means may not be statistically different.

analytic procedures in your daily life. For example, you may have had a feeling that you are spending more money than you have—a thought that is really an informal hypothesis in disguise. You may then balance your checkbook in an attempt to learn how much money you have, but you shouldn't stop there—you should also compare your monthly balance with your monthly income to see whether you are saving money or falling deeper into debt.

Once a scientific researcher has acquired enough data, he or she will begin to use the individual "trees" to see the overall "forest"—that is, to use the data to recognize broader patterns. Typically, this will be done either by putting the data in tables or by plotting it in graphs. In fact, tables and graphs are such an important part of the data analysis and presentation process that **Chapter 2** in Part 1 of this study guide is devoted entirely to the topic of scientific data presentation. In a curious way, analyzing scientific data is like solving a word puzzle in which the letters of words are scrambled and must be rearranged to form pieces of words before whole words begin to take shape. Scientific data alone are like the scrambled letters—a long and tedious process of interpreting the data, comparing it with previous experiments on similar topics, and fitting the outcome into the hypothesis formulated at the beginning of the experiments is required before the scrambled letters begin to resemble words. The truly great scientists of our era are able to put their data in the broadest possible context, so not only do their experiments allow them to accept or reject their own hypotheses, but their conclusions lead to advancements in the field of biology as a whole through insights gained from their experiments.

E. PUBLISHING DATA

The final step in experimentation is publishing the data and conclusions. Scientists receive financial and intellectual support for their work not just to indulge their own curiosity, but to encourage the results of their experiments to be distributed throughout the scientific community. Most journals subject submitted articles to rigorous peer review by other scientists in the same area of research. A manuscript is published only after intense scrutiny and often several rounds of revisions.

Scientific data has many outlets, most of them in the form of scientific journals. There are thousands of scientific journals, with titles and topics as broad as *Science* or as focused as *Pediatric Developmental Pathology*. Your institution's library probably has a large section devoted to scientific journals of all kinds. Of course, as electronic publishing on the Web has become commonplace in the last several years, most journals now publish in both traditional paper format and on the Web for subscribers.

From the inception of a question posed as a hypothesis through the gathering of resources, the completion of experiments, the analysis and interpretation of data, and finally the publication of results, the overall scientific endeavor can often take years and many thousands of dollars. The next time you are in your library, walk down the aisles of scientific journals and think of the enormous investment of time and resources that has produced our scientific knowledge base. Appreciate the fact that every page of Freeman's *Biological Science* is based on information gathered through careful scientific experimentation.

CHECK YOUR UNDERSTANDING

Go to your library's journal section and look up a biological research journal such as *Journal of Experimental Biology* or *Physiological and Biochemical Zoology*. Find an article about experimentation with "zebra fish." Now, use your Web browser to find that same journal on the Internet, and again try to find an article involving zebra fish. Which format do you find most convenient? Which is easier for finding articles? Which format is better for viewing figures (tables, graphs, etc.)? What do you think will be the future of the electronic versus paper publishing of scientific information? Ask your biology professor to comment about his or her preference for scientific information.

All science, including biological science, depends on the clear, concise presentation of data. Let's turn to that topic in the next chapter.

Chapter 2

Presenting Biological Data

*In **Chapter 1** of Part 1, we considered how experiments are set up, carried out, and to some extent how they are analyzed. In this chapter, we will focus on ways in which the information produced from such experimental procedures is presented and interpreted.*

A. INTRODUCTION TO SCIENTIFIC DATA

Scientists thrive on data, the information gathered from scientific observation or experimentation. They think about it all the time—what type of data they need, how much data they need, how they can get it, and what it might indicate once they acquire it. Whether they are taking the temperature of the lava from a smoldering volcano or measuring the rate of oxygen consumption in a growing zebra-fish embryo, all scientists rely on data.

Data are the currency of science, and learning to interpret the many ways it can be presented will be an important part of your developing your science literacy. But do not think that only scientists deal with data, or that its interpretation is a completely new skill, because you have been handling data in many everyday situations for years! Try to imagine the data you encounter regularly; you might think of the money in your checking account, your credit card balance, your grade point average, or even your Facebook friends. The data on the monthly statement from your credit card company, for example, tracks the date and amount of your purchases. Interpreting this data involves determining if you are over your limit and budgeting your future credit card purchases to reflect your income.

As our world grows more technologically advanced, it is becoming increasingly important to understand complex forms of data and to apply the knowledge gained from these data. Drawing accurate conclusions from various forms of data presentation will do more than help you correctly answer questions on your biology tests; the critical thinking and analysis skills you develop will help you excel in other aspects of your education as well as throughout your life.

This chapter introduces the typical ways in which biologists and indeed all scientists use, present, and interpret data—methods you are also likely to encounter in your classroom and laboratory.

Scientists generally present their data in one of several forms:

- Writing
- Tables and graphs
- Simulations and animations

Let's discuss each form in greater detail.

B. PRESENTING SCIENTIFIC DATA IN WRITING

You're probably familiar with the saying, "The pen is mightier than the sword," meaning that more can be achieved by effective written communication than by legions of armies. For thousands of years, learned people have recorded information (we might call it data) about their observations of the natural world. For example, Aristotle recorded his observations of the beating heart of a chick embryo, which he exposed by carefully making a small window in the shell of a chicken egg. Socrates left similar written observations of the chick embryo heart, as did Galileo. Since that time, other writings have left a rich legacy of knowledge about the anatomical and physiological development of the chick embryo, which now serves as an important animal mode for studying a variety of cardiovascular diseases. These contributions could not be understood without the clear, concise writing that describes the observations—the data—of each writer. Today, writing remains a mainstay of scientific communication. In *Biological Science*, Scott Freeman writes about the great wealth of biological information supplemented by a multitude of tables and graphs.

You have already used writing to convey scientific information—think of the lab reports, essays, and book reports you have produced so far in your academic career. While you may already understand the components of good scientific writing, **Chapter 4** in Part 1 of this study guide, "Reading and Writing to Understand Biology," explores these components in considerable detail. It is important to remember that the ability to write clearly is not limited to those who want to become scientists. Those of you studying for careers in business, education, or virtually any other discipline will be required to communicate your ideas and data to others, and your study of the presentation of scientific information will help you successfully communicate all types of data.

C. PRESENTING DATA IN TABLES AND GRAPHS

In the previous section, you recalled past lab reports from science courses such as biology, chemistry, and physics. Did you ever use tables or figures to help explain the data you produced? While the pen may be mightier than the sword, you are probably also familiar with the phrase, "a picture is worth a thousand words." A table or a graph is a picture of scientific data. Though you can describe your favorite summer vacation to your friends by talking to them, you may want to take out your photo album as well, using the pictures to supplement your stories. The combination of the pictures and your verbal description provides the most complete communication of your ideas. In the same way, a scientist explaining his or her experiments or an author describing biological principles will almost always resort to a combination of words and visual tools, such as tables and graphs.

Tables and graphs are similar in that they are visual, highly structured ways of looking at data, often displaying independent and dependent variables as discussed in **Chapter 1** of Part 1.

2.1 Tables in Scientific Writing

Tables are such a convenient way of communicating data that they can be found all around us. As you read this book, you may have a beverage or a snack close by. Take a minute to look at the table on the packaging. It is probably a relatively simple table showing nutrition information such as caloric content and the amount of fat, cholesterol, sodium, and so on found in the product.

While the packaging could describe this information with words, a simple table provides the same information more efficiently. To consider various table formats, let's return to the example of cactus seedlings introduced in **Chapter 1** of this part. We could describe the experiment in paragraph form, but consider the following table:

Cactus Group (5 pots/group)	Amount of Water Given to Seedlings (ml/week)	Average Seedling Growth (mm/week)
Group 1	10	3
Group 2	20	7
Group 3	30	13
Group 4	40	10
Group 5	50	5

A substantial amount of information about the experiment can be gathered from this table, such as:

- There are five cactus groups represented in the experiment.
- There are five pots in each group.
- A standard watering procedure is shown, in which the most-watered group of seedlings received five times more water than the least-watered group.
- The growth column indicates that all of the seedlings grew to some extent.
- Seedlings receiving a certain amount of water grew the most; seedlings receiving more or less water than this amount did not grow as well.

Tables can be relatively straightforward, as in this example, or relatively complex, with many columns and many rows. Regardless of layout, however, all tables will contain independent and dependent variables.

 CHECK YOUR UNDERSTANDING

Record some of your observations about this table. What does it show you about the experimental design, and what general trends do you notice?

2.2 Graphs in Scientific Writing

Although tables can be extraordinarily useful to organize data, graphs are also used extensively to communicate this information. If you pick up a newspaper (especially *USA Today*, which uses more graphics than many other national publications), you will probably find several examples of graphs (look in the "Business" section, where you may find the recent performance of the stock market presented both in a table and a graph).

Look through *Biological Science*, how many different types of graphs can you identify?

There are many different types of graphs, each with advantages and disadvantages that suit it for a particular kind of data. Most graphs have another feature in common besides dependent and independent variables. They all have an x-axis, also called an abscissa, and a y-axis, or ordinate. The x-axis runs horizontally along the bottom of the graph, while the *y*-axis runs vertically (**Figure 2.1**). Typically, the x-axis is used to plot the independent variable (for example, watering amounts for our seedlings), while the dependent variable (seedling growth) is plotted on the y-axis.

Let's identify the specific types of graphs you will encounter in *Biological Science*. We'll continue to use the example of cactus seedlings and growth, because you are now familiar with these data in tabular form.

Scatter Plots

A scatter plot is aptly named; each data point is individually represented on the graph (in contrast to graphs in which only the averages of several data points are recorded) (**Figure 2.2**). It may appear that data are simply scattered on the graph. The scatter plot in **Figure 2.2** presents the growth data from every pot in the cactus seedling experiment.

Looking at the scatter plot again, you should be able to see that, at each amount of water, there is variation between individual pots. Further, notice that the amount of variation differs for each amount of water, with peak growth when seedlings are given 30 ml of water per week. Scatter plots are very useful in that variation and trends in the data can be examined simultaneously. Sometimes,

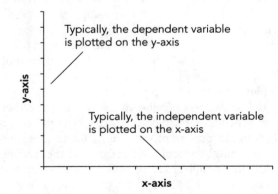

FIGURE 2.1 Graphs typically have an *x*-axis, showing the independent variable controlled by the researcher, and a *y*-axis, showing how dependent variable values change as the independent variable is altered.

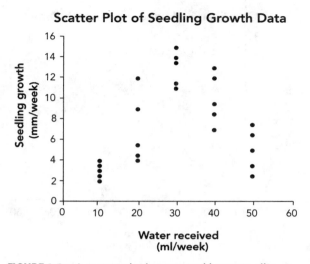

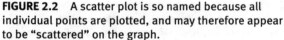

FIGURE 2.2 A scatter plot is so named because all individual points are plotted, and may therefore appear to be "scattered" on the graph.

What information or patterns can you identify from **Figure 2.2**? Did the amount of water received by a group make a difference? Did each individual plant respond the same in each group? Remember that the control group received the same amount of water as would be received in the wild (20 ml/week); did plants receiving different amounts of water display different responses than the control group? What was the range of weekly seedling growth at 30 ml of water per week, and what was the average response at that level?

however, scatter plots are awkward because they can contain too much data, which makes the detection of patterns in the data more difficult. For this reason, scientists sometimes choose to communicate their data using line graphs.

Line Graphs

A line graph does not show individual data points, but instead shows the averages of several data points, connected by one or more lines that best fit the data points. In other words, line graphs appear to "connect the dots" of a scatter plot. The line graph in **Figure 2.3** displays the average seedling growth for each group at each watering.

This line graph clearly shows the pattern that is only hinted at by the scatter plot; namely, that seedlings do poorly with little water, but they are also adversely affected by too much water. The graph indicates that an intermediate watering amount, slightly more than the seedlings would receive in their natural habitat, is just right for optimal growth.

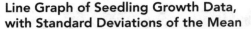

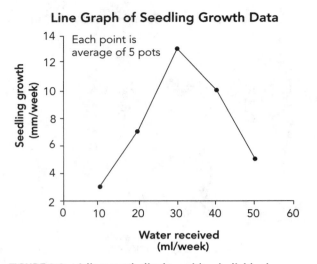

FIGURE 2.3 A line graph displays either individual or mean values of a dependent variable against the independent variables and, as the name implies, joins these points with a line to help reveal overall trends.

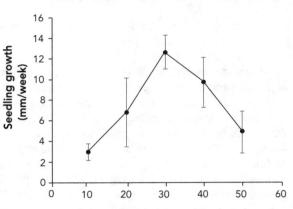

FIGURE 2.4 Line graphs may also show how the data varies about the plotted mean values, as in this line graph showing the standard deviation of the mean for each mean seedling growth value.

Look at some graphs in a printed version of a newspaper or on a news website. Which graph type is more common, scatter plots or line graphs? Look at examples of each—which format works better to communicate the data being presented?

While line graphs communicate trends very clearly, they often do not provide any sense of variation in the data. This problem can be minimized by plotting one of several kinds of statistical variables to describe variation in the data. The most common of these variables is the standard deviation of the mean, which is defined in **Chapter 1** in Part 1 of this study guide. **Figure 2.4** is a replicate of the first line graph we looked at (**Figure 2.3**) with the standard deviation now added. The vertical line drawn through each point represents the amount of variation in the independent variable at each dependent variable value.

CHECK YOUR UNDERSTANDING

Go back to your newspaper or news website. Find a simple line graph with no standard deviations. Without seeing any indication of the variability in the data, can you tell if the trends depicted are "trustworthy"?

Bar Graphs

Scientists often illustrate data with a bar graph in which the height of the bar represents the value of the dependent variable. Bar graphs can be plotted either horizontally or vertically, and there isn't much difference between the two—aesthetics and a sense for which communicates the data more effectively are often key factors when deciding whether to orient a bar graph horizontally or vertically. The two bar graphs presented plot the cactus seedling data in each form (**Figure 2.5**).

Just as line graphs can show variation in data, bar graphs can also be designed to show standard deviations (**Figure 2.6**).

Bar graphs are often most effective when showing the results of an experiment that compares several different variables, or when comparing several different experiments. On a line graph, separate lines often fall on top of one another, which can make data analysis more difficult. A bar graph, however, displays data by using different bars for each dependent variable value.

Let's consider a possible follow-up cactus seedling experiment for our biologist. In this experiment, not only does the amount of water differ for each batch of seedlings, but the amount of light given to the seedlings varies. One complete set of pots (5 pots in each of 5 groups) is given 16 hours of light per day, while another set is given 18 hours of light per day. The biologist concludes that the amount of light doesn't make much difference, but she still wants to communicate these findings graphically.

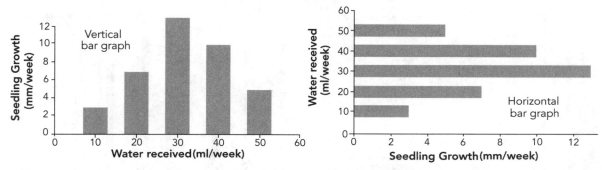

FIGURE 2.5 Bar graphs are often an effective means of showing the relationship between dependent and independent variables. In this figure, the same data set is plotted as a vertical bar graph (left) and a horizontal bar graph (right). Which do you think is most effective in presenting these data?

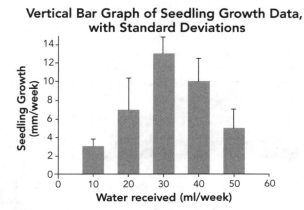

FIGURE 2.6 As with line graphs, bar graphs can also show variation in the data by displaying standard deviations of the mean.

Has the biologist added another dependent variable or another independent variable?

Now let's combine these data using both a line graph and a bar graph to highlight the differences between the two graph types (**Figure 2.7**).

Which do you think is more effective?

Pie Charts

Another common type of graph is a pie graph, or pie chart. As the name implies, a pie chart resembles a pie divided into more than one slice. A pie chart is ideally suited for presenting data expressed in the form of percentages of fractions of a whole. The size of each slice, when appropriately labeled, indicates the relative amount of the graphed variable. For example, a pie chart can effectively present data about the elemental composition of seawater (with each element shown as a percentage) or the fraction of each phylum represented in the total number of species in an ecosystem.

Can you think of a good way to present the cactus seedling data using a pie chart?

2.3 Putting Your Knowledge of Graphs to Work

Now that you've learned the common types of graphs, let's assess your understanding by asking you to try a little graphing of your own. For this exercise, you will need your copy of *Biological Science* and some graph paper (or a computer program that allows graphing).

1. Try to find an example of each type of graph in *Biological Science*. For each type, identify the x- and y-axis (except for a pie chart) and the dependent and independent variables.

2. Pick five chapters from *Biological Science* at random. For each chapter, record the total number of graphs, regardless of graph type. Let's consider the chapter number to be the independent variable and the number of graphs per chapter to be the dependent variable. Now, create these graphs:

   • Scatter plot, recording each individual data point

   • Line plot, using the average number of graphs for each chapter

   • Two bar graphs, one oriented vertically, the other horizontally

Which graph do you think is the most effective way to show the data you collected?

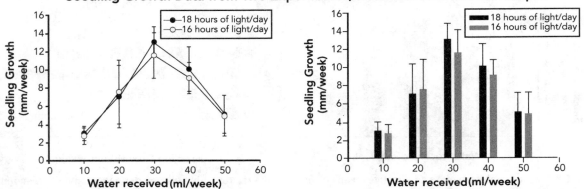

FIGURE 2.7 When the results of more than one experiment are plotted on a line graph, the data can be difficult to visualize. Sometimes a bar graph (right), with separate bars representing every dependent variable value, presents the results more clearly than a line graph with two plots (left).

3. Refer to the five chapters from *Biological Science* that you used to make the graphs in step 2. Create a worksheet so that you can keep track of the number of each type of graph (line graph, bar graph, etc.) as you flip through these chapters. Using a compass (or a coffee cup or another similar object), draw a circle on a piece of graph paper. Create a pie chart showing the fractions of the total number of graphs represented by each type.

2.4 Presenting Data with Simulations and Animations

Some of your favorite sites on the World Wide Web probably contain animations, which use iconic symbols or cartoon drawings that move in a lifelike way. Animations are becoming increasingly important in the communication of scientific data. They are especially effective when describing ongoing processes occurring over a period of time—for example, an animated graph that reveals its data in a series of stages rather than all at once.

Related to animations are simulations, often in the form of animated graphs or tables, that allow the recreation of a real-life situation. Rather than passively watching animation unfold, you can manipulate data (usually independent variables) in a model to see how the outcome (the dependent variable) changes in response to different data sets. An increasingly important tool for data communication and interpretation, simulations are used in everything from predicting how a forest fire might spread to anticipating a drug's effects on a patient's blood pressure.

CHECK YOUR UNDERSTANDING

Can you think of any times you have used a simulation in your personal life—that is, fed data into a model to see what the outcome would be? You might recall an example involving your personal finances, or perhaps you've calculated the effects of several potential test grades on your grade point average.

As you can see, there are many ways to effectively present scientific information, each with different characteristics that may suit it for specific types of data. Experiments may yield unexpected results, however, and careful interpretation is often necessary before data can be understood. In the next chapter we will explore the challenges of data interpretation, and you'll find techniques to help you understand scientific information and excel in your study of biology.

Chapter 3

Understanding Patterns in Biology and Improving Study Techniques

Learning background details about the form and function of life, from large organisms to microbes that cannot be seen without the help of a microscope, is a necessary prerequisite for a comprehensive understanding of life. This is an immense amount of information, and textbook authors are faced with the challenge of organizing it just as students are faced with the challenge of learning it. How can details about everything from molecules and viruses to ecosystems and evolution be compiled and presented so that they will be understood by even the most inexperienced student? Fortunately, biology is not just a haphazard accumulation of facts. Life is highly ordered and involves many ongoing processes that result in recognizable patterns—some simple, others more complex—that are easily presented in textbooks.

An understanding of detail is crucial to the pursuit of biology, because though individual structures and their functions may not seem important at first, each will play an important role in your discovery of the elaborate patterns found throughout biology. When used to complement your lectures and labs, your textbook is an important tool that will help you master details and discover a rich tapestry of biological patterns. This chapter will help you use the features of Freeman's *Biological Science* to gain full understanding of your course material and of biology as a whole.

✔ CHECK YOUR UNDERSTANDING

To understand how patterns and relationships in biology are organized, go to the section in your library that contains introductory biology textbooks such as Freeman's *Biological Science*. Look at the tables of contents in several different books. Do you see much variation in how the chapters are arranged in each book? What themes are present in every book? Looking at the publication dates of these books, can you see any ways in which the organization of biological information has evolved over the years?

A. DISCOVERING PATTERNS IN BIOLOGY

Researchers and students alike face the common challenge of gathering enough specific knowledge about biology to be able to recognize general patterns. Before attempting to understand complex biological concepts, you must learn about specific structures and processes that will explain biology in a broader context. In **Chapter 1** of Part 1, we considered the pitfalls of "not seeing the forest for the trees," which describes the idea that broad patterns and conclusions can be overlooked when attention is focused solely on detail. Let's apply this metaphor to the study of biology.

3.1 Seeing Both the Forest and the Trees—Use Details to Understand Concepts

As a beginning biology student, you will notice that there is no shortage of "trees"! Hundreds of terms, many of which may be unfamiliar, are used to describe organisms, their constituent parts, and the ways they interact. Learning each "tree" in the "biology forest" is a challenge that can be approached in several different ways. Understanding these details is necessary to do well in a biology course—for example, learning the basic structure of a plant leaf and the fundamental steps of cellular respiration are prerequisites for studying the more complex process of photosynthesis. Learning these details will be easier and more interesting if you try to understand *why* they are important. As you progress through this course, constantly question yourself about the material to assess your understanding: Why is a plant leaf so structurally complex? What are the functions of its separate components? How do a plant's leaves contribute to its survival? If you keep the *function* of an organism's structural components (the trees, to return to our metaphor) in mind, these details will be easier to learn and will help clarify a broader concept (the forest). As you read your textbook, it is important that you not just passively memorize terms; instead, continually ask yourself, "*What does this structure do, and how does it relate to the terms and concepts I've already mastered?*"

This study guide emphasizes the idea that the material you study will build upon your knowledge from previous chapters and will help you understand future topics. Each chapter in Part 2 of this Study Guide begins with a feature called "Looking Back—Concepts from Earlier Chapters." This features helps reinforce your knowledge by reminding you of important concepts you have already learned that will be useful in understanding the material of the current chapter. Similarly, each chapter concludes with "Looking Forward—Concepts in Later Chapters," that will indicate connections to concepts you will learn about in future chapters. If you look for general patterns and broad concepts as you learn the details of biology, you will eventually discover that the shadowy outline of the forest is starting to emerge from your study of the trees.

3.2 Look for Hierarchical Relationships

At this point in your academic career, you are probably familiar with creating outlines as a part of the process of writing. Every outline, whether it clarifies the structure of a college-level essay assignment you are writing or describes the layout of an entire textbook, follows the same basic format. Main ideas are identified by the major headings, and these topics are divided into more specific subheadings. While some ideas stand alone as key concepts, others need the support of more specific facts; key words or phrases can be included within subsections to indicate additional details that will be discussed in the essay or book.

In an outline, facts and ideas are presented in the form of a hierarchy. A hierarchy is a system of categorical groupings that is arranged according to relative importance. A good example of a hierarchy is a book's table of contents, which is essentially a complex outline; its hierarchical structure distinguishes it from an index, in which ideas and facts are listed alphabetically regardless of their relative importance. A good table of contents is more than a guide to help the reader locate information; it also establishes a hierarchical relationship between topics.

It is important to understand the hierarchy of information in *Biological Science* as well as in other textbooks you may encounter. To help you with this concept, a section entitled "Key Concepts" is included in each chapter in Part 2 of this study guide. This section presents the main ideas of the corresponding textbook chapter in a hierarchical format, allowing you to easily distinguish pivotal concepts from supporting details based on their relative position.

For example, consider this study guide excerpt from Part 2, Chapter 40, on plant reproduction:

An Introduction to Plant Reproduction

Sexual Reproduction
- Most plants reproduce sexually. **Sexual reproduction** is based on the reduction division known as **meiosis** and on **fertilization**, the fusion of haploid cells called **gametes**.
- **Sperm** are small cells from the male that contribute genetic information in the form of DNA but few or no nutrients to the offspring. Female gametes are called **eggs**, which contribute a store of nutrients to the offspring.
- **Outcrossing** occurs when male and female gametes are exchanged between individuals of the same species. When **self-fertilization** occurs, a sperm and an egg from the same individual unite to form a progeny.
- The primary advantage of **selfing** is that successful **pollination** is virtually assured.

The Land Plant Life Cycle
- Plants are the only organisms that have both a multicellular form that is diploid and a multicellular form that is haploid. This life cycle is called **alternation of generations**.
- The diploid phase is called the **sporophyte**, and the haploid phase is called the **gametophyte**.
- A **spore** is a cell that grows directly into an adult individual. A **gamete** is a reproductive cell that must fuse with another gamete before growing into a new individual.
- Meiosis occurs in sporophytes and results in the production of haploid spores, which are produced in structures called **sporangia**.
- Spores divide by mitosis to form multicellular, haploid gametophytes.
- Fertilization occurs when two gametes fuse to form a diploid zygote. The zygote grows by mitosis to form the sporophyte.
- Study the following diagram (**Figure 40.2**) to make sure you understand these life-cycle intricacies.

This example introduces several details about plant reproduction, and their relative positions in the outline indicate relationships between them. Based on the structure of the outline, you can determine that there are three important points under the heading of plant reproduction—sexual reproduction, plant life cycles, and asexual reproduction. Each of these topics is supported by specific details; for example, meiosis and fertilization involving gametes are both important components of sexual reproduction. Such details are also included under the topics of plant life cycles and asexual reproduction. Although this information could be discussed in paragraph form, a hierarchical outline allows relationships between topics to be defined clearly by their appearance on the page.

3.3 Accept Variability

To understand biology, it is necessary to accept that individual organisms, structures, or behaviors may not always fit into expected patterns. Variability is a natural part of biological science; occasional variations actually emphasize that most organisms match a standard pattern. While *Biological Science* concentrates on the typical components of general patterns, it also includes realistic variation. This variation is especially apparent in your book's many graphs, which often show individual data points as well as an overall pattern. Consider this graph from Chapter 25, page 442, which illustrates the relationship between human birth weight and mortality (**Figure 3.1**).

Logically, we might expect to observe lowest mortality at one ideal birth weight, with a greater mortality rate if birth weight is either higher or lower than this optimum. In fact, the actual data do support this hypothesis. However, if you examine the data points representing mortality (in **Figure 3.1**, the blue points connected by a U-shaped curve) and the actual distribution of body weights at birth (the red bars), you will see that while there are obvious trends, there is variability among individual data points. In this example, though variability is illustrated by any individual points lying off the indicated curve (drawn from the average of all data points), most points fall fairly close to this line. In some experiments, variability is far more pronounced, requiring sophisticated statistical analysis to identify a pattern in a widely distributed set of data.

Accepting variation will prevent you from being distracted by any data that do not fit the predicted pattern. Variation, though it may appear to disrupt experimental data, is not just a nuisance to be ignored. After all, it is variation among individual organisms that enables a species to change over time, and variation is therefore the foundation of evolution.

3.4 Be Comfortable with the Unknown

In addition to facing the challenges of variability, biology students must be prepared to encounter uncertainty. New facts emerge and new hypotheses are tested daily, making biology a field that is constantly evolving and reinventing itself. Though much is already known about biology, an even greater amount is waiting to be discovered and understood. *Biological Science* discusses both accepted facts and current theories, distinguishing what is proven from what is conjectured. You may pose questions to your professor and lab instructor that they may not be able to answer, because the answer is not yet known. This uncertainty simply reflects the rapid growth of the field and explains biologists' constant search for greater understanding.

B. THRIVING IN YOUR BIOLOGY COURSE

As with any course, you will need to thoroughly understand biology in order to succeed on quizzes and tests and achieve your desired grade. Several tools are at your disposal, many of them integrated directly into your textbook, that can make the learning process easier and more rewarding. Additionally, certain study habits may improve your ability to understand the material you are studying.

3.5 Budget Your Time Carefully

The amount of time you spend on your biology course outside of class—studying, preparing for tests, and working on your laboratory assignments—will be reflected in the grade you achieve. Your other classes also require that you take the time to study, read the material, and write papers. How can you be sure that you are devoting enough time to each class?

It is important to be realistic about your time commitments. Everyone in your biology class has different schedules, course loads, and activities, so it is important to develop a study pattern that suits your life. For example, let's say that your biology course involves two lectures and a lab each week; therefore, you will probably spend at least five hours a week in class for your biology course alone. You may need at least another five hours a

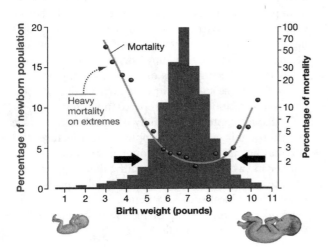

FIGURE 3.1 The histogram shows the percentages of newborns with various birth weights. The gray dots are datapoints indicating the percentage of newborns in each weight class that died, plotted on a logarithmic scale on the right. The gray line is a function that fits the datapoints.

week to complete assignments and prepare for exams, meaning that biology alone might take up 10 hours of your time each week! Considering that there are five classes in a typical course load, you may devote 50 hours each week to academics alone—leaving little time for family, friends, a part-time job, and relaxation! Realistically consider what you can expect to achieve in the time you have available. If you concentrate on what you hope to achieve in college and on your future goals, your courses will become a top priority.

3.6 Attend Your Lectures and Labs!

Lectures and labs are not optional—if they were, your course syllabus might say, "Please come to class occasionally." Instead, it probably lists the exact times for all lectures and labs, implying (or specifically stating) that they are mandatory. By attending every lecture and lab, taking careful notes, and participating in class while you are there, you can take an active role in the learning process. Going to class will give you the opportunity to question your instructor and classmates about material you may not understand and allow you to gain different perspectives and practical experience. Attending class regularly, therefore, may be the one thing you can do to most improve your performance in a biology course.

3.7 Take Good Notes in Class

Attending class is necessary for success, but it is also important to make the most of your class time by taking good notes. Though you may have personal note-taking methods that work well for you, try structuring your notes in the form of an outline to illustrate a hierarchy of ideas. A hierarchy, as introduced earlier in this chapter, arranges concepts according to relative importance. Try to mimic your textbook as you take notes in class, reflecting the hierarchy of facts outlined in the table of contents and in the chapters themselves.

✔ **CHECK YOUR UNDERSTANDING**

Examine the hierarchical format used in these sample notes on Chapter 44, "Gas Exchange and Circulation," and compare it with the format of notes you've recently taken for this class. Are they similar? Can you think of ways to improve your own note-taking technique?

Even if you do not choose to follow this method precisely, consider modifying your own note-taking technique to include a hierarchy of ideas. This format will help to solidify your understanding of the relationships between concepts.

3.8 Integrate Your Lecture and Laboratory Material

Lectures and laboratories are often taught at different times by different instructors, so it may sometimes seem that they are two separate courses. Remembering that the laboratory is designed to support the lecture, and vice versa, will help you understand the material you are studying. Though they may not fall on the same day, the scheduling of laboratories and lectures often coincides so that in a given week, the material discussed in each will be related. For example, in the same week as your lecture on animal diversity, you may have a laboratory examining the morphological diversity of a wide array of animal specimens. Similarly, your lectures on the role of water and ions in an animal's physiology may be accompanied by a laboratory about the structure and function of the mammalian kidney. If you don't immediately see a connection between your lectures and labs, ask your professor or lab instructor to explain how they are related.

3.9 Take Advantage of Your Textbook

While this study guide is designed to help maximize your understanding of your biology course, it is not meant to replace your textbook. The vast amount of information contained in its chapters makes your textbook an indispensable supplement to your biology lectures and laboratories.

First, familiarize yourself with the book's features—the table of contents, glossary, index, and useful learning features such as "Key Concepts" and "Check Your Understanding"—as well as with the structure and content of the chapters. Parts of the textbook can be used like a reference book if you need to review topics you recognize from previous courses. You can use the index to locate the page on which a particular term is presented, thus avoiding a tedious search of an entire chapter. The glossary can be a useful tool to test your knowledge: Write down any key terms in your lecture notes or reading assignment, and then define these concepts in your own words and compare your definitions to those offered in the glossary.

Chapter content is carefully designed to introduce and thoroughly explain each topic, because lectures and labs may not always provide enough information to fully understand a concept. For more difficult topics, you may need to refer to the book for additional explanations. A complex subject such as genetics, for example, might demand a careful reading of the entire chapter before you can fully understand the topic.

3.10 Take Advantage of the Media for Students

Be sure to take advantage of the media support available for your textbook. Use the Study Area in Mastering Biology (*www.masteringbiology.com*) for targeted and efficient use of valuable study time. Some of the many study tools include BioFlix™ 3-D movie-quality animations that focus on the toughest topics, engaging activities, and cumulative chapter quizzes that help prepare for exams. Animations and activities are called out in your textbook by the Mastering Biology icon (see, for example, page 77). The interactive eText is available 24/7 and enables you to highlight text, add study notes, and review personalized instructor notes at your convenience.

3.11 Share Your Knowledge

One of the best ways to learn is to tell others what you know. Find a study partner and meet regularly to discuss material. Take turns explaining topics to each other as thoroughly as possible while using any new terms and concepts you are learning. The partner who is listening can then critique the other's knowledge, identifying any missing information to ensure that both partners thoroughly understand the topic. In this way, you will be able to quickly identify your particular areas of weakness. You will eventually need to explain these concepts on laboratory and lecture exams (especially if these tests involve essay questions); any self-testing you do before an exam will build your confidence and ensure your mastery of the subject.

*Though biology can be complex at times, understanding principles of data interpretation and improving your study technique can make learning this subject both challenging and rewarding. Reading effectively and writing to clearly communicate information will also help you in this course, throughout your education, and in the "real world" as well. In **Chapter 4** of Part 1 of the study guide, we will discuss these skills and give you the opportunity to assess your abilities.*

Chapter 4

Reading and Writing to Understand Biology

In Part 1, **Chapter 2** *of this study guide, we noted that while scientific information is often conveyed visually with graphs and animations, writing remains the most common and effective means of scientific communication. This chapter will help you improve your ability to read and interpret information about biology, evaluate your note-taking skills, and learn to convey biological ideas in your own words.*

A. READING FOR UNDERSTANDING

Your success in this biology course (as well as in other classes) may depend on your ability to gather important information from what you read.

- Can you easily pick out the main topic of a paragraph?
- Do you put figure information together with key points from the text?
- When you read about an experiment, can you identify the hypothesis being tested?
- Do you satisfactorily remember information you read?
- Are you able to efficiently skim your text when reviewing for a test?

4.1 Tips for Effective Reading

Identifying key points in a text helps readers understand the big picture and remember the most important information. When a text is well written, each paragraph should have one or two easily identifiable main ideas. Often a topic sentence can be found near the beginning of the paragraph, followed by further support, elaboration, or explanation. The supporting information may describe examples or analogies, explain how the topic relates to a larger context, or present alternate views to help the reader understand the information.

To maximize the amount of information you gather from reading your textbook, try the following:

- First skim the section headings to get an overview of the subject covered.
- Next, read carefully, taking the time to examine all figures noted in the text.
- If you don't understand a topic, refer to your study guide; it may present the same information in different words or in a simplified manner.
- Look up any unfamiliar words—your textbook's glossary explains many new terms.
- As you read, stop periodically to ask yourself what you have learned; these mini-reviews will help you remember information later.
- Note any topics you have difficulty understanding so you can ask your professor, teaching assistant, or another student to explain the subject.

Several techniques may improve your memory of what you read. For example, you might try underlining topic sentences and other important points, but **be selective**. If you do not want to mark your book, making notes on sticky pads and attaching them to pages with key information will serve the same purpose. Note-taking is another effective way to increase the amount of information you retain while reading, but certain methods may work better than others. Taking notes as you read may distract you from learning the material and may tempt you to copy the text directly into your notes. Instead, take notes after you have finished the assigned reading. You learn and remember more from writing and reviewing your notes if you use your own words.

Last but not least, answer the questions provided in your text and in this study guide. Compare your answers to those given and note any discrepancies. The questions you did not answer correctly indicate the topics on which you need additional review.

When you review for a test,

- Skim the section headings again and identify topics that are not clear to you.

- Be sure you can accurately define any boldfaced vocabulary words.

- Refer to this study guide and to your notes for a summary of topics, focusing on those you feel unsure of.

- Go over the figures in your textbook, testing yourself to see if you can explain each figure without referring to the text.

- As a final review of important topics, re-read your class notes—or better yet, re-write them!

✔ CHECK YOUR UNDERSTANDING

(a) Practice your reading skills using the following passage from Chapter 1, page 2 of *Biological Science*. Underline or note in some other way the one or two key point(s) of each paragraph. Look up any unknown words that are not explained in the text. At the end of the passage, ask yourself what you have learned (see answer key at the end of this chapter for suggested answer).

*Two of the greatest unifying ideas in all of science laid the groundwork for modern biology: the cell theory and the theory of evolution by natural selection. Formally, scientists define a **theory** as an explanation for a very general class of phenomena or observations. The cell theory and theory of evolution address fundamental questions: What are organisms made of? Where do they come from?*

When these concepts emerged in the mid-1800s, they revolutionized the way biologists think about the world. They established two of the five attributes of life: Organisms are cellular, and their populations change over time.

Neither insight came easily, however. The cell theory, for example, emerged after some 200 years of work. In 1665 Robert Hooke used a crude microscope to examine the structure of cork (a bark tissue) from an oak tree. The instrument magnified objects to just 30 × (30 times) their normal size, but it allowed Hooke to see something extraordinary. In the cork he observed small, pore-like compartments that were invisible to the naked eye. These structures came to be called cells.

What have you learned?

Reading and writing are closely related. Taking notes and/or making flash cards as you read helps you understand and remember information and will make it easier for you to study later on.

B. WRITING FOR UNDERSTANDING

Practice writing clearly about biology when taking notes and answering short-answer questions in your text and study guide. Taking notes on the material you read will help you analyze, understand, and remember the information; answering short-answer questions will allow you to consider facts from a different angle and put your thoughts into your own words. Finally, getting in the habit of writing clearly helps you effectively communicate your mastery of the material on course tests and assignments.

4.2 Studying Is Easy with Good Notes

Your own notes will help you study better than someone else's because they emphasize the material you did not already know and are phrased in your own words. Although it is acceptable to use informal language—incomplete sentences, abbreviations, and so on—when writing notes and flash cards, it is essential to write with clarity, brevity, and logic. Include important facts, eliminate extraneous details, and make sure that your notes will still make sense to you months later. (Why are "extraneous details," that you probably won't need to know for quizzes and tests, included in your textbook? These details are meant to support important points in the text. Though it may not appear on exams, this information will help you understand the more complex topics you are required to know.)

Be sure your notes will allow you to distinguish key concepts from supporting details and examples when you review them later. Consider using the Cornell Note-Taking System. This method requires that you leave a large margin and some space at the bottom of each note-page. During lecture, take as complete notes as possible. Soon after, reduce your notes to brief questions or cues to print in the margin. Finally, summarize each page at the bottom. You can now easily quiz yourself by covering the note-taking space and reciting your question answers or cue explanations (for more about this method, see the Internet References section at the end of this chapter).

✔ **CHECK YOUR UNDERSTANDING**

(b) Read this passage called "Evolutionary Legacy" from David S. Goodsell's article, "Biomolecules and Nanotechnology," in *American Scientist* 88 (3), May–June 2000. First read for understanding, underlining the main points as you go. Then take notes on the passage using a separate piece of paper. Be selective, distinguishing main points from supporting examples or details.

The process of evolution by natural selection places strong constraints on the form that biological molecules may adopt. Because genetic information is passed directly from generation to generation, cells must maintain a living line back to the earliest primordial cells. If a cell fails to generate a living descendent, all of its biological discoveries will be lost. This is far more limiting than the technology of our familiar world. If we create machines that don't function, we scrap them and go back to the drawing board. But if a cell takes a gamble and changes a critical machine, it had better get it right the first time or the result will be disastrous.

The picture is not entirely grim, however, as cells have several levels of redundancy within which to develop new machines. First, the plans for a given machine may be duplicated, which allows the duplicate to be modified and ultimately perfected to perform a function different from the original. This is very common in the evolution of life. Hemoglobin, the protein that carries oxygen in our blood, is an example. Our cells contain information for building several different types of hemoglobin. One is optimized for carrying oxygen in the blood of adults, whereas another is found in the blood of a fetus. The fetal hemoglobin has a higher affinity for oxygen, allowing it to capture oxygen from the mother's blood. About 200 million years ago, a gene duplication allowed the fetal hemoglobin to be perfected separately.

Second, biology seldom involves a single cell. A population of cells—billions, trillions—is the biologically relevant entity. Within this population there exists ample room for experimentation. Millions of modifications may be tried, even if most are ultimately lethal. The population will still survive and individuals with rare improvements may grow to dominate in later generations. Human immunodeficiency virus (HIV) shows the benefits of evolutionary change, accelerated so that we can see the effects in months instead of millennia. HIV reverse transcriptase, the enzyme that copies the virus's genetic information, is particularly error-prone. Because of this, the population of viruses within an infected individual contains viruses with all possible single-site mutations—thousands of variants on the wild-type virus. The best of these will dominate, but even the weakest are continually created and recreated in

subsequent generations by the low-fidelity copying mechanism. Thus, when an infected individual is treated with anti-HIV drugs, the population has a wide range of different mutants to choose from, some of which may be resistant to the drug: The virus is made more efficient by its very inefficiency. The hallmark of biological evolution is the plasticity provided by mutation and genetic recombination. Within a population, or through genetic duplication within a single cell, a great many variants may be tested and the occasional improvement saved.

Evolution carries with it one important drawback, however: the problem of legacy. Once a key piece of machinery is perfected, it is difficult to replace it or make major modifications without killing the cell. This is particularly true for major molecular processes, such as protein synthesis, energy production and molecular machines. This leads to the remarkable uniformity of all earthly living things when observed at the molecular level. All are built of the same basic components.

Compare your notes with those shown in the answer key at the end of this chapter. Do your notes mention the same main points as those in the answer key? Do you prefer your notes or the sample notes? You probably find your own notes easier to understand because they are in your own words, phrased in a way that makes sense to you. If you prefer the sample notes, try to figure out why—were your notes too detailed? Too brief? Did you miss some main points or fail to make a distinction between major topics and supporting details? Use these observations to identify ways in which you can improve your note-taking skills.

4.3 Writing Essays and Other Assignments

Why Is Writing So Important?

Writing is essential in science! A scientist's work is not limited to research conducted in a laboratory; he or she must be able to effectively communicate research goals, procedures, and results in grant applications and scientific papers. As a biology student, you also need to effectively communicate your ideas—most often in laboratory reports and in answers to essay or short-answer questions. The ability to write clearly and concisely will improve your performance in all these areas.

Evaluating and improving your writing technique will benefit your study of biology in several ways:

- Taking coherent notes in your own words makes studying easier.

- Writing down information helps you recall it later on exams.

- Clearly written short-answer and essay responses effectively communicate your knowledge.

- Well-written laboratory reports demonstrate your understanding of the process studied.

(c) Compare these two short-answer responses to the question: How does what we know about artificial selection support the theory of evolution?

1. Artificial selection is when you select something good. so plants and animals change to get better buds or something quickly because their reproduction is helped by. This is like natural selection which Darwin said caused evolution and they both is types of selection therefore evolution must occur.

2. Artificial selection occurs when humans control the reproduction of domesticated organisms to favor certain variable, heritable traits. In response to this selection, populations change rapidly over a relatively short period of time. This supports the idea that natural selection could also cause changes over time (evolution).

Which answer more effectively communicates the material? Why?

_____

_____

_____

_____

Maintaining clarity and presenting ideas cohesively is especially challenging when completing an assignment requiring a longer response. What can you do to improve your essay-writing technique?

Think First

Before you start writing, be sure you understand your topic and the writing assignment.

- Decide on a thesis statement; this sentence introduces the topic to be addressed and identifies a particular position that your paper will defend.

- Identify examples or supporting arguments to support your thesis.

- Make an outline to organize your essay or lab report. Making an outline is worth the effort. It ensures that your argument is thoroughly developed and communicated in a way that is easily understood by other readers.

- Prepare a strong conclusion.

(d) You are applying to work with a professor on his or her summer research project. Make an outline for the cover letter you plan to attach to your résumé and application.

Write

After completing your outline, use it to write a first draft of your essay. Though it is not necessary to be a perfectionist at this stage—you will edit later—try to create a clear topic sentence for each of your paragraphs. Unless directed otherwise, use active verbs and avoid passive voice. For example, it is more effective to write, "We inadvertently contaminated our sample," than "Mistakes were made and the sample was contaminated." Be concise, but support each of your main points. Consider using metaphors, similes, analogies, and examples to explain your topic; these creative techniques encourage readers to make their own connections, enabling them to understand your position more clearly.

Look again at the previous sample essay by David Goodsell as an example of a well-written piece. It can be difficult to analyze good writing because while bad writing distracts you as you read, alerting you to problems, good writing may seem almost unnoticeable because it allows you to easily understand the information being discussed.

- In his first paragraph, Goodsell introduces his thesis statement (that evolution limits the forms of biological molecules) and explains it with a comparison between the evolution of life and the development of machines.

- He uses the next two paragraphs to explore ideas that contradict this thesis (suggesting that evolution's limitations are not strong enough to be substantial). He supports this alternate view by modifying his previous analogy between machinery and life to discuss a similarity between the two.

- The words *first* and *second* help the reader understand that these two points are meant to support this alternative argument. By presenting a position that opposes his thesis statement, Goodsell shows that he thoroughly understands the topic and provides an opportunity for direct refutation of arguments that contradict his thesis.

- Goodsell's final paragraph explains that his thesis remains relevant despite the conflicting arguments, and he concludes the essay by identifying a situation in which his thesis is particularly relevant.

Throughout his essay, Goodsell uses clear logic, explaining and supporting his statements appropriately. He employs many techniques of strong writing, such as varying the sentence structure and using repetition—in moderation—to emphasize important points. He avoids using the passive voice. However, remember that it is an excerpt from a longer article. Had he addressed the same topic in an essay of this length, Goodsell would probably have developed and supported his thesis while spending less time discussing opposing arguments.

How can you, a student writer, be sure that you are applying enough of these writing techniques to your own assignments? While drafting an outline will help you organize information before you begin writing, you may not always remember to vary sentence structure, use active verbs, or support your thesis with examples as you write. This is why editing is such an important step in the writing process.

Edit, Edit, Edit!

The best way to improve a paper you have written is to re-read and edit your work a day after you finish writing. Allowing time to pass before editing gives you a fresh perspective, permitting you to detect problems that you might not have noticed otherwise. A more immediate editing technique involves reading your paper aloud to yourself or a friend. Speaking your words out loud will help you catch any grammatical errors or awkward phrasings. You should also notice if certain words are used too frequently. If necessary, find replacements for overused words by looking up synonyms in a thesaurus. Last but not least, if your essay is on a computer, use the spell- and grammar-check programs. Misspelled words and poorly constructed sentences interrupt the flow of your argument, distract your reader, and leave a poor impression. Believe it or not, half of becoming a good writer is becoming a good self-editor.

✔ **CHECK YOUR UNDERSTANDING**

(e) Practice identifying effective writing techniques as you read this passage from Carl Zimmer's article entitled "Do Parasites Rule the World?" (*Discover* 21 [8], August 2000). You might first read the article for understanding and then take notes on the passage. Finally, look at the details of sentence structure, observing strengths of the author's writing style.

As scientists discover more and more parasites and uncover the extent and complexity of their machinations, they are fast coming to an unsettling conclusion: Far from simply being along for the ride, parasites may be one of nature's most powerful driving forces. At the Carpinteria salt marsh, Kevin Lafferty has been exploring how parasites may shape an entire regions' ecology. In a series of exacting experiments, he has found that a single species of fluke—Euhaplorchis californiensis—journeys through three hosts and plays a critical role in orchestrating the marsh's balance of nature.

Birds release the fluke's eggs in their droppings, which are eaten by horn snails. The eggs hatch, and the resulting flukes castrate the snail and produce offspring, which come swimming out of their host and begin exploring the marsh for their next host, the California killifish. Latching onto the fish's gills, the flukes work their way through fine blood vessels to a nerve, which they crawl along to the brain. They don't actually penetrate the killifish's brain but form a thin carpet on top of it, looking like a layer of caviar. There the parasites wait for the fish to be eaten by a shorebird. When the fish reaches the bird's stomach, the flukes break out of the fish's head and move into the bird's gut, stealing its food from within and sowing eggs in its droppings to be spread into marshes and ponds.

In his research, Lafferty set out to answer one main question: Would Carpinteria look the same if there were no flukes? He began by examining the snail stage of the cycle. The relationship between fluke and snail is not like the one between predator and prey. In a genetic sense, infected snails are dead, because they can no longer reproduce. But they live on, grazing on algae to feed the flukes inside them. That puts them in direct competition with the marsh's uninfected snails.

To see how the contest plays out, Lafferty put healthy and fluke-infested snails in separate mesh cages at sites around the marsh. "The tops were open so the sun could shine through and algae could grow on the bottom," says Lafferty. What he found was that the uninfected snails grew faster, released far more eggs, and could thrive in far more crowded conditions. The implication: In nature, the parasites were competing so intensely that the healthy snails couldn't reproduce fast enough to take full advantage of the salt marsh. In fact, if flukes were absent from the marsh, the snail population would nearly double. That explosion would ripple out through much of the salt marsh ecosystem, thinning out the carpet of algae and making it easier for the snails' predators, such as crabs, to thrive.

After you read and analyze the passage from Zimmer's article, answer the following questions:

1. What is Zimmer's thesis?

2. What hypothesis was Lafferty testing?

3. What is the larger context for Lafferty's study?

4. Note the verbs that Zimmer uses. Are they active or passive?

5. Can you identify examples of a simile and a metaphor in this passage?

6. Does sentence structure vary?

7. What other strengths and/or weaknesses can you identify in Zimmer's writing?

4.4 Cite Your Sources and Do Not Plagiarize!

Most important, do not plagiarize. Plagiarism, stealing someone else's ideas and passing them off as your own, is unethical. When you plagiarize, you sabotage your educational experience by throwing away an opportunity to explain your own ideas and perceptions. This does not mean you should not use material from outside sources. If you find relevant material on your topic, include it in your essay with a citation acknowledging its source. An author who cites many legitimate sources demonstrates high-quality scholarship. However, note that you rarely find verbatim or "direct" quotes in scientific writing. This is because the author is expected to write all his or her own sentences, using citations to properly attribute the source of the information.

✓ CHECK YOUR UNDERSTANDING

(f) Now that you can identify the characteristics of strong writing, take out an essay or lab report you have written recently. Re-read it, looking for the elements of good writing described in this chapter. Is your thesis well supported? Which are more common in your paper, active or passive verbs? Does the sentence structure in the essay show enough variation? How might you further edit the piece? Are your sources appropriately cited?

As you can see, there is more to good writing than simply identifying a topic and carelessly listing information about it. The ability to convey an idea or support a position thoroughly in writing is a skill that will improve your grade in this biology course, help you as your education continues, and be useful throughout your professional career.

REFERENCES

The following references contain further information about ways to improve writing technique and the fundamentals of language and grammar.

Text References

Pauk, Walter. 2001. *How to Study in College*, 7th ed. Houghton Mifflin.

Pechenik, J. 2009. *A Short Guide to Writing about Biology*, 7th ed. Longman.

Soloway, J. ed. 2001. *Grammar Smart: A Guide to Perfect Usage*. Princeton Review Publishing, L. L. C.

Strunk, W. Jr., and E. B. White. 2000. *The Elements of Style*, 4th ed. Longman.

Internet References

The Cornell Note-Taking System

http://lsc.sas.cornell.edu/Sidebars/Study_Skills_Resources/cornellsystem.pdf

http://www.gearfire.net/cornell-note-templates/

Plagiarism: Facts, Frequently Asked Questions, and How to Avoid It

http://www.plagiarism.org/

http://owl.english.purdue.edu/owl/resource/589/01/

INTRODUCTION CHAPTER 4— ANSWER KEY

Check Your Understanding

(a) The main points for these paragraphs are:

- A scientific theory explains many related observations and establishes general principles; examples are the cell theory and the theory of evolution.
- These theories explain that life is cellular and populations change.
- Robert Hooke was the first to observe cells in the late 1660s.

Words you might have wanted to look up in the glossary include: theory and cell.

(b) Notes taken using the Cornell Note-Taking System are shown in **Figure 4.1**. To learn more about this method, refer to the "Internet References" section of this chapter.

(c) The second answer communicates the information more clearly. Although the first response contains relevant information, the poor spelling, grammar errors, and misplaced punctuation in this answer are distracting. Its sentences are not structured well, forcing the reader to search for key information and piece it together. To what does the "they" in the last sentence refer? If this was a student answer from an exam, would you be convinced that the student knows what artificial selection is? Careful consideration of this response reveals that the student understands similarities between artificial and natural selection and that evolution is change over time, but a tired or distracted test grader might not take the time to draw these conclusions, deducting points because the response is hard to read.

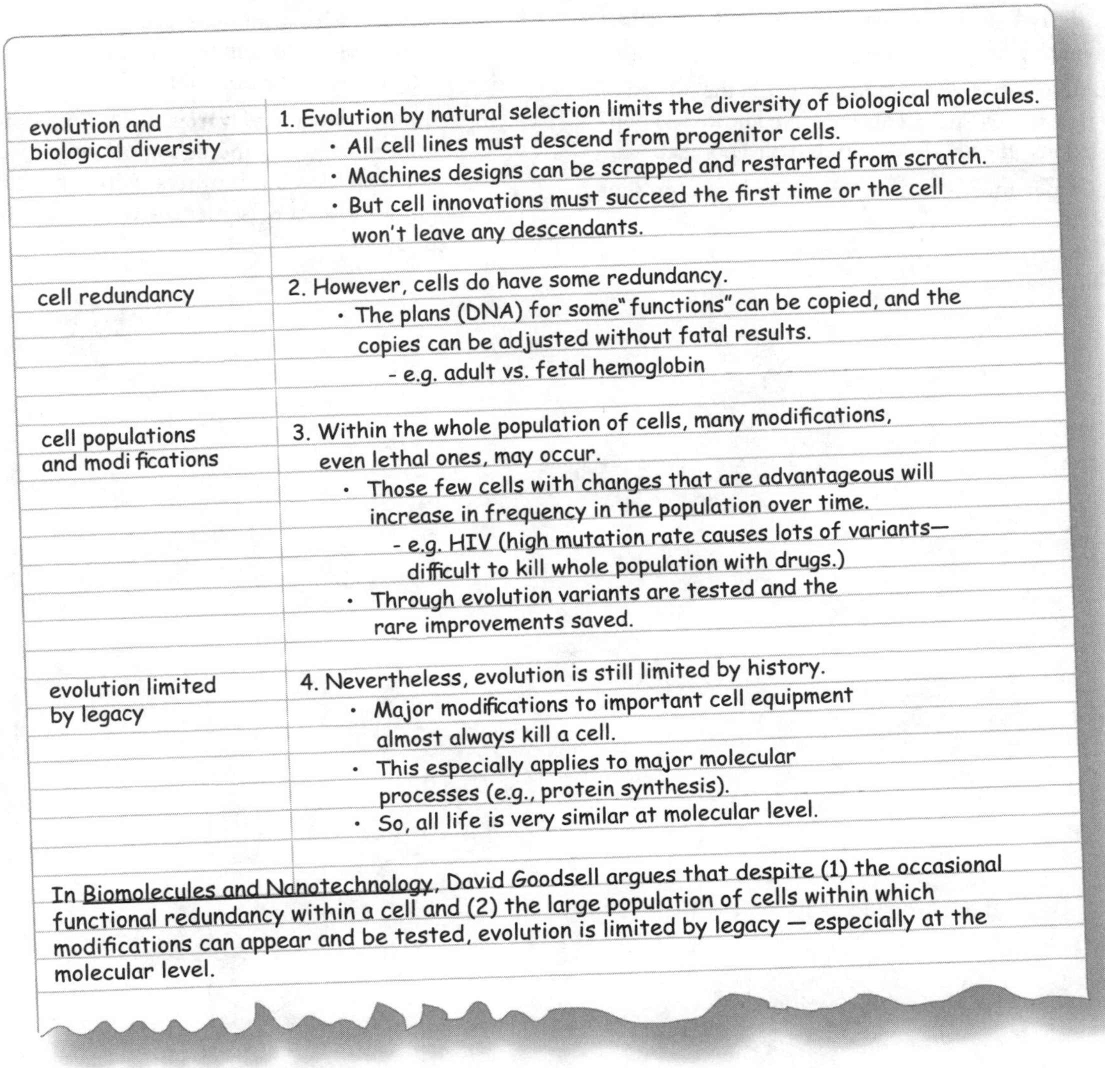

evolution and biological diversity	1. Evolution by natural selection limits the diversity of biological molecules.
	  • All cell lines must descend from progenitor cells.
	  • Machines designs can be scrapped and restarted from scratch.
	  • But cell innovations must succeed the first time or the cell
	    won't leave any descendants.

cell redundancy	2. However, cells do have some redundancy.
	  • The plans (DNA) for some "functions" can be copied, and the
	    copies can be adjusted without fatal results.
	     - e.g. adult vs. fetal hemoglobin

cell populations and modifications	3. Within the whole population of cells, many modifications,
	    even lethal ones, may occur.
	  • Those few cells with changes that are advantageous will
	    increase in frequency in the population over time.
	     - e.g. HIV (high mutation rate causes lots of variants—
	     difficult to kill whole population with drugs.)
	  • Through evolution variants are tested and the
	    rare improvements saved.

evolution limited by legacy	4. Nevertheless, evolution is still limited by history.
	  • Major modifications to important cell equipment
	    almost always kill a cell.
	  • This especially applies to major molecular
	    processes (e.g., protein synthesis).
	  • So, all life is very similar at molecular level.

In <u>Biomolecules and Nanotechnology</u>, David Goodsell argues that despite (1) the occasional functional redundancy within a cell and (2) the large population of cells within which modifications can appear and be tested, evolution is limited by legacy — especially at the molecular level.

FIGURE 4.1   Notes taken using the Cornell Note-Taking System.

(d) See following outline.

  I. Thesis: Hire me this summer

    A. Introduce myself—how did I hear about professor?

    B. Professor's research sounds interesting

    C. Want to get research experience

  II. I like biology

    A. Have done well in biology classes

    B. Interested in nature: hiking, gardening, animals

    C. Want to major in biology; interest in professor's research area

  III. I work hard and am responsible

    A. Held jobs for the past two summers (references available)

    B. Volunteer work at local animal shelter

  IV. Conclusion

    A. Hire me this summer

    B. Contact information

    C. Thanks for your consideration

(e) Possible answers for the end-of-article questions are as follows:

  1. Zimmer's thesis is that "parasites may be one of nature's most powerful driving forces."

2. Lafferty tested the hypothesis that flukes affect snail population dynamics.

3. Lafferty's study was conducted within the context of a marsh ecosystem; it also observed how snail populations affected algae, crab, and bird populations.

4. Zimmer mostly uses active verbs in the present or past tense, for example: "scientists *discover*," "Lafferty *found*," "birds *release*," "eggs *hatch*," and "flukes *castrate*."

5. An example of a simile is "looking like a layer of caviar," and an example of a metaphor is "that explosion would ripple out."

6. Sentence structure varies well; for example, Zimmer begins some sentences with prepositional phrases and others with the subject. Zimmer also uses sentences of many different lengths.

PART 2

Chapter Guides

Chapter 1

Biology and the Tree of Life

- Organisms obtain and use energy, are made up of cells, process information, replicate, and—as populations—evolve.

- All organisms are made of cells, and all cells come from preexisting cells.

- Species change over time because individuals with favorable heritable traits have more offspring, on average, than other individuals.

- Phylogenetic trees depict evolutionary relationships among species such that closely related species that share distinctive traits are placed close to each other.

- Biologists ask questions, develop hypotheses, and design experiments to test the predictions of competing hypotheses.

A. CHAPTER OUTLINE

1.1 What Does It Mean to Say That Something Is Alive?

- Life-forms obtain and use energy, are made of cells, use information from both their genes and from the environment, can replicate, and are both the result of evolution and members of populations that evolve.

1.2 The Cell Theory

- A scientific **theory** explains many related observations and establishes general principles for a field of study.

- Robert Hooke and Anton van Leeuwenhoek were the first to observe cells in the late 1660s (**Figure 1.1**).

- Additional observations made by many researchers confirmed that all organisms are made of cells.

Are All *Organisms Made of Cells?*

- Yes! Cells are the fundamental building blocks of life, from the smallest single-celled bacteria to the largest multicellular organisms.

- A **cell** consists of concentrated chemicals dissolved in water and enclosed by a plasma membrane. Most cells are highly organized and can divide to reproduce.

Where Do Cells Come From?

- Scientific theories usually have two parts: the description of a pattern observed in nature and an explanation of the process that causes the observed pattern.

- The cell theory states that all cells come from preexisting cells. This process explains why all organisms are made of cells (the pattern part of the cell theory).

Two Hypotheses

- A **hypothesis** is a proposed explanation for a specific observation or question.

- Before the cell theory, many biologists thought organisms could spontaneously appear in nonliving materials; the all-cells-from-cells hypothesis challenged the spontaneous generation hypothesis.

An Experiment to Settle the Question

- Useful hypotheses lead to **predictions**, statements describing the results that one should obtain if the hypothesis is true.

- The spontaneous generation hypothesis predicts that microorganisms can spontaneously appear in nutrient broth without being exposed to previously existing cells.

- Louis Pasteur tested this prediction by placing sterilized nutrient broth in a straight-necked flask exposed to preexisting cells in the air and in a swan-necked flask in which condensed water prevented exchange with the air (**Figure 1.2**).

- Pasteur's experiment was well designed because there was only one difference between his two treatments: the neck shape of the flask.

One Hypothesis Supported

- Cells only appeared in the straight-necked flask; these results supported the all-cells-from-cells hypothesis and opposed the spontaneous generation hypothesis.

- All cells come from preexisting cells, so single-celled organisms in an isolated population are genetically related; cells in multicellular animals also descend from a single ancestral cell, the fertilized egg.

- Charles Darwin and Alfred Russel Wallace realized that all species are also related.

✓ CHECK YOUR UNDERSTANDING

(a) Why didn't biologists discover earlier that all life is made of cells?

_____

_____

1.3 The Theory of Evolution by Natural Selection

What Is Evolution?

- **Evolution** occurs when a population's characteristics changes over time.

- Species change over time ("descent with modification") and are connected by shared ancestry (the pattern part of the theory of evolution).

What Is Natural Selection?

- Natural selection explains *how* evolution occurs (the process part of the theory).

Two Conditions of Natural Selection

- Individuals of the same species living in the same area at the same time are called a **population**.

- Natural selection occurs when (1) individuals differ from each other for one or more heritable trait(s) and (2) in a particular environment, individuals with certain traits tend to survive and/or reproduce better than individuals with other traits.

- Natural selection's effect on individuals increases the frequency of favorable traits in populations over time; selection acts on individuals but it is the population that evolves.

✓ CHECK YOUR UNDERSTANDING

(b) If you were a hunter-gatherer without any modern technology, what heritable traits might help you survive and/or reproduce?

_____

_____

Selection on Maize as an Example

- **Artificial selection** occurs when humans (instead of the environment) determine which individuals will produce the most offspring.

- A long-term selection experiment on maize started in 1896. The original maize population showed variation in the heritable trait of kernel protein content.

- Plants with the highest kernel protein content were chosen to be the parents of the next generation and thus produced many more offspring than plants with low kernel protein content.

- **Figure 1.3** shows how the average percent protein content changed in this maize population over 100 generations. See **BioSkills 2** and Chapter 2 of Part 1 of this study guide for additional information on making and reading graphs.

Fitness and Adaptation

- **Fitness**, in biology, refers to an individual's ability to produce offspring. Individuals with high fitness have more offspring than those with low fitness.

- A trait that increases an individual's fitness in a particular environment is an **adaptation**. In the maize example, high kernel protein content became adaptive once researchers started artificially selecting for it.

- The cell theory and the theory of evolution are two central, unifying ideas of biology.

✓ CHECK YOUR UNDERSTANDING

(c) How does what we know about artificial selection support the theory of evolution by natural selection?

_____

_____

_____

1.4 The Tree of Life

- Natural selection can cause populations of one species to diverge and form new species. This is called **speciation**.

- Evolution by natural selection implies that all species are descended from a single common ancestor such that a family tree of all organisms—the **tree of life**—could be drawn.

Using Molecules to Understand the Tree of Life

- Carl Woese and colleagues attempted to discover the evolutionary relationships, or **phylogeny**, of all organisms by studying small subunit ribosomal RNA (rRNA).

- Small subunit rRNA is found in all organisms because it is required for cell growth and reproduction. RNA is made of four ribonucleotides (symbolized by the letters A, U, G, and C) connected in a linear sequence (**Figure 1.4**).

Analyzing rRNA

- Ribosomal RNA ribonucleotide sequence is a trait that has changed, or evolved, over time. Closely related species share a recent common ancestor and should have more similar small subunit rRNA than distantly related species do.

- Small subunit rRNA sequences can be analyzed to produce a **phylogenetic tree** showing the probable evolutionary relationships of the organisms studied.

- On a phylogenetic tree, species that share distinctive traits are closely related and are placed close to each other.

The Tree of Life Estimated from an Array of Genes

- An analysis of small subunit rRNA and other genes from many different species can be used to produce a tree of life (**Figure 1.5**; see **BioSkills 3** for additional information on reading phylogenetic trees).

- This tree shows three major groups of organisms: the eukaryotes (Eukarya), the Bacteria, and the Archaea.

- Eukaryotic cells have a nucleus whereas prokaryotes (the Bacteria and the Archaea) do not (**Figure 1.6**). Most prokaryotes are unicellular whereas many eukaryotes are multicellular.

- The tree of life work led to recognition (1) that the Archaea are a major domain of life, (2) that fungi are more closely related to animals than to plants, and (3) that traditional classification systems did not accurately represent evolutionary history.

The Tree of Life Is a Work in Progress

- Researchers continue to analyze additional molecules and other types of data. New data and new analysis techniques cause continuing debate about the placement of certain branches, improving our understanding of the tree of life.

CHECK YOUR UNDERSTANDING

(d) Archaea are more closely related to Eukaryotes than to Bacteria. Draw a simple phylogenetic tree depicting this relationship. What hypothesis might you make about RNA sequence similarity among these groups?

_____

_____

How Should We Name Branches on the Tree of Life?

- **Taxonomy** is the naming and classification of organism. Any named group is a **taxon** (plural: **taxa**).

- Modern taxonomy attempts to accurately portray the evolutionary history, or phylogeny, of organisms.

- The **domain** is a taxonomic category created to accurately describe the three major branches of life: the Bacteria, the Archaea, and the Eukarya.

- The term **phylum** (plural: **phyla**) refers to major domain lineages (e.g., the chordates).

Scientific (Species) Names

- Carolus Linnaeus created the naming system used today where each organism is given a unique two-part **scientific name** based on Latin or Greek word roots.

- The first part is the **genus**, which includes one to several closely related species. The second name is the **species**. Individuals that breed together are included in the same species.

- Scientific names are italicized, and genus names are capitalized. Each two-part name is unique (e.g., _Homo sapiens_ for humans).

Scientific Names Are Often Descriptive

- The Latin or Greek word roots used in scientific names typically describe the organism in some way (e.g., _Homo_ means "man" and _sapiens_ means "wise"). Thus, learning common word roots may be useful (see **BioSkills 4**).

1.5 Doing Biology

The Nature of Science

- Scientists ask questions about the natural world that can be addressed by collecting data. Questions that cannot be answered in this way cannot be addressed by science.

- Teaching evolution is controversial in some public schools although most biologists and most religious leaders do not think their faith conflicts with the scientific theory of evolution because religion and science answer different types of questions.
- Science involves formulating hypotheses and evaluating them based on data-based evidence whereas religious faith addresses questions that cannot be answered by data.

Why Do Giraffes Have Long Necks? An Introduction to Hypothesis Testing

- Giraffes must have long necks so they can better compete for food that is high up—right? But it turns out food competition is only part of the story.
- Data collected by Robert Simmons and Lue Scheepers suggest that long necks in giraffes may be favored because they are useful in fights rather than because they allow giraffes to reach food that is not available to other mammals.
- Alternative hypotheses must be evaluated by testing their predictions.
- These predictions can be tested with carefully designed observational or experimental studies. A hypothesis is supported if its predictions are correct but must be modified or replaced with an alternative explanation if its predictions are not accurate.

The Food Competition Hypothesis: Predictions and Tests

- The hypothesis that giraffes evolved long necks in order to reach high food sources leads to three predictions, each of which can be separately evaluated.
- First, neck length is variable. Studies in zoos and in nature confirm this prediction.
- Second, neck length is heritable. This prediction is still untested, because giraffe breeding experiments are difficult to conduct.
- Third, giraffes typically feed high in trees, especially when food is scarce. Data do not support this prediction; giraffes normally feed with bent necks (**Figure 1.7**).

The Sexual Competition Hypothesis: Predictions and Tests

- An alternative hypothesis proposes that giraffes evolved long necks because longer-necked males win more fights than shorter-necked males do. Winners gain access to estrous females and presumably father more offspring.
- Research shows that long-necked males win more fights than shorter-necked males.

- Long necks may primarily improve the fitness of male giraffes competing for matings with females rather than affecting giraffe ability to compete for food.

CHECK YOUR UNDERSTANDING

(e) Write out the predictions you would make based on the sexual competition hypothesis, and compare your predictions to the data given to support this hypothesis.

_____

_____

_____

_____

- Experiments with giraffes are difficult, so for an in-depth analysis of experimental design, we turn to ant navigation.

How Do Ants Navigate? An Introduction to Experimental Design

- Experiments allow scientists to test the effect of one defined factor on some natural process.
- For example, Saharan desert ants search for food to bring back to their colony in scorching heat. Ants take a wandering path while foraging, but return straight to their nest via the shortest route—a straight line.

The Pedometer Hypothesis

- Initial studies showed that ants know the approximate direction of their nest relative to the Sun. However, they must also know how far to go in order to successfully find the colony.
- These ants do not use landmarks to find their way, so Matthias Wittlinger and colleagues hypothesized that they use leg movement information to determine how far they have gone.
- This pedometer hypothesis predicts that ants measure distance by combining information about the number of steps they have taken with knowledge of their stride length and the angles of their turns to find their colony.

Testing the Hypothesis

- Ants travelled 10 m through a channel to a feeder, where they were caught and assigned to one of three test groups.
- **Stumps:** twenty-five ants had their lower legs cut so that their legs were shorter than normal. **Normal:** twenty-five ants did not have any leg modification. **Stilts:** twenty-five ants had bristles glued to their legs to give them longer legs.

- Ants were placed in a different channel for their return trip and researchers measured the distance they travelled before looking for their nest (**Figure 1.8**).

- Ants with stumps stopped short, normal ants travelled the correct distance, and ants with stilts walked too far before searching for their nest opening.

- When these same ants were recaptured several days later, ants from all three groups travelled the correct distance, demonstrating that the manipulation in and of itself did not affect ant behavior.

Interpreting the Results

- The results support the pedometer hypothesis because the stride length and step number did affect ant nest-finding ability.

- If the pedometer hypothesis had been wrong, stride number and length should have had no effect on ant homing ability. A statement of what we should observe if the tested hypothesis is not correct is called a **null hypothesis**.

✓ CHECK YOUR UNDERSTANDING

(f) Consider the earlier giraffe example. What would be the null hypothesis for the sexual competition hypothesis and what results would have supported this null hypothesis?

_____

_____

_____

Important Characteristics of Good Experimental Design

- Control groups should be included to check for other factors that could affect experimental results (e.g., inclusion of the "normal" ant group checked whether switching ants to a new channel affected their behavior).

- Experimental conditions not being tested should be held constant (e.g., each group contained the same species of ant from the same nest at the same time of day, etc.).

- Tests should be repeated; larger sample sizes minimize the influence of random variation on the results.

- Good experimental design minimizes the possibility of alternative interpretations of results. In our ant example, researchers were able to conclude that desert ants do use stride length and number to measure distance travelled from their nest.

- Questions about organisms lead to hypotheses that are evaluated using evidence-based decision making.

B. MASTERING CHALLENGING TOPICS

This chapter presents several key unifying concepts in biology. Many topics introduced in this chapter are covered in greater depth later on in this text. At this point you do not need to be too concerned about the details of evolutionary processes or what rRNA does. Do, however, spend some time thinking about how biologists use hypotheses to generate predictions that can be tested in order to discover information about the natural world. You might find it useful to practice generating your own hypotheses and testable predictions. For example, if plants seem to do better in one window than in another, what hypothesis might explain this difference? What testable predictions follow from this hypothesis, and how could you test them? Can you design an experiment that controls all variables except the one of interest? What kind of sample size would you use? Throughout the following chapters, description of important experiments will often be used to explain key biological concepts. If you master the ability to learn about biology through evaluating experiments, you will not have to resort to rote memory because you will *understand* the topic.

C. ASSESSING WHAT YOU'VE LEARNED

1. Which statement is *false*?
   a. All living organisms are made of cells.
   b. All living organisms are multicellular.
   c. Cells arise from preexisting cells.
   d. All life is organized in compartments surrounded by thin flexible membranes.

2. In Pasteur's experiment, what different results would have supported the spontaneous generation hypothesis?
   a. No cell growth in either flask
   b. Cell growth only in the swan-necked flask
   c. Cell growth in the straight-necked flask
   d. Cell growth in both flasks

3. Pasteur's experiment with a swan-necked flask rejected the spontaneous generation hypothesis by demonstrating:
   a. cells can grow in broth after it has been boiled.
   b. cells grow in broth unless the broth is sealed off from the air.
   c. cells grow only in broth exposed to a source of preexisting cells.
   d. bacteria, but not fungi, can grow in broth.

4. A group of individuals of the same species living in an area at the same time constitutes a:
   a. community.
   b. population.
   c. genus.
   d. family.

5. In addition to being made of cells, all organisms (choose all that apply):
   a. process information.
   b. reproduce sexually.
   c. are the products of evolution.
   d. acquire and use energy.
   e. have a nucleus.

6. The theory of evolution by natural selection states:
   a. all species are descended from a common ancestor.
   b. no traits are heritable.
   c. natural selection makes individuals more fit.
   d. species remain unchanged over time.

7. You observe a population in which organisms vary greatly in size. Larger organisms leave more offspring on average, yet over many generations there is no change in the average size of individuals in the populations. Which of the following most likely explains this result?
   a. Fitness does not affect natural selection in this population.
   b. Smaller organisms are more fit even though they leave fewer descendents.
   c. Size is not heritable in this population.
   d. This population does not evolve.

8. A moth learns to visit white flowers, and the good food source lets it lay lots of eggs. What do you conclude?
   a. Evolution has occurred.
   b. The moth will definitely pass this adaptation on to its offspring.
   c. Artificial selection has occurred.
   d. The moth has a high fitness.

9. You obtain a small sample of maize. Every kernel in your sample happens to have exactly 10 percent protein content. Can you use this sample to repeat the maize artificial selection experiment? Choose the best answer.
   a. No, the average percent protein was 11 percent at the start of the maize artificial selection experiment.
   b. Yes, but you are likely to get different results.
   c. Yes, but it will take a long time.
   d. No, you must have a starting sample that is variable for percent protein.

10. Researchers did a second maize selection experiment where they selected individuals with the lowest protein content. Which of the following statements about these two selection experiments is true? (Choose all that apply.)
    a. Both of these experiments are examples of artificial selection.
    b. It is impossible to predict the outcome of this second experiment.
    c. High protein content was an adaptation in the first experiment, but low protein content is an adaptation in the second experiment.
    d. Both experiments should show an increase in average protein content over time.

11. Modern taxonomy attempts to describe the _____, or historical relationships, among organisms.
    a. variation
    b. natural selection
    c. phylogeny
    d. physical similarity

12. If dogs have an RNA sequence that starts with G-U-G-A-C-U-G-A, but the same piece of RNA starts with G-A-G-A-C-C-G-A in cats and with U-U-G-A-C-U-G-A in bears, which of these groups appear to be more closely related based on this information?
    a. Dogs and bears
    b. Dogs and cats
    c. Bears and cats
    d. These sequences suggest that dogs, bears, and cats are equally related.

13. Use the tree of life (**Figure 1.5**) to determine which of the following groups is most closely related to animals.
    a. Archaea
    b. Slime molds
    c. Red algae
    d. Land plants

14. If the ant experiment described earlier had just included one ant in each group and the stump ant was observed to travel the same distance as the normal ant, which of the following might you conclude?
    a. Cutting the ant's leg injured it so that it walked less far than it might have otherwise.
    b. Individual variation among ants is too great to conclude much from tests with single ants.
    c. Something happened to the normal ant to make it walk less far than usual.
    d. Ants do not use number of steps and stride length to calculate distance from nest.

15. Which of the following steps is *not* vital to a well-designed experiment?
    a. A control group is included.
    b. All variables other than those being tested are controlled or kept the same between treatments.
    c. Experiments are repeated on many individuals (large sample size).
    d. The steps described in a, b, and c are all important.

CHAPTER 1—ANSWER KEY

Check Your Understanding

(a) Prior to the mid-1600s, microscopes powerful enough to let researchers see cells were not available.

(b) A keen sense of smell and taste might have helped a hunter-gatherer survive better by improving his or her plant identification skills.

(c) Artificial selection shows that differential survival and reproduction cause change in a population over time because those heritable traits that are selected for increase in frequency in each succeeding generation.

(d)

Hypothesis: Archaea RNA is more similar to Eukaryote RNA than to Bacteria RNA sequence. Note: because phylogenetic trees show the same hypothesized relationships when branches are rotated around nodes, it doesn't matter which of the two shorter branches above is labeled Archaea and which is labeled Eukarya. The same evolutionary history is shown either way.

(e) **Prediction 1:** Males with longer necks win more fights—supported by data.

Prediction 2: Males that win fights have more offspring; this is indirectly supported by data showing that winners of fights gain access to estrous females.

Prediction 3: Neck length is heritable.

Predictions 2 and 3 require further testing to confirm the sexual competition hypothesis as proposed by Simmons and Scheepers.

(f) The null hypothesis for the sexual competition hypothesis would be that neck length has no effect on fight outcome. Observations of no relationship between neck length and number of fights won would have supported this null hypothesis.

Assessing What You've Learned

1. b; 2. d; 3. c; 4. b; 5. a, c, d; 6. a; 7. c; 8. d; 9. d; 10. a, c; 11. c; 12. a; 13. b; 14. b; 15. d

Looking Forward—Key Concepts in Future Chapters

Cells—Chapter 6 and Unit 2, especially Chapter 7

Evolution of the first cells is discussed in **Chapter 6**, followed by Unit 2 (**Chapters 7–11**), which covers cell structure and function, cell–cell interactions, some chemical pathways found in many cells, and cell replication (how cells make more cells).

Evolution and Natural Selection—Chapter 4, Unit 5 (Chapters 24–27)

Chapter 4 investigates hypotheses on how first life might have evolved from nonliving macromolecules. Evolutionary processes in general, including natural selection, are more thoroughly examined in Unit 5.

Phylogeny and Taxonomy—Unit 6 (Chapters 28–35)

Chapter 1 introduces phylogeny, taxonomy, and the major groups of organisms. Unit 6 describes major groups of organisms and how they are related in greater detail.

Water and Carbon: The Chemical Basis of Life

Looking Back—Key Concepts in Earlier Chapters

Evolution—Chapter 1

Having learned that species change over time and that all organisms descend from a single common ancestor, in Chapter 2 you explore the conditions under which this common ancestor (the first life) evolved.

KEY CONCEPTS

- Atoms can bond together to form molecules. Electron sharing in these chemical bonds varies from equal sharing in nonpolar covalent bonds to complete transfer in ionic bonds.

- Life depends on water, which is an extremely efficient solvent due to its polarity and its ability to form hydrogen bonds.

- Active motion and stored potential (e.g., chemical bonds) are both types of energy and can be used to do work or supply heat.

- Chemical reactions spontaneously occur only if they lower potential energy and/or increase disorder; nonspontaneous reactions require energy input.

- Most key compounds found in organisms contain carbon. Some important carbon-containing molecules formed early in Earth's history.

A. CHAPTER OUTLINE

Chapter Introduction

- The theory of **chemical evolution** states that complex carbon-containing compounds formed from simpler molecules in early Earth's oceans (the pattern) as energy from sunlight and heat was converted to chemical energy in molecular bonds (the process explaining the pattern).

- Eventually this process created a compound that could replicate itself. When this compound became surrounded by a membrane, life began.

2.1 Atoms, Ions, and Molecules: The Building Blocks of Chemical Evolution

- Organisms are primarily made of hydrogen, carbon, nitrogen, and oxygen. Examining the structure of these atoms and simple molecules they form will improve our understanding of their function in chemical evolution and in cells today.

- Structure's effect on function is a major theme in biology.

Basic Atomic Structure

- An atom's nucleus contains protons and neutrons. Negatively charged electrons orbit the nucleus and are attracted to the protons. An atom is electrically neutral when the number of protons equals the number of electrons (**Figure 2.1**).

- Each **element** is defined by its number of protons (its **atomic number**), which is indicated by a subscript on the periodic table (**Figure 2.2**).

- Superscripts indicate an element's **mass number**, the sum of the atomic nucleus' protons and neutrons.

- Atoms of an element that differ in number of neutrons are called **isotopes**. Neutrons and proton masses are similar and each equal about 1 dalton, whereas the mass of an electron is so small that it is typically ignored.

- Electron **orbitals** hold up to two electrons, and **electron shells** are groups of orbitals numbered by distance from the nucleus. Electrons fill the inner shells first (**Figure 2.3**).

- Elements commonly found in organisms have at least one unpaired electron in their outer (**valence**) shell. When electron orbitals are not filled, **chemical bonds** may form to fill the valence shell.

How Does Covalent Bonding Hold Molecules Together?

- Atoms are more stable when each orbital has two electrons. Orbitals with one electron each can overlap so that the nuclei share the two electrons, forming a **covalent bond** (**Figure 2.4**). A **molecule** is formed when atoms are joined by covalent bonds.

Nonpolar and Polar Bonds (Figure 2.5)

- A **nonpolar covalent bond** forms when electrons are evenly shared between two atoms (e.g., a C–H bond). If one atom holds onto the shared electrons more tightly than the other does, a **polar covalent bond** forms.

- Oxygen's affinity for electrons, its **electronegativity**, is very high. Carbon and hydrogen have approximately equal electronegativities, and nitrogen's electronegativity is greater than C and H but less than oxygen's.

Polar Bonds Produce Partial Charges on Atoms

- Water (H_2O) has two polar covalent bonds; electrons stay closer to the oxygen atom, giving it a partial negative charge ($\delta-$) and leaving the hydrogen atoms with partial positive charges ($\delta+$).

- A water molecule's partial charges make life possible.

Ionic Bonding, Ions, and the Electron-Sharing Continuum

- **Ionic** bonds form when electrons are completely transferred from one atom to another (**Figure 2.6**) to give the resulting **ions** (charged atoms) a full outermost electron shell.

- An atom that loses an electron becomes positively charged (a **cation**), and an atom that gains an electron becomes negatively charged (an **anion**).

- Opposite charges attract, so cations and anions form salts (e.g., table salt, NaCl).

- Electron sharing in chemical bonds exists in a continuum from equal sharing (nonpolar covalent bonds) to partial sharing (polar covalent bonds) to complete transfer or no sharing (ionic bonds). See **Figure 2.7**.

- Molecules found in organisms mostly have the stronger covalent bonds.

✔ CHECK YOUR UNDERSTANDING

(a) Examine **Figure 2.3**. What type of bond is found in lithium fluoride and why?

_____

_____

In glucose ($C_6H_{12}O_6$), what types of bonds are found between: C and H?

_____

C and O?

_____

Why?

_____

_____

Some Simple Molecules Formed from C, H, N, and O

- The number of unpaired electrons in the valence shell determines the number of bonds an atom can make.

- Thus, C bonds with four H to form methane (CH_4), N bonds with three H to form ammonia (NH_3), and O bonds with two H to form H_2O (**Figure 2.8**).

- When there are two unpaired electrons in the valence shell, two nuclei can share four electrons in a double bond (e.g., carbon dioxide, CO_2). A triple bond can form if there are three unpaired electrons (e.g., molecular nitrogen, N_2).

The Geometry of Simple Molecules

- Molecular function is influenced by shape, and molecular shape is determined by bond geometry.

- Methane is tetrahedral because the electrons in the four C–H bonds push the bonds as far apart as they can get.

- Water is bent (V-shaped) in one plane because only two of the four electron orbitals in oxygen's valence shell participate in bonds (**Figure 2.9**).

Representing Molecules (Figure 2.10)

- The **molecular formula** states the numbers and types of atoms in a molecule.

- **Structural formulas** show which atoms are bonded together, and indicate single, double, or triple bonds.

- **Ball-and-stick models** show three-dimensional geometry and indicate the relative sizes of atoms.

- **Space-filling models** show spatial relationships between atoms more accurately but are harder to read. See **BioSkills 6** in your text for additional information.

 **CHECK YOUR UNDERSTANDING**

(b) Draw H_2CO and HCN, showing the electrons and orbitals involved in the bonds as in **Figure 2.8** (*Hint:* C forms a double bond with O and a triple bond with N in these two molecules).

Basic Concepts in Chemical Reactions

- Earth's ancient atmosphere contained CH_4, NH_3, H_2O, CO_2, and N_2. These molecules combined to form more complex compounds in chemical reactions. Chemical reactions can form, break apart, or rearrange chemical bonds.

- Molecules bond to each other in whole-number combinations. Researchers use the **mole**, 6.022×10^{23} molecules, to measure relative quantities of molecules.

- The **molecular weight** of a molecule equals the sum of the mass numbers of all its atoms, and the mass of one mole equals a molecule's molecular weight in grams.

- The concentration of a substance dissolved in a liquid (a **solution**) is typically expressed as **molarity** (M)—the number of moles present per liter of solution.

- Chemical evolution is thought to have occurred in an aqueous (water-based) solution.

CHECK YOUR UNDERSTANDING

(c) A nitrogen atom has 7 protons and 7 neutrons.
      What is its atomic number?

_____

      What is its mass number?

_____

What is the molecular weight of molecular nitrogen (nitrogen gas, N_2)?

_____

What type of bond holds N_2 together (be specific)?

_____

Is this bond weaker or stronger than the bond holding NaCl together?

_____

How much would a mole of N_2 weigh?

_____

2.2 The Early Oceans and the Properties of Water

- Life is based on water because water is a great **solvent**; substances (**solutes**) dissolve easily in water.

Why Is Water Such an Efficient Solvent?

- The O side of a water molecule has a partial negative charge ($\delta-$) and the H atoms carry partial positive charges ($\delta+$). The molecule has a bent shape so the partial negative charge sticks out away from the H atoms (**Figure 2.12**).

- Partial charges on the H and O in different H_2O molecules attract each other and form weak **hydrogen bonds**. These bonds are weaker than covalent or ionic bonds, but many together have important biological effects.

- Hydrogen bonds also form between H_2O and other polar molecules or ions dissolve in water. These bonds help these **hydrophilic** substances stay in solution.

- Uncharged and nonpolar or **hydrophobic** substances interact with each other and avoid water (**Figure 2.13**).

CHECK YOUR UNDERSTANDING

(d) Compare and contrast hydrogen bonds and ionic bonds.

_____

_____

_____

_____

How Does Water's Structure Correlate with Its Properties?

Cohesion, Adhesion, and Surface Tension

- Water molecules stick together due to hydrogen bonding; this is called **cohesion**. When water molecules hydrogen-bond with a polar or charged surface, the process is called **adhesion**.

- Cohesion and adhesion explain many properties of water, including the formation of a concave meniscus when water adheres to the sides of a glass container and surface tension when water sticks together at the surface (**Figure 2.14**).

Water Is Denser as a Liquid than as a Solid (Figure 2.15)

- Hydrogen bonds in ice connect molecules in a crystal-like pattern. Liquid water has fewer hydrogen bonds, so its molecules pack more tightly. Thus, ice is less dense than liquid water and large bodies of water rarely freeze through because ice floats.

Water Has a High Capacity for Absorbing Energy

- The amount of energy it takes to raise the temperature of 1 gram of a substance 1°C is called its **specific heat**. Water has a very high specific heat because hydrogen bonds must be broken in order for water molecules to move faster.

- Water also has a high **heat of vaporization** (the amount of energy required to change 1 gram of liquid to a gas). This is why the evaporation of water in sweat cools you.

- Compounds important in chemical evolution dissolved in the ocean, where water's temperature-buffering capacity protected them from energy that could have broken them apart.

✔ CHECK YOUR UNDERSTANDING

(e) Ocean temperatures have increased about 1/10th of a degree in the past 10 years. This does not sound like much—so why is it viewed as strong evidence for global warming?

_____

_____

_____

_____

Acid–Base Reactions Involve a Transfer of Protons

- A water molecule can lose a proton, a **hydrogen ion** (H^+), which tends to associate with another water molecule to make H_3O^+. This dissociation reaction can be written: $H_2O + H_2O \rightarrow H_3O^+ + OH^-$ (a **hydroxide ion**).

- **Acid–base reactions** are chemical reactions in which one molecule or ion gives up protons (the **acid**), and another accepts protons (the **base**). Water can act as both an acid and a base. Water is a very weak acid because only a few molecules dissociate.

- Strong acids and bases can greatly increase or decrease (respectively) a solution's proton concentration.

pH Indicates the Concentration of Protons

- Pure water at 25°C has a proton concentration, symbolized as $[H^+]$, of 1.0×10^{-7} M. The **pH scale** indicates $[H^+]$: $pH = -\log[H^+]$ (**Figure 2.16**, see **BioSkills 7** for a review of logarithms).

- Because pH is a logarithmic scale, one unit increase in pH represents a tenfold increase in hydrogen ion concentration.

Buffers Protect against Damaging Changes in pH

- Most living cells have a pH of about 7 and are very sensitive to changes in pH. The hydrogen ion concentration affects acid–base reactions and the structure and function of polar and charged substances.

- **Buffers** are compounds that minimize the pH changes of a solution. They help maintain relatively constant conditions (**homeostasis**) within organisms.

- Most buffers are weak acids. For example, acetic acid (CH_3COOH) gives up a proton if the pH starts to increase but acetate ions (CH_3COO^-) accept protons to form acetic acid if pH decreases.

The pH Scale Measures the Acidity or Alkalinity of a Solution

- Pure water has a pH of 7 and is considered neutral. Solutions with a pH lower than 7 are acidic (more likely than water to give up protons); those with a pH greater than 7 are basic and tend to accept protons.

2.3 Chemical Reactions, Chemical Evolution, and Chemical Energy

- The chemical evolution theory proposes that the simple molecules present on ancient Earth reacted with one another to create larger, more complex molecules.

- These reactions could have occurred in the atmosphere in the presence of the volcanic gases H_2, CH_4, and CO or in deep-sea vents where the same volcanic gases exist along with iron, nickel, and other metals.

How Do Chemical Reactions Happen?

- Chemical reactions are written out with the initial **reactants** on the left and the **product(s)** on the right. A double arrow means that the reaction is reversible and *g, l,* and *aq* indicate whether the molecules are gas, liquid, or an aqueous solution.

- Chemical reactions should be written out so that they are balanced and have the same number of each atom on each side of the equation.

- **Chemical equilibrium** occurs when forward and reverse reactions proceed at the same rate. Temperature and concentration affect chemical equilibriums. For example, adding more reactants will cause more product(s) to be made until a new equilibrium is achieved.

- Temperature also affects chemical equilibriums. **Endothermic** reactions must absorb heat to proceed, whereas **exothermic** reactions release heat.

What Is Energy?

- **Energy** is the ability to do work or supply heat, and **potential energy** is stored energy.

- Because electrons are attracted to the protons in the nucleus, electrons in outer shells have more potential energy than those in inner shells and will fall into an inner shell if possible (**Figure 2.17**).

- **Kinetic energy** is movement energy. Molecules are always moving; this type of kinetic energy is called **thermal energy**. **Temperature** measures thermal energy. If an object is "cold," its molecules are moving slowly.

- When a cold object touches a hot object (with faster-moving molecules), thermal energy (**heat**) is transferred from the hot to the cold object.

- One type of energy can change into another type of energy, but is not created or destroyed. For example, as an electron in an outer shell falls to an inner shell, potential energy is converted to kinetic energy (**Figure 2.18**).

- The **first law of thermodynamics** states that energy is conserved.

- Energy transformation was important in chemical evolution. Molecules on ancient Earth were exposed to a lot of heat and unscreened radiation from the Sun.

Chemical Evolution: A Model System

- Computer models that simulate chemical reactions were used to determine whether formaldehyde (H_2CO) and hydrogen cyanide (HCN) could have been produced on ancient Earth.

- H_2CO and HCN are intermediates in the creation of more complex molecules, but their formation is not spontaneous; energy must be added to the reactant gases CO_2 and H_2.

What Makes a Chemical Reaction Spontaneous?

- Reactions proceed spontaneously without any energy input when products have less potential energy than reactants and/or when products are less ordered than reactants (**Figure 2.21**).

- In exothermic reactions, reactants have more potential energy than products and the difference is released as heat. The difference between reactant and product potential energy is symbolized by ΔH (Δ means "change"). Exothermic reactions have a negative ΔH.

- **Entropy**, symbolized by S, is the amount of disorder in a group of molecules. When entropy increases, ΔS is positive (products less ordered than reactants). The **second law of thermodynamics** states that entropy increases over time in an isolated system.

- You can determine whether a reaction will proceed spontaneously by assessing the overall change in potential energy (ΔH) and disorder (ΔS). Changes in entropy are more important at higher temperatures.

- A reaction is spontaneous when the **Gibbs free-energy change** (ΔG) is negative. $\Delta G = \Delta H - T\Delta S$, where T is temperature in degrees Kelvin. Reactions proceed in the direction that lowers the overall free energy.

- Reactions that proceed spontaneously ($\Delta G < 0$) are **exergonic** and release energy. Reactions that require energy input to occur ($\Delta G > 0$) are **endergonic**. When $\Delta G = 0$, a reaction is at equilibrium and no net change in reactants or products occurs.

CHECK YOUR UNDERSTANDING

(f) Consider the following reaction: $C_6H_{12}O_6 + 6\ O_2 \rightarrow 6\ CO_2 + 6\ H_2O$.

Is ΔH positive or negative?

_____

Is ΔS positive or negative?

_____

How about ΔG?

_____

Will this reaction proceed spontaneously?

_____

The Roles of Temperature and Concentration in Chemical Reactions

- Spontaneous reactions $(-\Delta G)$ may nonetheless occur extremely slowly. High temperatures and high concentrations cause more reactant collisions and increase reaction rates (**Figure 2.22**).

CHECK YOUR UNDERSTANDING

(g) In the experiment shown in **Figure 2.22**, why did the students test temperature and concentration separately instead of together (for example, by increasing temperature and concentration at the same time)?

_____

_____

What would happen to the reaction rate if temperature and reactant concentration were increased at the same time?

_____

_____

What might happen to the reaction rate if the temperature was increased but the reactant concentration was decreased?

_____

_____

Energy Inputs and the Start of Chemical Evolution

- A computer model of chemical reactions among CH_4, NH_3, H_2O, CO_2, H_2, and N_2 molecules included both spontaneous reactions and endergonic reactions that occur in the presence of sunlight energy (photons).

- High-energy photons reached ancient Earth because there was little ozone (O_3) in the atmosphere. These photons can knock electrons away from valence shells, breaking apart molecules and forming **free radicals**—highly reactive atoms with unpaired electrons (**Figure 2.23**).

- Reactions involving these free radicals were included in the computer simulation of reactions that could occur on ancient Earth.

The First Reactions in Chemical Evolution

- Researchers measured the actual rates of reactions that form H_2CO and other molecules and used these rates in their computer models.

- These computer models show that significant amounts of H_2CO and HCN form under temperature and concentration conditions likely found in early Earth's atmosphere.

- Atmospheric H_2CO and HCN would have rained into the ocean. These compounds could also have formed in deep-sea vents where extremely hot water provides energy to drive endergonic reactions.

How Did Chemical Energy Change during Chemical Evolution?

- Electrons in H_2CO bonds have more potential energy than do the electrons in CO_2 and H_2.

- The formation of H_2CO and HCN is a critical step in chemical evolution because energy from sunlight has been converted to **chemical energy** (potential energy in chemical bonds).

- The equation for formaldehyde formation can be written to include the necessary energy input as follows:

$$CO_2(g) + 2\ H_2(g) + sunlight \rightarrow H_2CO(g) + H_2O(g).$$

2.4 The Importance of Carbon

- Molecules containing carbon are called **organic molecules**.

- Carbon has four unpaired valence electrons, allowing it to form four covalent bonds. This flexibility in bonding makes possible an incredible diversity of carbon-based molecular structures (**Figure 2.24**).

Linking Carbon Atoms Together (Figure 2.25)

- Larger organic molecules with carbon–carbon bonds easily form when CH_2O and HCN are heated.

Functional Groups

- Groups of H, N, or O atoms (**functional groups**) bonded to C determine the behavior of organic compounds (**Table 2.3**). Carbon provides the structural framework, but the functional groups determine a molecule's chemistry.

- For example, amino groups attract protons but carboxyl groups tend to donate them, and carbonyl groups often participate in reactions that link molecules together.

- Molecules with hydroxyl groups are weak acids, whereas phosphate groups give molecules a negative charge and sulfhydryl groups link different molecules together.

- During chemical evolution, further reactions among organic molecules with various functional groups produced the common large molecules still found in cells today: proteins, nucleic acids, and carbohydrates.

✔ CHECK YOUR UNDERSTANDING

(h–m) Label the following functional groups:

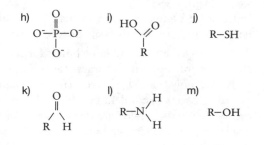

B. MASTERING CHALLENGING TOPICS

Chapter 2 is a whirlwind tour of basic chemistry. You need to know these basics in order to understand (1) molecules and chemical reactions that are essential to life, and (2) how life could have evolved from nonlife. You may want to look at an introductory chemistry text to ensure you understand concepts such as **electron shells**, **potential** and **kinetic energy**, different **bond types**, **pH**, and how **acid–base reactions** work. You may also want to go back and carefully examine the figures in this chapter. Make a concept map for yourself, showing how the concepts introduced in this chapter relate to our best hypotheses about the chemical reactions that could have led to life.

C. ASSESSING WHAT YOU'VE LEARNED

1. Two isotopes of an element vary in the number of:
   a. electrons.
   b. protons.
   c. neutrons.
   d. atoms.

2. The number of _____ within an atom never varies among atoms of a specific element.
   a. electrons
   b. protons
   c. neutrons

3. A silicon atom has four unpaired electrons in its valence shell, two paired electrons in its inner shell, and 14 electrons in all. How many hydrogens will it form covalent bonds with?
   a. None, it will form ionic bonds instead of covalent bonds.
   b. 2
   c. 4
   d. 14

4. Which group of atoms determines a molecule's behavior?
   a. Functional group
   b. Carbon skeleton
   c. Hydrogen atoms
   d. Ion group

5. Ninety-six percent of matter found in living organisms is composed of these four elements:
   a. hydrogen, carbon, oxygen, and sulfur.
   b. carbon, sulfur, nitrogen, and iron.
   c. sulfur, hydrogen, oxygen, and nitrogen.
   d. hydrogen, carbon, oxygen, and nitrogen.

6. A large boulder is balanced on top of a hill. As it starts to roll down the hill, _____ energy is transferred into _____ energy.
   a. kinetic; potential
   b. potential; kinetic
   c. kinetic; thermal
   d. potential; chemical

7. Which of these factor(s) determine(s) whether a chemical reaction will occur spontaneously?
   a. Amount of entropy (disorder in a group of molecules)
   b. Temperature
   c. Amount of potential energy differences between products and reactants
   d. All of the above

8. Which phase of matter has the highest entropy?
   a. Solid
   b. Liquid
   c. Gas

9. Which of the following compounds is most likely to form hydrogen bonds with water when in an aqueous solution?
   a. H_2
   b. CH_4
   c. NH_3
   d. None of these, only water can form hydrogen bonds with water.

10. Substances that give up protons during acid–base reactions are:
    a. isotopes.
    b. acids.
    c. bases.
    d. ions.

11. How many protons are present in 1 liter of an aqueous solution with pH 6? (Hint: See definition of molarity.)
    a. 1.0×10^6
    b. 6.022×10^{17}
    c. 6.022×10^{23}
    d. No protons are present.

12. Which of the following correctly orders bond types from least electron sharing to most electron sharing?
    a. Ionic, polar covalent, and covalent
    b. Polar covalent, ionic, and covalent
    c. Covalent, polar covalent, and ionic
    d. Covalent, ionic, and polar covalent

13. The bonds between the oxygen and hydrogen atoms within a water molecule are _____ bonds.
    a. hydrogen
    b. ionic
    c. polar covalent
    d. nonpolar covalent

14. Which of the following is part of the theory of chemical evolution?
    a. On ancient Earth, nonpolar covalent bonds became polar covalent bonds as energy was absorbed from the Sun.
    b. Elements evolve through time as their bonds change.
    c. Natural selection led to formation of complex molecules found in living cells today.
    d. On ancient Earth, simple chemical compounds combined to form larger complex molecules.

15. Which of the following functional groups can act as a base?
    a. Carboxyl
    b. Amino
    c. Hydroxyl
    d. Carbonyl

CHAPTER 2—ANSWER KEY

Check Your Understanding

(a) Lithium fluoride has ionic bonds; lithium, like sodium, tends to transfer one electron to fluoride, which, like chlorine, accepts one electron to fill its outer electron shell. In glucose, C and H form nonpolar covalent bonds because their electronegativities are similar, whereas C and O form polar covalent bonds because O has a higher electronegativity than C.

(b)

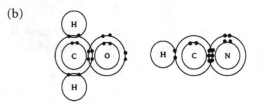

(c) The atomic number for nitrogen is 7, and its mass number is 14.

 The molecular weight of N_2 is 28. A triple covalent bond holds N_2 together, and a mole of N_2 would weigh 28 g.

 Covalent bonds are stronger than ionic bonds. For example, ionic bonds easily dissolve in water.

(d) Both hydrogen bonds and ionic bonds form due to attraction of opposite charges for each other. However, only partial charges (from unequally shared electrons in covalent bonds) are involved in hydrogen bonds, while ions have gained or lost whole electrons. Hydrogen bonds are much weaker than ionic bonds.

(e) A rise in ocean temperatures is strong evidence for global warming because water has such high specific heat. It takes a *lot* of energy to warm all the ocean water even a tenth of a degree.

(f) ΔH is negative because more bonds are created than broken; the products have less potential energy than the reactants. ΔS is positive because there are more molecules and thus more entropy in the products than in the reactants. ΔG must thus be negative, and this reaction can proceed spontaneously, although it might proceed slowly in the absence of any enzymes, as you will see in Chapter 9.

(g) Good experimental design varies just one variable at a time. If the students had changed temperature and concentration at the same time, they would not have been able to tell which variable caused the reaction rate change. However, if temperature and reactant concentration were increased at the same time, we would expect the reaction rate to increase even more than if only one of these variables were changed. It is hard to predict what would happen to the reaction rate if the temperature were increased, but the reactant concentration decreased since these effects oppose each other. The end reaction rate would depend on the amount of temperature increase and reactant concentration decrease.

(h) phosphate

(i) carboxyl

(j) sulfhydryl

(k) carbonyl

(l) amino

(m) hydroxyl

Assessing What You've Learned

1. c; 2. b; 3. c; 4. a; 5. d; 6. b; 7. d; 8. c; 9. c; 10. b; 11. b; 12. a; 13. c; 14. d; 15. b

Looking Forward—Key Concepts in Later Chapters

Chemical Evolution—Chapters 3–6

Chapter 2 sets the stage for the major players in chemical evolution. **Chapters 3 through 6** continue this story and describe hypotheses about how life evolved from nonlife in parallel with descriptions of key organic molecules: proteins, nucleic acids, carbohydrates, and lipids.

Chemical Reactions—Chapters 9 and 10

Respiration, photosynthesis, and other important metabolic pathways involve a series of chemical reactions.

Water Is Special—Chapters 37, 42, 54, and others

Water has some unusual properties that make it essential to life. **Chapters 37** and **42** discuss how plants and animals deal with their water requirements, and **Chapter 54** describes how water cycles through the ecosystem.

Functional Groups—Chapters 3, 4, 5, 9, 38, 43, and others

Functional groups introduced in this chapter are discussed in greater detail later, with reference to specific chemical pathways and nutrition.

Chapter 3

Protein Structure and Function

Looking Back—Key Concepts in Early Chapters

Chemical Evolution—Chapter 2

The start of chemical evolution and basic chemistry needed to understand the chemical reactions discussed in this chapter are described in **Chapter 2**. For example, Chapter 2 discussed how formaldehyde (H_2CO) and hydrogen cyanide (HCN) can form from reactants such as H_2, CO_2, CH_4, and NH_3 present on ancient Earth.

Chemical Bonds, Chemical Reactions, and Chemical Energy—Chapter 2

Chapter 2 also provides background information on how chemical bonds form and how the change in Gibbs free energy (ΔG) determines whether a reaction can occur spontaneously. In this chapter you will learn how enzymes speed reactions by lowering the free energy required for transition states.

KEY CONCEPTS

- Most cell functions depend on proteins.

- Proteins are made of amino acids that vary in side-chain composition. Side chains affect amino acid structure and function.

- Protein structure is amazingly diverse and can be analyzed at four levels ranging from amino acid sequence to subunit interactions.

- Enzymes are protein catalysts that speed up specific chemical reactions by bringing reactants together in a way that facilitates the reaction.

A. CHAPTER OUTLINE

Chapter Introduction

- Alexander I. Oparin and J. B. S. Haldane independently proposed the theory of chemical evolution in the 1920s.

- This theory proposes that complex carbon compounds formed in ancient Earth's atmosphere and ocean (pattern component) as energy from sunlight and other sources was converted into chemical energy in covalent bonds (process component).

- Chemical evolution involves four steps: (1) production of small molecules such as H_2CO and HCN; (2) creation of a **prebiotic soup** in shallow ocean waters as larger organic subunit molecules formed; (3) linkage of subunits to form the large organic molecules important in cells today; and (4) evolution of a self-replicating molecule.

- This chapter focuses on proteins: one of the four major classes of large organic molecules.

3.1 Early Origin-of-Life Experiments

- Stanley Miller (1953) simulated chemical evolution in the laboratory in order to test whether the first stages of chemical evolution could have occurred on ancient Earth.

- Miller put methane (CH_4), ammonia (NH_3), and hydrogen (H_2) gases in a large flask (the "atmosphere"). He connected this flask to another flask with boiling water (the "ocean"). Water vapor circulated from the "ocean" to the "atmosphere."

- A condenser cooled the water vapor on its return from the "atmosphere" flask to the "ocean" flask. As the vapor rained down, it carried molecules with it so that molecules dissolved in the water also circulated through the system (**Figure 3.1**).

- Miller added sparks (simulated lightning) to the "atmosphere" as an energy source.

- Presto! HCN, H_2CO, and other more complex organic compounds, including amino acids, were made and accumulated in the "ocean" flask.

3.2 Amino Acids and Polymerization

- Experiments similar to Miller's show that amino acids and other organic molecules easily form under conditions that more accurately simulate conditions on ancient Earth (e.g., using CO, CO_2, and H_2 in the "atmosphere").
- Amino acid precursors are also made when experimental conditions mimic deep-ocean volcanoes and when sunlight is used as an energy source instead of electric sparks.
- Meteorites sometimes contain amino acids, demonstrating that these organic molecules are also produced in outer space.

✔ CHECK YOUR UNDERSTANDING

(a) How would you change Miller's experimental setup to test whether chemical evolution of complex organic molecules can occur in an atmosphere of CO, CO_2, and H_2 using energy from sunlight?

_____

_____

_____

The Structure of Amino Acids

- The huge diversity of proteins in living organisms is made from just 20 or so amino acid subunits.
- All **amino acids** have a central carbon atom bonded to H, an amino functional group (NH_2), a variable side chain (the "R-group") and a carboxyl functional group (COOH) (**Figure 3.2**).
- In water (pH 7), the amino group attracts a proton to form NH_3^+, and the carboxyl group loses a proton (acting as a weak acid) to form COO−. This polarity helps amino acids stay in solution and makes them more reactive.

The Nature of Side Chains

- The 20 major amino acids differ only in the variable side chain (R-group) attached to the central carbon. R-groups determine amino acid function because they differ in their size, shape, reactivity, and interactions with water.

Functional Groups Affect Reactivity

- Amino acids can be grouped as nonpolar, polar, or electrically charged (**Figure 3.3**) and include many of the functional groups discussed in Chapter 2.

The Polarity of Side Chains Affects Solubility

- Nonpolar **hydrophobic** R-groups cannot form hydrogen bonds with water and tend to group together in solution, whereas polar or charged **hydrophilic** R-groups interact readily with water (**Table 3.1**).

✔ CHECK YOUR UNDERSTANDING

(b) Look at the structures of valine and serine in **Figure 3.3**. Why is serine more likely to interact with water than valine?

_____

_____

_____

How Do Amino Acids Link to Form Proteins?

- **Polymerization** is the process of linking molecular subunits (**monomers**) to make **polymers** (macromolecules of attached monomers; **Figure 3.4**).
- Proteins macromolecules (large polymers) are made from amino acid monomers. These and other polymers presumably formed in the prebiotic soup.
- Monomers do not form polymers spontaneously because polymers have less entropy ($-\Delta S$) and higher potential energy ($+\Delta H$) than do unlinked monomer subunits, making ΔG positive at all temperatures for polymer formation.

Could Polymerization Occur in the Energy-Rich Environment of Early Earth?

- Monomers polymerize through **condensation reactions** (also called **dehydration reactions**) that release a water molecule as the polymer bond forms. In the reverse reaction, **hydrolysis**, water reacts with the polymer to release a monomer subunit (**Figure 3.5**).
- Condensation reactions require energy input since ΔG is positive for these reactions (whereas hydrolysis is energetically favorable).
- Experiments show that polymers can form on tiny mineral particles like those found in clay or mud. Adsorption to these tiny particles protects a growing macromolecule from hydrolysis.
- Amino acid polymerization can also occur in hot, metal-rich environments like those found near undersea volcanoes. Polymers can form in cooler environments if a carbon- and sulfur-containing gas is present.

The Peptide Bond

- Amino acid condensation reactions bond the carboxyl group of one molecule to the amino group on another, forming a **peptide bond** (**Figure 3.6**). A chain of linked amino acids (called residues) is a **polypeptide**.

- Peptide bonds are especially strong covalent bonds, having some double-bond qualities.
- Side chains stick out from linked amino acid residues (**Figure 3.7**) and can interact with water and/or other R-groups.
- Polypeptides are flexible and have directionality. The N-terminus has an unlinked amino group, and the C-terminus has a carboxyl group. While the peptide bond does not rotate, the single bonds on each side of it can (**Figure 3.8**).
- By convention, polypeptides are shown with the N-terminus on the left and the free carboxyl on the right. Amino acids are numbered starting from the N-terminus.
- **Oligopeptides** have less than 50 amino acids, whereas **proteins** have 50 or more linked amino acids.

3.3 Proteins Are the Most Versatile Large Molecules in Cells

- Proteins perform most cell work.
- Proteins called enzymes speed up cell reactions (**catalysis**).
- Antibodies and complement proteins defend cells against disease.
- Motor and contractile proteins move cells and cell cargo.
- Proteins carry and receive signals between cells.
- Structural proteins provide mechanical support for cells and form accessory structures such as hair.
- Transport proteins move molecules in and out of cells and throughout the body.

3.4 What Do Proteins Look Like?

- Proteins have incredibly diverse sizes, shapes, and chemical properties that make possible the many different protein functions.
- Protein structure can be analyzed at four levels.

Primary Structure

- The **primary structure** of a protein is its unique sequence of amino acids. With 20 amino acids, a 10-amino-acids-long polypeptide has billions of possible sequences! Specifically, a polypeptide of length n has 20^n possible sequences.
- Because the amino acid R-groups affect a polypeptide's size, shape, chemical reactivity, and interactions with water, just one amino acid change can radically alter protein function (**Figure 3.10**).

(c) Your text explains that a change from glutamate to valine in one of hemoglobin's amino acids causes sickle-cell disease. Which structural levels were likely affected and why?

_____

_____

_____

Would you expect the effects to be more or less serious if glutamate were changed to aspartate? Why?

_____

Secondary Structure

- Hydrogen bonding between the carbonyl O of one amino acid residue and the amino H of another creates protein **secondary structure**.
- Partial charges on these atoms attract each other only when a polypeptide is bent such that these functional groups lie close to each other (**Figure 3.11**).
- These hydrogen bonds can cause the polypeptide to coil into an α-helix or to fold into a β-pleated sheet (like the folds in a paper fan).
- Ribbon diagrams show α-helices as coils and β-pleated sheets as arrows lying next to each other with alternating directionality.
- Primary structure affects the secondary structures that form; different amino acids facilitate formation of different secondary structures.
- A protein's secondary structure increases its stability; although one hydrogen bond is fairly weak, many hydrogen bonds contribute to each of these secondary structures.

Tertiary Structure

- Interactions among amino acid R-groups and between R-groups and the peptide-bonded backbone determine a polypeptide's final three-dimensional shape or **tertiary structure** (**Figure 3.12**).
- Hydrogen bonds can form between H atoms and the carbonyl group in the peptide backbone and between H atoms and R-groups with partial negative charges.
- Hydrophobic side chains group together in water and are stabilized by weak electrical attractions called **van der Waals interactions**. Each van der Waals interaction is weaker than a hydrogen bond, but the summed effect can be significant.
- Covalent **disulfide bonds** or bridges can form between sulfur atoms of two cysteines.

- Ionic bonds can form between side chains with opposite charges.
- A protein's primary and secondary structure determine its tertiary structure.

Quaternary Structure

- Some proteins have several polypeptide subunits; bonds or other interactions between subunit R-groups or peptide backbones produce **quaternary structure (Figure 3.13)**.
- Quaternary structure depends on subunit tertiary structure, which in turn depends on primary and secondary structure.
- The combined effects of primary, secondary, tertiary, and sometimes quaternary structure (**Table 3.2**) allow for amazing diversity in protein form and function.
- Overall structure depends on the way in which polypeptide chains fold.

Folding and Function

- Polypeptides in solution automatically fold so that hydrophobic side chains group together away from water and hydrophilic side chains face water.
- This folding is often spontaneous and exergonic because the bonds and van der Waals interactions stabilize the molecule.

Folding in Ribonuclease

- Christian Anfison and colleagues showed that the enzyme ribonuclease was **denatured**, or unfolded, when treated with chemicals that break hydrogen bonds and disulfide bonds (**Figure 3.14**; proteins can also be denatured by heat).
- The folded three-dimensional shape is basic to protein function, so a denatured protein cannot function normally. However, when Anfison removed the denaturing compounds, ribonuclease spontaneously refolded and was able to function normally.
- Proteins called **molecular chaperones** help proteins fold correctly in cells. Many of these molecular chaperones are heat-shock proteins, produced after a cell experiences high temperature or other conditions that disrupt protein folding.

Folding in Prions

- **Prions** are misfolded forms of normal proteins that are infectious. They can cause normal proteins to misfold into the infectious form; yet again protein shape determines protein function.
- A misfolded form of the prion protein (**Figure 3.15**) causes the spongiform encephalopathies. These diseases are typically transmitted when individuals consume tissue containing prions.

3.5 Enzymes: An Introduction to Catalysis

Enzymes Help Reactions Clear Two Hurdles

- Chemical reactions occur when molecules collide with the correct orientation and have enough kinetic energy to break and/or form chemical bonds despite the repulsion of the electrons involved in the bond.
- Enzymes bring reactant molecules (**substrates**) together in the right positions and decrease the necessary kinetic energy for a reaction by facilitating the formation of a **transition state** in which a combination of old and new bonds is present.
- **Activation energy** is the free energy necessary to achieve this transition state. Reactions occur when reactants have enough kinetic energy to reach the transition state.
- Temperature is a measure of kinetic energy. Chemical reactions occur faster at higher temperatures than at lower temperatures because more molecules have enough energy to reach the transition state (**Figure 3.16**).

Graphing the Energy of the Transition State

- **Figure 3.17** graphically represents free-energy changes during a chemical reaction. ΔG is negative (the reactants have more free energy than the products), so this reaction is exothermic and spontaneous.
- However, the activation energy (E_a) needed to form the transition state is high, so this reaction would occur very slowly.
- Transition states have high free energy states because the substrate bonds are destabilized, but the new bonds are not yet complete. The less stable the transition state, the higher the activation energy and the slower the reaction.

Enzymes Lower the Activation Energy

- Stabilizing the transition state can speed up reactions because when activation energy is lower, less kinetic energy is necessary for the reaction to proceed.
- Electrons in the transition state can be stabilized by interactions with other ions, atoms, or molecules. A **catalyst** is a molecule that lowers the activation energy of a reaction by stabilizing the transition state (**Figure 3.18**).
- Catalysts are not used up in chemical reactions and do not change ΔG or the energy of the reactants or products. Only E_a, the amount of energy required to achieve the transition state, is affected.
- **Enzymes** are protein catalysts that typically catalyze only one reaction. They speed up reactions by

bringing reactants together in the correct orientation and stabilizing the transition state.

- Most biological chemical reactions occur only at meaningful rates in the presence of an enzyme; many enzymes allow reactions to proceed a million times faster than they would without the enzyme.

✔ **CHECK YOUR UNDERSTANDING**

(d) Under identical conditions, which of the following reactions will proceed faster and why? (Hint: You may wish to refer to **Figures 3.17** and **3.18** in your text and label ΔG and E_a on these graphs before answering.)

Reaction 1

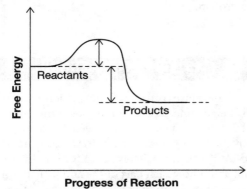

Reaction 2

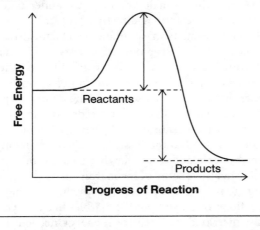

_____

_____

_____

How Do Enzymes Work?

- Fischer's 1894 **lock-and-key model** of enzyme action described enzymes as being like locks into which substrate "keys" fit.

- This model correctly identified that enzymes (1) bring substrates together in specific positions that facilitate reactions and (2) are very specific as to which reaction(s) they catalyze since substrates only bind to sites with the right shape and chemical properties.

- The substrate-binding site on an enzyme is called the **active site** (**Figure 3.19**). It is usually a fold or hole within a globular protein. Catalysis occurs at the active site.

- The lock-and-key model has been revised because enzymes are not rigid like locks; instead, they are flexible and often change shape as substrates bind to the active site. This is referred to as an **induced fit**.

- Interactions between the substrate and active-site R-groups include hydrogen bonding, formation of temporary covalent bonds involved in atom transfer, and gain or loss of protons aided by acidic or basic R-groups.

- Substrate interaction with the enzyme active site stabilizes the transition state, causing a decrease in the reaction's activation energy.

- Reaction products have a low affinity for the active site and are therefore released from the enzyme, restoring the enzyme to its original shape.

- Enzyme action has three steps: **initiation**, when reactants are correctly oriented as they bind to the active site; **transition state facilitation**, as interactions between the substrate and active-site R-groups lower the reaction's activation energy; and **termination** when products are released (**Figure 3.20**).

Do Enzymes Act Alone?

- Enzymes often require **cofactors**—atoms or molecules not included in an enzyme's primary structure. Some cofactors are metal ions and others are small organic molecules called **coenzymes**.

- Enzyme cofactors often bind to the active site and are involved in transition state stabilization.

- Many vitamins are necessary because they are precursors for enzyme cofactors. Vitamin deficiency diseases occur when there are not enough enzyme cofactors to allow normal enzyme function.

Most Enzymes Are Regulated

- Enzymes associate with molecules that change protein structure in ways that affect enzyme activity.

- **Competitive inhibition** occurs when a regulatory molecule similar to the substrate competes with the substrate for active-site binding (**Figure 3.21a**).

- **Allosteric regulation** occurs when a molecule changes enzyme conformation by binding at a location other than the active site. Allosteric regulation is more common than competitive inhibition and can either increase or decrease enzyme activity (**Figure 3.21b**).

What Limits the Rate of Catalysis?

- Enzyme kinetics describe how various factors affect enzyme-catalyzed reaction rates.

- **Figure 3.22** shows that when substrate concentrations are low, the speed of product formation increases linearly with substrate concentration. The rate of product formation slows and then levels out at higher substrate concentrations.

- Although the rate of product formation is much lower for uncatalyzed reactions, these uncatalyzed reactions show linear increases in reaction rate across all ranges of substrate concentration.

- Initially, each increase in substrate concentration allows the available enzymes to facilitate more reactions.

- As substrate concentration increases, active sites become filled for all enzymes and no further increase in reaction rate is possible. This is known as saturation kinetics.

How Do Physical Conditions Affect Enzyme Function?

- Enzymatic activity is strongly influenced by temperature (effect on enzyme movement and substrate kinetic energy) and pH (effect on R-group charge and proton- and electron-transfer ability of the active site).

- Most enzymes function best at some particular temperature and pH. Because natural selection favors organisms with enzymes that function efficiently, temperature and pH optima usually reflect an enzyme's environment (**Figure 3.23**).

- Primary structure variations that increase enzyme activity at the temperatures and pH the enzyme typically experiences help organisms adapt to their environment.

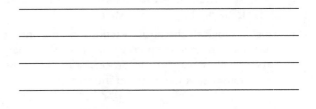

CHECK YOUR UNDERSTANDING

(e) The enzymes shown in **Figure 3.23** have distinct peaks at which their activities are highest. Why aren't these enzymes able to maintain their maximum activity over a wide range of temperatures and pH? Provide an example of how a temperature or pH change could affect enzyme action.

_____

_____

_____

_____

_____

_____

Was the First Living Entity a Protein Catalyst?

- A self-replicator would need to catalyze the polymerization of its copy, and enzymes are the most efficient catalysts known.

- Proteins could have formed in the prebiotic soup, but the first self-replicator likely also needed a copy-making template—something not found in proteins. Consequently, the first living entity was probably not a protein.

B. MASTERING CHALLENGING TOPICS

This chapter gives you lots of information about proteins and their amino acid subunits—how they fit together and the chemical reactions involved. Be sure to carefully study the text figures. Practice drawing a generic amino acid until you can reliably reproduce it. As you draw, think about how each part of the amino acid functions. Draw a second amino acid and show how they would link together.

Sometimes analogies help us understand processes that we cannot directly observe. For example, the end-products of spontaneous chemical reactions typically have less **potential energy** (ΔH) than the starting reactants do. Similarly, an object on the floor has less potential energy than an object on the table does, because it is closer to Earth's center of gravity. But the object on the table needs **kinetic energy** to reach the **transition state** of passing the table edge before it will fall to the floor. If you have many marbles on the table and a fan blowing in the room, air currents may give some of the marbles enough kinetic energy to reach the table edge, at which point they fall to the floor. But if you put the marbles in a box, you have just increased the **activation energy** needed to reach the transition state of reaching the table edge and fewer (if any) marbles will fall to the floor. Similarly, even spontaneous chemical reactions may proceed so slowly as to be hardly detectable if the activation energy needed to reach the transition state is high and/or the kinetic energy of the molecules is low.

C. ASSESSING WHAT YOU'VE LEARNED

1. Stanley Miller showed experimentally that complex, carbon-containing compounds could form from simple molecules present in early Earth's atmosphere and oceans. In Miller's experiment, _____ was the energy source that permitted these complex molecules to form.
   a. heat
   b. light
   c. electricity
   d. wind

2. Selenocysteine is an extremely rare but naturally occurring amino acid. Its R-group is CH_2-SeH, where Se stands for selenium. Selenium has the same electronegativity as carbon. Based on this information, how would you describe this side chain?
   a. Nonpolar
   b. Polar
   c. Electrically charged

3. All amino acids have a(n) _____ functional group and a(n) _____ functional group, but vary in their _____ group.
   a. carboxyl; amino; R-
   b. R-; carboxyl; amino
   c. phosphate; R-; amino
   d. amino; hydroxyl; R-

4. Monomers join to form polymers via:
   a. acid–base reactions.
   b. hydrolysis.
   c. condensation reactions.
   d. exergonic.

5. Some proteins have sections rich in the amino acid proline that assume a spiral structure stabilized by hydrogen bonding. What level of structure does this represent?
   a. Primary
   b. Secondary
   c. Tertiary
   d. Quaternary

6. Which of the following statements regarding proteins is correct?
   a. Alpha helices and beta sheets are part of a polypeptide's tertiary structure.
   b. Nucleotides join via condensation reactions to form polypeptides.
   c. The words *polypeptide* and *multienzyme complex* have identical meaning.
   d. Protein structure is correlated with protein function.

7. Which of the following side-chain interactions are important in determining tertiary structure?
   a. Covalent bonds
   b. Ionic bonds
   c. Hydrogen bonds and van der Waals interactions
   d. a and c
   e. b and c
   f. a, b, and c

8. Why do proteins have so many different cell functions?
   a. They are the simplest cell molecules.
   b. They are structurally diverse.
   c. They were the first life-form.
   d. They have strong peptide bonds.

9. Which condition(s) speed(s) the rate of chemical reactions? (Choose all that apply.)
   a. High kinetic energy of reactants
   b. High pH
   c. Presence of a catalyst
   d. High concentration of products

10. How do catalysts increase reaction rates?
    a. By increasing the kinetic energy of reactants
    b. By providing chemical energy
    c. By decreasing ΔG
    d. By stabilizing the transition state

11. Further research led to modifications in the lock-and-key model of enzyme action. Which assumption was *not* correct?
    a. Enzymes are rigid structures analogous to a lock.
    b. Enzymes bring substrates together so that reactions are more likely to occur.
    c. Enzymes are very specific as to the reaction they catalyze.

12. Why is lysine more likely to interact with water than valine?
    a. Lysine is more hydrophobic than valine.
    b. Lysine has more hydrogen than valine does.
    c. Lysine is charged at cell pH.
    d. Lysine's central carbon bonds to an amino group and a carboxyl group.

13. A molecule is a competitive inhibitor of an enzyme if:
    a. it changes enzyme shape by binding to a site other than the active site.
    b. it can bind in an enzyme's active site because it is chemically similar to the substrate.
    c. it competes with the enzyme for substrate binding.

14. At low substrate concentrations, enzyme-catalyzed reaction rates _____ as substrate concentration increases, and at high substrate concentrations, enzyme-catalyzed reaction rates _____.
    a. show an exponential increase; increase linearly
    b. increase linearly; increase linearly
    c. decrease linearly; plateau at a maximum speed
    d. increase linearly; plateau at a maximum speed

15. Which of the following statements is correct?
    a. All enzymes function best in the neutral pH range.
    b. Different versions of enzymes may have different temperature and pH optima.
    c. Most enzymes are not affected by temperature or pH.
    d. Cofactors help enzymes by providing energy for chemical reactions.

CHAPTER 3—ANSWER KEY

Check Your Understanding

(a) I would place CO, CO_2, and H_2 in the large flask and expose the experimental setup to sunlight instead of to sparks. Better yet, I might use a light source that mimics the sunlight of ancient Earth (before the buildup of UV-absorbing ozone).

(b) Serine has an OH group, which is polar due to oxygen's high electronegativity; valine just has C and H in its R-group, and these atoms form nonpolar covalent bonds. Thus, serine is more likely to interact with water.

(c) All structural levels were likely affected because all are influenced by primary structure. However, the effects would have been less serious if glutamate were changed to aspartate because these two R-groups are similar (both are charged and soluble in water).

(d) Reaction 1 will proceed faster than reaction 2 because it has a lower activation energy. Activation energy, not ΔG, determines the speed at which a reaction takes place.

(e) The degree of hydrogen bonding changes with temperature and the state of basic and acidic R-groups changes with pH. Because the shape of the active site and its ability to stabilize a reaction's transition state depends on hydrogen bonding and R-group state, the active site optimally catalyzes a reaction only over a certain temperature and pH range. For example, an R-group that is able to accept a proton from the substrate at a certain pH might pick up a proton from the surrounding solution at a lower pH. This would prevent the R-group from being able to accept the substrate's proton and the enzyme would no longer be able to effectively catalyze the reaction.

What You've Learned

1. c; 2. a; 3. a; 4. c; 5. b; 6. d; 7. f; 8. b; 9. a, c; 10. d; 11. a; 12. c; 13. b; 14. d; 15. b

---

Looking Forward—Key Concepts in Future Chapters

Proteins are the workhorse molecules of cells, so these molecules will appear in many later chapters. Here, a few protein functions are highlighted.

Structural Proteins and Transport Proteins—Chapters 6–8, 37, 42, 43, and others

Proteins form most of the major structures in cells and are responsible for transporting molecules into and out of cells.

Enzymes—Chapters 9, 10, 16, and others

Chapters **9** and **10** focus on some of the most important cell reactions. These reactions are, of course, catalyzed by enzymes. You will learn more about enzyme regulation in these chapters. In **Chapter 16** you will learn about how enzymes are involved in transcription and translation—the processes by which cells make enzymes and other proteins.

Other Protein Functions—Chapters 39, 47, and 48 (hormones) and Chapter 49 (defense)

Nucleic Acids and the RNA World

KEY CONCEPTS

🔑 Nucleotides consist of a sugar (ribose or deoxyribose), a phosphate group, and a nitrogen-containing base. Nucleotides polymerize to form ribonucleic acid (RNA) and deoxyribonucleic acid (DNA).

🔑 A sequence of nitrogen-containing bases forms DNA's primary structure. DNA's secondary structure is a double helix: two twisted DNA strands running in opposite directions held together by base pairing. DNA's structure easily allows organisms to store and replicate information.

🔑 RNA's primary structure is similar to that of DNA, but RNA is mostly single-stranded with short regions of hairpins and double helices where RNA folds back on itself.

🔑 The first self-replicating molecule was probably made of RNA because RNA can both carry information and catalyze chemical reactions.

A. CHAPTER OUTLINE

Chapter Introduction

- The theory of chemical evolution suggests that the first step in the evolution of life was the formation of a self-replicating molecule. Chance errors in this molecule's copy-making would then create variations that could undergo natural selection.

- Most biologists think the first self-replicator was a molecule of ribonucleic acid (RNA). This chapter's discussion of nucleic acid structure and function will also introduce this RNA world hypothesis.

4.1 What Is a Nucleic Acid?

- A **nucleic acid** is a polymer of nucleotides (**Figure 4.1**). Each **nucleotide** is phosphate group bonded to a sugar (via its 5′ carbon), which is, in turn, bonded to a nitrogenous base.

- A **sugar** is an organic compound with a carbonyl group and several hydroxyl groups.

- **Ribonucleotides** contain ribose and **deoxyribonucleotides** contain deoxyribose. These are both five-carbon sugars (pentoses), but deoxyribose has an H instead of an −OH on its 2′ carbon.

- Nitrogenous bases include adenine (A) and guanine (G) from the two-ring purine structural group and cytosine (C), uracil (U), and thymine (T) from the single-ring pyrimidine group.

- Adenine, guanine, and cytosine are found in both types of nucleotides, but ribonucleotides contain uracil whereas deoxyribonucleotides contain thymine.

Could Chemical Evolution Result in the Production of Nucleotides?

- Chemical evolution simulations have not yet produced nucleotides.

- Sugars and purines are easily made, but ribose would have had to dominate in an environment for nucleic acids to form, and pyrimidines are not easily synthesized under conditions thought to exist on ancient Earth.

How Do Nucleotides Polymerize to Form Nucleic Acids?

- Nucleotides polymerize to form nucleic acids when a condensation reaction forms a **phosphodiester linkage** between the phosphate group of one nucleotide and the –OH group on the 3′ carbon of another nucleotide (**Figure 4.2**).

- **Ribonucleic acid** (RNA) is made from nucleotides containing ribose, and **deoxyribonucleic acid** (DNA) is made from nucleotides containing deoxyribose.

DNA and RNA Strands Are Directional

- Sugar-phosphate linkages form the "backbone" of a nucleic acid; this chain is directional because one end has an unlinked 5′ carbon and the other end has an unlinked 3′ carbon (**Figure 4.3**).

- The nucleotide base sequence creates a nucleic acid's primary structure, which is written out from the 5′ to the 3′ direction using the first letter of each nucleotide.

Polymerization Is an Endergonic Process

- In cells, enzyme-catalyzed polymerization only occurs after nucleotide free energy has been raised by adding two extra phosphate groups per nucleotide (**Figure 4.4**). These **phosphorylated** nucleotides are said to be "activated."

Could Nucleic Acids Form in the Prebiotic Soup?

- Phosphorylated nucleotides can form RNA when incubated for several days with tiny mineral particles. (See **gel electrophoresis** and **autoradiography** descriptions in **BioSkills 9** for a description of techniques used to identify these RNA molecules.)

- If all the nucleotides were able to form during chemical evolution, they could have polymerized on clay particles to form RNA and DNA.

✔ CHECK YOUR UNDERSTANDING

(a) Why can't the condensation reaction illustrated in **Figure 4.2** happen exactly as shown?

_____

_____

_____

_____

4.2 DNA Structure and Function

- Both nucleic acids and proteins have a backbone-like primary structure formed via condensation reactions. Phosphodiester bonds connect nucleotides in DNA and RNA and peptide bonds connect amino acids in proteins.

- DNA and RNA secondary structure is due to hydrogen bonding between nitrogenous bases that extend outward from the nucleic acid backbone.

What Is the Nature of DNA's Secondary Structure?

- James Watson and Francis Crick discovered DNA's secondary structure in 1953. Their model built on the following results from other laboratories.

Early Data Provide Clues

- Chemists had determined the structure of nucleotides and the nature of phosphodiester linkages, so Watson and Crick knew that DNA had a sugar-phosphate backbone.

- Chargaff's rules stated that in DNA the number of purines equals the number of pyrimidines, the numbers of T's and A's are equal, and the number of C's and G's are equal.

- Rosalind Franklin and Maurice Wilkins used DNA **X-ray crystallography** (**BioSkills 10**) to calculate distances between atom groups that regularly repeated; they concluded that the structure of DNA was either helical or spiral.

DNA Strands Are Antiparallel

- Using models, Watson and Crick concluded that two antiparallel DNA strands form a **double helix** with the hydrophilic sugar-phosphate backbone facing the exterior and with closely packed purine-pyrimidine nitrogenous base pairs on the interior.

- *Antiparallel* means that if one DNA strand starts with the 5′ end and ends with a 3′ end, the other strand is oriented such that its 3′ end matches up with the first strand's 5′ end.

- Franklin and Wilkins's measurements fit DNA geometry as follows: purine-pyrimidine pairing makes the width of the double helix equal to 2.0 nm. The distance between adjacent bases is 0.34 nm and the length of one complete helical turn is 3.4 nm (**Figure 4.8**).

- Within the helix, adenine hydrogen-bonds with thymine and guanine hydrogen-bonds with cytosine: this is known as **complementary base pairing** or **Watson–Crick pairing** (**Figure 4.6**).

- The guanine-cytosine pair is stronger than the adenine-thymine pair because this pair forms three, instead of two, hydrogen bonds.

CHECK YOUR UNDERSTANDING

(b) Why did Watson and Crick conclude that adenine bonds to thymine and guanine bonds to cytosine within the double helix?

_____

_____

_____

_____

_____

The Double Helix

- DNA's double helix is like a twisted ladder where the complementary base pairs form the rungs. Twisting occurs to maximize hydrogen bonding between the pairs (**Figure 4.7**).
- Each turn of the helix has 10 complementary base pairings, and the double helix has a major and minor groove.
- DNA is water soluble due to its hydrophilic sugar-phosphate backbone, but it is extremely stable due to its tightly packed hydrophobic interior.

DNA Functions as an Information-Containing Molecule

- The nitrogenous base sequence of DNA stores information required for the growth and reproduction of all living cells.
- Complementary base pairing provides a simple mechanism for DNA replication since each strand can serve as a mold or template for formation of a new complementary strand (**Figure 4.9**).
- DNA's double helix can be pulled apart by enzymes or heating.
- The exposed bases on the original **template strand** can then pair with free deoxyribonucleotides, which join together by phosphodiester linkages to create a new **complementary strand** with an antiparallel directionality.
- Through complementary base pairing, two daughter molecules are produced, each identical to the original strand.

- DNA replication requires many enzymes in today's cells—DNA cannot self-replicate.

Is DNA a Catalytic Molecule?

- DNA's stability makes it a reliable store for genetic information; DNA has few exposed chemical groups that could participate in chemical reactions, and the presence of an H instead of OH on the sugar's 2′ carbon makes DNA less reactive than RNA.
- Stable molecules like DNA are poor catalysts because stabilization of the transition state requires interaction between the substrate and the catalyst. Enzymes, for example, have exposed functional groups that interact with substrates.
- Proteins have incredible structural diversity, leading to a similar diversity of specific binding sites. DNA, on the other hand, has an extremely simple primary and secondary structure and does not appear to be able to catalyze any chemical reactions in nature.
- Thus, biologists think the first life-form was RNA, not DNA.

4.3 RNA Structure and Function

Structurally, RNA Differs from DNA

- DNA only has primary and secondary structure, but RNA can have four levels of structure, just like proteins.

Primary Structure

- RNA's primary structure is similar to DNA's, except: (1) RNA contains uracil instead of thymine, and (2) ribose, instead of deoxyribose, is the sugar in the sugar-phosphate backbone.
- The hydroxyl group on ribose's 2′ carbon can participate in reactions that degrade RNA, making RNA more reactive and less stable than DNA.

Secondary Structure

- RNA's secondary structure, like DNA's, is due to complementary base pairing between purine and pyrimidine nitrogenous bases. In RNA, two hydrogen bonds form between adenine and uracil, and three form between guanine and cytosine.
- However, RNA bases typically pair with complementary bases on the *same* strand: the RNA strand folds over, forming a hairpin structure where the bases on one side of the fold align with an antiparallel segment on the other side of the fold (**Figure 4.10**).

✓ CHECK YOUR UNDERSTANDING

(c) An RNA strand folds into a hairpin structure. One side of the base of the hairpin has the sequence (listed as always in 5′ to 3′ order): U-U-A-G-U-C-A. What is the nucleotide order on the complementary strand listed in 5′ to 3′ order?

_____

_____

A second RNA strand has the sequence C-A-G-G-G-U-C on one side of its hairpin stem. Which hairpin structure will be more stable? Why?

_____

_____

_____

- Other RNA secondary structures also occur spontaneously, and all are stabilized by hydrogen bonds. Hydrogen bond formation is exothermic; thus hairpin formation is exergonic even though entropy is decreased.

Tertiary and Quaternary Structures

- Sometimes hairpins and other RNA secondary structures have additional folds or are attached to other RNA strands, giving some RNA molecules tertiary and quaternary structure (**Table 4.1**).
- RNA molecules vary in their shapes and chemical properties; their structural and chemical complexity is intermediate between DNA's simplicity and protein diversity.

RNA's Structure Makes It an Extraordinarily Versatile Molecule

- RNA's function is also intermediate between that of proteins and DNA. For example, RNA molecules help process information but also catalyze some important reactions.

✓ CHECK YOUR UNDERSTANDING

(d) List four structural differences between DNA and RNA.

_____

_____

_____

_____

RNA Is an Information-Containing Molecule

- A RNA molecule could theoretically make a copy of itself; an initial template strand could be used to make a complementary strand that, once separated, could serve as a mold to form an RNA strand identical to the original template.
- RNA would be a rather unstable information-containing molecule. But, it might have survived long enough in the prebiotic soup to replicate itself.

RNA Can Function as a Catalytic Molecule

- RNA's structure is less variable than that of proteins, but it does have some secondary and tertiary structure as well as some chemical reactivity. These qualities allow RNA to stabilize some transition states.
- Altman and Cech discovered that, in the single-celled *Tetrahymena*, RNA **ribozymes** catalyze the hydrolysis and condensation of phosphodiester linkages. Other ribozymes catalyze other reactions.
- Perhaps an RNA molecule could catalyze the reactions necessary for its own replication!

4.4 The First Life-Form

- RNA can both provide a template for copying itself and catalyze the polymerization reaction that links monomers to make the (complementary) copy. These are the essential qualities of a self-replicator.
- Researchers propose that the first life-form was RNA. This is known as the RNA world hypothesis. Because no self-replicating molecules exist today, researchers test the RNA world hypothesis using laboratory simulations.
- Wendy Johnston and others in David Bartel's lab used artificial selection to create an RNA replicase that could catalyze the addition of ribonucleotides to a growing RNA strand.
- Researchers made many variable large RNA molecules. These were incubated with a small RNA template that was marked with a molecular tag.
- A few of the large RNAs used themselves as a template and catalyzed, albeit inefficiently, the addition of a few ribonucleotides to the starting template.
- Researchers used the molecular tag to isolate the round-1 ribozymes and created the next round of large RNA molecules by copying these ribozymes in a way that caused a few random changes to the primary sequence.
- The above process was repeated 18 times, each time selecting the best ribozymes as templates for the next round.

- At each step, some ribozymes showed improved catalyst function, and the final population included ribozymes that were good catalysts for the addition of ribonucleotides to a growing complementary RNA strand.

✔ CHECK YOUR UNDERSTANDING

(e) The researchers copied the selected RNA molecules in a way that introduced a few random changes to their sequence of bases. Most of the modified ribozymes worked worse than the original ones. Why, then, did the research team introduce these random changes?

_____

_____

_____

Do you think most natural mutations are also detrimental?

_____

_____

_____

- These results support the RNA world hypothesis; researchers may soon be able to produce a self-replicating molecule similar to the one that might have been our world's first life-form.

✔ CHECK YOUR UNDERSTANDING

(f) Give three arguments in support of the hypothesis that the first self-replicator was an RNA molecule.

_____

_____

_____

_____

_____

B. MASTERING CHALLENGING TOPICS

This chapter continues the structure-function theme from the previous chapter. Linking structures with functions is a great way to organize many biology concepts. Look at Table 1, where DNA and RNA structures are compared. Make your own table with additional columns for the functions related to the listed structures. Then add two more columns for protein structure and function. This will help you consider the similarities and differences between the different levels of structure for proteins and nucleic acids. A good understanding of how structure relates to function for these molecules will also help you master the details of the debate about which macromolecule might have been the first life-form.

C. ASSESSING WHAT YOU'VE LEARNED

1. The three components of a nucleotide are:
   a. a phosphorous atom, a 6-carbon sugar, and a purine.
   b. a sugar, a pyrimidine, and a carbonyl group.
   c. a phosphate group, a ribose, and an amino group.
   d. a sugar, a nitrogenous base, and a phosphate group.

2. Polymerization of nucleotides only becomes energetically favorable when:
   a. nucleotides are placed with small mineral particles.
   b. nucleotides are phosphorylated.
   c. the right enzyme is present.
   d. bases line up correctly so hydrogen bonds can form.

3. Which of the following issues are major challenges to the theory of chemical evolution?
   a. That living organisms use DNA for their genetic code and proteins as their catalysts
   b. That no self-replicating molecules exist today, so researchers can't study the question
   c. The ribose problem and the origin of pyrimidine bases
   d. That RNA cannot catalyze reactions and proteins cannot serve as templates for replication

4. What type of bond is a phosphodiester linkage?
   a. Covalent
   b. Polar covalent
   c. Ionic
   d. Hydrogen

5. If one strand of DNA has a 5′ to 3′ sequence of A-A-G-T-C-G, what will be the sequence on the complementary strand? Recall that nucleic acid sequences are always written out in the 5′ to 3′ direction.
   a. G-G-A-C-T-A
   b. T-T-C-A-G-C
   c. A-T-C-A-G-G
   d. C-G-A-C-T-T

6. The _____ structure of a nucleic acid is the sequence of nitrogenous bases.
   a. primary
   e. secondary
   f. tertiary
   g. quaternary

7. Chargaff's rules state that in DNA, the
   a. number of purines = number of pyrimidines, the number of A's = number of G's, and the number of T's = number of U's.
   b. number of purines = number of pyrimidines, the number of T's = number of A's, and the number of C's = number of G's.
   c. sugar-phosphate backbone is formed by phosphodiester linkages and is helical in structure.
   d. DNA strands are antiparallel and have complementary base pairing.

8. Which statement about DNA is *not* correct?
   a. DNA contains deoxyribose sugars.
   b. DNA has a double-helix structure.
   c. DNA contains uracil.
   d. DNA has a sugar-phosphate backbone.

9. Each nucleotide has one nitrogenous base: either a purine with a double-ring structure or a pyrimidine with a single ring. Which statement is true regarding complementary base pairing?
   a. Purines always pair with other purines, and pyrimidines always pair with other pyrimidines.
   b. The pyrimidine base cytosine (C) always pairs with the purine base guanine (A).
   c. In DNA, the purine base adenine (A) always pairs with the purine base guanine (G).
   d. Watson and Crick discovered complementary base pairing.

10. DNA is water soluble because:
    a. the nitrogenous bases form hydrogen bonds with water.
    b. the exterior-facing sugar-phosphate backbones are negatively charged and hydrophilic.
    c. the molecule has a hydrophobic interior.
    d. cells are about 70 percent water.

11. Could DNA catalyze its own replication?
    a. No, its primary and secondary structures are too simple and unreactive to catalyze reactions.
    b. Yes, all living organisms store their genetic information in their DNA sequences.
    c. Yes, DNA and RNA are structurally and functionally similar.
    d. No, DNA cannot serve as a template for its own replication.

12. _____ can catalyze chemical reactions as well as carry information to copy itself (themselves).
    a. DNA
    b. Proteins
    c. RNA
    d. Nitrogenous bases

13. Which of the following statements is (are) true about RNA? (Choose all that apply.)
    a. RNA primarily functions as a catalyst in modern cells.
    b. RNA is more reactive than DNA.
    c. RNA and protein structure are equally complex.
    d. RNA could theoretically serve as a template for its own synthesis.
    e. RNA is less stable than DNA.

14. Due to complementary base pairing, _____ form(s) a secondary structure called the hairpin loop.
    a. protein
    b. DNA
    c. RNA
    d. sugars

15. Wendy Johnston and co-workers created RNA molecules that catalyzed:
    a. the formation of hairpin loops.
    b. the formation of nucleotides from sugar, phosphate, and nitrogenous base subunits.
    c. DNA replication.
    d. the addition of ribonucleotides to a growing RNA strand.

CHAPTER 4—ANSWER KEY

Check Your Understanding

(a) **Figure 4.2** cannot happen as shown because it is an endergonic reaction; the product has both less entropy and more potential energy than the reactants. Enzymes only lower activation energy—they cannot provide energy. The energy for this reaction is provided by "activating" a nucleotide through the addition of two more phosphate groups. This increase in reactant potential energy makes the formation of the phosphodiester linkage exergonic, allowing this reaction to occur.

(b) Watson and Crick concluded that adenine bonds to thymine and cytosine to guanine because (1) only a purine-pyrimidine pair would fit inside the double helix based on Franklin and Wilkins' X-ray diffraction measurements, and (2) this pairing would explain Chargaff's observation that the number of T's always equals the number of A's in an organism's DNA.

(c) The complementary strand's sequence is U-G-A-C-U-A-A (remember that the strands are antiparallel and that you list nucleic acid sequences in 5′ to 3′ order). The second hairpin structure will be more stable because G and C form three hydrogen bonds, but U and A form only two. Thus, the second hairpin stem will have more hydrogen bonds stabilizing it.

(d) RNA differs from DNA in that it (1) has ribose, not deoxyribose, as its nucleotide sugar; (2) uses uracil instead of thymine; (3) is single stranded, not double stranded; and (4) can have tertiary structure due to the folding of secondary structures into complex shapes (more complex/less stable structure).

(e) Most changes impaired ribozyme function, because it is much easier to break something than to fix it. For example, if you *randomly* made changes to your car engine, most of the changes would cause problems; only a very few would improve the engine's function. The researchers introduced random changes to get the very few ribozymes with improved catalyst function. In nature, most mutations are also detrimental.

(f) Arguments supporting the hypothesis that the first self-replicator was an RNA molecule: (1) RNA can serve as a template for its own synthesis (although it takes two rounds of replication to get back to the original template molecule); (2) RNA can catalyze some chemical reactions, including phosphodiester bond formation; (3) none of the other macromolecules can do both 1 and 2; and (4) researchers have already successfully created RNA replicases that catalyze the addition of ribonucleotides to a growing RNA strand.

Assessing What You've Learned

1. d; 2. b; 3. c; 4. b; 5. d; 6. a; 7. b; 8. c; 9. d; 10. b; 11. a; 12. c; 13. b, d, e; 14. c; 15. d

Looking Forward—Key Concepts in Later Chapters

DNA—Chapters 11–20, 25, and others

DNA encodes the genetic information of all cells. Any discussion of genetics therefore necessitates reference to DNA. DNA is equally essential to descriptions of cell reproduction and natural selection (because traits must be heritable in order to be selected).

RNA—Chapters 16 and 35

RNA's main function in extant cells is to use the information contained in DNA to produce proteins. This process is described in **Chapter 16**. **Chapter 35** describes how viruses use RNA.

An Introduction to Carbohydrates

An Introduction to Carbohydrates

Looking Back—Key Concepts from Earlier Chapters

Chemical Evolution—Chapters 2–4

The environment of ancient Earth and basic chemistry needed to understand the chemical reactions that cause chemical evolution are described in **Chapter 2**. **Chapters 3** and **4** discuss chemical evolution with respect to proteins and nucleic acids.

Structure-Function Correlation—Chapters 3 and 4

You should be familiar with the ubiquitous correlation between structure and function from the discussion in **Chapters 3** and **4** of this relationship for proteins and nucleic acids. In Chapter 5 you learn about how the molecular structure of carbohydrates relates to their function.

KEY CONCEPTS

- Sugars and other carbohydrates vary a lot in structure.

- Condensation reactions form glycosidic linkages that join sugar monomers (monosaccharides) to form polymers called polysaccharides.

- Carbohydrates can form fibrous structures, indicate cell identity, and store chemical energy.

A. CHAPTER OUTLINE

Chapter Introduction

- Single-sugar monomers (monosaccharides), small polymers (oligosaccharides), and large polymers (polysaccharides) with the generalized chemical formula $(CH_2O)_n$ are called **carbohydrates**.

- Carbohydrate monomers contain a carbonyl group, several hydroxyl functional groups, and many carbon-hydrogen bonds.

5.1 Sugars as Monomers

- Sugars provide chemical energy to cells and serve as molecular building blocks.

How Monosaccharides Differ

- Monosaccharides vary in functional group placement; when the carbonyl group is at the end of the molecule, the sugar is called an aldehyde sugar or *aldose*, and when the carbonyl is in the middle of the carbon chain, it is a ketone sugar or *ketose* (**Figure 5.1**).

- Sugar carbonyl and hydroxyl functional groups can participate in many different chemical reactions.

- The number of carbons in monosaccharides varies. Three-carbon sugars are **trioses**; five-carbon sugars, like ribose, are **pentoses**; and six-carbon sugars, like glucose, are **hexoses**. Carbons are numbered starting at the end closest to the carbonyl group.

- Sugars can have the same chemical formula but different structures (e.g., glucose and galactose, $C_6H_{12}O_6$, **Figure 5.2**). As always, structure affects function—only glucose can be directly used by the cell in ATP-producing reactions.

- Sugars in aqueous solution tend to form ring structures. When glucose forms a ring, carbon-1 (the carbonyl carbon) can bond with the fifth carbon's hydroxyl group in two different orientations, forming either α-glucose or β-glucose (**Figure 5.3**).

(a) Draw two ketose pentose sugars that have the same chemical formula but different structures.

- Many different monosaccharide structures are possible due to the variety of ring forms, carbon numbers, and functional group placements. Large structural variation permits a diversity of functions.

Monosaccharides and Chemical Evolution

- Monosaccharides, including ribose, are easily made under conditions simulating the prebiotic soup.

- A 3-C ketose and other sugar-like molecules were found on a meteorite, suggesting that simple sugars can be synthesized on space debris and might have rained down on ancient Earth.

- It is, however, unlikely that polymerization of monosaccharides occurred during chemical evolution, and researchers do not understand why ribose appears to have been the most common monosaccharide in the prebiotic soup.

5.2 The Structure of Polysaccharides

- **Polysaccharides** form when a condensation reaction between hydroxyl groups of two monosaccharides creates a **glycosidic linkage** (**Figure 5.4**). The simplest polysaccharide is a **disaccharide** of two linked monomers.

- Monomers joined by glycosidic linkages can be identical or different, and these linkages can form between any hydroxyl groups. Each sugar has at least two hydroxyl groups, so the location and geometry of these bonds varies widely.

- The α- and β-1,4-glycosidic linkages are both common. The numbers refer to the carbons on each side of the bond and α- and β- refer to the linkage structure.

- Enzymes can more easily break α-glycosidic linkages than β-glycosidic linkages.

Starch: A Storage Polysaccharide in Plants

- Plants join many glucose monomers together with α-1, 4-glycosidic linkages to store sugars as **starch**. The angle of these bonds causes the polysaccharide to coil into a helix.

- Starch is a mix of unbranched and branched helices (amylose and amylopectin, respectively; **Table 5.1**). Starch branches when a glycosidic bond forms between carbon 1 on one strand and carbon 6 on another (α-1,6-bonds).

Glycogen: A Highly Branched Storage Polysaccharide in Animals

- Animals store sugars as **glycogen**. Branches (via α-1, 6-glycosidic linkages) occur in glycogen about once every 10 glucose monomers, whereas amylopectin branches occur about once every 30 monomers (**Table 5.1**).

- Humans store glycogen in the liver and in muscles. During exercise, glycogen is broken down, releasing glucose for use by working muscles.

(b) Identify two similarities and two differences between starch and glycogen.

_____

_____

_____

_____

Cellulose: A Structural Polysaccharide in Plants

- Most organisms have a protective barrier called a **cell wall** outside their cell membranes. This cell wall is made of polysaccharides in algae, plants, bacteria, fungi, and other groups.

- Plant cell walls are largely made of **cellulose,** a β-glucose polymer with β-1,4-glycosidic linkages. These linkages flip every other monomer upside down, causing a geometry that allows extensive hydrogen bonding between side-by-side strands of cellulose (**Table 5.1**).

- Each hydrogen bond is relatively weak, but many together provide great strength in cellulose structures.

Chitin: A Structural Polysaccharide in Fungi and Animals

- **Chitin** forms the cell walls in fungi. It is also found in some algae and animals—it is the primary component of insect and crustacean exoskeletons. Although structurally similar to cellulose, chitin is composed of *N*-acetylglucosamine monomers.

- Chitin monomers, like those of cellulose, are joined by β-1,4-glycosidic linkages such that every other monomer is flipped over, allowing hydrogen bonding between adjacent strands and creating a tough, stiff, protective sheet (**Table 5.1**).

Peptidoglycan: A Structural Polysaccharide in Bacteria

- Bacterial cell walls are primarily composed of **peptidoglycan**, which is made of two alternating monosaccharides joined by β-1,4 glycosidic linkages.

- One of the sugar monomers is linked to a chain of amino acids, and peptide bonds link the amino acid chains of adjacent strands (**Table 5.1**).

- Cellulose, chitin, and peptidoglycan all consist of long parallel strands linked to each other. The resulting structure easily withstands forces that push or pull it, and it functions well as a protective sheet.

Polysaccharides and Chemical Evolution

- Polysaccharides are important to organisms today, but they probably were not involved in the origin of life.

- Glycosidic linkages appear to form only in the presence of enzymes, so polysaccharides probably did not form in the prebiotic soup. No known ribozymes catalyze these reactions.

✔ CHECK YOUR UNDERSTANDING

(c) Explain how structural polysaccharides differ from storage polysaccharides and how these differences in structure relate to their different functions.

_____

_____

_____

_____

_____

_____

- Monosaccharides have few functional groups and fairly simple secondary structures. They lack the structural and chemical complexity required for catalysis, and no known reactions are catalyzed by polysaccharides.

- Polysaccharide monomers cannot base pair; they do not provide the information necessary for copy-making.

5.3 What Do Carbohydrates Do?

- Carbohydrates are important building blocks in the synthesis of other molecules. Ribose and deoxyribose are nucleotide subunits; and other sugars furnish "carbon skeletons" for the synthesis of many important molecules, including amino acids.

- Carbohydrates also form fibrous structures, indicate cell identity, and store chemical energy.

The Role of Carbohydrates as Structural Molecules

- The structural polysaccharides—cellulose, chitin, and peptidoglycan—all form long strands with bonds between adjacent strands. These strands can be organized into fibers or layered in sheets to give cells and organisms great strength and elasticity.

- The β-1,4-glycosidic linkages of these structural molecules are difficult to hydrolyze. Very few enzymes have active sites that can accommodate their geometry. Molecules with these bonds are great structural molecules because they are so resistant to degradation.

- Although humans cannot digest cellulose, it aids digestion (as dietary fiber) by absorbing water and adding bulk and moisture to feces.

The Role of Carbohydrates in Cell Identity

- Polysaccharides are unable to *store* information in a way that can be replicated, but they do *display* information on the outer surface of cells.

- **Glycoproteins**—proteins with a covalent bond to a carbohydrate (usually an oligosasccharide)—project outward from cell membranes and are essential in cell–cell recognition and cell–cell signaling (**Figure 5.5**).

- In multicellular organisms, each cell type has different glycoproteins, each of which is recognized by the immune system as "self" and distinguished from foreign cells from other species (such as infecting bacteria) or individuals.

- Monosaccharide structural diversity makes possible the incredible diversity of glycoproteins necessary to indicate cell identity. Once again, structure correlates with function.

The Role of Carbohydrates in Energy Storage

- Carbohydrates store and provide chemical energy for cells.

(d) Why are polysaccharides good at cell identification but unable to effectively store information or catalyze chemical reactions?

_____

_____

_____

_____

Carbohydrates Store Sunlight as Chemical Energy

- Today, most sugars are produced via **photosynthesis**: $CO_2 + H_2O + sunlight \rightarrow (CH_2O)_n + O_2$, a key process that transforms sunlight's kinetic energy into carbohydrate bond chemical energy.

- Carbohydrates have more free energy than CO_2 does, because C–H and C–C bond electrons are shared more equally and held less tightly than those in C–O bonds (**Figure 5.6**).

- Because carbohydrate electrons are held less tightly, they are farther from the nucleus and have higher potential energy than tightly held electrons like those in CO_2.

- Fatty acids have even more C–H bonds than carbohydrates do and thus store even more free energy than do carbohydrates (**Figure 5.7**). Both molecules are important sources of chemical energy for cells.

Enzymes Hydrolyze Carbohydrates to Release Glucose

- Starch and glycogen are convenient glucose stores because enzymes easily catalyze reactions that release glucose subunits from α–glycosidic linkages.

- **Phosphorylase** catalyzes the hydrolysis of α-glycosidic linkages in glycogen, and **amylase** catalyzes the hydrolysis of these linkages in starch.

- Very few enzymes can hydrolyze β-glycosidic linkages like those found in the structural polysaccharides.

(e) Termites have bacteria in their guts that enable them to eat wood. What about these bacteria might be responsible for enabling termites to eat wood? (Be specific.)

_____

_____

Energy Stored in Glucose Is Transferred to ATP

- The free energy in carbohydrate C–H and C–C bonds can be used to produce **adenosine triphosphate (ATP)** via the reaction: $CH_2O + O_2 + ADP + P_i \rightarrow CO_2 + H_2O + ATP$.

- The easily accessed free energy stored in ATP drives endergonic reactions and performs all kinds of cell work.

- Carbohydrates store energy, but this energy must be converted to ATP before it can power cell work.

B. MASTERING CHALLENGING TOPICS

Much of this chapter relates to important structural and functional differences between α- and β-1,4-glycosidic bonds. The *best* way to get a feel for the structural differences is to make models of each bond type that joins two sets of glucose molecules. If you do not have a chemical model set, you could use cardboard and toothpicks. Seeing the three-dimensional structures will drive home the importance of whether the C-1 hydroxyl and the C-6 carbon are on the same or opposite sides of the ring and how this structural difference affects glycosidic bond geometry. You can also draw your own pictures of starch, glycogen, cellulose, chitin, and peptidoglycan to improve your understanding of their structural differences. Finally, make a table summarizing the major structural and functional differences of polysaccharides made from monomers that are joined by these two types of bonds.

	α-1,4-glycosidic bonds	β-1,4-glycosidic bonds
Bond structure		
Polysaccharides containing this bond and the monomers thus linked		
Cell function		
Relationship between cell function and bond structure		

C. ASSESSING WHAT YOU'VE LEARNED

1. Examine the ring form of fructose shown below. Which type of sugar is this?

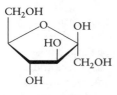

a. A pentose aldose
b. A pentose ketose
c. A hexose aldose
d. A hexose ketose

2. In α-glucose, the C-1 hydroxyl group is (refer to **Figure 5.3** if necessary):
a. gone.
b. turned into a carbonyl group.
c. on the same side of the ring as C_6.
d. on the opposite side of the ring from C_6.

3. Which of the following statements regarding sugars is correct?
a. Amino acids undergo condensation reactions to form polysaccharides.
b. Monosaccharides undergo condensation reactions to form polysaccharides.
c. Monosaccharides undergo hydrolysis to form polysaccharides.
d. Glucose is the monomer for all polysaccharides.

4. Why is glucose better than galactose as a source of chemical energy for cells?
a. Glucose's structure permits it to more easily be converted to usable chemical energy.
b. Glucose has more carbons than galactose.
c. Glucose has more C–H bonds than galactose.
d. Glucose has less entropy than galactose.

For questions 5 through 7, identify the cell-wall polysaccharide for each type of organism using the following lettered choices:
a. amylopectin
b. cellulose
c. chitin
d. glycogen
e. peptidoglycan
f. starch

5. Bacteria _____

6. Fungi _____

7. Plants _____

8. The monomer in chitin is:
a. α-glucose.
b. β-glucose.
c. amylase.
d. N-acetylglucosamine.

9. Researchers conclude that polysaccharides played little or no role in the origin of life because:
a. ribozymes cannot catalyze the formation of glycosidic linkages.
b. polysaccharide monomers are unable to form complementary base pairs.
c. polysaccharides cannot catalyze reactions.
d. all of the above are correct.

10. Which of the following describe ways in which carbohydrates function in organisms? (Choose all that apply.)
a. Cell protection
b. Storage of chemical energy
c. Transport of molecules
d. Cell-cell recognition
e. Catalysis

11. Proteins covalently bonded to carbohydrates are called:
a. glycoproteins.
b. amyloproteins.
c. cephalosporins.
d. saccharoamides.

12. In carbohydrates, electrons are:
a. closer to the carbon atom than they were in CO_2.
b. farther from the carbon atom than they were in CO_2.
c. shared less evenly in C–H bonds than in C–O bonds.
d. more often found in polar covalent bonds than when in CO_2.

13. Fats have more free energy than carbohydrates because:
a. they are made during photosynthesis.
b. they have more kinetic energy.
c. they have more C–H bonds.
d. they have more C–O bonds.

14. One thing that cellulose, chitin, and peptidoglycan all have in common is:
a. hydrogen bonds between adjacent strands.
b. covalent bonds between adjacent strands.
c. α-1,4-glycosidic linkages.
d. β-1,4-glycosidic linkages.

15. Based on this chapter's discussion of why carbohydrates have high free energy, which of the following is likely to be true of the electrons in the bond between ATP's second and third phosphates?
    a. The electrons are shared fairly equally, giving the bond a high free energy.
    b. The electrons are not shared equally, giving the bond a high free energy.
    c. The electrons are shared fairly equally, giving the bond a low free energy.
    d. The electrons are not shared equally, giving the bond a low free energy.

CHAPTER 5—ANSWER KEY

Check Your Understanding

(a)
```
      H                    H
      |                    |
   H–C–OH              H–C–OH
      |                    |
   H–C–OH                C=O
      |                    |
    C=O                 H–C–OH
      |                    |
   H–C–OH              H–C–OH
      |                    |
   H–C–OH              H–C–OH
      |                    |
      H                    H
```

(b) Similarities: Both made of glucose monomers; both have α-1,4-glycosidic linkages; both are used to store chemical energy. Differences: Starch is made by plants, glycogen by animals; glycogen is more branched than starch (more β-1,6-glycosidic bonds).

(c) The β-glycosidic bonds in structural polysaccharides are not easily broken down (making these carbohydrates hard to digest) and cause every other monomer to be flipped upside down. This produces a more linear structure and allows hydrogen bonding between adjacent strands. The many hydrogen bonds between strands make these polysaccharides strong yet flexible. In contrast, α-glycosidic bonds are more easily broken down, making these polysaccharides a good source of energy. However, their helical structure is not as strong as the parallel sheets of the structural polysaccharides.

(d) Polysaccharides do not form a template that can be copied and passed on to the daughter cells, so they are poor molecules for information storage. Polysaccharides also lack the complexity of secondary and tertiary structure necessary to catalyze reactions, yet are structurally diverse enough to allow each cell type and species to display different polysaccharides.

(e) The bacteria in termite guts have an enzyme that can hydrolyze the β-1,4-glycosidic linkage found in cellulose. Very few organisms have such enzymes, which is why so few organisms can eat wood.

Assessing What You've Learned

1. d; 2. d; 3. b; 4. a; 5. e; 6. c; 7. b; 8. d; 9. d; 10. a, b, d; 11. a; 12. a; 13. c; 14. d; 15. a

---

Looking Forward—Key Concepts in Later Chapters

Sugars—Chapters 9, 10, 37, and 43

Chapter 9 describes how organisms break down sugars to release and use the energy contained in their C–H bonds; **Chapter 10** describes how some organisms can use the energy in sunlight to make sugars (photosynthesis). **Chapters 37** and **43** describe the importance of sugars to plants and animals at the organismal level.

Glycoproteins—Chapter 49

Chapter 49 describes the important role that glycoproteins and proteoglycans have in forming a physical shield for a membrane, helping to prevent the invasion of microbial pathogens.

Lipids, Membranes, and the First Cells

The Cell Theory—Chapter 1

Chapter 1 introduced the cell theory as one of the unifying themes of biology: All life is made of cells. In Chapter 6, you learn about the membranes that create a barrier between an organism and the external world and define the cell.

Entropy—Chapter 2

Chemical reactions can proceed spontaneously only when entropy, a measure of disorder, increases and/or potential energy decreases. Entropy also determines the movement of molecules and ions across membranes in diffusion and osmosis.

Proteins—Chapter 3

Membrane channels, transporters, and pumps are all proteins. You may wish to review **Chapter 3**'s discussion of protein structure when learning how proteins insert into membranes and bind ions or molecules. Yet again structure and function are closely related.

KEY CONCEPTS

- Phospholipids have both hydrophilic and hydrophobic regions. This makes them amphipathic and allows phospholipids to spontaneously form selectively permeable bilayers in solution.

- Solutes spontaneously diffuse from regions of high concentration to regions of low concentration. Water can diffuse across lipid bilayers in a process called osmosis.

- Proteins determine the rate at which ions, polar molecules, and large molecules cross membranes. Many membrane proteins are channels, transporters, or pumps.

A. CHAPTER OUTLINE

Chapter Introduction

- The mostly lipid **cell membrane** or **plasma membrane** surrounds a cell, separating the life inside the cell from the external environment.

- Development of a plasma membrane around a self-replicator created the first cell.

- Cells could now begin to control their internal environment by keeping harmful compounds out and allowing useful compounds in, greatly increasing the efficiency of cell chemical reactions.

6.1 Lipids

- Molecules that contain only carbon and hydrogen atoms are called **hydrocarbons**; these molecules are nonpolar and consequently hydrophobic, because electrons are evenly shared in C–H bonds.

- **Lipids** have a significant hydrocarbon component; they are nonpolar, hydrophobic organic compounds that do not dissolve in water.

- **Fatty acids** (hydrocarbon chains bonded to a carboxyl COOH group) and isoprene are the key building blocks for the synthesis of lipids in organisms (**Figure 6.1**).

A Look at Three Types of Lipids Found in Cells

- Amino acids, nucleotides, and sugars are defined by their chemical structure, but lipids are defined by their solubility. Lipid structure is highly variable.

Fats

- **Fats**, also known as a *triacylglycerols* or *triglycerides*, are made when three fatty acids are linked to a three-carbon **glycerol**.

- Condensation reactions between the glycerol hydroxyl groups and fatty-acid carboxyl groups form **ester linkages**. Although this bond is similar to those formed between other macromolecule subunits, fats are not polymers of repeating units (**Figure 6.2**).

Steroids

- All **steroids** have a four-ring structure but can differ in the functional groups or side groups attached to these rings (**Figure 6.3a**).
- One steroid, cholesterol, is an important part of plasma membranes. It has a long hydrocarbon tail attached to the general steroid ring structure.

Phospholipids

- **Phospholipids** are formed from a glycerol linked to a phosphate group PO_4^{2-}, which in turn bonds to a small charged or polar organic molecule (**Figure 6.3b**).
- The glycerol's other two OH groups form bonds to either two isoprene chains (in Archaea) or two fatty acids (in Bacteria and Eukarya).
- Phospholipids are critical components of plasma membranes.

The Structures of Membrane Lipids

- Membrane lipids are **amphipathic**—they contain both a polar, hydrophilic region and a nonpolar, hydrophobic region. The polar region interacts with water, but the nonpolar end does not.
- Phospholipids form membranes because they are amphipathic; their polar or charged head region can interact with water molecules, but their nonpolar tails cannot.

> ✔ **CHECK YOUR UNDERSTANDING**
>
> (a) Examine the structure of fats in (**Figure 6.2**). Why don't these fats form membranes?
>
> _____
>
> _____
>
> _____

6.2 Phospholipid Bilayers

- Phospholipids with short tails tend to form **micelles** (tiny droplets), and those with longer tails tend to form **lipid bilayers** (two layers) in water. The polar heads face outward toward water and the hydrophobic tails face inward toward other tails (**Figure 6.4**).
- Micelles and lipid bilayers form spontaneously without any energy input. These structures are more stable (have lower potential energy) because free phospholipids disrupt water's hydrogen bonds.
- The lowered potential energy more than makes up for the decreased entropy of micelles and lipid bilayers, so the overall free energy of the system decreases and formation of these structures is exergonic.

Artificial Membranes as an Experimental System

- Lipid bilayers break and form small spherical compartments (vesicles) when shaken. Water is located both outside and inside these spherical bilayers. When these vesicles are made artificially, they are called liposomes (**Figure 6.5**).
- Hypotheses about membrane function are tested using liposomes and planar bilayers—artificial membranes across a hole in a wall between two aqueous solutions (**Figure 6.6**).
- **Permeability**, the ease with which substances cross a barrier, is crucial to plasma membrane function because differentiation between an internal and external environment occurs when some molecules or ions cross a barrier more easily than others.
- Researchers investigate membrane permeability by varying either the solute added to one side of the bilayer or the bilayer composition. They can control which factors change from one treatment to another, making this a powerful experimental system.

Selective Permeability of Lipid Bilayers

- Experiments with liposomes and planar bilayers show that small, nonpolar molecules move across phospholipid bilayers quickly, but charged substances and large molecules cross slowly if at all.
- Phospholipids bilayers have **selective permeability**; some molecules cross these bilayers more easily than others (**Figure 6.7**).
- Charged compounds, large polar molecules, and ions cannot pass through lipid bilayers because they are more stable in an aqueous solution than among the nonpolar, hydrophobic phospholipid tails.
- Water crosses membranes fairly rapidly, even though it is polar, because it is such a small molecule.

> ✔ **CHECK YOUR UNDERSTANDING**
>
> (b) Compare the permeability scale in **Figure 6.7** with the pH scale shown in **Figure 2.16**. Why are powers of 10 used for these scales?
>
> _____
>
> _____

How Does Lipid Structure Affect Membrane Properties?

- Hydrocarbon chain length and the number of double bonds affect membrane permeability. This is another example of structure affecting function.

Bond Saturation Is an Important Aspect of Lipid Structure

- A hydrocarbon chain without double bonds is **saturated**; it has the maximum number of hydrogen atoms bonded to its carbon skeleton.

- A hydrocarbon chain with one or more double bonds is **unsaturated**. Double bonds prevent the two carbons from rotating freely and place a "kink" in a hydrocarbon chain (**Figure 6.8**).

- Saturated fats contain more C—H bonds than unsaturated fats and thus store more free energy than do unsaturated fats.

Bond Saturation and Hydrocarbon Chain Length Change Membrane Fluidity and Permeability

- The kinks in unsaturated fats disrupt hydrophobic interactions within a lipid bilayer. Hydrophobic interactions are stronger among saturated than among unsaturated fats.

- Hydrophobic interactions are also stronger among longer hydrocarbon tails than among shorter tails.

- Bilayers with unsaturated hydrocarbons and/or shorter hydrocarbon tails are more fluid and permeable than those with saturated hydrocarbons and/or longer tails (**Figures 6.9 and 6.10**).

- For example, saturated fats are solid at room temperature; those with extremely long hydrocarbon tails form stiff **waxes**. In contrast, unsaturated fats are liquid **oils** at room temperature.

Cholesterol Reduces Membrane Permeability

- Cholesterol also affects membrane permeability (**Figure 6.11**); it fills spaces between phospholipids, increasing hydrophobic interactions and reducing bilayer permeability.

How Does Temperature Affect the Fluidity and Permeability of Membranes?

- Experiments with tagged phospholipids show their lateral movement through membranes (**Figure 6.12**).

- Cell membranes are fluid at room temperature because the bilayer molecules move relatively fast. At lower temperatures the molecules move more slowly and the hydrophobic tails pack more tightly, making the membrane less fluid.

- Decreased fluidity causes decreased permeability. Fluidity and permeability are affected by both temperature and hydrocarbon structure.

CHECK YOUR UNDERSTANDING

(c) Based on the preceding discussion, what kind of lipid composition would you expect to find in a shortening like Crisco?

_____

Examine the labels on shortening and olive oil, and use the information you find there to explain why olive oil is fluid at room temperature but shortening is not.

_____

_____

_____

Why is shortening softer at room temperature than in the refrigerator?

_____

_____

6.3 Why Molecules Move across Lipid Bilayers: Diffusion and Osmosis

Diffusion

- Dissolved molecules and ions (**solutes**) are in constant random motion due to their thermal or kinetic energy. Movement caused by this random kinetic motion is called **diffusion**.

- Due to random molecular motion, solutes spontaneously move down their **concentration gradient**. Molecules bump into each other more often in areas of high concentration, so they tend to move toward areas of low concentration.

- Diffusion down a concentration gradient is spontaneous because it increases entropy.

- Molecules or ions on one side of a phospholipid bilayer tend to diffuse toward the other side over time (**Figure 6.13**).

- Equilibrium (no *net* movement) is achieved when molecules or ions are randomly distributed throughout the solution. Molecules still move, but they are equally likely to move in any direction.

Osmosis

- Diffusion of water is called **osmosis**. For example, when a selectively permeable membrane separates two solutions, water spontaneously moves toward the solution with the higher solute concentration (i.e., lower water concentration).

- Osmosis is spontaneous because it increases entropy.

- Osmosis can shrink or swell (and even burst) membrane-bound vesicles and cells (**Figure 6.15**).

- If a solution outside a membrane has a higher concentration of solutes than the interior does, and if the solutes are unable to diffuse across the membrane, water will leave the vesicle, causing it to shrivel. Solutions that shrink cells are **hypertonic**.

- **Hypotonic** solutions are less concentrated than the vesicle or cell interior. In this case, if the membrane is not permeable to the solute, water will flow into the vesicle or cell, which will then swell and perhaps burst.

- A solution with the same solute concentration as the inside of a vesicle or cell is **isotonic** and does not affect membrane shape. These -*tonic* terms always apply to relative concentrations.

- Osmosis and diffusion always reduce the differences between solutions on two sides of a phospho–lipid bilayer. Thus, liposomes in the prebiotic soup probably did not provide a significantly different internal environment for self-replicating molecules.

- However, prebiotic soup liposomes could have grown due to the addition of new lipids and they could have divided when shaken by wave action.

CHECK YOUR UNDERSTANDING

(d) Using your knowledge of how the concentration of solutions outside cells affects cell shape, explain why salting meat is an effective way to preserve it (and reduce bacteria spoilage).

_____

_____

6.4 Membrane Proteins

- Amphipathic proteins can be added into lipid bilayers (**Figure 6.16**).

- Amino acid R-groups vary in their affinity for water. Nonpolar amino acids are stable in membranes due to hydrophobic interactions with hydrocarbon tails, and polar or charged amino acids are stable next to the polar heads and in the surrounding water.

- By mass, cell membranes contain as much protein as phospholipids.

Evolution of the Fluid-Mosaic Model

- H. Davson and J. Danielli, in an early model of membrane structure, proposed that proteins coated a pure lipid bilayer.

- The more recent **fluid-mosaic model** (S. J. Singer and G. Nicolson) suggests many proteins are inserted within the lipid bilayer, making the membrane a fluid, dynamic mosaic of phospholipids and proteins (**Figure 6.17**).

- Development of freeze-fracture preparations for a **scanning electron microscope** provided key evidence for the fluid-mosaic model. In this technique, membranes are frozen and then split to reveal the middle of the bilayer (**Figure 6.18**).

- Membrane proteins look like mounds sticking out of half of the bilayer, leaving pits in the other half. Membrane-spanning proteins are called **integral membrane proteins or transmembrane proteins**.

- Other **peripheral membrane proteins** are located on the membrane surface, and they are often attached to the integral membrane proteins (**Figure 6.19**).

- Membrane proteins participate in many cell functions, but this chapter focuses on how integral membrane proteins transport ions and molecules across the plasma membrane.

Systems for Studying Membrane Proteins

- To separate integral membrane proteins from cell membranes, researchers treat the membranes with small, amphipathic molecules called **detergents** (**Figure 6.20**).

- Detergents have short hydrophobic tails that interact with membrane phospholipid tails, breaking apart the bilayer. Detergents then surround hydrophobic parts of membrane proteins, forming water-soluble detergent-protein complexes.

CHECK YOUR UNDERSTANDING

(e) Pour some oil into a pan or bowl of water. What happens? Why?

_____

_____

Now add some dishwasher or laundry detergent and stir. What happens? Why?

_____

_____

- The now-soluble membrane proteins can be isolated and purified using **gel electrophoresis**, which separates the proteins based on their size. The purified proteins can be inserted into planar bilayers or liposomes for experimental analysis.

- Three types of **transport proteins** affect membrane permeability: channels, transporters, and pumps.

Protein Transport I: Facilitated Diffusion via Channel Proteins

- Ions can only cross phospholipid bilayers through proteins called **ion channels** that form tunnels across the lipid bilayer.

- Ions move from areas of like charge to areas of unlike charge and from areas of high concentration to areas of low concentration—they move in response to the **electrochemical gradient** (**Figure 6.22**).

✔ **CHECK YOUR UNDERSTANDING**

(f) Some steroids function as hormones. Based on what you know about the selective permeability of lipid bilayers, predict whether steroid hormones can cross membranes on their own, or whether they require a transport protein. Explain your reasoning.

_____

_____

_____

_____

Is CFTR an Ion Channel?

- Researchers measured charge flow (electrical current) across membranes with and without CFTR (cystic fibrosis transmembrane conductance regulator) to test the effect of this protein on chloride ion (Cl^-) flow.

- Chloride ions move across planar bilayers with CFTR, but no current flows across membranes without CFTR. These results supported the hypothesis that CFTR is a chloride channel (**Figure 6.22**).

Protein Structure Determines Channel Selectivity

- Cells have a wide variety of membrane channels. **Channel proteins** are highly selective—their structures allow only a certain type of ion or molecule to cross the membrane.

- **Aquaporins**, for example, increase the rate at which water crosses membranes tenfold and are highly selective for water.

- Hydrophobic R-groups on the exterior side of the tunnel interact with the membrane hydrocarbon tails, whereas the interior of the pore is lined with polar amino acid R-groups (**Figure 6.23a**). Once again, structure correlates with function.

Movement through Membrane Channels Is Regulated

- Aquaporins, CFTR, and many other ion channels are **gated channels** that open only when bound to a specific molecule or in the presence of a certain electrical charge.

- For example, one type of potassium channel lets potassium ions through only when the outside of the membrane is positive relative to the inside. The positive charge causes a shape change that opens the channel (**Figure 6.23b**).

- Channels allow membranes to exert some control over transmembrane flow of molecules and ions and provide a mechanism for allowing small charged compounds into the cell.

- However, these substances always diffuse through channels down their electrochemical gradients. This process is also called **facilitated diffusion** or **passive transport**, since the cell expends no energy.

Protein Transport II: Facilitated Diffusion via Carrier Proteins

- Some proteins that facilitate diffusion change shape to allow the ion or molecule to cross the membrane. These proteins are called **carrier proteins** or **transporters**.

The Search for a Glucose Transporter

- Glucose is a subunit for important macromolecules and a great energy source, but lipid bilayers are not very permeable to glucose.

- Researchers obtained intact cell membranes by bursting red blood cells in a hypotonic solution to form red blood cell "ghosts." These plasma membranes are more permeable to glucose than are pure phospholipid bilayers.

- Researchers isolated a glucose carrier named GLUT-1, which increases membrane permeability to glucose. Glucose enters liposomes with GLUT-1 proteins at the same high rate observed in cell membranes.

How Does Glut-1 Work?

- Researchers think glucose binds to GLUT-1 on the cell's exterior surface, causing a conformational change (just like enzymes change shape during substrate binding) that transports glucose to the cell interior (see **Figure 6.25**).

- Like channels, carrier proteins reduce differences between the inside and the outside of a cell or vesicle.

When glucose concentrations are equal, no net glucose movement occurs by GLUT-1.

✓ CHECK YOUR UNDERSTANDING

(g) Cells often break down glucose as a source of energy. How does this process alter the concentration gradient for glucose?

_____

_____

How does this process explain why many cells can rely on transporters like GLUT-1 to bring glucose into the cell, when the direction of transport is determined by the electrochemical gradient?

_____

_____

_____

Protein Transport III: Active Transport by Pumps

- Cells extant today can also transport molecules or ions *against* their electrochemical gradient; this process requires energy and is called **active transport**.

- Adenosine triphosphate (**ATP**) provides energy in a form that proteins can easily use. When a phosphate group leaves ATP and binds to a protein, it increases the protein's potential energy. The other product, ADP, has less potential energy than ATP.

- The phosphate's negative charges interact with the protein's amino acids, causing a shape change. Another conformational change occurs when the phosphate group leaves.

The Sodium-Potassium Pump

- The **sodium-potassium pump**, Na^+/K^+-ATPase, uses energy from ATP to transport Na^+ and K^+. Ion movement against an electrochemical gradient occurs when the pump's shape changes as a phosphate group from ATP first binds and then leaves.

- Three Na^+ ions bind to Na^+/K^+-ATPase on the inside of a cell. A phosphate from ATP binds to the pump, changing its shape so that the Na^+ ions are released outside the cell. The pump now binds two K^+, which are released inside the cell when the phosphate group drops off, returning the pump to its original shape (**Figure 6.26**).

- Other pumps move other ions or molecules across cell membranes against electrical or concentration gradients.

Secondary Active Transport

- The sodium-potassium pump creates a chemical and electrical gradient across the membrane because it returns only two positive K^+ charges for every three Na^+ charges removed from the cell.

- A chemical and/or electric gradient is itself a form of potential energy. This energy can be used to move other molecules affected by the gradient. This is known as **secondary active transport** or **cotransport**.

- For example, gut cells bring glucose into cells along with Na^+ ions. Energy in the Na^+ gradient created by Na^+/K^+-ATPases powers the glucose transport against a glucose concentration gradient.

Plasma Membranes and the Intracellular Environment

- Selective permeability combined with the specificity of passive and active transport protein action allow cells to create and maintain an internal environment very different from the external environment (**Figure 6.27**).

- From the earliest life, natural selection would have favored cells with membrane proteins that helped the cell develop a favorable internal environment.

B. MASTERING CHALLENGING TOPICS

One good way to review and test your comprehension of **diffusion**, **osmosis**, **fluidity**, **concentration gradients**, and **electrochemical gradients** is to draw figures like that in **Figure 6.6b** for different experiments you could do. Then explain your expected results based on information given to you in this chapter.

If you put hydrophobic molecules like CO_2 on one side of a lipid bilayer, what would happen? How about if you put glucose on one side? What could you add to the membrane to increase glucose's permeability? In general, how could you modify the composition of the lipid bilayer to make it more or less permeable? What happens when the membrane is impermeable to a solute on one side (but permeable to water)? What happens if you add more water to one side? Ions carry a charge, and opposite charges attract; this is why ions respond to electrical gradients as well as to chemical gradients. What variation on this technique can researchers use to measure membrane permeability to ions?

Answers to all these questions are in the text, but doing your own thought experiments helps you master the information in this chapter and better understand how researchers test hypotheses about membranes.

C. ASSESSING WHAT YOU'VE LEARNED

1. What type of molecule is represented in this figure?

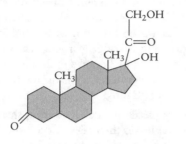

   a. A fat
   b. A steroid
   c. A phospholipid
   d. A fatty acid

2. Why are fatty acids called *acids*?
   a. Because they have lots of hydrogen atoms
   b. Because they form ester bonds with glycerol
   c. Because the carboxyl group tends to lose a proton in aqueous solution
   d. Because they like solutions that have a low pH

3. Liposomes and planar bilayers are powerful experimental tools because
   a. the phospholipid bilayers are selectively permeable and only certain molecules can pass through them.
   b. they form spontaneously so it is easy to make them in the laboratory.
   c. researchers can simulate conditions found on ancient Earth.
   d. they allow precise control over which factor changes from one experiment to the next.

4. Which of the following molecules will most easily cross a lipid bilayer?
   a. N_2
   b. Ribose
   c. Ca^{2+}
   d. Water

5. Why is the cell membrane less permeable to glucose than to glycerol?
   a. Glucose is larger than glycerol.
   b. Glucose contains oxygen; glycerol does not.
   c. Glycerol is nonpolar; glucose is polar.
   d. Glucose is a charged molecule; glycerol is not.

6. Lipid bilayers are most permeable when they are made of _____ fatty acids and _____ hydrocarbon chains.
   a. saturated; long
   b. unsaturated; long
   c. saturated; short
   d. unsaturated; short

7. At room temperature, a cell membrane has the consistency of olive oil. Cooling the membrane _____ its fluidity and _____ its permeability.
   a. increases; decreases
   b. decreases; increases
   c. decreases; decreases
   d. increases; increases

8. Spontaneous movement of any molecule from an area of high concentration to an area of low concentration is:
   a. osmosis.
   b. diffusion.
   c. equilibrium.
   d. active transport.

9. Distilled water is _____ relative to a solution containing dissolved salts.
   a. hypertonic
   b. hypotonic
   c. isotonic
   d. none of the above

10. Which statement best describes the fluid-mosaic model of membrane structure?
    a. Two layers of proteins are interspersed with phospholipids.
    b. A phospholipid bilayer has a few peripheral proteins associated with it.
    c. A phospholipid bilayer is coated on both sides with hydrophilic integral proteins.
    d. A phospholipid bilayer contains diverse proteins, including embedded amphipathic proteins.

11. Detergents are:
    a. amphipathic.
    b. hydrophobic.
    c. hydrophilic.
    d. osmotic.

12. Based on this chapter's discussion of channel proteins and CFTR, which of the following will cause a cross-membrane current when added to one side of a planar bilayer with CFTR proteins inserted in it?
    a. H_2O
    b. Glucose
    c. Na^+
    d. None of the above

13. You place human red blood cells in isotonic salt solution. Will glucose move across the cell membrane, and if so, in which direction?
    a. Yes, glucose will move into the cells.
    b. Yes, glucose will move out of the cells.
    c. No, glucose will not cross the membrane at any appreciable rate.
    d. It is impossible to tell from the information given in this chapter.

14. A protein binds to an ion, diffuses across a membrane, and releases it on the other side. No ATP or other chemical energy is used. This protein is a(n):
    a. aquaporin.
    b. gated channel.
    c. transport protein.
    d. ion pump.

15. Certain bacteria live in the salty environment of the Dead Sea and use the Na^+/K^+-ATPase to pump out sodium ions (potassium ions are present, but at a much lower concentration). Working with these bacteria in the lab, you identify one with a sodium-potassium pump mutation that allows it to bind three instead of only two K^+ ions. Which of the following is the most likely outcome for a cell with this mutated pump in its natural habitat?
    a. The cell is not able to control other ion gradients as effectively because its ability to establish an electrochemical gradient is impaired.
    b. The cell produces many offspring because it is more efficient and uses less ATP for ion transport.
    c. The outside of the cell becomes more positively charged with respect to the cell interior.
    d. More Na^+ cotransport can occur.

CHAPTER 6—ANSWER KEY

Check Your Understanding

(a) The fats don't form membranes because fats are not amphipathic. Fats are entirely nonpolar without any hydrophilic parts.

(b) It is useful to show these scales as powers of 10, because a wide range of permeabilities or concentrations can easily be shown in the same figure.

(c) Oil has more unsaturated fats than shortening does. Oil is fluid at room temperature since double bonds in its unsaturated fats make kinks in the hydrocarbon chains and disrupt hydrophobic interactions. Shortening is softer at room temperature than in the refrigerator because molecules move slower at lower temperatures, causing hydrocarbon tails to pack more tightly.

(d) Bacteria reaching the salted meat shrivel up and die as water leaves their cells due to osmosis. The salted meat is hypertonic to bacterial cells.

(e) The oil floats on top of the water and does not mix with the water. When the detergent is mixed in, the oil splits up into smaller and smaller droplets.

(f) The four-ring hydrocarbon structure of steroids is hydrophobic. Thus, lipid bilayers are permeable to almost all steroids, and these hormones can cross membranes without transport proteins (although certain side groups can decrease the speed with which a steroid crosses a lipid bilayer).

(g) As cells use glucose, the concentration of glucose in a cell decreases. This maintains a lower concentration of glucose in the cell than in, for example, the blood. Glucose moves into the cell down its concentration gradient via transporters such as GLUT-1.

Assessing What You've Learned

1. b; 2. c; 3. d; 4. a; 5. a; 6. d; 7. c; 8. b; 9. b; 10. d; 11. a; 12. d; 13. b; 14. c; 15. a

---

Looking Forward—Key Concepts in Later Chapters

Cells—Unit 2 (Chapters 7–11), Chapter 21, Chapter 22, and other chapters

Because cells are the fundamental units of life, they reappear throughout your text; the next unit discusses the basics of cell structure, metabolism, and replication (division). Later on, you will learn how cells differentiate to form different tissue types in multicellular organisms.

Membranes and Membrane Proteins—Chapters 7–10, 37, 38, 42, 45–47, and 49

Membranes and their proteins determine what substances can cross the membrane to enter the cell. They thus have a critical role in maintaining homeostasis and transmitting signals to the cell. Membrane proteins also can act as enzymes and make physical connections between cells. You'll learn more about these processes in the units on how plants and animals work. In **Chapters 7–10**, you will learn about internal cell membranes and their functions.

Chapter 7

Inside the Cell

Prokaryotes and Eukaryotes—Chapter 1

Chapter 1 first introduced you to the differences between prokaryotes and eukaryotes and showed how these groups are related on the tree of life. In this first chapter you also learned about cell theory—that cells are the basic unit of all life, and all cells come from preexisting cells.

The Cell Membrane—Chapter 6

In the last chapter you learned about the structure and function of the plasma membrane. In this chapter you learn more about internal cell membranes—the endomembrane system. Remember that the structure (phospholipid bilayer with proteins) and function of these membranes is similar to that of the cell membrane.

KEY CONCEPTS

- Cell structure is closely tied to cell function, both for cell parts and for overall cell shape and composition.
- Molecular "zip codes" ensure proper transport of cell materials.
- Overall cell structure relies on a cytoskeletal framework that is actively involved in cell division, movement, and transport.
- Carbohydrates can form fibrous structures, indicate cell identity, and store chemical energy.
- Cells are busy, organized structures.

A. CHAPTER OUTLINE

Chapter Introduction

- The cell is the fundamental unit of life; a selectively permeable **plasma membrane** surrounds the cell, creating a special internal environment.

7.1 Bacterial and Archaeal Cell Structures and Their Functions

- According to **phylogeny**, there are three main groups of organisms: Bacteria, Archaea, and Eukarya. Because Bacteria and Archaea share a similar **morphology**, they are called prokaryotes. Eukaryotes have a nucleus, but prokaryotes do not.

A Revolutionary New View

- New research reveals that bacterial cells are highly organized, with many distinctive structures, only some of which are visible under low magnification (**Figure 7.1**).

Prokaryotic Cell Structures: A Parts List

The Chromosome Is Organized in a Nucleoid

- The **chromosome**, a large DNA molecule associated with a few proteins, is the biggest prokaryotic cytoplasmic structure. Most prokaryotic species have one circular chromosome made of DNA and associated proteins.
- DNA's nitrogenous base sequence carries the organism's genetic information. A **gene** is a section of DNA containing directions to build one polypeptide or RNA molecule.
- Because the chromosome may be hundreds of times longer than the cell, it is "supercoiled" (**Figure 7.2**).
- The chromosome is located in the **nucleoid** region of the cell.
- Prokaryotes often also contain additional small supercoiled DNA pieces called **plasmids**. Plasmids usually carry genes needed only under rare environmental conditions.

Ribosomes Manufacture Proteins

- All prokaryotes synthesize proteins on ribosomes made of RNAs and proteins.

Internal Membrane Complexes in Photosynthetic Species

- Some prokaryotes have extensive internal membranes for photosynthesis (**Figure 7.3**). The plasma membrane folds in, sometimes pinching off to form vesicles. These membranes contain the pigments and enzymes necessary for photosynthetic reactions.

Organelles Perform Specialized Functions

- Certain prokaryotes have membrane-bound storage containers. Many of these qualify as **organelles** because they contain enzymes or structures specialized for a specific function.

The Cytoskeleton Structures the Cell Interior

- Prokaryotic cells have a supportive structure of thin protein filaments. These proteins help cell division occur, and in some species they help maintain cell shape.

The Plasma Membrane Separates Life from Nonlife

- The prokaryotic plasma membrane surrounds the **cytoplasm**, a highly concentrated solution containing all the molecules that make up the inside of the cell.

Flagella Enable Some Species to Swim

- Some prokaryotes have one or more rotating tail-like **flagella** on their cell surface; each is made of many proteins that work together to spin the flagellum and move the bacterium.

The Cell Wall Forms a Protective "Exoskeleton"

- The cytoplasm is typically hypertonic to the outside environment, so water tends to enter the cell by osmosis, causing it to expand.
- The influx of water is resisted by a tough, fibrous **cell wall** surrounding the cell membrane. The cell wall is usually made of peptidoglycan, which may, in turn, be surrounded by **glycolipids** (**Figure 7.4**).
- Cell walls protect cells and give them shape and structure.

7.2 Eukaryotic Cell Structures and Their Functions

- On average, eukaryotes are significantly larger than prokaryotes (about 5 to 100 μm diameter versus less than 10 μm diameter), and many are multicellular.
- Large eukaryotes can consume smaller prokaryotes, but their size makes it difficult for molecules to diffuse across the entire cell.

The Benefits of Organelles

- Eukaryotic cells break up their large cell volume into smaller compartments called organelles.

- Compartmentalization increases chemical reaction efficiency by separating incompatible chemical reactions.
- Compartmentalization also groups together enzymes and substrates. Groups of enzymes that work together may also be located close to each other on or within internal membranes.
- Eukaryotes can be distinguished from prokaryotes because their chromosomes exist within a membrane-bound nucleus; they are typically larger than prokaryotes and have extensive internal membranes and a more dynamic cytoskeleton (**Table 7.1**).

Eukaryotic Cell Structures: A Parts List

- Cell structure correlates with cell function for eukaryotes too. Typical plant and animal cells are shown in **Figure 7.6**.

The Nucleus

- All eukaryotes store and process genetic information in a large nucleus surrounded by a double-membrane **nuclear envelope** that connects with the endoplasmic reticulum (**Figures 7.7** and **7.8**).
- **Nuclear pores** provide passage through the envelope, and nuclear lamina proteins organize the linear chromosomes and help stiffen and shape the nucleus.
- Ribosome synthesis and assembly occurs in the **nucleolus**.

Rough Endoplasmic Reticulum

- The **rough endoplasmic reticulum** (**rough ER**) is a ribosome-studded network of tubes and sacs. These ribosomes make proteins destined for the plasma membrane, organelles, or cell secretion.
- Proteins made by rough ER ribosomes move to the inside of the rough ER as they are made, where they are then folded and otherwise modified in the rough ER **lumen**.

Smooth Endoplasmic Reticulum

- Endoplasmic reticulum without ribosomes is called the **smooth endoplasmic reticulum** or **smooth ER**. Lipid synthesis and breakdown occurs here, so this organelle contains enzymes that catalyze these reactions.
- The smooth ER may also store calcium ions (Ca^2) that act as cell signals.
- The smooth and rough ER, together with the Golgi apparatus and the lysosomes, make up the **endomembrane system**. In each case, differences in organelle structure relate to organelle function.

(a) What is the difference between smooth and rough ER? Why does it make sense that rough ER is more important than smooth ER in the synthesis and distribution of proteins?

_____

_____

_____

_____

Golgi Apparatus

- The **Golgi apparatus** is a series of flat membrane sacs (**cisternae**) stacked together.

- The Golgi's *cis* side faces the rough ER and receives products from it; the *trans* side faces the plasma membrane and sends finished, packaged products to the cell surface in small vesicles (**Figure 7.10**).

Ribosomes

- Many ribosomes are found free in the **cytosol**—the fluid part of the cytoplasm. The cytoplasm includes everything inside the plasma membrane except for the nucleus.

- As in prokaryotes, eukaryotic ribosomes are made of RNA and protein and are molecular machines for protein synthesis. Ribosomes do not have a membrane and thus are not organelles.

Peroxisomes

- **Peroxisomes** are single-membrane, globular organelles that bud off from the ER.

- Peroxisomes are oxidation reaction centers. The specific reactions and enzymes may vary from cell to cell, but many produce hydrogen peroxide (H_2O_2) as a by-product.

- The enzyme catalase quickly degrades the dangerously reactive H_2O_2 to harmless water and oxygen, so that the cell is not harmed.

- Plant leaves have specialized peroxisomes, called **glyoxisomes,** that oxidize fats to make an energy storage compound.

Lysosomes

- **Lysosomes**, found in animal cells, vary in size and shape, but they are all single-membrane-bound centers for cell storage and/or waste processing.

- Lysosomes have an acid interior (pH 5.0) and contain digestive enzymes called acid hydrolases that work best at a low pH. These enzymes break macromolecules into their monomer subunits.

- The lysosome receives packages that originate when damaged organelles are engulfed by a membrane so they can be digested and recycled (**autophagy**) or when the plasma membrane engulfs a smaller cell or food particle (**Figure 7.14**).

- Lysosomes also receive materials via **receptor-mediated endocytosis (Figure 7.15)**; macromolecules outside the cell bind to protein receptors, causing the membrane to fold inward and pinch off to form an **early endosome.**

- This membrane-bound vesicle matures into a **late endosome** and eventually into a lysosome as it receives digestive enzymes from the Golgi and has its pH lowered by proton pumps.

- After macromolecules are hydrolyzed, their subunits are transported out of the lysosome by proteins in the organelle's membrane. The exported subunits can then be reused to make new macromolecules.

- Phagocytosis and receptor-mediated endocytosis are types of **endocytosis,** a process whereby a cell membrane pinches off a vesicle to bring outside material into the cell. **Pinocytosis** brings fluid into the cell, but these vesicles do not go to lysosomes.

- The smooth and rough ER, together with the Golgi apparatus and the lysosomes, make up the **endomembrane system.**

Vacuoles

- Plants and fungi have large **vacuoles** that function primarily in water and/or ion storage and help the cell maintain its normal volume. Sometimes these vacuoles store other molecules (proteins, toxins, pigments) or contain digestive enzymes.

Mitochondria

- **Mitochondria** have two membranes; the inner one is folded into sac-like **cristae** that contain the **mitochondrial matrix (Figure 7.17)**. Most enzymes involved in making ATP are in the inner mitochondrial membrane or in the matrix solution.

- ATP production is a mitochondrion's core function.

- Mitochondria grow and divide independently of cell division; they make their own ribosomes and contain their own supercoiled DNA in a small, circular chromosome similar to those found in bacteria.

Chloroplasts

- Most plant and algal cells have **chloroplasts** that convert light energy to chemical energy via photosynthesis.

- Like mitochondria, chloroplasts have a double membrane, grow and divide independently, and contain their own circular chromosome. Via photosynthesis, chloroplasts convert light energy to chemical energy.

- Inside the chloroplast, flat vesicles called **thyla–koids** are stacked to form **grana**. Everything required for photosynthesis is located either in the thylakoid membranes or out in the chloroplast **stroma** (**Figure 7.18**).

The Cell Wall

- Fungi, algae, and plants have an outer **cell wall** typically made of carbohydrate fibers in a stiff matrix of polysaccharides and proteins (**Figure 7.19**). This cell wall protects the cell.

- Some plants produce yet another protective barrier of **lignin**, a tough molecule that, when combined with cellulose, forms wood.

Cytoskeleton

- Eukaryotes have a complex, interconnected system of protein fibers that form the **cytoskeleton**. The cytoskeleton gives the cell shape; it also aids cell movement and the transport of materials within the cell.

✓ **CHECK YOUR UNDERSTANDING**

(b) Label the organelles in the following diagrams. Are these plant, animal, or bacterial cells?

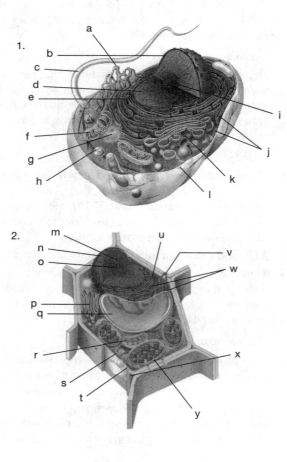

7.3 Putting the Parts into a Whole

- The structures of cell components are closely related to their functions (**Table 7.2**). A cell's size, shape, and composition also correlate with its function.

Structure and Function at the Whole-Cell Level

- **Figure 7.21** shows examples of how cell structure correlates with cell function; cells that export proteins have extensive rough ER and Golgi, cells that synthesize steroids have lots of smooth ER, and plant cells that store starch have many vacuoles.

The Dynamic Cell

- Microscopes reveal the basic structure of cells, and **differential centrifugation** (**BioSkills 11**) allows the isolation and chemical analysis of each cell component. However, neither of these techniques gives a complete picture of living cells.

- Live cells are dynamic, living things with interacting parts and constantly moving molecules.

- Cells take in substances, synthesize molecules, get rid of wastes, and reproduce. Every second within your cells, phospholipids move within membranes, millions of ATP are hydrolyzed, and each enzyme might catalyze 25,000 or more reactions.

- Within a cell, gravity has little effect whereas kinetic energy and electrostatic attractions are major forces. Events occur in nanoseconds and over micrometers.

- Certain modern imaging techniques (**BioSkills 10**) are able to capture cell dynamism by labeling and tracking molecules and organelles within living cells.

7.4 Cell Systems I: Nuclear Transport

- The nucleus is a highly organized information center for the cell. The nuclear lamina anchors each chromosome and maintains the overall shape and structure of the nucleus.

- Ribosomes are produced in the nucleolus, and other centers make mRNA. Many enzymes work together to properly decode and process genetic information.

Structure and Function of the Nuclear Envelope

- Molecules shuttle to and from the nucleus through thousands of **nuclear pores** that connect the nucleus with the cytoplasm (**Figure 7.22**). More than 50 proteins form the **nuclear pore complex**, which extends through both nuclear membranes.

- Initial research on the nuclear pore complex involved injecting gold particles into cells and determining their location after various time intervals using electron microscopy.

- After 2 minutes, gold particles were associated with nuclear pores; after 10 minutes, many had entered the nucleus. These data supported the idea that pores regulate the transfer of molecules into the nucleus.

- Traffic through nuclear pores is selective. DNA never leaves the nucleus, but RNAs are exported into the cell. The nuclear pore complex allows nucleotide triphosphates and proteins used in the nucleus to enter.

How Are Molecules Imported into the Nucleus?

- **Viruses** use a host cell's proteins to replicate themselves. To do this, certain viral proteins must enter the nucleus. Researchers noticed that certain changes in viral proteins prevented them from entering the nucleus through the nuclear pore.

- Researchers thus hypothesized that proteins destined for the nucleus have a molecular address tag or "zip code" allowing them to enter the nucleus. Viral proteins have access to the nucleus if they contain this **nuclear localization signal (NLS)**.

- The NLS was studied using nucleoplasmin, which helps assemble chromosomes. This protein has a globular core surrounded by several long "tails." Radioactively labeled nucleoplasmin quickly enters the nucleus when injected into a cell's cytoplasm.

- Researchers separated the nucleoplasmin core from its tails using proteases, and radioactively labeled first one and then the other part.

- When injected into cells, isolated tail fragments were transported to the nucleus; but core fragments remained in the cytoplasm. Thus, the NLS must be part of the tail region (**Figure 7.23**).

- Further analysis demonstrated that the NLS, the molecular zip code, is a 17-amino-acid sequence. Other proteins destined for the nucleus have similar localization signals.

- Movement of proteins and other large molecules into and out of the nucleus requires energy and involves transport proteins called importins and exportins.

✔ CHECK YOUR UNDERSTANDING

(c) Why did researchers investigating nucleoplasmin's nuclear localization signal (**Figure 7.23**) inject radiolabeled tails and cores into different cells?

_____

_____

_____

_____

7.5 Cell Systems II: The Endomembrane System Manufactures and Ships Proteins

- Small molecules like ions, ATP, and amino acids travel throughout the cell via diffusion, but the movement of large molecules uses energy and is highly organized and regulated.

- Some proteins made on ribosomes in the cytosol have signal sequences that cause transport to either mitochondria or chloroplasts. Other proteins are made in the rough ER and travel through the endomembrane system.

Studying the Pathway through the Endomembrane System

- The "secretory pathway" hypothesis proposes that secreted proteins are made and processed in a multistep pathway involving the rough ER and Golgi. These organelles are especially prominent in cells that secrete proteins.

The Logic of a Pulse-Chase Experiment

- George Palade and co-workers did **pulse-chase experiments** to reveal the movements of newly synthesized proteins within pancreatic cells that secrete enzymes. These cells were grown in culture (in vitro).

- A cell was first given a "pulse"—a high concentration of a labeled molecule for a short time—followed by the "chase"—lots of unlabeled molecules for a longer time period.

- Palade's group gave cells a 3-minute pulse of radiolabeled leucine. This amino acid became part of proteins produced during that 3-minute period, thus labeling those proteins. After the pulse, nonradioactive leucine was supplied.

Results of a Pulse-Chase Experiment

- When cells were killed and examined by electron microscopy immediately after the pulse, all labeled protein was inside the rough ER.

- Researchers examined cells after increasingly longer chase periods and found that the labeled proteins proceeded from the rough ER to the Golgi, into secretory vesicles on the *trans* side of the Golgi, and then out of the cell (**Figure 7.25**).

- These experiments supported the secretory pathway hypothesis and led researchers to conclude that the rough ER and Golgi apparatus function as an integrated endomembrane system.

(d) Why is the chase a necessary step in pulse-chase experiments?

_____

_____

_____

Entering the Endomembrane System: The Signal Hypothesis

- The **signal hypothesis** predicts that proteins bound for the endomembrane system have a molecular zip code just as proteins bound for the nucleus bear the nuclear localization signal.

- Günter Blobel and colleagues proposed that as ribosomes make secretory proteins, the first few amino acids signal the growing polypeptide to enter the ER.

- This hypothesis is supported by the observation that secreted proteins made in a test tube are 20 amino acids longer than are the proteins secreted by cells.

- Blobel proposed that a 20-amino-acid **ER signal sequence** must direct the proteins to the ER and then be removed before the protein is secreted. His group went on to identify this sequence.

- After a ribosome synthesizes the ER signal sequence, the sequence binds to an RNA–protein complex in the cytosol called the **signal recognition particle (SRP)**.

- The SRP-signal sequence-ribosome complex then binds to a SRP receptor in the ER membrane. The SRP is then released, the signal sequence is removed, and the ribosome completes protein synthesis (**Figure 7.26**).

- The new protein enters either the rough ER lumen (if it will be shipped to an organelle or secreted) or the rough ER membrane (if it is a membrane protein) and is folded with the help of chaperone proteins.

- Within the ER lumen, many proteins gain a carbohydrate made of 14 sugar subunits. After this **glycosylation**, the protein is called a **glycoprotein** (**Figure 7.27**).

Moving from the ER to the Golgi

- Palade and colleagues proposed that proteins are transported from the ER to the *cis* face of the Golgi apparatus in vesicles that bud off of the ER. Differential centrifugation of vesicles containing labeled proteins supported this interpretation.

What Happens Inside the Golgi Apparatus?

- Within the Golgi apparatus, cisternae (flattened vesicles) form at the *cis* face and then move toward the *trans* face, where they eventually break apart.

- Golgi cisternae mature as they move from *cis* to *trans* face, containing different glycosylation enzymes at different stages. Proteins within the cisternae thus undergo continual modification as they pass through the various sacs.

- Near the *cis* Golgi face, some proteins get sugar-phosphate groups added. Later, the 14-sugar ER carbohydrate is removed. Finally, near the *trans* face, proteins may gain carbohydrates that protect the protein or help it attach to surfaces.

How Do Proteins Reach Their Destination?

- A few proteins remain within the endomembrane system, but most proteins that pass through the rough ER and Golgi are sent to lysosomes, the plasma membrane, chloroplasts, or outside the cell.

- Molecular zip codes serve as protein shipping tags that ensure they are transported to the correct destinations. For example, proteins bound for lysosomes have mannose-6-phosphate on their surface. If this sugar is removed, the proteins are not correctly shipped.

- Mannose-6-phosphate binds to a protein in certain vesicles. These vesicles have surface proteins that interact with lysosome membrane proteins, ensuring that those proteins are correctly delivered to a lysosome.

- Similar processes direct other proteins to vesicles destined for other locations. Each protein exits the Golgi with a molecular tag for a specific type of transport vesicle. These vesicles have tags that ensure their own transport to a specific cell location.

- Some proteins are sent to the cell surface in vesicles that fuse with the plasma membrane, releasing their contents to the exterior of the cell (**exocytosis**).

- Protein sorting and transport from the Golgi apparatus is highly organized and efficiently gets proteins to the proper compartments (**Figure 7.28**). Proteins made in the cytoplasm have their own molecular zip codes directing their transport.

(e) What are the possible outcomes if a cell is unable to make mannose-6-phosphate?

_____

_____

_____

7.6 Cell Systems III: The Dynamic Cytoskeleton

- A complex network of fibers makes up the cytoskeleton, which provides structural support for cells. The cytoskeleton is dynamic; it moves and changes in order to alter the cell's shape, transport materials within the cell, or move the cell itself.
- The three types of cytoskeletal elements are actin filaments, intermediate filaments, and microtubules (**Table 7.3**).

Actin Filaments

- The smallest-diameter cytoskeletal elements are **actin filaments**, also known as **microfilaments**. Actin, a globular protein, polymerizes to form two long, fibrous strands twisted around each other.

Structure

- The globular protein actin polymerizes to form two long, fibrous strands twisted around each other. Actin filaments grow and shrink as actin subunits are added or removed.
- Actin monomers and the filaments made from them are asymmetrical, or polar, with a plus and a minus end. Actin filaments grow faster at the plus end than at the minus end because polymerization occurs fastest there.
- Actin filaments are grouped into long fibers or dense networks, which are most often found just inside the plasma membrane. These filaments, linked by other proteins, help define the cell's shape and resist tension (pull).

Microfilament Function

- Actin is also involved in cell movement through its interactions with **myosin**—a **motor protein** that can convert ATP's chemical energy into mechanical work.
- A conformational change of myosin's head region uses the chemical energy released by ATP hydrolysis to slide bound actin filaments (**Figure 7.29a**).
- Actin–myosin movement causes (1) contraction of a ring of filaments that pinches an animal cell in two during **cytokinesis** at the end of cell division, and (2) **cytoplasmic streaming**, which moves cytosol and organelles through plant and fungal cells (**Figure 7.29b**).
- Some cells use actin filaments for **cell crawling** by using the filaments to push the plasma membrane outward.

Intermediate Filaments

- **Intermediate filaments** are defined by size (about 10-nm diameter) instead of by composition; all are nonpolar cytoskeletal elements that function as structural support for the cell.
- Keratins in our skin help resist pressure and abrasion, and secreted keratins form hair and nails. **Nuclear lamins** of the nuclear lamina anchor chromosomes and give shape to the nucleus.

- Some intermediate filaments reach from the nucleus to the cell membrane and form a flexible skeleton, which contributes to cell shape and holds the nucleus in place.

Microtubules

- Microtubules are large, hollow tubes made of α- and β-tubulin **dimers** (two joined monomers).
- Microtubules grow or shrink as subunits are added or removed and are polar; the two ends differ, and the molecule is more likely to grow from one end than from the other.
- Microtubules grow outward from a **microtubule organizing center**. This center is called the **centrosome** in animals. An animal centrosome has two microtubule bundles called **centrioles** (**Figure 7.30**).
- Plants have many microtubule organizing centers in each cell, whereas animals and fungi have just one. Microtubules radiating from these centers move chromosomes to daughter cells during cell division.
- Microtubules help the cell resist compression and create a structural framework for organelles. They also help move the cell and materials within the cell.

✔ **CHECK YOUR UNDERSTANDING**

(f) You expose animal cells in culture to radiation to obtain mutants. You find one mutant cell that looks like it has gone through most of mitosis but will not/cannot separate into two new cells. You suspect that one of the cytoskeletal filaments is involved. Which one do you suspect and why?

_____

_____

_____

Studying Vesicle Transport

- The giant axon is an extremely large neuron that runs the length of a squid's body. If the squid is disturbed, the axon carries a signal that contracts muscles that propel the animal away from danger.
- Ronald Vale and co-workers studied vesicle transport in this axon because it was easy to see and work with. They could squeeze the cytoplasm out of the cell and observe vesicle transport without being hindered by the plasma membrane.

Microtubules Act as "Railroad Tracks"

- Researchers using video microscopy saw vesicles moving along a filamentous track (**Figure 7.31**). The diameter and chemical composition of the track indicated that it was made of microtubules.

- Vesicle transport required energy because movement stopped if ATP was removed from the cytoplasm.

- The movement of vesicles from the ER to the Golgi apparatus also appears to depend on microtubules because vesicle movement is abnormal after treatment with a drug that disrupts microtubules.

A Motor Protein Generates Motile Forces

- Vale and co-workers produced microtubules from purified α- and β-tubulin. They added ATP and transport vesicles isolated by differential centrifugation, but no transport occurred.

- Eventually they found a cell fraction that allowed vesicle movement. Further purification led to isolation of the motor protein **kinesin**, which generates vesicle movement by converting the chemical energy of ATP into mechanical work.

- X-ray diffraction of kinesin shows three distinct regions: (1) a head with two globular pieces, each of which can bind to ATP and to a microtubule; (2) a tail that binds to a transport vesicle; and (3) a stalk that connects the head and tail regions (**Figure 7.32**).

- The kinesin head region steps along the microtubule as it binds and hydrolyzes ATP, while the tail region carries the transport vesicle. Different kinesins carry different types of vesicles.

Flagella and Cilia: Moving the Entire Cell

- Bacterial flagella are made of flagellin and rotate like a propeller to move a cell, whereas eukaryotic flagella are made of microtubules and wave back and forth. These two types of flagella evolved independently.

- Eukaryotic flagella are encased by the plasma membrane, but bacterial flagella are not.

- Structurally, eukaryotic flagella and cilia are very similar. Eukaryotes with flagella typically have only one or two, whereas those with cilia have many shorter structures. Many eukaryotes do not have either structure.

How Are Cilia and Flagella Constructed?

- Cilia and flagella have two central microtubules surrounded by nine microtubule doublets (one of which is incomplete). This 9 + 2 structure is called the **axoneme**.

- Spokes connect doublets to the central microtubules, molecular bridges connect doublets to one another, and each doublet has two "arms" projecting toward an adjacent doublet (**Figure 7.34**).

- The axoneme attaches to the cell at a **basal body**, which is structurally identical to a centriole. The basal body is important in axoneme growth.

What Provides the Force Required for Movement?

- Ian Gibbons isolated *Tetrahymena* axonemes by using a detergent to remove the plasma membrane and then performing differential centrifugation. This cell-free system was easy to manipulate and perform experiments on.

- The isolated axonemes beat only in the presence of ATP, showing that cilia movement requires energy.

- Furthermore, axonemes could not beat or use ATP when protein binding was disrupted. Electron microscopy showed that the "arms" between doublets were missing. These arms are a motor protein called **dynein**.

- Dynein changes shape when a phosphate group from ATP attaches to it so as to "walk" along microtubules.

- Cilia and flagella bend instead of elongating because the spokes and bridges constrain movement; this makes the microtubules slide past one another when the dynein arms on just one side of the axoneme walk (**Figure 7.35**).

- Cells are dynamic integrated structures!

✔ **CHECK YOUR UNDERSTANDING**

(g) Name three research techniques described in this chapter, and briefly state their usefulness in cell research.

_____

_____

_____

B. MASTERING CHALLENGING TOPICS

Chapter 7 has introduced you to many new terms and research techniques. Consider making note cards for yourself for all these new terms and techniques. If you have trouble memorizing, be sure to give yourself plenty of time before the test. It will be easier if you can study a few note cards at a time. You might even try tapping up note cards around your room—above the light switch, on your computer, on your sock drawer, on your mirror. This way you can think about these new terms a little bit each day as you go through your regular daily routine.

Once you know some terms, put up new cards to learn. You can also try explaining vesicle transport to family or friends not familiar with biology. If you can explain it clearly (and correctly) to them, you know you have mastered the material.

C. ASSESSING WHAT YOU'VE LEARNED

1. How are photosynthetic prokaryotes similar to chloroplasts?
   a. They both have extensive membranes containing pigments and enzymes.
   b. They both contain a nucleus surrounded by a double membrane.
   c. They both store calcium.
   d. Their flagella have similar structures.

2. Which of the following is true of DNA in bacterial cells?
   a. All bacterial DNA is part of a chromosome in the cell's nucleoid region.
   b. Bacterial DNA usually forms linear chromosomes.
   c. Bacterial DNA does not contain genes.
   d. Bacterial DNA is supercoiled.

3. The nuclear envelope is:
   a. a single membrane surrounding the nucleus.
   b. continuous with the Golgi apparatus.
   c. supported by the nuclear lamina.
   d. made of chromatin.

For questions 4–8, match the given functions with the correct organelle.

For extra review, note how the structures of these organelles correlate with their functions.

a.	Nucleus	e.	Peroxisomes
b.	Rough ER	f.	Lysosomes
c.	Smooth ER	g.	Mitochondria
d.	Golgi apparatus	h.	Chloroplasts

4. _____ processing and packaging proteins for secretion

5. _____ isolation of dangerous oxidation reactions

6. _____ lipid processing and phospholipid manufacture

7. _____ synthesis of plasma membrane proteins

8. _____ hydrolysis of macromolecules

9. After the activation of proton pumps, this vesicle receives digestive enzymes from the Golgi apparatus.
   a. Peroxisome
   b. Lysosome
   c. Endosome
   d. Chromosome

10. Which of these molecules is most likely to contain the NLS?
    a. mRNA for protein synthesis
    b. An enzyme that attaches carbohydrates to proteins
    c. RNA polymerase, the enzyme that catalyzes RNA synthesis
    d. The receptor for a water-soluble hormone

11. The glucocorticoid receptor is only found in the cytoplasm until the hormone glucocorticoid binds to it. The receptor then moves to the nucleus where it regulates the gene transcription. If a certain amino acid sequence on the receptor is deleted, the receptor still binds glucocorticoid but does not enter the nucleus. Now suppose that if the mutant receptor is fused with the tail fragment of nucleoplasmin, the receptor enters the nucleus regardless of whether glucocorticoid is bound. Furthermore, fusion of the core fragment of nucleoplasmin with the sequence deleted from the glucocorticoid receptor results in accumulation of the core fragment in the nucleus. Which conclusion might be drawn from these observations?
    a. The binding of glucocorticoid to the receptor normally exposes a nuclear localization signal.
    b. The glucocorticoid receptor does not have a nuclear localization sequence.
    c. Glucocorticoid is the nuclear localization signal for the receptor.
    d. The glucocorticoid receptor enters the nucleus by a different mechanism than that used by nucleoplasmin.

12. The signal hypothesis predicted that:
    a. nuclear localization signals bind to signal recognition particles in cytosol.
    b. proteins bound for the endomembrane system have a molecular zip code.
    c. only proteins destined for the nucleus need a molecular zip code.
    d. secreted proteins will be 20 amino acids shorter when synthesized in a test tube.

13. Microtubules, intermediate filaments, and microfilaments are made of _____, respectively.
    a. actin, keratins and other nonpolar filaments, and α- and β-tubulin
    b. α- and β-tubulin, lamins, and myosin
    c. actin, myosin, and α- and β-tubulin
    d. α- and β-tubulin, keratins and other nonpolar filaments, and actin

14. The mixing of purified microtubules with transport vesicles and ATP does not result in movement of the vesicles. Why?
    a. Actin is missing.
    b. Kinesin is missing.
    c. Glucose is missing.
    d. Intermediate filaments are missing.

15. Suppose the plasma membrane around a flagellum opens to reveal the axoneme inside. The radial spokes connecting peripheral microtubule doublets to the central pair are then broken by chemical

treatment. ATP is then added. What is the expected observation?

   a. The flagellum will undergo normal bending.

   b. No movement will occur, because the plasma membrane is not present.

   c. The axoneme will elongate.

   d. No movement will occur, because the ability to use ATP is lost.

CHAPTER 7—ANSWER KEY

Check Your Understanding

(a) Rough ER has ribosomes attached, and smooth ER does not. Because ribosomes are the site of protein synthesis, it makes sense that rough ER would be responsible for the synthesis and distribution of many cell proteins.

(b) 1. Animal cell: (a) smooth ER, (b) nuclear envelope, (c) flagellum, (d) chromosomes, (e) rough ER, (f) mitochondrion, (g) centrioles, (h) lysosome or peroxisome, (i) nucleolus, (j) ribosomes, (k) Golgi apparatus, and (l) plasma membrane. 2. Plant cell: (m) chromosomes, (n) nuclear envelope, (o) nucleolus, (p) golgi apparatus, (q) vacuole, (r) mitochondrion, (s) peroxisome, (t) plasma membrane, (u) rough ER, (v) smooth ER, (w) ribosomes, (x) cell wall, and (y) chloroplast

(c) If radiolabeled tails and cores had been injected into the same cell, the researchers would not have been able to determine which had entered the nucleus by monitoring radiation—because the cores and tails themselves cannot be detected, only the radiation.

(d) Without the chase, radioactive proteins would have been found throughout the endomembrane system during the later time periods, and it would have been impossible to determine whether the radioactivity observed in the ER and Golgi was due only to new proteins or whether some previously made protein remained in the ER and Golgi.

(e) Inability to make mannose 6-phosphate (known as a congenital disorder of glycosylation) causes lysosomal problems because many lysosomal enzymes are not correctly packaged and delivered. This can sometimes be detected due to abnormally high plasma and serum concentrations of these enzymes. Organism-level effects can be quite severe.

(f) You suspect that actin filaments are involved because they are responsible for cytokinesis in animals.

(g) Radiolabeling (helps localize proteins within cells), pulse-chase (tracks movement of labeled proteins within cells), subcellular fractionation/differential centrifugation (helps isolate cell components for chemical testing and identification), and cell-free extracts (e.g., the squid giant axon and isolated axonemes; allow easier experimental manipulation of cell components).

Assessing What You've Learned

1. a; 2. d; 3. c; 4. d; 5. e; 6. c; 7. b; 8. f; 9. c; 10. c; 11. a; 12. b; 13. d; 14. b; 15. c

Looking Forward—Key Concepts in Later Chapters

Mitochondria and Chloroplasts—Chapters 9 and 10

You will learn much more about mitochondria and chloroplast structure and function when you read about cell respiration and photosynthesis in **Chapters 9** and **10** respectively.

Cell Division—Chapters 11 and 12

During cell division, whole chromosomes are moved around the cell by microtubules. Much of what you learned about intracellular transport in this chapter is thus relevant to your understanding of mitosis and meiosis.

DNA and Genes—Chapters 13–Chapter 18

Because DNA contains all the information for protein synthesis, these chapters on DNA and genes will aid your understanding of cell function. For example, this chapter mentions that ribosomes are the site of protein synthesis, but the process itself (translation) is described in **Chapter 16**.

Prokaryotes—Chapter 28

Most of this chapter describes eukaryotic cell structure and function, but much of life is prokaryotic. You will learn more about Bacteria and Archaea in **Chapter 28**.

Microfilaments—Chapters 11, 12, 29, and 46

In Chapter 7, you learned how microfilaments help give shape to the cell. These actin filaments also play active roles in cell division (**Chapters 11** and **12**), pseudopodia motility (**Chapter 29**), and muscle contraction (**Chapter 46**).

Cell-Cell Interactions

Looking Back—Key Concepts from Earlier Chapters

Enzyme Function—Chapter 3

Hormone-receptor interactions are similar to the enzyme-substrate interactions described in **Chapter 3**; both rely on reversible conformational changes. Chapter 8 also introduces you to some important groups of enzymes: kinases and phosphatases.

The Cell Membrane and Its Associated Proteins—Chapter 6

Cadherins, gap junction proteins, hormone receptors, and G proteins are just some of many membrane proteins mentioned in Chapter 8. Remember from **Chapter 6** that the cell membrane is dynamic—these proteins are always moving around, and their motion ensures that activated G proteins interact with the enzymes they activate.

KEY CONCEPTS

- ◖━ Extracellular material provides structural support and helps bind cells together.

- ◖━ Neighboring cells stick together using cell-cell connections and communicate with each other via cell-cell gaps.

- ◖━ Thousands of cells can form an integrated whole in multicellular species because they respond in a coordinated manner to intercellular signals.

- ◖━ Signals between cells must be received, processed, responded to, and deactivated. Those with cell-surface receptors must produce an intracellular signal.

A. CHAPTER OUTLINE

Chapter Introduction

- Cells constantly interact with and respond to other cells and their environment. In multicellular organisms, a cell's external environment consists primarily of other cells. How do cells communicate and cooperate with each other in multicellular organisms?

8.1 The Cell Surface

- The plasma membrane separates the cell interior from its outer environment.

- The phospholipid bilayer has peripheral proteins on its surfaces and integral proteins within it. Some of these proteins attach to the cell's cytoskeleton and/or to extracellular structures.

The Structure and Function of an Extracellular Layer

- Most cells secrete extracellular material that glues them to other cells, defines cell shape, and/or defends the cell from external threats.

- Extracellular material typically has a "fiber composite" structure of long cross-linked filaments surrounded by a stiff ground substance.

- The rods or filaments protect against stretching forces (tension), while the ground substance protects against compression.

- Different groups of organisms use different molecules to form this extracellular fiber composite, and many extracellular layers are flexible as well as sturdy.

The Cell Wall in Plants

- Plant cell walls are dynamic—they respond to changes in their environment and can signal other cells and respond to signals received from other cells.

Primary Cell Walls

- The extracellular material secreted by plant cells forms a **primary cell wall** with a fiber composite structure.

- Enzymes in the plasma membrane secrete cellulose, which becomes cross-linked with other polysaccharides and is bundled into filaments called microfibrils.

- The microfibrils form a crisscrossed network that becomes filled with a gelatinous polysaccharide ground substance (**Figure 8.2**).

- **Pectin**s are common ground substances in plant cell walls. They keep cell walls moist because they attract and hold water. Pectins and similar polysaccharides are made in the rough endoplasmic reticulum and Golgi apparatus and secreted to the extracellular space.

- The primary cell wall determines cell shape. Normally, the solute concentration is higher inside the cell than outside. Consequently, water enters the cell by osmosis, fills up the cell volume, and pushes the plasma membrane against the cell wall.

- **Turgor pressure** is the force with which a filled cell pushes against its cell wall.

- In young cells, enzymes called expansins catalyze reactions that permit microfibrils to slide past one another. Turgor pressure then causes cell wall elongation, allowing cell growth.

Secondary Cell Walls

- Mature plant cells may secrete a **secondary cell wall** inside the primary cell wall. Secondary cell wall structure varies and relates to cell function. For example, cells that form wood have a secondary cell wall containing tough, rigid **lignin**.

The Extracellular Matrix in Animals

- Most animal cells secrete a fiber composite called the **extracellular matrix** (**ECM**). As in plants, the extracellular material strengthens cells and helps bind them together.

- Animals also use gelatinous polysaccharides as a ground substance, but a protein called **collagen** is the predominant fibrous filament (**Figure 8.3**).

- Collagen and other protein fibers are synthesized in the rough ER, processed in the Golgi apparatus, and secreted via exocytosis. These proteins are more elastic than cellulose or lignin and form a flexible extracellular layer.

- The composition and amount of ECM correlates with cell function, varying widely among cell types. For example, bone has more ECM than skin.

- The ECM also helps cells stick together. Cytoskeletal actin filaments connect to transmembrane **integrins**, which bind to ECM **fibronectin**s, which bind to collagen. These protein-protein attachments link cell cytoskeletons to the ECM (**Figure 8.4**).

- ECM breakdown can cause major problems. For example, when cancer cells break away from their ECM connections, they can travel throughout the body (**metastasis**).

✔ CHECK YOUR UNDERSTANDING

(a) List similarities and differences between plant and animal extracellular material. How do the differences relate to differences in the function of extracellular material in animals and plants?

_____

_____

_____

_____

_____

8.2 How Do Adjacent Cells Connect and Communicate?

- Cell-cell physical connections are the basis of **multicellularity**.

- In most multicellular organisms, similar cells with similar functions group together to form **tissues**. These cells must attach to each other and exchange materials and information.

Cell-Cell Attachments in Eukaryotes

- Plant cell walls are glued together by gelatinous pectins and other molecules that form the middle lamella (**Figure 8.6**). When enzymes break down this layer, the surrounding cells separate.

- A layer of gelatinous polysaccharides also exists between cells in many animal tissues.

- Integrins connect cell cytoskeletons to the ECM, and some tissues have additional proteins that more strongly connect neighboring cells.

- Stronger protein attachments are especially found in **epithelia**, the tissues covering external and internal body surfaces.

Tight Junctions

- **Tight junctions** are cell-cell attachments in which the proteins of neighboring cell membranes line up and bind to one another, stitching the two cells together to form a watertight seal between the two plasma membranes (**Figure 8.7**).

- The exact structure and function of tight junctions depends on the type of epithelia in which they are found. For example, tight junctions in the small intestine allow small ions to pass, whereas those in the bladder do not.

- Tight junctions are dynamic: Some tissues loosen or tighten these junctions in response to environmental cues.

Desmosomes

- Epithelial cells and some muscle cells are held together by **desmosomes** made of proteins that link the cytoskeletons of adjacent cells. These proteins bind to each other and to the proteins that anchor cytoskeletal intermediate filaments (**Figure 8.8**).

Selective Adhesion

- When adult sponges are treated with chemicals that cause cells to separate from each other, the resulting mixed-up cell mass reconnects when the chemicals are removed.

- Reconnecting sponge cells sort themselves by species and tissue type. This is known as **selective adhesion** (**Figure 8.9**). This result suggested that cell-cell connections are species- and tissue-specific.

The Discovery of Cadherins

- In animals, each major cell type has its own cell adhesion proteins. The **cadherins** are cell adhesion proteins found in plasma membranes. These proteins only bind to other cadherins of the same type. The attachment molecules in desmosomes are cadherins.

- Researchers studying these cell adhesion proteins had to isolate membrane proteins from a particular type of cell.

- Pure preparations of each protein were injected into rabbits, one at a time, causing each rabbit to produce **antibodies**—proteins that bind specifically to one part of the injected protein.

- Researchers added one type of antibody at a time to mixtures of dissociated cells and observed whether the cells could reconnect normally.

- Antibodies that prevented cell-cell attachment must bind to adhesion proteins, blocking their normal binding function (**Figure 8.10**).

(b) In the cadherin research, why was it necessary to isolate the membrane proteins before injecting them into the rabbits for antibody production?

_____

_____

_____

_____

_____

_____

_____

- The cell-type specificity of cadherins and other cell adhesion proteins allows animal cells to selectively attach to other animal cells of the same tissue type within the same individual.

- Cell-cell connections help adjacent cells stick together; cell-cell gaps allow adjacent cells to communicate.

Cells Communicate via Cell-Cell Gaps

- Direct connections between cells allow tissue cells to work together. In both plants and animals, cells typically keep their own organelles and macromolecules, but share ions and/or small molecules in their cytoplasm.

Plasmodesmata

- Plant cells communicate via cell-wall gaps called **plasmodesmata**, where the plasma membranes and cytoplasms of two cells connect such that smooth ER actually runs between one cell and the next (**Figure 8.11a**).

- Like the nuclear pore complex, plasmodesmata contain proteins that regulate protein traffic. Some proteins passing through plasmodesmata help coordinate the activity of neighboring cells.

Gap Junctions

- Most animal tissues have specialized proteins in adjacent cells that match up to form a channel. These channels are called **gap junctions** (**Figure 8.11b**).

- Water, ions, and small molecules flow through these channels and help coordinate activities of the connected cells so that a tissue can act as an integrated whole.

8.3 How Do Distant Cells Communicate?

- Cells that are not in physical contact can also communicate with each other.

Cell-Cell Signaling in Multicellular Organisms

- **Hormones** are intercellular signaling molecules secreted from cells that travel through the organism to act on distant target cells.

- Hormones are usually small molecules present in minute concentrations. They nevertheless have a major impact on target cells and the organism as a whole.

- The functions and chemical structures of plant and animal hormones vary widely (**Table 8.1**).

- Hormone chemical structure affects the manner in which its signal is received. Steroids and other lipid-soluble hormones can diffuse across a cell's plasma membrane, but hydrophilic hormones must be recognized at the cell surface.

Signal Reception

- Hormones and other cell-cell signals must bind to a receptor molecule in order to affect a target cell.

- When several tissues and organs respond to the same signal, it is because they all have appropriate receptors. Presence of identical receptors in distant cells and tissues permits the coordination of cell activities throughout a multicellular organism.

- Proteins that change their conformation or activity when a hormone binds to them are **signal receptors**.

- Receptors are dynamic and may increase or decrease in number over time. For example, receptor numbers tend to decline following prolonged hormonal stimulation. Changes in receptor number affect cell sensitivity to that hormone.

- Receptors can be blocked. Many drugs act by binding to and blocking specific receptors.

- Lipid-soluble signals can diffuse across the plasma membrane, so their receptors are located inside cells. Other signal receptors are in the plasma membrane.

Signal Processing

- Steroid hormones directly initiate cell response; they enter a cell, bind to a receptor protein, and either directly affect membrane pump activity or travel to the nucleus as a hormone-receptor complex that directly alters gene expression (**Figure 8.13**).

- Hormones that cannot diffuse across the plasma membrane rely on **signal transduction**—a series of events that converts the hormone signal from its original extracellular chemical form to a new intracellular form (**Figure 8.14**).

Signal Amplification

- Hormone messages are also amplified during signal transduction. Signal amplification allows small hormone concentrations to cause large cell responses.

- After a receptor binds to a plasma membrane receptor, it causes a secondary signal involving many ions or molecules. This amplifies the original signal.

- G-proteins and enzyme-linked receptors are two important signal transduction systems. G-proteins trigger the production of intracellular "second" messengers, whereas enzyme-linked receptors trigger protein activation via phosphorylation.

Signal Transduction via G-Proteins

- **G-protein**s are peripheral membrane proteins located inside the cell. They are closely associated with transmembrane signal receptors and are named for their ability to bind guanosine triphosphate (GTP) and guanosine diphosphate (GDP).

✔ CHECK YOUR UNDERSTANDING

(c) Why are steroid receptors located inside cells, whereas most other hormone receptors are located on the plasma membrane?

_____

_____

_____

- GTP, like ATP, has high potential energy. G-proteins are turned on, or activated, when they bind GTP and are inactivated when they hydrolyze GTP, forming GDP.

- When a hormone binds to a G-protein-associated plasma membrane receptor, the receptor changes shape. The receptor's conformational change activates a G-protein on the membrane's inner surface.

- The activated G-protein releases GDP, binds GTP, and then splits into two parts, one of which activates a nearby enzyme in the plasma membrane (**Figure 8.15**).

- G-protein-activated enzymes catalyze the production of small, nonprotein signaling molecules called **second messengers**. These intracellular signals spread the hormone message throughout the cell.

- Second messengers can be quickly produced in large quantities and rapidly diffuse throughout the cell. One hormone molecule stimulates the production of many second messenger molecules, greatly amplifying the original signal.

- Second messengers (**Table 8.2**) often activate **protein kinase** enzymes that activate or inactivate other proteins by adding a phosphate group to them.

- The same second messenger may cause different effects in different cell types, and more than one type of second messenger may be involved in a cell's response to any given signal.

Signal Transduction via Enzyme-Linked Receptors

- **Enzyme-linked receptors**, such as **receptor tyrosine kinases** (**RTKs**), directly catalyze a cell reaction when activated.

- RTKs are transmembrane proteins that form dimers after binding a hormone signal. The dimers become active after a phosphate is transferred to them from ATP.

- A protein bridge forms between the activated RTK and a peripheral membrane protein called **Ras**. This bridge allows Ras to exchanges its GDP for GTP and become active.

- Active Ras causes the phosphorylation and activation of yet another enzyme that phosphorylates yet more proteins. This **phosphorylation cascade** activates many enzymes, amplifies the original signal, and directs the cell response (**Figure 8.16**).

- Phosphorylation cascade proteins are often kept close to each other by scaffolding proteins; this increases the speed of the reaction sequence.

- Some G-proteins trigger phosphorylation cascades and some phosphorylation cascades produce second messengers.

- Both G-protein and receptor tyrosine kinase signal transduction convert an easily transmitted extracellular message into an amplified intracellular message that carries information throughout the cell, inducing the cell response.

✔ **CHECK YOUR UNDERSTANDING**

(d) Compare signal reception by a G-protein receptor versus a receptor tyrosine kinase.

_____

_____

_____

_____

_____

_____

Signal Response

- Second messengers and/or phosphorylation cascades can affect cell activity by altering gene expression and/or by activating or deactivating particular enzymes, membrane channels, or other cell proteins (e.g., wheat seeds and GA_1).

Signal Deactivation

- Cells have automatic mechanisms for signal deactivation. Rapid response, followed by equally rapid response elimination, ensures that cells remain sensitive to future hormone signals and to any change in the number or activity of signal receptors.

- For example, activated G-proteins can hydrolyze the GTP bound to them. This reaction alters the G-protein's conformation, deactivating it so that it no longer activates the enzyme producing the second messenger.

- Second messengers are either degraded (for example, phosphodiesterase deactivation of cAMP and cGMP) or pumped back into storage (Ca^{2+}).

- Phosphorylation cascades are turned off by phosphatases—enzymes that remove phosphate groups from proteins. Phosphatases are always present in cells, ready to shut off phosphorylation cascades as soon as hormone stimulation stops.

- Problems occur when signaling pathways are not properly shut down following stimulation. For example, defective Ras proteins that do not properly deactivate can contribute to the development of cancer.

✔ **CHECK YOUR UNDERSTANDING**

(e) Why is signal deactivation important? Generally speaking, what might happen if a signal was not deactivated?

_____

_____

_____

_____

Cross-Talk: Synthesizing Input from Many Signals

- From reception to processing to response to deactivation, cell-cell signaling allows organisms to produce an integrated whole-organism response to changing conditions within and outside of the multicellular organism (**Figure 8.17**).

Quorum Sensing in Bacteria

- Cell-cell communication in bacteria is called **quorum sensing**. Some bacteria change their activity once their population size crosses some threshold.

- Species-specific signaling molecules are secreted by cells and diffuse through the environment. The response to these signals varies from species to species.

- Some bacteria respond to a signaling molecule by secreting a hard, polysaccharide-rich substance called a **biofilm** that attaches cells to the surface on which they live.

- For example, dental plaque is a biofilm encasing oral bacteria, and *Pseudomonas aerunginosa* bacteria secrete a biofilm that attaches them within the lung and protects them from immune system response and from antibiotics.

- Quorum sensing permits bacteria to communicate and coordinate activity. Like cells in multicellular organisms, cell-cell signaling helps bacteria gain information about and appropriately respond to a changing environment.

B. MASTERING CHALLENGING TOPICS

Antibodies are a key research tool for the cell biologist. The experiments shown in **Figure 8.10** are quite complex. If you did not completely understand these experiments after one reading, consider the following analogy.

You are told that antibodies are proteins that specifically bind to a section of another protein (animals make antibodies that bind to foreign proteins as part of their immune response). Pretend that the proteins isolated from a specific cell type and injected into the rabbit are like various-shaped objects that you place on a table for a 5-year-old to use in making play-dough molds. After awhile, you have many molds of an AA battery, a Lego, some children's scissors, and various other objects. Let's pretend that these molds, when left out, eventually find and bind to the object they fit with.

For your first experiment (Experiment A), you leave out the battery molds and discover that they bind to all the batteries in the house, causing your mini flashlights to go out, etc., but having no effect on Lego attachment. For Experiment B, you put away the battery molds and take out the Lego molds. Now your mini-flashlights work, but the Legos won't connect anymore, because they are all covered with the molds. If you put the Lego molds away and take out the scissors molds (Experiment C), the Legos and flashlights work, but kids in the house are no longer able to cut paper.

In Experiments A and C, the molds/antibodies did not affect Lego attachment because the objects used to make the molds are not involved in that process. In Experiment B, Lego attachment no longer works, because the mold/antibody blocks the attachment/binding site. Taking this analogy one step further, you are told that cadherin proteins are cell-type specific. In our analogy, this is comparable to discovering that the molds that block regular Legos in some houses have no effect on Duplo Legos found in other houses. Each house/cell type has its own attachment proteins that are blocked only by their specific molds/antibodies.

C. ASSESSING WHAT YOU'VE LEARNED

1. Extracellular material can function to: (Choose all that apply.)
   a. glue cells together.
   b. actively transport nutrients into cells.
   c. protect and support cells.
   d. define cell shape.
   e. catalyze chemical reactions.

2. The primary cell wall of plants is made of:
   a. pectins and expansins.
   b. cytoskeletal elements and peripheral proteins.
   c. cellulose and gelatinous polysaccharides.
   d. lignin and collagen.

3. Turgor pressure occurs when:
   a. solute concentration is the same inside the cell and outside, so that water pushes on the cell wall.
   b. solute concentration is higher inside the cell than outside, so that water leaves the cell by osmosis.
   c. solute concentration is lower inside the cell than outside, so that water enters the cell by osmosis.
   d. solute concentration is higher inside the cell than outside, so that water enters the cell by osmosis.

4. Which of the following statements is (are) true about the animal cell extracellular matrix (ECM)? (Choose all that apply.)
   a. Composition and amount of ECM varies widely among different cell types.
   b. Protein-protein attachments link the cell cytoskeleton to the ECM.
   c. The ECM helps maintain cell turgor pressure in animal cells.
   d. Most ECM components are made in the smooth ER.
   e. Collagen is the most common ECM protein fiber.

5. Only multicellular organisms:
   a. have quorum sensing.
   b. have organelles.
   c. have extracellular layers.
   d. have long-term cell-cell physical connections.

For questions 6 to 8, use the following options to identify each structure correctly.

    a. Found between plant cell walls, contains gelatinous pectins

    b. Found inside the primary cell wall, may incude lignin

    c. Found between epithelial cells in animals, form watertight seals

    d. Common in certain muscle cells, links cytoskeletons of adjacent cells

    e. Made by the immune system, bind to one specific section of another protein

6. _____ tight junctions

7. _____ middle lamella

8. _____ desmosomes

9. What evidence suggests that cells have selective adhesion?

    a. Electron micrographs show cell-cell attachments.

    b. Randomly mixed sponge cells sorted themselves by tissue and species.

    c. Chemical treatments are unable to separate sponge cells of the same tissue type.

    d. Rabbits produce antibodies to injected membrane proteins.

10. Plant cells communicate via cell-wall gaps called

    a. desmosomes.

    b. tight junctions.

    c. gap junctions.

    d. plasmodesmata.

11. Which of the following are shared by cells connected by gap junctions? (Choose *all* that apply.)

    a. Ions

    b. RNA

    c. Amino acids

    d. Smooth ER

    e. Nucleotides

12. **Table 8.1** notes that thyroxine's signal is received inside the cell and regulates metabolism in animals. From this information, what can you deduce about this hormone?

    a. It probably causes a large second messenger response.

    b. It is probably lipid soluble.

    c. It probably binds to a receptor tyrosine kinase.

    d. Beta-blockers can probably block thyroxine receptors.

13. Which of the following events occurs third in signal transduction via G-proteins?

    a. The G-protein splits into two parts.

    b. An enzyme produces a second messenger.

    c. Hormone binding changes the shape of a membrane receptor.

    d. The G-protein releases GDP.

14. In some cells, activation of a hormone receptor causes an increase in Ca^{2+} levels, which causes activation of an enzyme that produces nitric oxide. Nitric oxide then activates another enzyme that makes cGMP. Based on this description, nitric oxide could be considered a:

    a. steroid-hormone.

    b. G protein.

    c. signal receptor.

    d. second messenger.

15. How are receptor tyrosine kinase signals deactivated?

    a. Phosphodiesterases remove phosphate groups from ATP.

    b. Phosphatases remove phosphate groups from activated proteins.

    c. Membrane pumps return calcium ions to storage.

    d. cAMP and cGMP are converted to inactive AMP and GMP.

CHAPTER 8—ANSWER KEY

Check Your Understanding

(a) *Similarities*: Both use gelatinous polysaccharides as a ground substance, and both function as structural support. *Differences*: They have different cross-linked fibers (polysaccharides vs. collagen). The fiber differences make animal cell connections more flexible. This flexibility is critical for organisms that move around more than plants do.

(b) If the proteins had not been isolated before injection into the rabbit, the rabbit would have produced antibodies to all the proteins. These antibodies would have been all mixed together. Subsequent addition of these mixed antibodies into the cell culture might have allowed researchers to determine that a cell adhesion protein was present, but researchers would not have been able to determine which of the proteins injected into the rabbit was the one involved in cell adhesion.

(c) Steroid hormones are lipid soluble and can diffuse through the phospholipid bilayer, but most other hormones cannot easily cross the plasma membrane.

(d) In general, signal reception occurs when a hormone binds to a receptor, causing a conformational change. In the case of G-protein receptors, the receptor's conformational change causes an associated G-protein to release GDP, bind GTP, and split into two parts (one of which will in turn activate the next enzyme in this signal transduction pathway). In the case of receptor tyrosine kinases, hormone binding causes the receptor itself to form a dimer, bind ATP, and become phosphorylated (and able to activate the next part of this signal transduction pathway).

(e) Signal deactivation is important because if signals were not turned off, cells would be unable to respond the next time a signal was sent. Automatic signal deactivation allows cells to remain responsive. Thus, cells keep responding only if the signal continues to be sent and will stop responding if the signal dissipates.

Assessing What You've Learned

1. a, c, d; 2. c; 3. d; 4. a, b, e; 5. d; 6. c; 7. a; 8. d; 9. b; 10. d; 11. a, c, e; 12. b; 13. a; 14. d; 15. b

---

Looking Forward—Key Concepts in Later Chapters

Gene Expression—Chapter 17 and 18

You learned in this chapter that some signal transduction events end by affecting gene expression. Further information on how gene expression is controlled in Bacteria and Eukaryotes is given in **Chapters 17** and **18**. **Chapter 18** also discusses how defects in gene regulation can cause cancer (a topic also touched on in **Chapter 11**).

Tissues and Organs—Chapters 36 and 41

Chapter 8 describes how cells physically connect to form tissues and organs. You will learn more about tissue and organ structure and function in **Chapters 36** and **41** (plant and animal form and function, respectively).

Chemical Signals in Plants and Animals—Chapters 39, 47, and others

You may want to come back to this chapter later when you read about specific hormones used as chemical signals in plants and animals. These hormones affect organisms via the signal transduction processes described in this chapter. **Chapter 45** looks at electrical signals—another important type of signal transduction.

Chapter 9

Cellular Respiration and Fermentation

KEY CONCEPTS

- ☞ Exergonic reactions, such as the breakdown of ATP, help power the endergonic reactions required for life.

- ☞ Cellular respiration makes ATP by gradually releasing and using the high potential energy stored in molecules such as glucose.

- ☞ The four steps of cellular respiration are (1) glycolysis, (2) pyruvate processing, (3) the citric acid cycle, and (4) electron transport and chemiosmosis.

- ☞ Biochemical processes such as cellular respiration and fermentation are highly regulated, often using feedback inhibition.

- ☞ Fermentation converts NADH back to NAD^+ so that glycolysis can continue when electron transport chains shut down due to lack of an electron acceptor.

A. CHAPTER OUTLINE

Chapter Introduction

- Hydrolysis of **adenosine triphosphate (ATP)** provides the chemical energy that powers most cell work. ATP is the major energy currency of the cell.

- Energy to make ATP comes from sugars and other compounds with high potential energy. Addition of a third phosphate to **adenine diphosphate (ADP)** makes ATP.

- This chapter introduces you to **metabolism**—the chemical reactions that occur in cells.

9.1 The Nature of Chemical Energy and Redox Reactions

- Chemical energy is a type of potential energy. In cells, electrons store chemical energy; they have high potential energy when they are close to other negative charges and/or far from the positively charged protons in atom nuclei.

The Structure and Function of ATP

- ATP has high potential energy because its four negative charges are close together. During ATP hydrolysis, the bond holding ATP's outermost phosphate breaks, releasing an inorganic phosphate ($H_2PO_4^-$, abbreviated as P_i) and ADP (**Figure 9.1**).

- ATP hydrolysis is highly **exergonic** and releases 7.3 kilocalories (kcal) of energy per mole of ATP under standard temperature and pressure conditions.

Why Does ATP Hydrolysis Release Energy?

- The products (ADP and P_i) have more entropy and less potential energy than the reactant (ATP). The negative charges on ADP and P_i experience less electrical repulsion and are stabilized by interactions with the surrounding water.

What Happens When Proteins Are Phosphorylated by ATP?

- Cells use the energy released during ATP hydrolysis to do cell work. Often the P_i released from ATP is added to a substrate (**phosphorylation**).

- Phosphorylation reactions are exergonic because ADP and the attached phosphate have less potential energy than did ATP.

- Phosphate addition changes protein shape. Movement during this shape change can do cell work (**Figure 9.2**).

How Does ATP Drive Endergonic Reactions?

- Endergonic reactions cannot occur on their own in cells. These reactions must be coupled to an exergonic reaction like ATP hydrolysis in order to proceed (**energetic coupling, Figure 9.3**).

- For example, ATP hydrolysis phosphorylates B in an exergonic reaction. Phosphorylated B (an "activated" substrate) can now react with A to form AB because phosphorylated B's free energy is high enough so that the production of AB in this manner is exergonic.

What Is a Redox Reaction?

- In **reduction–oxidation** or **redox reactions**, an **electron donor** loses electrons (is **oxidized**), and an **electron acceptor** gains electrons (is **reduced**).

- If one reactant is oxidized, another must be reduced because the "lost" electron must go somewhere. The "loss" of an electron may just mean that it moved farther away from its atom's nucleus as its relative position shifts within a covalent bond.

An Example of Redox in Action

- For example, during photosynthesis carbon's electrons move closer to their atom's nucleus, but oxygen's electrons move further away from their nucleus. Thus, carbon is reduced and oxygen is oxidized (**Figure 9.4**).

- Electrons in the reduced products of photosynthesis are held more loosely (further from the nucleus) and thus have more potential energy. Photosynthetic products also have less entropy, so the overall reaction is endergonic and only proceeds with energy input from sunlight.

✓ CHECK YOUR UNDERSTANDING

(a) Compare the energy transformations in the endergonic reaction A + B → AB just discussed and in the overall reaction for photosynthesis.

_____

_____

_____

_____

_____

Another Approach to Understanding Redox

- During electron (e^-) transfers, a proton (H^+) may move with the e^- such that the molecule with the reduced atom gains a hydrogen atom (since H^+ plus $e^- = H$).

- Molecules with many C—H bonds (e.g., carbohydrates and fats) have high potential energy because electrons are shared equally in C—H bonds. When molecules are oxidized in cells, they often lose a proton along with their electron.

- Oxidized molecules have low potential energy and tend to have many C—O bonds and few, if any, C—H bonds. Electrons in C—O bonds are held tightly and thus have low potential energy.

- Pay attention to the "H's" in a reaction to figure out which molecules are reduced and which are oxidized. For example, the reduced form of **nicotinamide adenine dinucleotide** (NAD^+) is NADH.

- NADH can act as an electron donor and is an important **electron carrier** during cellular respiration (**Figure 9.5**).

What Happens When Glucose Is Oxidized?

- When glucose burns (**Figure 9.6**), its potential energy is released as kinetic energy (heat). When cells oxidize glucose, this energy is released slowly during many controlled redox reactions so that it can be used to make ATP from ADP and P_i.

9.2 An Overview of Cellular Respiration

- ATP is continually made and hydrolyzed; at any one time a cell contains only enough ATP for a few minutes of work. Glucose, primarily made by photosynthesis, is broken down to make ATP via cellular respiration or fermentation (**Figure 9.7**).

- In addition to using glucose to make ATP, organisms use glucose to build fats, carbohydrates, and other compounds. These compounds can, in turn, be broken down to retrieve glucose and make ATP.

- Most ATP is made from glucose via cellular respiration: a four-step process in which each step consists of a series of chemical reactions with distinctive initial reactants and end products.

- The four steps of cellular respiration are: (1) **glycolysis**—glucose is broken down to two pyruvates; (2) **pyruvate processing**—forms acetyl-CoA; (3) **citric acid cycle**—acetyl-CoAs are oxidized to CO_2; (4) **electron transport and chemiosmosis**—electrons from NADH and $FADH_2$ provide energy to make a proton gradient.

- Some ATP is made during glycolysis and the citric acid cycle, but most is made in step 4 when protons flow back across a membrane (**Figure 9.8**). Any suite of reactions that produces ATP using an electron transport chain is called **cellular respiration**.

9.3 Glycolysis: Processing Glucose to Pyruvate

- The common ancestor of all living organisms probably used glycolysis to make ATP because practically all organisms have glycolytic enzymes.

- Hans and Edward Buchner discovered glycolysis when they added sucrose to yeast extracts and found that the sugar was broken down and fermented, producing alcohol. This discovery showed that cell metabolism could be studied in vitro (outside the cell).

- The Buchners and others soon found that the fermentation reactions lasted much longer when inorganic phosphate was added.

- All intermediate compounds of glycolysis are phosphorylated (e.g., fructose bisphosphate); only glucose and pyruvate lack phosphate.

- Glycolytic reactions stopped if the yeast extract was boiled. This observation suggested the involvement of enzymes, because enzymes can be inactivated by heat. In fact, a different enzyme catalyzes each step of glycolysis.

✔ CHECK YOUR UNDERSTANDING

(b) Why did the fermentation reactions observed by the Buchners last longer when inorganic phosphate was added?

_____

_____

Glycolysis Is a Sequence of 10 Reactions

- The 10 reactions of glycolysis occur in the cytosol (**Figure 9.9**).

- The first reaction of glycolysis uses ATP to phosphorylate glucose, forming glucose-6-phosphate. A rearrangement creates fructose-6-phosphate, which gains another phosphate group from ATP to make fructose-1,6-bisphosphate.

- The initial energy investment is more than paid off in subsequent reactions. The sixth reaction reduces 2 NAD^+, the seventh reaction produces 2 ATP, and the last reaction produces another 2 ATP.

- The net yield of glycolysis is 2 NADH, 2 ATP, and 2 pyruvates per glucose.

- In glycolysis, ATP is produced by **substrate-level phosphorylation**, because enzymes catalyze the direct transfer of phosphates from a phosphorylated intermediate to ADP (**Figure 9.10**).

✔ CHECK YOUR UNDERSTANDING

(c) What would happen to glycolysis reactions if a cell ran completely out of ATP?

_____

_____

_____

How Is Glycolysis Regulated?

- High levels of ATP inhibit the enzyme **phosphofructokinase**, which catalyzes step 3 of glycolysis. The products of steps 1 and 2 can be used in other metabolic pathways, but step 3's product is only used in glycolysis, so this step is a good regulatory point.

- Addition of a substrate typically increases reaction rate. However, ATP is also one of the end-products of glycolysis. When an enzyme in a pathway is inhibited by the pathway's product, it is called **feedback inhibition** (**Figure 9.11**).

- Cells with lots of ATP do not need to make more. Thus, natural selection should favor individuals with phosphofructokinase enzymes that are inhibited at high concentrations of ATP.

- ATP inhibition of phosphofructokinase allows cells to stop glycolysis when ATP is plentiful and conserve glucose stores for another time.

- Phosphofructokinase has two ATP binding sites: an active site, where ATP is hydrolyzed, and a **regulatory site** (**Figure 9.12**), where ATP acts as an allosteric regulator. When ATP binds to the regulatory site, the enzyme's shape changes in a way that decreases enzyme activity.

9.4 Pyruvate Processing

- Eukaryotes transport the pyruvate made by glycolysis to the mitochondria.

- Mitochondria have two membranes. The inner membrane is folded into sac-like **cristae**. Most Krebs cycle enzymes are in the **mitochondrial matrix**—the aqueous area inside the cristae (**Figure 9.13**).

- Pyruvate crosses the outer membrane through small pores and is actively transported (energy from ATP-hydrolysis) into the mitochondrial matrix by an inner membrane protein called the pyruvate carrier.

- Inside the mitochondria, pyruvate reacts with **Coenzyme A (CoA)** to produce **acetyl CoA**. CoA is a common enzyme cofactor that transfers acetyl $(-COCH_3)$ groups to substrates.

- The reaction between pyruvate and CoA is catalyzed by an enzyme complex called **pyruvate dehydrogenase**, which is found in the mitochondrial matrix in eukaryotes (and in the cytosol of prokaryotes).

- The reaction between pyruvate and CoA results in the reduction of NAD^+ to NADH and the oxidation of one of pyruvate's carbons to CO_2. The remaining two carbons form the acetyl group transferred to CoA (**Figure 9.14**).

- Pyruvate dehydrogenase is also sensitive to feedback inhibition. When ATP is plentiful, pyruvate dehydrogenase becomes phosphorylated and changes shape such that catalysis is inhibited.

- High concentrations of acetyl CoA and NADH also promote pyruvate dehydrogenase phosphorylation. In contrast, high concentrations of NAD^+, CoA, or AMP increase pyruvate dehydrogenase's enzymatic activity.

- Because pyruvate processing is under both positive and negative control, it is another key regulatory point in glucose oxidation. High concentrations of products inhibit pyruvate dehydrogenase, whereas large supplies of reactants stimulate it.

9.5 The Citric Acid Cycle: Oxidizing Acetyl CoA to CO_2

- Biologists discovered eight small **carboxylic acids** (R-COOH) involved in redox reactions that produced CO_2, the endpoint of glucose metabolism. Researchers proposed that these carboxylic acids were associated with glucose metabolism.

- Researchers observed that adding any of these carboxylic acids to cells caused an increase in cell respiration, but the added molecules were not used up.

- The eight carboxylic acids oxidize rapidly and can react in sequence (from least to most oxidized). Hans Krebs realized that the carboxylic acid reactions could occur in a cycle tied to the breakdown of pyruvate.

- The cycle's start point, citrate, forms when pyruvate and the cycle's endpoint, oxaloacetate, are added to cells. The cycle is called the **citric acid cycle** (also the Krebs cycle or the tricarboxylic acid cycle) because citrate becomes citric acid when it gains a proton.

- Experiments with radioactive carbon isotopes confirmed that the cycle occurred as Krebs proposed (**Figure 9.15**).

- The energy released by the oxidation of one acetyl CoA is used to make 3 NADH, 1 $FADH_2$, and 1 **guanosine triphosphate (GTP)** or **ATP** via substrate-level phosphorylation.

- Citric acid cycle reactions occur in the cytoplasm of prokaryotes and primarily in the mitochondrial matrix of eukaryotes. Since glycolysis produces two pyruvate molecules, the cycle turns twice per glucose molecule processed in cellular respiration.

How Is the Citric Acid Cycle Regulated?

- The citric acid cycle is regulated; low ATP causes high reaction rates whereas high ATP concentrations inhibit citric acid cycle reactions (**Figure 9.16**). For example, ATP inhibits the enzyme that converts acetyl CoA to citrate (feedback inhibition).

- Feedback inhibition also occurs at two other points in the Krebs cycle. In one case, NADH binds to the enzyme's active site as a competitive inhibitor; in the other case, ATP binds to an allosteric site.

- The citric acid cycle is carefully regulated and can be inhibited at several points using different feedback inhibition mechanisms.

- The citric acid cycle starts with acetyl CoA, ends with CO_2, and occurs in the mitochondrial matrix. The potential energy released from acetyl CoA is used to produce NADH, $FADH_2$, ATP, and GTP.

What Happens to the NADH and $FADH_2$?

- Two pyruvates are formed for each glucose molecule that enters glycolysis. After each pyruvate has been fully oxidized in the citric acid cycle, the cell has produced 10 NADH, 2 $FADH_2$, and 4 ATP (**Figure 9.17**):

$$C_6H_{12}O_6 + NAD^+ + 2\,FAD + 4\,ADP + 4P_i \rightarrow 6\,CO_2 + 10\,NADH + 2\,FADH_2 + 4\,ATP.$$

- The four ATP are made by substrate-level phosphorylation and can be used for cell work, the CO_2 is a waste gas that you exhale when you breathe, and NADH and $FADH_2$ are used to make more ATP in a reaction that could be written as:

$$NADH + FADH_2 + O_2 + ADP + P_i \rightarrow$$
$$NAD^+ + FAD + H_2O + more\ ATP.$$

- Thus, most of the energy produced during glucose oxidation ($C_6H_{12}O_6 + 6\ O_2 \rightarrow 6\ CO_2$) comes NADH and $FADH_2$. The free-energy changes associated with these reactions are shown in **Figure 9.18**.

- Glycolysis, pyruvate processing, and the citric acid cycle transfer electrons from glucose to NAD^+ and FAD, making NADH and $FADH_2$, which carry the electrons to the final electron acceptor in eukaryotes: oxygen.

9.6 Electron Transport and Chemiosmosis: Building a Proton Gradient to Produce ATP

- Eukaryotes oxidize NADH in their inner mitochondrial membranes and prokaryotes oxidize NADH in their plasma membrane. Analysis of these membranes revealed molecules that switch between a reduced and an oxidized state during respiration.

Components of the Electron Transport Chain

- Molecules in the inner mitochondrial membrane that are involved in the oxidation of NADH and $FADH_2$ make up the **electron transport chain (ETC)**.

- Energy released by sequential redox reactions is used to pump protons across the inner mitochondrial membrane, creating a proton gradient. Protons flow back into the mitochondrial matrix through **ATP synthase**, driving ATP production from ADP and P_i.

- This method of ATP production is linked to oxidation of NADH and $FADH_2$, so it's called **oxidative phosphorylation**. When oxygen accepts electrons at the end of the ETC, water forms, and glucose oxidation is complete.

- Most ETC molecules are proteins with distinctive chemical groups (flavins, iron-sulfur complexes, heme groups, etc.) that are easily reduced or oxidized.

- **Ubiquinone**, also called **coenzyme Q** or **Q**, is a carbon-containing ring attached to a hydrophobic tail of isoprene subunits. Q is lipid soluble and easily moves

throughout the inner mitochondrial membrane, whereas all but one of the ETC proteins are anchored in the membrane.

CHECK YOUR UNDERSTANDING

(d)   The inner mitochondrial membrane is highly folded. Why might this be advantageous?

_____

_____

- ETC molecules differ in their electronegativity—their tendency to become reduced or oxidized.

- Electrons proceeding down an ETC are held more and more tightly (have less and less potential energy) as they are passed from molecule to molecule (**Figure 9.19**). Each reaction releases a small amount of energy.

- Researchers used poisons to determine the ETC sequence. A poison that inhibits a ETC protein causes ETC molecules that receive electrons before the poisoned protein to stay in a reduced state, whereas those "downstream" of the poisoned protein remain oxidized.

- NADH donates electrons to a flavin-containing protein at the top of the chain, but $FADH_2$ donates electrons to an iron-sulfur protein that passes electrons directly to Q.

- Oxygen is the final electron acceptor in eukaryotes, and the total difference in potential energy from NADH to oxygen is 53 kcal/mol.

The Chemiosmotic Hypothesis

- No enzymes could be found that directly used the energy released by electron transport chain redox reactions to make ATP. Peter Mitchell proposed an indirect connection between the redox reactions and ATP production.

- Mitchell suggested that the ETC pumps protons from the matrix to the intermembrane space. This would create an electrochemical gradient favoring the movement of protons back into the matrix. An enzyme in the inner membrane could use this **proton-motive force** to make ATP.

- Mitchell used the term **chemiosmosis** to describe the production of ATP via a proton gradient. The following experiment supported his hypothesis.

- Researchers made vesicles from artificial membranes containing an ATP-synthesizing enzyme found in mitochondria. They also inserted bacteriorhodopsin, a protein that uses light energy to pump protons.

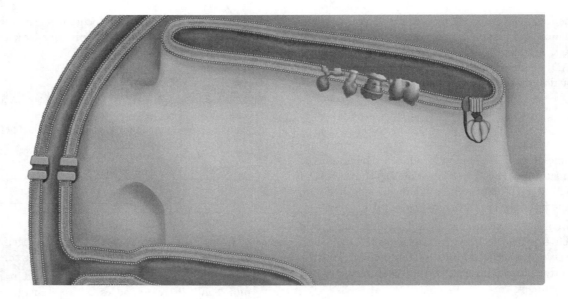

(e) On the figure above, label the locations of glycolysis, pyruvate processing, the citric acid cycle, and the electron transport chain. Also label the mitochondrial matrix and indicate which part of the cell shown would have the lowest pH during cellular respiration.

- When lit, bacteriorhodopsin pumped protons out of the vesicles, creating a proton gradient, and ATP was produced inside the vesicles (**Figure 9.20**).

- This experiment showed that as long as there was a proton gradient, ATP could be produced even in the absence of an electron transport chain.

How Is the Electron Transport Chain Organized?

- Electron transport chain proteins are organized into four protein–cofactor complexes.

- Lipid-soluble Q and the protein **cytochrome *c*** transfer electrons between complexes. In complexes I and IV, protons pass through the electron carriers as they are pumped out of the matrix (**Figure 9.21**).

- Q also carries protons across the membrane; when Q accepts electrons from complex I or II, it also gains protons. Q then diffuses through to the outer side of the membrane, where its electrons are transferred to complex III and its protons are released into the intermembrane space.

The Discovery of ATP Synthase

- Efraim Racker observed that inside-out mitochondrial membrane vesicles had large stalk- and knob-shaped

(f) A compound called dinitrophenol (DNP) can be used to poke holes in the inner mitochondrial membrane. What would this do to ATP synthesis by oxidative phosphorylation?

_____

_____

_____

proteins on their surfaces, attached to a base within the membrane. The stalks and knobs fell off when shaken or treated with urea.

- Isolated stalks and knobs hydrolyzed ATP, but vesicles with just the membrane-inserted base could not make ATP even though proton transport occurred normally. Thus, the base appeared to be a proton channel.

- Racker hypothesized that the stalk-and-knob component was an enzyme that both hydrolyzes and synthesizes ATP. When stalk-and-knob components were added back to vesicles, the vesicles regained their ability to synthesize ATP.

- The knob-and-stalk protein is **ATP synthase**. Protons flow through the F_o unit (the base), causing the stalk connecting the two units to spin. As the F_1 unit (the knob) spins with the rod, its subunits are deformed in a way that catalyzes the phosphorylation of ADP to ATP (**Figure 9.22**).

(g) Why wasn't it enough for Racker to show that ATP synthesis stopped when the knobs and stalks were removed? That is, why was his explanation more convincing after he added back the isolated stalk-and-knob components, causing the vesicles to regain the ability to synthesize ATP?

_____

_____

_____

_____

_____

_____

Organisms Use a Diversity of Electron Acceptors

- Electrons from glucose are transferred to NADH and FADH$_2$, passed down the ETC, and accepted by oxygen during cellular respiration in our cells. Proton pumping by the ETC creates the proton-motive force that drives ATP synthesis (**Figure 9.23**).

- ATP synthase produces about 25 of the 29 ATP molecules produced per glucose during cell respiration, so most of the usable energy from the oxidation of glucose comes from oxidative phosphorylation.

Aerobic versus Anaerobic Respiration

- All eukaryotes and many prokaryotes use oxygen as the final electron acceptor of electron transport chains. Cellular respiration in these organisms is called **aerobic respiration**.

- Many other bacteria and archaea, especially those in oxygen-poor environments, use other electron acceptors (e.g., NO_3^{2-} and SO_4^{2-}) in **anaerobic** respiration. Some prokaryotes also use electron donors other than glucose (e.g., H_2, H_2S, CH_4).

- All these cells use electron transport chains to create a proton-motive force that drives ATP synthesis.

Aerobic Respiration Is Efficient

- Oxygen, with its high electronegativity, is the best electron acceptor. Its electrons have very low potential energy. The large energy difference between NADH and O_2 electrons permits the generation of a large proton-motive force for ATP production.

- Cells that use other electron acceptors cannot make as much ATP per glucose. These cells grow more slowly than aerobic respirators do. Thus, aerobic respirators typically outcompete anaerobic respirators in environments with oxygen.

- But what happens when oxygen or other electron acceptors are temporarily unavailable and NADH electrons have no place to go? The ETC stops, yet cells must have NAD$^+$ to produce ATP from glycolysis. What do cells do?

9.7 Fermentation

- **Fermentation** pathways convert NADH back to NAD$^+$ and allow glycolysis to continue producing some ATP via substrate-level phosphorylation when the ETC electron acceptor is not available (**Figure 9.24**).

- During fermentation, pyruvate—or a molecule derived from pyruvate—accepts electrons from NADH, regenerating NAD$^+$. The molecule formed from the reduction of pyruvate is often excreted from the cell as waste.

Many Different Fermentation Pathways Exist

- In humans, pyruvate accepts electrons from NADH in **lactic acid fermentation** when muscles metabolize glucose faster than oxygen can be supplied. This type of fermentation produces lactate in addition to regenerating NAD$^+$ (**Figure 9.25a**).

- Yeast deprived of oxygen use **alcohol fermentation**, in which pyruvate loses CO_2 and is converted to acetylaldehyde, which accepts an electron from NADH to produce ethanol and NAD$^+$ (**Figure 9.26b**).

- Other fermentation pathways also exist, and some bacteria and archaea exclusively use fermentation to break down glucose (e.g., in your small intestine; in the cow's rumen).

Fermentation as an Alternative to Cellular Respiration

- Because the electron acceptors in fermentation are less electronegative than oxygen, only 2 ATP are produced per glucose—compared to about 29 ATP per glucose in aerobic respiration.

- Consequently, organisms capable of cellular respiration never use fermentation if an appropriate electron acceptor is available for their ETC. Organisms that normally use oxygen as an electron acceptor only use fermentation when oxygen is not available.

- Organisms that can switch between fermentation and aerobic cell respiration are called **facultative aerobes**.

✔ CHECK YOUR UNDERSTANDING

(h) What would happen if a winemaker followed the industrial process for making wine, but then bubbled oxygen through the vats of grape juice instead of letting it sit undisturbed?

_____

_____

9.8 How Does Cellular Respiration Interact with Other Metabolic Pathways?

- Cell metabolism involves thousands of different chemical reactions. Types and amounts of molecules inside cells are constantly changing (**Figure 9.26**).

- Energy and carbon are the fundamental requirements of all cells.

- **Catabolic pathways** involve the breakdown of molecules and the production of chemical energy in the form of ATP, whereas **anabolic pathways** result in the synthesis of larger molecules from smaller components.

- Together, these pathways allow cells to survive, grow, and reproduce, supplying chemical energy and carbon-containing building blocks for synthesis of the molecules necessary for life.

Processing Proteins and Fats as Fuel

- Cells use enzyme-catalyzed reactions to break down glycogen, starch, and most simple sugars into glucose or fructose, which can enter the glycolytic pathway.

- Fats can be broken down into glycerol, which can be oxidized and phosphorylated to form an intermediate of glycolysis (glyceraldehyde-3-phosphate), and acetyl CoA, which enters the citric acid cycle.

- Carbon compounds also remain after proteins are broken down into amino acids and amino groups are removed and excreted. The carbon compounds can be converted into pyruvate, acetyl CoA, and/or other intermediates of glycolysis or the citric acid cycle.

- Carbohydrates, fats, and proteins are broken down in different ways, but glycolysis and the citric acid cycle are central to ATP production in these catabolic pathways (**Figure 9.27**). For ATP production, cells first use carbohydrates, then fats, and finally proteins.

Anabolic Pathways Synthesize Key Molecules

- Molecules found in carbohydrate metabolism are used in anabolic metabolic pathways too (**Figure 9.28**).

- About half of our amino acids can be synthesized from citric acid cycle compounds.

- Acetyl CoA is the carbon building block for fatty-acid synthesis.

- The second compound in glycolysis can be used to synthesize a key intermediate in nucleotide production for RNA and DNA synthesis.

- Pyruvate and lactate can be used to synthesize glucose when sufficient ATP is present to power these reactions. Glucose can, in turn, be converted to glycogen (in animals) or starch (in plants) for storage.

- Molecules may contribute to many cell functions, and metabolic pathways are interconnected and tightly regulated.

B. MASTERING CHALLENGING TOPICS

The chemiosmotic hypothesis can be difficult to understand. This is partly why some researchers opposed this theory for so long. But it really just proposes a slightly different type of energy transformation than those we are more familiar with. For example, in **Chapter 2** you learned how potential energy can be converted to kinetic energy. In the electron transport chain, the potential energy of electrons is simply converted to the potential energy in an electrochemical gradient. The electrochemical proton gradient can be compared to the water and gravity gradient observed in a dam. Cells use redox reactions and the potential energy of electrons to pump the protons in the same way a water pump could burn fuel to pump water up a dam. Then the protons flow back across the membrane through ATPase, and the energy in the proton gradient is converted into energy used to phosphorylate ATP. Similarly, water flowing through a dam can be used to turn turbines that then produce electricity.

C. ASSESSING WHAT YOU'VE LEARNED

1. Which of the following compounds discussed in this chapter could be considered an activated substrate?
   a. Glucose
   b. GTP
   c. Fructose 1, 6-bisphosphate
   d. Pyruvate

2. Why do reduced molecules tend to have a lot of C—H bonds?
   a. Reduced molecules hold electrons tightly, and C—H bonds hold electrons tightly.
   b. Reduced molecules like glucose can't have any oxygen atoms.
   c. Carbohydrates and fats have lots of C—H bonds.
   d. Reduced molecules usually gain protons along with electrons.

3. The burning of methane can be written as $CH_4 + O_2 \rightarrow CO_2 + H_2O$. Which of the following is true of this reaction?
   a. Oxygen is oxidized and carbon is reduced.
   b. Carbon is oxidized and oxygen is reduced.
   c. Both carbon and oxygen are reduced.
   d. This is not a redox reaction because no electrons are completely transferred.

4. What is the purpose in having several steps in glycolysis or the citric acid cycle rather than a single step from glucose and oxygen to carbon dioxide and water?
   a. The multistep approach is the only way to convert glucose to carbon dioxide.
   b. The multistep approach increases the amount of potential energy in the reaction.
   c. The multistep approach makes better use of the available potential energy.
   d. The multistep approach increases the amount of heat produced in the reaction.

5. Under which of the following conditions is pyruvate dehydrogenase likely to be most active in a cell? (See if you can answer using logic, without looking back at the text.)
   a. High acetyl CoA, high NAD^+
   b. High AMP, low NAD^+
   c. High ATP, high NADH
   d. High NAD^+, low AMP

6. In the Buchner experiment, why did boiling the yeast extract prevent the conversion of ethanol from glucose?
   a. Phosphate required for the reactions was vaporized.
   b. Sucrose was destroyed.
   c. Proteins were denatured.
   d. Yeast cells were killed.

7. Why is a different enyzme involved in each step of glycolysis?
   a. Each step occurs in a different subcellular location.
   b. Each step occurs in a different cell.
   c. Each step involves a different cofactor.
   d. Each step involves a different chemical reaction.

8. Nearly every living organism uses glucose as a nutrient source of energy. Why?
   a. Glucose is the only molecule capable of providing the energy to produce ATP.
   b. The ability to harvest energy from glucose appeared very early in biological evolution.
   c. Glucose contains more potential energy than any other nutrient molecule.
   d. The structure of glucose is very similar to ATP.

9. The reactions summarized below are a type of fermentation. Why?

$$2 \text{ pyruvate} + \text{NADH} \rightarrow\!\!\rightarrow\!\!\rightarrow 2\, CO_2 + NAD^+ + \text{2,3-butanediol}$$

   a. The reactions use the potential energy present in NADH.
   b. CO_2 is released.
   c. A type of alcohol is produced.
   d. Pyruvate is reduced and NADH is oxidized to NAD^+.

10. If glucose is labeled with ^{14}C, what molecule will become radioactive as glycolysis and the citric acid cycle are completed?
    a. Water
    b. Carbon dioxide
    c. ATP
    d. NADH

11. At the end of the citric acid cycle, but before the electron transport chain, the oxidation of glucose has produced a net yield of:
    a. 3 CO_2, 5 NADH, 1 $FADH_2$, and 2 ATP.
    b. 6 CO_2, 10 NADH, 2 $FADH_2$, and 4 ATP.
    c. 6 CO_2, 10 NADH, 2 $FADH_2$, and 6 ATP.
    d. None of the above are correct.

12. What would have been the result of Krebs's experiment if the citric acid cycle were linear instead of circular?
    a. Oxaloacetate + pyruvate would have produced citric acid.
    b. Oxaloacetate + pyruvate would have produced acetyl CoA.
    c. Oxaloacetate + pyruvate would have produced no products.
    d. Oxaloacetate + pyruvate would have produced glucose.

13. If oxygen gas is labeled with ^{18}O, what molecule will become radioactive as cellular respiration occurs?
    a. Water
    b. Carbon dioxide
    c. ATP
    d. NADH

14. Cyanide poisons an electron transport chain protein that has less free energy than Q. Which of the following would occur during cyanide poisoning?
    a. Q would stay reduced and stop moving protons across the mitochondrial membrane.
    b. Q would stay oxidized and stop moving protons across the mitochondrial membrane.
    c. Q would not be affected, but the proton gradient would dissipate.
    d. The electron transport chain would stop functioning because the cell would run out of NADH.

15. Why does NADH donate electrons to the beginning of the electron transport chain, whereas $FADH_2$ donates electrons to the middle of the chain?
    a. $FADH_2$ is more rapidly oxidized than NADH.
    b. NADH has more potential energy than $FADH_2$.
    c. $FADH_2$ has more reducing potential than NADH.
    d. NADH has more electrons to donate than $FADH_2$.

CHAPTER 9—ANSWER KEY

Check Your Understanding

(a) Both photosynthesis (really a whole pathway, as you will see in Chapter 10) and A + B → AB are endergonic reactions. In both of these reactions, the products have less entropy than the reactants; and in photosynthesis, at least, the product has more potential energy than the reactants due to the increased distance of the electrons from their carbon nuclei. The key difference in energy transformations between these two reactions is that while photosynthesis is powered by light energy, A + B → AB is powered by chemical energy. As the text discusses, this reaction occurs only when it is coupled to ATP hydrolysis.

(b) Inorganic phosphate is one of the substrates for glycolysis. Thus, glycolysis and fermentation stop when no more phosphate is available.

(c) If a cell ran completely out of ATP, it could not perform glycolysis because the first and third steps require ATP input before ATP is made in the final steps.

(d) The folds provide more surface area—more area for electron transport chain components and ATPases.

(e) See figure below.

(f) When DNP puts holes in the inner mitochondrial membrane, no ATP can be produced by ATPase because the proton gradient disappears.

(g) Something else that was responsible for ATP synthesis might have been knocked off along with the knobs.

(h) You would produce carbon dioxide, water, and lots more yeast, but little if any alcohol because the yeast would use aerobic respiration instead of fermentation.

Assessing What You've Learned

1. c; 2. d; 3. b; 4. c; 5. a; 6. c; 7. d; 8. b; 9. d; 10. b; 11. b; 12. c; 13. a; 14. a; 15. b

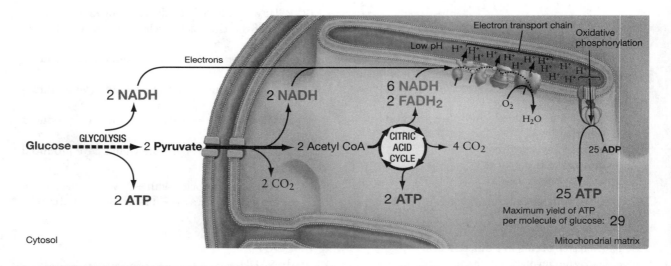

Looking Forward—Key Concepts in Later Chapters

Many later chapters discuss cell work powered by ATP hydrolysis.

Photosynthesis—Chapter 10

In the next chapter, you learn about another ATP-producing process that uses an electron transport chain—photosynthesis.

Bacteria and Archaea—Chapter 28

Bacteria and Archaea are groups that show incredible metabolic diversity. Chapter 9 just touches on this diversity by mentioning that many of these species use anaerobic respiration.

Photosynthesis

Use of Radioactive Isotopes—Chapters 7 and 9

Chapter 7 described "pulse-chase" experiments involving the use of radioactive isotopes to discover how molecules are transported throughout the cell. **Chapter 9** mentions that radioactive isotopes were used to discover intermediate compounds in the Krebs cycle. Here in Chapter 10, you learn how isotopes were used in research identifying the metabolic pathways involved in carbon fixation.

Redox Reactions and Electron Transport Chains—Chapter 9

Chapter 9 introduced you to redox reactions, electron transport chains, and chemiosmosis—all key components of photosynthesis. As detailed in that chapter, mitochondria also use an electron transport chain to convert the energy of high-potential-energy electrons to more easily usable chemical bond energy (e.g., ATP).

KEY CONCEPTS

🔑 Photosynthesis converts light energy to chemical energy using two linked sets of reactions.

🔑 Light-capturing reactions use excited electrons and electron-transport chains to produce ATP (by chemiosmosis) and the electron-carrier NADPH.

🔑 Calvin cycle reactions use ATP and NADPH to make an intermediate used in carbohydrate production. Rubisco catalyzes the initial fixation of CO_2.

🔑 Carbon dioxide enters plants through openings called stomata. Some plants use CAM or C_4 photosynthetic pathways to increase leaf CO_2 concentration and minimize photorespiration.

A. CHAPTER OUTLINE

Chapter Introduction

- **Photosynthesis** is the ability to convert energy from sunlight into chemical energy. Photosynthetic organisms are **autotrophs**, organisms that make all their own food. Organisms that derive their food from other organisms are called **heterotrophs**.

- Photosynthesis is the ultimate food source for most organisms, and the evolution of this metabolic pathway is one the great events in the history of life.

10.1 Photosynthesis Harnesses Sunlight to Make Carbohydrate

- Photosynthesis occurs in the green parts of plants. It requires sunlight, carbon dioxide, and water and produces oxygen. Photosynthesis allows plants to convert the electromagnetic energy of sunlight into chemical energy in C–C and C–H bonds.

- The overall reaction when glucose is the carbohydrate made is:

$$6CO_2 + 12H_2O + \text{light energy} \rightarrow C_6H_{12}O_6 + 6O_2 + 6H_2O$$

- Photosynthesis reduces carbon dioxide to glucose (or other sugars). These endergonic reactions are approximately the reverse of cellular respiration, the exergonic reactions that oxidize glucose to carbon dioxide in order to make ATP.

Photosynthesis: Two Linked Sets of Reactions

- Cornelius van Niel studied photosynthesis in purple sulfur bacteria. These grow even when starved of sugars if exposed to sunlight and hydrogen sulfide (H_2S).

- Sulfur (S), instead of O_2, is released as a by-product: $CO_2 + 2H_2S + \text{light energy} \rightarrow (CH_2O)_n + H_2O + 2S$.

- Water is a product, but not a reactant in this reaction. Because CO_2 is used in this reaction but no O_2 is made, van Niel concluded that the O_2 released in most photosynthesis must come from H_2O, not CO_2 (just as the 2S comes from the $2H_2S$).

- The hypothesis that photosynthetic O_2 comes from H_2O was supported by the observation that isolated chloroplasts make O_2 when exposed to sunlight, even if no CO_2 is present.

- Algae and plants exposed to water containing the heavy isotope of oxygen, ^{18}O, emit $^{18}O_2$ gas in the presence of sunlight. This result confirmed that photosynthetic O_2 comes from H_2O.

- What about the fate of the carbon? Melvin Calvin fed $^{14}CO_2$ to algae and identified the order in which products became ^{14}C-labeled. He found that $^{14}CO_2$ was used to make carbohydrates even in the dark.

- This light-independent part of photosynthesis, the **Calvin cycle**, reduces CO_2 and produces sugar. While the Calvin cycle does not require light itself, it depends on products from the light-dependent reactions.

- Photosynthesis involves two linked sets of reactions. The first set of reactions is triggered by light and produces oxygen from water; the second set of reactions (the Calvin cycle) uses the products of the first set of reactions to produce sugar from CO_2.

- These two sets of reactions are linked by electrons. As H_2O splits and forms O_2, electrons are released and transferred to $NADP^+$, forming NADPH, an electron carrier similar to NADH. NADPH's electrons are used in the Calvin cycle reduction of CO_2.

- ATP made by the light-dependent reactions is also used in the Calvin cycle (**Figure 10.1**).

Photosynthesis Occurs in Chloroplasts

- Photosynthesis occurs in green plant parts—specifically, in small green organelles called **chloroplasts**. Chloroplast membranes release O_2 when exposed to sunlight.

- Chloroplasts have an outer and an inner membrane and are filled with vesicle-like structures called **thylakoids**, which form interconnected stacks called **grana**. The surrounding fluid is the **stroma**, and the space inside the thylakoids is its **lumen**. (**Figure 10.2**).

- Thylakoid membranes contain lots of **pigments**. Pigments absorb certain wavelengths of light and transmit others.

- Chlorophyll is the most abundant thylakoid pigment. Chlorophyll absorbs blue and red light and transmits green light, so most plants as well as many algae and bacteria are green.

10.2 How Does Chlorophyll Capture Light Energy?

- Electromagnetic radiation is described by its **wavelength** (**Figure 10.3**); the part of the **electromagnetic spectrum** that humans see is called **visible light** and encompasses wavelengths from 400 to 710 nanometers (nm).

- Shorter wavelengths (e.g., blue) have more energy than do longer wavelengths (e.g., red).

- Light has particle-like as well as wavelike properties. Packets of light are called **photons**, and each photon or wavelength has a specific amount of energy. Pigments can absorb this energy.

Photosynthetic Pigments Absorb Light

- Photons can be absorbed, transmitted, or reflected by molecules. White light is a mix of all visible wavelengths; when pigments absorb some, but not all wavelengths, they appear to us to be the color of the reflected or transmitted wavelengths.

- Pigments can be extracted and separated via a technique called **thin layer chromatography** (**Figure 10.4**). Raw leaf extract separates into components as a solvent travels up a coated support. Extract molecules travel at different rates based on their size and/or solubility.

- Leaves contain several pigments. Researchers can isolate individual pigments and determine the wavelengths each absorbs. A graph of light absorbed versus wavelength gives the **absorption spectrum** for a given pigment (**Figure 10.5a**).

- The major leaf pigments are the **chlorophylls** (chlorophyll *a* and chlorophyll *b*), which absorb red and blue light and transmit green light; and the **carotenoids**, which absorb blue and green light and thus appear yellow, orange, or red.

- Algae on a slide lit with a spectrum of colors produce oxygen (measured by bacterial growth) in areas exposed to blue and red light. These wavelengths make up the **action spectrum** for photosynthesis—the wavelengths that drive light-capturing reactions (**Figure 10.5b**).

- Chlorophylls absorb red and blue wavelengths, so the action spectrum data suggests that they are the primary photosynthetic pigments.

CHECK YOUR UNDERSTANDING

(a) A thousand years from now, biologists discover life on a planet circling another star. They find plant-like organisms on this planet that can transform the electromagnetic energy of light into chemical energy.

However, these "plants" are yellow and red. Why aren't they green, and does this surprise you?

_____

_____

_____

_____

_____

What Is the Role of Carotenoids and Other Accessory Pigments?

- Carotenoids absorb light and pass the energy on to chlorophyll.

- The two classes of carotenoids found in plants are the carotenes (e.g., beta-carotene, **Figure 10.6a**) and the xanthophylls (e.g., zeaxanthin in corn). These molecules are similar and are both found in chloroplasts.

- Carotenoids absorb wavelengths of light not absorbed by chlorophyll, thus extending the range of wavelengths that can be used in photosynthesis.

- Carotenoids also stabilize free radicals produced by electromagnetic radiation, protecting chlorophyll from degradation.

Flavonoids Are a Natural Sunscreen

- Flavonoids also protect chlorophyll and other plant molecules from destructive radiation by absorbing high-energy ultraviolet (UV) light.

The Structure of Chlorophyll

- Chlorophyll _a_ and _b_ have similar structures but slightly different absorption spectra.

- Both chlorophylls have a long tail of isoprene subunits that stays embedded in the thylakoid membrane, and a large ring structure (the "head") containing a magnesium atom where light is absorbed (**Figure 10.6b**).

When Light Is Absorbed, Electrons Enter an Excited State

- A photon striking chlorophyll can transfer its energy to an electron if the photon's energy is similar to the energy required to raise the electron to a higher energy state.

- An unexcited chlorophyll electron has an energy state of 0 (its ground state). Red photons have the right amount of energy to raise a chlorophyll electron's energy state up to level 1, and blue photons can raise its energy state up to level 2 (**Figure 10.7**).

- Photons of intermediate energy (e.g., green light) cannot easily be absorbed by chlorophyll.

- Ultraviolet wavelengths have so much energy that they can eject electrons from pigment molecules, and infrared wavelengths may heat atoms but have too little energy to raise them to a higher energy level.

- Electrons in high-energy states may fall back to their ground state, releasing heat and lower-energy photons (**fluorescence; Figure 10.8**). Only about 2 percent of the electrons excited in chloroplast chlorophyll release their energy as fluorescence.

- In the thylakoid membrane, 200 to 300 chlorophyll molecules and accessory pigments group together in complexes called **photosystems**. Each photosystem has an antenna complex and a reaction center that help capture and process excited electrons.

The Antenna Complex

- When **antenna complex** electrons absorb photon energy and become excited, they pass the energy, but not the electron, to a nearby chlorophyll molecule.

- The original electron falls back to its ground state, but the energy transfer excites the next chlorophyll molecule's electron. This phenomenon is called resonance; it explains how energy is transmitted from chlorophyll to chlorophyll to the reaction center.

The Reaction Center

- In the **reaction center**, excited electrons are actually passed to a specialized chlorophyll that acts as an electron acceptor. The reduction of this electron acceptor completes the transformation of electromagnetic energy into chemical energy.

- The reaction center couples a light-induced exergonic reaction to an endergonic redox reaction.

- Excited electrons in photosynthetic pigments can (1) fluoresce as they drop back down to a low energy level, (2) excite adjacent pigment electrons, causing resonance in the antenna complex, or (3) be transferred to an electron acceptor in the reaction center (**Figure 10.9**).

✔ **CHECK YOUR UNDERSTANDING**

(b) Chlorophyll can fluoresce; so why aren't plants noticeably fluorescent?

_____

_____

_____

_____

10.3 The Discovery of Photosystems I and II

- Researchers found that green algae had similar photosynthetic responses to far-red (700 nm) and red (680 nm) light. But the rate of photosynthesis was **much** higher (more than double) when algae were exposed to both wavelengths at once (**Figure 10.10**). This result was called the enhancement effect.

- Robin Hill and Faye Bendall proposed that green algae and plants have two types of reaction centers: **photosystem II** and **photosystem I**. Photosynthesis is more efficient when both photosystems work together, causing the enhancement effect.

- Many bacteria have photosystems similar to those found in algae and plants, and several bacteria have only one of the two photosystems. Researchers could thus study each photosystem in isolation before trying to understand how they work together.

How Does Photosystem II Work?

- Photosystem II is similar to the single photosystem found in purple nonsulfur and purple sulfur bacteria.

Pheophytin Accepts Electrons and Transfers Them to an Electron Transport Chain

- When the antenna complex transmits energy to photosystem II's reaction center, a high-energy electron is donated to **pheophytin**—a molecule similar to chlorophyll that lacks magnesium in its head region (**Figure 10.11**).

- Reduced pheophytin passes an electron to an electron transport chain (ETC) in the thylakoid membrane. This ETC is similar to the mitochondrial ETC in both structure and function.

- Both the mitochondrial and photosystem II's electron transport chain contain cytochromes and quinones. Electrons participate in redox reactions and are gradually stepped down in potential energy.

- These redox reactions lead to proton pumping, triggering ATP synthesis via chemiosmosis in the chloroplast.

The ETC Sets up a Proton Gradient That Drives ATP Synthase

- **Plastoquinone (PQ)**, a small hydrophobic molecule, receives electrons from pheophytin and transports them across the thylakoid membrane to a cytochrome complex.

- Within the cytochrome complex, electrons from PQ are passed through a series of iron- and copper-containing proteins. The potential energy released is used to add protons to other plastoquinones (**Figure 10.12**).

- These PQs release the protons on the lumen side of the thylakoid. This proton transport increases the proton concentration inside the thylakoid, causing a thousand-fold difference in proton concentration (thylakoid interior at around pH 5).

- This proton motive force then drives ATP production when protons flow through ATP synthase back into the stroma. Proton flow is an exergonic process that drives the endergonic synthesis of ATP from ADP and P_i.

- The capture of light energy by photosystem II to produce ATP chemical energy is called **photophosphorylation**.

Photosystem II Obtains Electrons by Oxidizing Water

- Recall that radioisotope experiments showed that the oxygen in O_2 produced in photosynthesis comes from water. Part of photosystem II "splits" water to replace its lost electrons: $2 H_2O \rightarrow 4 H^+ + 4e^- + O_2$.

- When excited electrons leave to enter the ETC, photosystem II becomes so electronegative that enzymes can remove electrons from water, leaving protons and oxygen.

Oxygenic Photosynthesis and the Oxygen Atmosphere

- Cyanobacteria, algae, and plants produce oxygen as a by-product of photosynthesis. Organisms that split water in photosystem II during photosynthesis are said to perform **oxygenic** photosynthesis.

- Purple sulfur and purple nonsulfur bacteria cannot oxidize water. They perform **anoxygenic** photosynthesis.

- Photosystem II is the only known protein complex capable of splitting water. All oxygen we breathe originates from splitting of water in photosystem II. Addition of O_2 to the atmosphere by oxygenic photosynthesis had a huge impact on the history of life.

- Oxygen is toxic to anaerobic organisms. Furthermore, organisms that evolved the ability to use O_2 as an electron acceptor in cell respiration could make ATP more efficiently than organisms using other electron acceptors. Aerobic organisms became dominant on our planet.

✔ CHECK YOUR UNDERSTANDING

(c) Purple sulfur bacteria were commonly used to research photosystem II. Why were purple sulfur bacteria a better model system than trees for studying these processes?

_____

_____

_____

_____

_____

How Does Photosystem I Work?

- Heliobacteria, which produce NADH from NAD^+ in the presence of sunlight, were used to study photosystem I. In cyanobacteria, algae, and land plants, this photosystem reduces $NADP^+$ to produce the electron-carrier NADPH.

- Excited electrons from specialized chlorophyll molecules within the reaction center of photosystem I are passed down an ETC of iron- and sulfur-containing molecules to **ferredoxin** (**Figure 10.13**).

- Ferredoxin passes electrons to the enzyme ferredoxin/$NADP^+$ oxidoreductase (also called $NADP^+$ reductase). This enzyme transfers a proton and two electrons to $NADP^+$, forming NADPH.

- The entire photosystem is based in the thylakoid membrane.

- Like NADH from the citric acid cycle, the electrons from NADPH can be used to reduce other compounds.

- Thus, photosystem I produces chemical energy stored in NADPH's reducing power, and photosystem II produces chemical energy stored in ATP.

✔ CHECK YOUR UNDERSTANDING

(d) Which photosystem's electron transport chain is more similar to that in the mitochondria, and why?

_____

_____

_____

The Z Scheme: Photosystems II and I Work Together

- The model for how photosystems I and II interact is called the **Z scheme** because an electron's path and potential energy changes can be graphed in a Z shape (**Figure 10.14**).

- The Z scheme starts when photons excite electrons in photosystem II's antenna complex. Chlorophylls transmit the energy to the reaction center, where a special chlorophyll pair called P680 passes an excited electron to pheophytin.

- Pheophytin passes these high-energy electrons to an ETC, where the electrons are gradually stepped down in potential energy. Energy released by these redox reactions is used by PQ to carry protons across the thylakoid membrane.

- Energy in the proton gradient is used by ATP synthase to phosphorylate ADP.

- At the end of photosystem II's electron transport chain, a small protein called **plastocyanin** (PC) gains an electron from the cytochrome complex and carries it back across the thylakoid membrane to photosystem I.

- Plastocyanin physically links photosystem II and photosystem I. One plastocyanin can shuttle over 1000 electrons per second between photosystems. This electron flow replaces electrons that leave the photosystem I reaction center.

- A chlorophyll called P700 in the photosystem I reaction center passes its excited electrons to the ETC containing ferredoxin. These electrons are used to reduce $NADP^+$ to NADPH.

- Photosystem II's electrons are replaced by the splitting of water, and photosystem I's electrons are replaced with the electrons that plastocyanin shuttles from photosystem II.

Understanding the Enhancement Effect

- The Z scheme explains the enhancement effect; photosynthesis is more efficient when both 680-nm and 700-nm wavelengths are available, allowing both photosystems to run at maximal rates.

Cyclic Photophosphorylation Produces Additional ATP

- In green plants and algae, photosystem I sometimes transfers electrons to the electron transport chain of photosystem II to increase ATP production instead of using them to reduce $NADP^+$; this is called **cyclic photophosphorylation** (**Figure 10.15**).

Where Are PS II and PS I located?

- Surprisingly, photosystem I and ATP synthase are most commonly found in the exterior, unstacked thylakoid membranes, while photosystem II is most common in the interior stacked membranes of grana (**Figure 10.16**).

✔ CHECK YOUR UNDERSTANDING

(e) Explain how photosystems I and II are similar and how they differ.

_____

_____

_____

_____

_____

10.4 How Is Carbon Dioxide Reduced to Produce Glucose?

- Photosystems I and II can produce ATP and NADPH only in the presence of light. The reactions that produce sugar from carbon dioxide do not require light—but they do require the ATP and NADPH produced by the light-dependent reactions.

The Calvin Cycle Fixes Carbon

- **Carbon fixation** adds carbon dioxide to an organic compound in a redox reaction (CO_2 is reduced). This converts or "fixes" CO_2 in a biologically useful form.

- Melvin Calvin led the research team that used the pulse-chase technique to discover the intermediate compounds produced when carbon dioxide is reduced to sugar.

- Calvin's group fed labeled carbon dioxide ($^{14}CO_2$) to green algae and then isolated and identified product molecules containing ^{14}C.

- By waiting different amounts of time between adding the $^{14}CO_2$ and killing the algae, they determined the order in which compounds are produced during the reduction of CO_2 to sugar.

- Product molecules from algal extract were separated using chromatography and identified by laying X-ray film over the chromatography surface. Radioactively labeled products left a black spot on the X-ray film (**Figure 10.17**) and could then be isolated.

- Cells killed right after exposure to $^{14}CO_2$ contained a radiolabeled three-carbon compound called 3-phosphoglycerate, an intermediate in glycolysis.

- Calvin's pathway produces sugar and glycolysis breaks down glucose, so it makes sense that the pathways share some intermediates.

RUBP Is the Initial Reactant with CO_2

- A five-carbon compound called **ribulose bisphosphate (RuBP)** reacts with CO_2 to produce a six-carbon compound that immediately splits in half to form two molecules of 3-phosphoglycerate.

The Calvin Cycle Is a Three-Step Process

- The cycle that reduces CO_2 to produce 3PG (and eventually glucose) is called the Calvin cycle. This cycle of reactions occurs in the chloroplast stroma and has three phases (**Figure 10.18**).

- *Fixation phase:* CO_2 reacts with RuBP, producing two 3-phosphoglycerate molecules.

- *Reduction phase:* 3-phosphoglycerate is phosphorylated by ATP and reduced by NADPH to produce **glyceraldehyde 3-phosphate (G3P)**. Some G3P is drawn off to make glucose and fructose.

- *Regeneration phase:* The remaining G3P is used to regenerate RuBP.

- Each turn of the Calvin cycle fixes one molecule of CO_2, so it takes three turns to produce one molecule of G3P.

- The Calvin cycle uses the substantial chemical energy stored in ATP and NADPH from the light-dependent reactions to power the reduction of CO_2 to carbohydrate $(CH_2O)_n$.

- Most Calvin cycle reactions also occur as part of glycolysis and/or another metabolic pathway, but the reaction between CO_2 and RuBP is unique to the Calvin cycle. This is the key reaction responsible for the fixation of atmospheric CO_2.

✓ CHECK YOUR UNDERSTANDING

(f) Compare and contrast the citric acid cycle of cell respiration (**Chapter 9**) with the Calvin cycle. How are they similar and how are they different?

_____

_____

_____

_____

The Discovery of Rubisco

- Arthur Weissbach and colleagues purified proteins from spinach extracts and tested whether they could catalyze the incorporation of $^{14}CO_2$ into 3PG. Eventually they isolated the carbon-fixing enzyme: ribulose 1, 5-bisphosphate carboxylase/oxygenase (**rubisco**).

- Rubisco is cube-shaped and has eight active sites for CO_2 fixation; it is very abundant in leaf tissue.

- Despite its importance, rubisco is slow and inefficient. Oxygen competes with CO_2 at the active site—rubisco catalyzes the addition of O_2 to RuBP as well as the addition of CO_2. This trait appears to be **maladaptive** because it decreases RuBP's efficiency.

- When O_2 and RuBP react in rubisco's active site, one product undergoes reactions that use up ATP and produce CO_2 (**Figure 10.19**). These reactions are called **photorespiration** because they consume O_2 and produce CO_2.

Carbon Dioxide Enters Leaves through Stomata

- Plants are covered with a waxy cuticle that prevents water loss via evaporation from the plant. This cuticle also prevents atmospheric CO_2 from directly entering surface leaf cells.

- **Stomata** (singular = **stoma**) are the leaf structures where gas exchange occurs. They consist of two **guard cells** that change shape when leaf CO_2 concentration is low.

- Chemical signals activate proton pumps in guard cell membranes. These pumps create a charge gradient; K^+ then flows into guard cells, water follows by osmosis, the guard cells swell and pull apart from each other, and create a **pore**.

- When leaf pores open, atmospheric CO_2 diffuses into the extracellular fluid and on into cells down its concentration gradient (**Figure 10.20**). The Calvin cycle continually uses CO_2, maintaining a steep concentration gradient favoring CO_2 diffusion into cells.

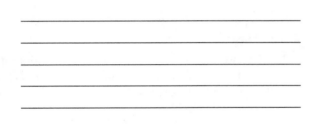

CHECK YOUR UNDERSTANDING

(g)   Why is rubisco considered an inefficient enzyme?

_____

_____

_____

_____

_____

Mechanisms for Increasing CO_2 Concentration

- Photosynthetic rate declines during photorespiration. When cell CO_2 concentration is high and O_2 concentration is low, carbon fixation is favored over photorespiration. But, the atmosphere contains much more oxygen than carbon dioxide.

- Furthermore, open stomata cause leaf water loss because the atmosphere is usually much drier than the leaf interior. Thus, when conditions are hot and dry, many plants must keep their stomata closed to prevent water loss, causing photosynthesis to stop.

- Even worse, cellular respiration continues when photosynthesis stops, producing yet more O_2—a condition that favors photorespiration.

C_4 photosynthesis

- Research groups using Calvin's pulse-chase approach found that in sugar cane (Kortschak's group) and corn (Karpilov's group), $^{14}CO_2$ was initially incorporated into four-carbon organic acids (e.g., malate, aspartate) instead of into 3PG (**Figure 10.22**).

- Carbon dioxide can be added to three-carbon compounds by **PEP carboxylase**. This type of CO_2 fixation is called **C_4 photosynthesis** because the initial

product has four carbons. The reaction catalyzed by Rubisco is part of **C_3 photosynthesis**.

- PEP carboxylase is common in **mesophyll cells** near the leaf surface. The four-carbon organic acids produced by C_4 photosynthesis travel to **bundle-sheath cells** surrounding the plant's **vascular tissue**, where water and nutrient transport occurs.

- Bundle-sheath cells of C_4 plants contain rubisco. Here the four-carbon organic acids release CO_2, which can then be processed in the Calvin cycle (**Figure 10.22**; model initially proposed by Hatch and Slack).

- The C_4 pathway has independently evolved several times in plants in hot, dry habitats. This pathway uses energy (ATP) but increases the concentration of CO_2 in cells where the Calvin cycle operates. C_4 photosynthesis limits photorespiration in these plants.

CAM Plants

- Flowering plants called the Crassulaceae evolved a different solution to the problem of photorespiration in hot, dry environments. These plants keep their stomata closed all day to minimize water loss. At night they open their stomata and temporarily fix CO_2 to organic acids using a pathway called **crassulacean acid metabolism (CAM)**.

- CAM plants store the organic acids they make at night in the central vacuoles of photosynthesizing cells. During the day, CO_2 is released from the stored organic acids and used by the Calvin cycle (**Figure 10.23**).

- CAM and C_4 photosynthesis both function as CO_2 pumps. C_4 photosynthesis stockpiles CO_2 in cells where rubisco is not active, and CAM plants store CO_2 at a time when rubisco is not active. Both pathways minimize photorespiration when stomata are closed.

CHECK YOUR UNDERSTANDING

(h)   Although our world is likely experiencing a period of overall global warming, climate change researchers think that certain areas might experience local cooling. Suppose you are studying a grassland that appears to be experiencing cooler, wetter summers on average. Considering only the differences in photosynthesis, what do you predict will happen to the ratio of C_4 to C_3 plants in the area over time and why?

_____

_____

_____

_____

How Is Photosynthesis Regulated?

- Light triggers the production of proteins for photosynthesis, but high sugar concentrations inhibit the production of these proteins. High sugar concentrations stimulate the production of proteins that process and store sugars.

- Regulatory molecules produced in the presence of light activate rubisco. Rubisco is inhibited when CO_2 is at low concentrations.

- Low P_i levels stimulate the production of Calvin cycle proteins. When sugar is produced and processed, P_i is released from NADPH and can be used in light-capturing reactions.

What Happens to the Sugar That Is Produced by Photosynthesis?

- G3P from the Calvin cycle is used to make glucose and fructose, which can combine to form **sucrose** in the cytosol (**Figure 10.24**).

- Water-soluble sucrose is easily transported throughout the plant; it can be broken down to fuel cell respiration or be converted to starch for storage.

- In rapidly photosynthesizing cells, when sucrose is abundant, glucose is temporarily stored as **starch** in the chloroplast. Because starch is not water soluble, it is broken down at night and used to make more sucrose for transport throughout the plant.

- Herbivores eat plants and use sugars and starch from photosynthesis for their own growth and reproduction. When carnivores eat herbivores, they gain chemical energy that ultimately came from the conversion of sunlight energy into chemical energy via photosynthesis.

B. MASTERING CHALLENGING TOPICS

Three new major metabolic pathways are described in this chapter: photosystem I, photosystem II, and the Calvin cycle. Connections between these and other metabolic pathways are also discussed. One good strategy for learning this information is to make your own lists of the order in which events occur, drawing flowcharts or concept maps depicting the connections between pathways.

For example, you might start a list describing photosynthesis as follows: (1) Photon hits chlorophyll and excites electron; (2) excited electron is passed to other chlorophyll molecules until it reaches a reaction center; (3) at the reaction center, the electron reduces an electron acceptor. Another study strategy is to look at the text figures and describe out loud to yourself or a study partner the names of structures and what happens where. In assessing the importance of each step along a pathway, think about what would happen if that part of the pathway failed.

C. ASSESSING WHAT YOU'VE LEARNED

1. How would the experiments of van Niel with purple sulfur bacteria have been different if oxygen came from carbon dioxide during photosynthesis?
   a. Elemental sulfur would be produced from carbon dioxide and hydrogen sulfide.
   b. Oxygen would be produced from carbon dioxide and hydrogen sulfide.
   c. Oxygen would be produced from carbon dioxide and water.
   d. Elemental sulfur would be produced from carbon dioxide and water.

2. If green plant cells are incubated with ^{18}O-labeled carbon dioxide, what molecule will become radioactive as the cells are exposed to light? (Choose all that apply.)
   a. Water
   b. Oxygen
   c. ATP
   d. Sugar

3. What two high-energy molecules produced during the light-dependent reactions of photosynthesis are used to drive the Calvin cycle?
   a. ATP and oxygen
   b. NADPH and carbon dioxide
   c. ATP and NADPH
   d. ATP and carbon dioxide

4. Which of the following is true about the overall reaction of photosynthesis as it occurs in plant cells: $6CO_2 + 12H_2O + $ light energy $\rightarrow C_6H_{12}O_6 + 6O_2 + 6H_2O$? (Choose all that apply.)
   a. Carbon dioxide is reduced.
   b. Carbon dioxide is oxidized.
   c. Sugar is oxidized.
   d. The overall reaction (as written above) is endergonic.
   e. The overall reaction (as written above) is exergonic.

5. What is the purpose of chemical reduction of an electron acceptor in the photosynthetic reaction center?
   a. It allows the energy of absorbed light to be trapped and converted to chemical energy.
   b. It adjusts the energy level of chlorophyll electrons to match the energy of light illuminating the leaf.
   c. It changes the wavelength of chlorophyll fluorescence so that it will not interfere with absorption of light.
   d. It provides electrons to be excited by the absorbed light.

6. Which of the following is evidence for there being two photosystems?
   a. Microscopy reveals two colors of chloroplast within the same leaf cells.
   b. The combination of light at 680 nm and 700 nm is much more effective at stimulating photosynthesis than at either wavelength alone.
   c. Chlorophyll absorbs light of two different wavelengths: blue and red.
   d. Two different high-energy molecules are produced during the light-dependent reactions of photosynthesis: ATP and NADPH.

7. What happens to the ATP synthesis rate if the pH of the thylakoid lumen decreases?
   a. The rate of ATP synthesis will not change.
   b. The rate of ATP synthesis will decrease.
   c. The rate of ATP synthesis will increase.

8. If a cell is very low on ATP, which of the following is most likely to occur?
   a. Decreased cellular respiration
   b. Increased starch synthesis
   c. Increased rate of Calvin cyle reactions
   d. Cyclic photophosphorylation

9. In an experiment conducted by Jagendorf, isolated thylakoids were incubated in an acidic solution at pH 4.0 until the pH was equilibrated across the thylakoid membrane. The thylakoids were then transferred to a buffer at pH 8.0 with ADP and inorganic phosphate. ATP was synthesized. Did this experiment require light to generate the ATP?
   a. Yes; ATP synthase requires high-energy electrons from photosystem II.
   b. No; ATP synthesis is dependent only on the presence of a hydrogen ion gradient and does not require light directly.
   c. Yes; the hydrogen ion gradient is generated via transfer of electrons in photosystem II.
   d. No; ATP synthesis is part of the light-independent reactions of photosynthesis.

10. Electrons excited by absorption of light in photosystem II are transferred to plastoquinone and so must be replaced. The replacements come from:
    a. photosystem I.
    b. water.
    c. oxygen.
    d. cytochrome.

11. Electrons excited by absorption of light in photosystem I are transferred to iron-sulfur electron acceptors and must be replaced. The replacements come directly from:
    a. photosystem II.
    b. ATP.
    c. NADPH.
    d. water.

12. The biochemical function of photosystem I is to:
    a. phosphorylate ADP.
    b. hydrolyze ATP.
    c. oxidize NADPH.
    d. reduce $NADP^+$.

13. Rubisco is an important enzyme because it
    a. catalyzes the splitting water, which provides electrons for photosynthesis.
    b. catalyzes carbon fixation.
    c. catalyzes the reaction that allows CAM plants to store CO_2.
    d. is one of the fastest enzymes known.

14. Why is it critical for plants to maintain a high concentration of carbon dioxide in the leaves?
    a. It helps to prevent photorespiration.
    b. It is the only substrate for rubisco.
    c. Oxygen cannot be produced without it.
    d. It is necessary for regeneration of RuBP from glyceraldehyde-3-phosphate.

15. How does the sequestering of carbon dioxide in CAM plants help them to survive?
    a. It keeps carbon dioxide away from places in the leaves where it would be toxic.
    b. It allows the plants to produce sugars in the winter when leaves are dead.
    c. It allows the light-dependent and light-independent reactions to occur in different places.
    d. It allows carbon dioxide to be gathered and used at different times of the day.

CHAPTER 10—ANSWER KEY

Check Your Understanding

(a) These new plant-like organisms are yellow and red instead of green because their pigments absorb blue and green light but reflect or transmit yellow and red light. This is not too surprising because these "plants" do not share an evolutionary history with life on our world. We would thus expect them to have different pigments and pathways for capturing light energy.

(b) Plant chlorophyll molecules are part of a photosystem. They pass the energy of excited electrons to other chlorophyll molecules or to the electron acceptor in the reaction center. Therefore, this energy is not normally given off as electromagnetic radiation and does not cause fluorescence.

(c) Some advantages to working with bacteria: They are simpler, faster-growing, and easy to keep lots of in the lab. Trees are more complex, take up a lot of space, and take a long time to grow and reproduce—making them, in many ways, more difficult to work with.

Purple sulfur bacteria were especially useful because they have only one of the two photosystems (II).

(d) Photosystem II's electron transport chain is more similar to that in the mitochondria than photosystem I's because photosystem II also creates a proton motive force that is used by ATP synthase to make ATP.

(e) Photosystems I and II are similar in that they both use electrons excited by photons to reduce molecules at the beginning of electron transport chains. Both are located in the thylakoid membrane. On the other hand, photosystems I and II most easily use light of different wavelengths, and they use the potential energy of the excited electrons to produce different compounds (II makes ATP and I makes NADPH). They occur in different parts of the thylakoid membrane, and they replace electrons lost from their chlorophylls in different ways (II splits water and I gets electrons from II).

(f) Both are biochemical cycles where the starting compound is regenerated at the end of the cycle. Both take place inside eukaryotic organelles. However, whereas the Krebs cycle occurs in the mitochondria, photosynthesis occurs in the chloroplast; the Krebs cycle makes ATP and NADH (and $FADH_2$), but the Calvin cycle uses up ATP and NADPH; and the Krebs cycle gives off CO_2, but the Calvin cycle uses up and/or fixes CO_2.

(g) Rubisco is inefficient because (1) it is slow, and (2) it catalyzes the toxic photorespiration reaction when cell CO_2 concentration is low and O_2 concentration is high. It evolved when there was little O_2 in Earth's atmosphere.

(h) You predict that the ratio of C_4 to C_3 plants should decrease. As the local climate becomes cooler and wetter, C_3 plants should be able to obtain plenty of CO_2 for photosynthesis through open stomata. Previously, C_4 plants had the advantage of being able to concentrate CO_2 where rubisco is active to minimize photorespiration. However, it uses energy to transport the 4-carbon organic acids into the bundle sheath cells. Thus, under cooler and wetter conditions, C_4 photosynthesis has a cost but little gain, whereas C_3 plants avoid the transport cost and will not suffer too much from photorespiration.

Assessing What You've Learned

1. b; 2. a, d; 3. c; 4. a, e; 5. a; 6. b; 7. c; 8. d; 9. b; 10. b; 11. a; 12. d; 13. b; 14. a; 15. d

Looking Forward—Key Concepts in Later Chapters

Oxygen Produced by Photosynthesis—Chapter 28

Photosynthetic bacteria (some similar to those used in experiments described in Chapter 10) were responsible for producing the oxygen that shapes our current world. Find out more about these organisms in **Chapter 28**.

C_4 and CAM Pathways for Carbon Fixation—Chapter 30

Chapter 30 gives you more information about green plants that use the C_4 and CAM strategies for carbon fixation.

Water and Sugar Transport in Plants—Chapter 37

Plants must take up water to replace water lost via stomata and transport fructose and other nutrients throughout the organism. These transport processes are detailed in **Chapter 37**.

Global Warming and CO_2 Levels—Chapter 54

You'll learn more about global warming and rising CO_2 levels in **Chapter 54**, on ecosystem ecology.

Chapter 11

The Cell Cycle

Cell Theory—Chapter 1

See **Chapter 1** for further discussion of the cell theory, its importance, and some supporting evidence.

Use of Radioactive Isotopes—Chapters 7, 9, and 10

Radioactive isotopes are extremely useful research tools. Several pulse-chase experiments and other experiments involving the use of radioactive isotopes are described in **Chapters 7, 9,** and **10.** The use of radioactive thymidine to identify the timing of DNA synthesis is described here in Chapter 11.

Microtubules—Chapter 7

Microtubules were introduced in their role as cytoskeletal elements in **Chapter 7.** Here in Chapter 11, you discover that microtubules also make up the mitotic spindle, which pulls apart sister chromatids during mitosis. The way that spindle fibers move chromosomes is somewhat similar to the way that cytoskeletal microtubules function in vesicle transport.

Signal Transduction—Chapter 8

Chapter 8 introduces the importance of cell-cell signaling via hormonal initiation of phosphorylation cascades. For example, cancer may develop when defective Ras proteins (**Figure 8.19**) prevent signal deactivation.

KEY CONCEPTS

- Dividing eukaryotic cells have a four-phase cell cycle that includes mitosis (M phase) and the three parts of interphase: G_1, S, and G_2.

- Chromosomes are copied during S phase and move to the middle of the cell during M phase. The chromosome copies then separate so that each daughter cell receives a copy of each chromosome and is genetically identical to the parent cell.

- Cell-cycle checkpoints control whether or not a cell can divide.

- Cell-cycle-checkpoint failure can cause uncontrolled cell division, which causes cancer in multicellular organisms.

A. CHAPTER OUTLINE

Chapter Introduction

- The cell theory states that all organisms are made of cells and that all cells come from other cells.

- Rudolf Virchow proposed that new cells arise through **cell division**. This hypothesis was confirmed by microscopic observations of **embryos** (developing organisms).

- There are two types of cell division in multicellular eukaryotes. In animals, **meiosis** produces **gametes**—the sperm or eggs (Chapter 12).

- The second type of eukaryotic cell division is **mitosis.** Mitosis produces **somatic cells** (all the non-gamete cells in your body). Daughter cells resulting from mitosis are genetically identical to the parent cell.

- Mitosis and meiosis are usually followed by **cytokinesis**—the division of the cytoplasm. These cell divisions are responsible for one of the fundamental attributes of life: reproduction.

- Mitosis produces genetically identical daughter cells.

11.1 Mitosis and the Cell Cycle

- In multicellular eukaryotes, mitosis and cytokinesis are responsible for growth, wound repair, and **asexual reproduction**. Offspring produced this way are genetically identical to the parent.

What Is a Chromosome?

- Certain dyes stain threadlike structures in the nuclei of dividing eukaryotic cells. In 1879 Walther Flemming published his observations of salamander embryo cell division (**Figure 11.1**). Paired threads split apart in a process he called mitosis.

- Roundworm studies revealed that the total number of threads per cell remains constant, division after division. W. Waldeyer named these threads **chromosomes** in 1888.

- We now know that each chromosome contains a long double helix of deoxyribonucleic acid (DNA) wrapped around proteins. DNA carries the cell's genetic information; a gene is a length of DNA that codes for a particular protein or ribonucleic acid.

- Chromosomes are replicated before mitosis, condense to a more compact form at the start of mitosis, and then one copy of each chromosome is moved to each of the two daughter cells (**Figure 11.2**).

- Each DNA copy in a replicated chromosome is called a **chromatid**. Chromatids are joined along their length and at a specialized region called the **centromere**. Attached chromatids are called **sister chromatids** and form one chromosome.

Cells Alternate between M Phase and Interphase

- Growing eukaryotic cells alternate between **mitotic** or **M phase**, when cells divide, and **interphase**, when no division occurs. Cells spend most of their time in interphase.

- Chromosomes are uncoiled during interphase and, consequently, are not visible using light microscopy.

The Discovery of S Phase

- Alma Howard and Stephen Pelc used radioactive isotopes to determine when chromosome replication occurs. They supplied growing cells with radioactive phosphorus or thymidine to label DNA as chromosomes were copied.

- Extra radioactive isotope was washed away and then the newly synthesized (and labeled) DNA could be visualized using X-ray film (autoradiography, **BioSkills 9**).

- Newly synthesized DNA, indicated by black dots, was found only in interphase nuclei. This indicated that DNA replication occurs during interphase.

- The DNA-replication part of interphase is called the **synthesis** (or **S**) **phase**. Howard and Pelc referred to the entire cell division sequence from formation of a new daughter cell, to DNA duplication, through cell division, as the **cell cycle**.

The Discovery of the Gap Phases

- To determine the length of S phase, researchers used a pulse-chase approach. They briefly exposed growing cells to radioactive thymidine (the pulse), flooded the cells with nonradioactive thymidine (the chase), and examined the cells after various time intervals.

- Only cells in S phase (synthesizing DNA) incorporated the label. Researchers used cultured cells because they could be manipulated more easily than cells inside living organisms (**BioSkills 12**).

✔ **CHECK YOUR UNDERSTANDING**

(a) Before radioactive isotopes were available, could researchers have determined when DNA synthesis occurred? Explain.

_____

_____

_____

_____

_____

_____

- It took 4 hours for labeled mitotic cells to appear, suggesting that there was a 4-hour gap in the cell cycle between the end of S phase and the beginning of M phase. This part of the cell cycle is now called G_2 **phase**, for second gap.

- Researchers determined that S phase lasts 6 to 8 hours because labeled mitotic nuclei can be observed for 6 to 8 hours. These cells must have been in S phase when radiolabeled thymidine was available.

- Researchers subtracted G_2-, S-, and M-phase times from total cell-cycle time (24 hours in these cells) to conclude that another cell-cycle gap must exist after M phase. This first gap, or G_1 **phase**, lasted about 7 to 9 hours.

The Cell Cycle

- The cell cycle comprises M phase and the three parts of interphase: G_1, S, and G_2 (**Figure 11.3**).

- During the gap phases, cells make additional cytoplasm and organelles replicate. Cells also perform other normal cell functions. A cell must grow large enough and have enough organelles before mitosis can occur.

✔ CHECK YOUR UNDERSTANDING

(b) What might happen to cells that repeatedly divided without any gap phases?

_____

_____

_____

11.2 How Does Mitosis Take Place?

- **Figure 11.4** shows how chromosomes change during interphase and mitosis. The number of chromosomes varies widely among species.
- **Chromatin**, the DNA–protein complex in eukaryotes, contains **histones**.
- Chromosomes consist of two sister chromatids at the start of mitosis. Chromatids are attached to each other at the centromere, and each contains the same genetic information, encoded in one long DNA double helix.

Events in Mitosis

- Mitosis begins when chromatin condenses and becomes visible.
- The two joined chromatids separate during mitosis to form independent chromosomes. Each daughter cell receives a copy of each chromosome, and thus a complete copy of the parental cell's genetic information.
- Mitosis (M phase) is a continuous process, but biologists identify several subphases (prophase, prometaphase, metaphase, anaphase, telophase) based on specific events occurring at different stages in the process (**Figure 11.5**).

Prophase

- During **prophase**, the previously replicated chromosomes become visible in the light microscope as they condense.
- In the cytoplasm, microtubules begin to form the **mitotic spindle**. Later on in mitosis, spindle fibers will pull chromosomes into the daughter cells and push cell poles apart.
- **Polar microtubules** extend from each spindle and overlap in the middle of the cell; **kinetochore microtubules** attach to chromosomes.
- Spindle fibers radiate from microtubule organizing centers. In animals these centers contain **centrioles** and are known as **centrosomes**.

Prometaphase

- The nucleolus disappears and the nuclear envelope breaks down during **prometaphase**.
- Kinetochore microtubules from each mitotic spindle attach to one of the two sister chromatids at structures called **kinetochores**, located near chromosome centromeres.
- Kinetochore microtubules now start moving chromosomes toward the middle of the cell, even as, in animals, centrosomes continue to move toward opposite poles of the cell.

Metaphase

- In **metaphase**, centrosomes reach opposite poles of the cell (in animals) and chromosomes, pulled by kinetochore microtubules, reach the middle of the cell.
- The imaginary plane where the chromosomes line up is called the **metaphase plate**. The mitotic spindle is now complete—stretching from each kinetochore to one of the two cell poles.

Anaphase

- The centromeres split at the start of **anaphase**. Kinetochore spindles shorten, pulling apart sister chromatids to create separate chromosomes that are pulled toward microtubule organizing centers at opposite cell poles.
- Motor proteins also push apart the two cell poles. Chromatid separation ensures that each daughter cell receives one copy of each parent chromosome.

Telophase

- A new nuclear envelope begins to form around each set of chromosomes during **telophase**. The spindle disintegrates, and the chromosomes become less compact and less visible as they de-condense.

✔ CHECK YOUR UNDERSTANDING

(c) Why might there be occasions when a researcher observing a mitotic cell under a microscope would have difficulty determining whether the cell is in prophase or prometaphase?

_____

_____

_____

_____

Cytokinesis Results in Two Daughter Cells

- Organelles replicate during interphase and the cell contents grow.

- The division of the cytoplasm to form two daughter cells (cytokinesis) typically occurs immediately after mitosis.

- Plant cell cytokinesis occurs when vesicles are transported from the Golgi apparatus to the middle of the dividing cell. These vesicles fuse to form a **cell plate** (**Figure 11.6**). Microtubules and other proteins help organize this process.

- Cytokinesis in animals, fungi, and slime molds occurs when a ring of action and myosin filaments inside the cell membrane contracts, causing the cell membrane to pinch inward and form a **cleavage furrow**.

- Bacteria divide by a mechanism similar to that used by animal cells. The constricting ring in bacteria is made of cytoskeletal elements called FtsZ fibers.

- Refer to **Table 11.1** for definitions of key mitotic structures.

How Do Chromosomes Move during Mitosis?

Mitotic Spindle Forces

- Microtubules are asymmetric and grow at their plus end. Kintochore microtubules grow until their plus ends attach to a kinetochore.

- Because microtubule length depends on its number of α- and β-tubulin dimers, researchers hypothesized that spindle microtubules shorten as tubulin subunits are lost.

- Fluorescently labeled tubulin subunits were added to prophase or metaphase cells. The whole mitotic spindle could then be seen using a fluorescent microscope.

- After the start of anaphase, biologists marked a region of the spindle with a laser light that stops fluorescence in the exposed region. This area remained stationary as anaphase progressed and the chromosomes moved toward the nonfluorescing region (**Figure 11.8**).

- Researchers concluded that chromosome movement occurs because tubulin subunits are lost from the kinetochore ends. The rest of the microtubules remain stationary.

✔ **CHECK YOUR UNDERSTANDING**

(d) In the experiment on spindle shortening, why was labeled tubulin added during prophase or metaphase instead of during anaphase?

_____

_____

_____

A Kinetochore Motor

- Dyneins and other kinetochore motor proteins appear to attach to the kinetochore's fibrous crown. These kinetochore proteins then "walk" down the microtubules as they shorten, similar to the way kinesin "walks" down microtubules during vesicle transport (**Figure 11.9**).

- Other kinetochore proteins catalyze the loss of tubulin subunits from the kinetochore end of the microtubules. The chromosome is pulled toward the cell pole as the kinetochore motor proteins walk down the shortening microtubule.

11.3 Control of the Cell Cycle

- Cell-cycle length varies greatly from tissue to tissue within the same individual. G_1 phase is practically eliminated in rapidly dividing cells. Other cells get permanently stuck in G_1 phase; this arrested stage is called the G_0 state.

- The rate of cell division may also respond to changes in conditions. These variations in cell-cycle length suggest that the cell cycle is regulated and that regulation varies among cells and organisms.

The Discovery of Cell-Cycle Regulatory Molecules

- Certain chemicals, viruses, or electric shocks can fuse two cells, forming a hybrid cell with two nuclei.

- When two cells in different cell-cycle stages were fused, one of the two nuclei sometimes changed phase. For example, fusion of an M phase cell with an interphase cell caused the interphase nucleus to enter M phase (**Figure 11.10a**).

- Researchers hypothesized that the M-phase cell cytoplasm contains a regulatory molecule that causes interphase cells to start mitosis. This hypothesis was confirmed by experiments with *Xenopus* frog eggs.

- Maturing *Xenopus* eggs start as **oocytes** and develop into mature M-phase eggs. These eggs are very large, making it feasible to purify large amounts of cytoplasm. Purified cytoplasm can be injected into eggs to test for the presence of mitotic regulatory factors.

- When purified M-phase cytoplasm was injected into the cytoplasm of frog oocytes arrested in the G_2 phase, the immature oocytes entered M phase. Cells injected with cytoplasm from interphase cells remained in G_2 phase (**Figure 11.10b**).

- Some factor is present in the cytoplasm of M-phase cells (but not in interphase cells) that induces mitosis.

This factor has now been purified and is called **mitosis promoting factor (MPF)**. MPF induces mitosis in all eukaryotes.

MPF Contains a Protein Kinase and a Cyclin

- MPF contains two polypeptide subunits. One, a **protein kinase**, catalyzes a protein phosphorylation reaction that transfers a phosphate group from ATP to a target protein.

- Many protein kinases are regulatory molecules because adding a phosphate group usually alters a protein's shape and activity, either activating or inactivating it. But, MPF protein kinase concentration does not change much during the cell cycle.

- The other MPF subunit, a **cyclin,** does change in concentration throughout the cell cycle; MPF cyclin concentration increases during interphase, peaks in M phase, and then decreases again (**Figure 11.11**).

- The MPF protein kinase can be active only when it is bound to the cyclin subunit; it is a **cyclin-dependent kinase (Cdk)**.

How Is MPF Activated?

- MPF's Cdk subunit is in an inactive phosphorylated form when it binds to cyclin. Late in G_2 phase, enzymes remove one of the phosphate groups from Cdk, causing a shape change that activates MPF.

- MPF then triggers several events that initiate chromosome condensation and the formation of the mitotic spindle.

How Is MPF Deactivated?

- MPF is controlled via **negative feedback** (the halting or slowing of a process by its products); MPF initiates mitosis, which activates an enzyme complex that degrades MPF's cyclin subunit, returning MPF's kinase subunit to an inactive conformation.

- Many cell processes are regulated by protein degradation. The enzyme complex responsible for MPF's degradation attaches small ubiquitin proteins to MPF's cyclin subunit during anaphase. This causes the proteasome to destroy the subunit.

- After cyclin is destroyed, it slowly builds up again during interphase.

Cell-Cycle Checkpoints Can Arrest the Cell Cycle

- A different cyclin and protein kinase control the transition from G_1 to S phase, and several regulatory proteins maintain the G_0 state of quiescent cells.

- Leland Hartwell and Ted Weinert studied **cell-cycle checkpoints** (critical points in cell-cycle regulation)

by analyzing mutant yeast cells that kept dividing when normal cells stopped.

- In our bodies, cells without effective cell-cycle checkpoints grow uncontrollably and form a **tumor**.

- Three checkpoints have been identified (**Figure 11.12**). At each one, interactions between regulatory molecules determine whether a cell proceeds with division.

G_1 Checkpoint

- Four factors affect whether cells pass the G_1 checkpoint: (1) *cell size*—cells must be large enough to split into two functional daughter cells; (2) *availability of nutrients*—in unicellular organisms; (3) *social signals*—in multicellular organisms; (4) *damage to DNA*.

- The protein **p53** activates genes that stop the cell cycle if a cell's DNA is damaged. p53 can also initiate **apoptosis** (programmed cell death). If p53 is defective, DNA is not repaired and more mutations may arise.

- Regulatory proteins such as p53 are called **tumor suppressors**.

G_2 Checkpoint

- Improper chromosome replication or DNA damage can cause cells to arrest between the G_2 and M phases. The activation of MPF via dephosphorylation is blocked, so cells remain in G_2 phase.

✔ **CHECK YOUR UNDERSTANDING**

(e)   Why did Hartwell and Weinert use yeast instead of bacteria to study cell-cycle regulation?

_____

_____

_____

_____

Metaphase Checkpoint

- Cells arrest during M phase if the chromosomes are not properly attached to the mitotic spindle. This mechanism prevents incorrect chromosome separation that could give daughter cells the wrong number of chromosomes.

- All three cell-cycle checkpoints prevent the division of damaged cells. Cells that are in G_0 state also do not pass the G_1 checkpoint.

11.4 Cancer: Out-of-Control Cell Division

- **Cancer** is a common, often lethal disease (**Figure 11.13**). All cancers originate from cell-cycle checkpoint failures. These checkpoint failures allow cells to grow uncontrollably.

- When cancerous cells spread throughout the body, they use nutrients and space needed by normal cells.

- Cancer cells have two types of cell-cycle defects: (1) defects that cause continuous activation of proteins required for cell growth (e.g., defects in Ras), and (2) defects that prevent tumor suppressor genes from stopping the cell cycle (e.g., p53).

Properties of Cancer Cells

- Uncontrolled cell division produces a mass of cells called a tumor. **Benign tumors** are noninvasive and can be cured by surgery if they can be removed without excessive organ damage. Cancer cells also are invasive.

- **Malignant tumors** are cancerous and can spread throughout the body via the blood or lymph and initiate new tumors (**Figure 11.14**).

- Detachment from the original tumor and invasion of other tissues is called **metastasis**. Once a cancer has metastasized, it cannot be cured by surgery alone.

Cancer Involves Loss of Cell-Cycle Control

- Mature cells often remain in the G_0 state; but if they pass through the G_1 checkpoint, they continue through mitosis and divide. Biologists hypothesize that most cancers involve defects in the G_1 checkpoint.

- Researchers must understand normal control mechanisms in order to understand the molecular nature of the disease.

Social Control

- Unicellular organisms pass the G_1 checkpoint when nutrients are available and cell size is sufficient.

- Cells of multicellular organisms instead respond to signals from other cells, so that they divide only when their growth benefits the whole organism. This is called *social control*.

- Normal cell cultures will not grow, even if provided with nutrients, unless polypeptides or small proteins called **growth factors** are present. Growth factors were first found in **serum**, the liquid that remains after blood clots and the cells are removed from the blood.

- **Platelet-derived growth factor (PDGF)** is a serum growth factor released into the blood by **platelets** at wound sites. Cells near the wound divide in response to PDGF and help heal the injury. PDGF can also be made by other cell types.

- Many growth factors exist, produced by many types of cells. Different cells divide in response to different combinations of growth factors, and specific growth factors must be present for a particular type of cell culture to grow.

✓ CHECK YOUR UNDERSTANDING

(f)   Why is there no social control in unicellular organisms?

_____

_____

_____

_____

- Cancer cells, however, do not need growth factors in order to divide—they are no longer subject to social control at the G_1 checkpoint.

How Does the G_1 Checkpoint Work?

- Growth factors initiate cell division by triggering production of cyclins and E2F. E2F is a regulatory protein that functions at the G_1 checkpoint a bit like MPF does at the G_2 checkpoint.

- E2F can trigger the expression of genes required for S phase. However, newly produced E2F normally binds to a tumor suppressor protein called Rb. While E2F is bound to **Rb protein**, it remains inactive and the cell remains in G_0.

- If growth factors continue to arrive, they can override the inhibitory effects of Rb. Growth factors also stimulate additional G_1 cyclin production. The G_1 cyclins bind to a Cdk, which is phosphorylated and inactive initially (**Figure 11.15**).

- When dephosphorylation activates G_1 cyclin-Cdk complexes, they catalyze the phosphorylation of Rb. Rb changes its shape in response and releases E2F. The freed E2F can now activate target genes, initiating the production of S phase proteins.

Why Do Social Controls and Cell-Cycle Checkpoints Fail?

- Cells can become cancerous when social controls fail and cells begin dividing in the absence of growth factors.

- In some human cancers the G_1 checkpoint cyclin is overproduced, permanently activating Cdk, which then continuously phosphorylates Rb. E2F remains in its active free form, sending the cell into S phase.

- Either the presence of excessive growth factors and/or cyclin production in the absence of growth factors can cause cyclin overproduction. This pathway includes the Ras protein discussed in Chapter 8—it is common to find defective Ras in cancerous cells.

- In other human cancers the G_1 checkpoint fails due to defective Rb. If Rb is absent or does not bind to E2F, the free E2F will trigger the gene expression that initiates S phase.

Cancer Is a Family of Diseases

- Many different types of defects can cause the G_1 checkpoint to fail, and most cancers result from multiple defects in cell-cycle regulation. Each type of cancer is caused by a unique combination of errors.

B. MASTERING CHALLENGING TOPICS

One effective way to learn different phases of the cell cycle is to simulate the events, using thread for the microtubules, shoelaces for chromosomes, and a plastic bag for the nuclear envelope. Recite aloud to yourself names of the phases and major events as you act out the process. You may also want to chart the different cell-cycle phases, listing their major events and any cell-cycle checkpoints. Finally, use your own mnemonic device for remembering the mitotic phases. For example, you could remember interphase, prophase, prometaphase, metaphase, anaphase, telophase, cytokinesis (IPPMATC) as "I Pet Purple Moose At The Circus."

C. ASSESSING WHAT YOU'VE LEARNED

1. You use Howard and Pelc's pulse-chase approach to study the cell cycle of a newly discovered unicellular eukaryote. First you determine that, in this species, mitosis takes about 4 hours and its cells typically divide every 20 hours under laboratory conditions. The black dots appear immediately in interphase cells, and in mitotic cells after 3 hours. You continue to find labeled mitotic cells for 6 hours. Which of the following is the best interpretation of these data?
   a. G_1 phase = 6 hours, S phase = 4 hours, and G_2 phase = 7 hours
   b. G_1 phase = 4 hours, S phase = 7 hours, and G_2 phase = 6 hours
   c. G_1 phase = 7 hours, S phase = 6 hours, and G_2 phase = 3 hours
   d. G_1 phase = 3 hours, S phase = 7 hours, and G_2 phase = 3 hours

2. Chromatin is:
   a. the structure that joins sister chromatids.
   b. the microtubule organizing center in animals.
   c. one strand of a replicated chromosome.
   d. the DNA–protein complex that makes up eukaryotic chromosomes.

3. Mitotic spindle fibers are composed of:
   a. microtubules.
   b. centrosomes.
   c. centromeres.
   d. kinetochores.

4. During mitosis, it is necessary for the nuclear envelope of the parent cell to dissolve. This is accomplished, at least partly, by phosphorylation of proteins associated with the nuclear envelope. If the enzyme responsible for that phosphorylation event is inhibited, at which phase of mitosis are cells likely to arrest?
   a. Prophase
   b. Prometaphase
   c. Metaphase
   d. Anaphase
   e. Telophase

5. Which of the following processes relies on a constricting ring of fibers? (Choose all that apply.)
   a. Chromosome movement toward the cell poles
   b. Cleavage furrow formation in fungi
   c. Bacterial fission
   d. Cell plate formation

6. Although the process of chromosome partitioning during mitosis is visible through the light microscope, the process of DNA replication is not. Why?
   a. Chromosomes form only during mitosis.
   b. Chromosomes are visible only after DNA has been duplicated.
   c. Chromosomes are too extended during S phase to be seen by light microscopy.
   d. Chromosomes do not contain protein until mitosis.

7. A certain species of animal has six pairs of chromosomes. How many molecules of DNA do the nuclei of these animals have during G_2 phase?
   a. 6
   b. 12
   c. 24
   d. 48

8. In animals, cytokinesis occurs:
   a. when microtubules pull the cell membrane into a cleavage furrow.
   b. when vesicles from the Golgi build up to divide the cell along the cell plate.
   c. when G_2 phase is complete.
   d. when a ring of actin and myosin filaments contracts.

9. The experiment where biologists marked spindle fibers made with fluorescent tubulin showed that microtubules shortened from the kinetochore ends. What results would have indicated that microtubules shorten from the microtubule organizing center end?
   a. Movement of the chromosomes toward a stationary marked area
   b. Movement of the marked region toward the cell pole
   c. Rapid disappearance of the marked microtubules
   d. None of the above

10. Some of the earliest experiments in control of the cell cycle were cell fusion experiments. Fusion of cells in M phase with cells in interphase caused the nuclei from the interphase cells to behave as if they were in mitosis. These results could have been obtained by adding just a part of the M-phase cells to the interphase cells. What portion of the M-phase cell would suffice to cause the interphase nuclei to mimic the process of mitosis?
    a. The nucleus
    b. Ribosomes
    c. The cell membrane
    d. Cytoplasm

11. Active MPF-Cdk phosphorylates proteins that cause:
    a. M-phase initiation.
    b. G_1 cell-cycle checkpoint.
    c. Rb protein.
    d. meiosis.

12. Cells that are too small to successfully produce daughter cells fail to pass:
    a. the G_1 checkpoint.
    b. the G_2 checkpoint.
    c. the metaphase checkpoint.
    d. all checkpoints.

13. Can surgery successfully cure a cancer that has metastasized?
    a. No; all body cells are dividing uncontrollably.
    b. Yes; it could remove all cells with defective cell-cycle regulation.
    c. No; cancer cells are no longer localized in one spot.
    d. Yes, if the tumor is benign.

14. What would happen if a cell had defective E2F?
    a. The cell would immediately enter the S phase.
    b. The cell would be unable to enter the S phase.
    c. The cell would become cancerous.
    d. The Rb protein would become activated.

15. One cause of cancer is mutation of genes that encode for proteins in the biochemical pathway regulated by PDGF. How might such mutations cause cancer?
    a. If the mutations caused the biochemical pathway to become permanently active, the cells might become cancerous.
    b. If the mutations caused the biochemical pathway to become permanently inactive, the cells might become cancerous.
    c. If PDGF stimulated cell division, the cells might become cancerous.
    d. Mutation of any gene will probably cause cancer.

CHAPTER 11—ANSWER KEY

Check Your Understanding

(a) Before radioisotopes became available in the mid-twentieth century, researchers could not determine when DNA synthesis occurred because there was no way to "see" when DNA synthesis occurred (especially since interphase chromosomes are not visible via light microscopy). When radioactively labeled nucleotides became available, researchers could use them to mark the incorporation of new nucleotides into DNA during S phase.

(b) Cells that divided without any gap phases would probably get smaller and smaller and have fewer and fewer organelles with each succeeding division. Eventually they would not have enough organelles or cytoplasm to function effectively.

(c) Recall that mitosis is a continuous process. It may thus be difficult to determine whether the nuclear envelope has completely disintegrated and when exactly the kinetochore microtubules have attached to the chromatids (at the kinetochore).

(d) Because the mitotic spindle is complete by the start of anaphase, labeled tubulin added in anaphase would not be incorporated into the mitotic spindle.

(e) Recall that bacteria replicate very differently from eukaryotes. They do not undergo mitosis, because they have circular chromosomes instead of linear chromosomes. Yeast, however, are eukaryotes that divide quickly and live well in culture—two qualities that make them useful in research.

(f) Natural selection favors unicellular organisms that leave the most offspring. Thus, unicellular organisms should divide whenever enough nutrients are available to allow sufficient growth for cell division (asexual reproduction). Only in multicellular organisms is it favorable to the organism to allow other cells (in the same organism) to influence cell division.

Assessing What You've Learned

1. c; 2. d; 3. a; 4. a; 5. b & c; 6. c; 7. c; 8. d; 9. b; 10. d; 11. a; 12. a; 13. c; 14. b; 15. a

Looking Forward—Key Concepts in Later Chapters

Meiosis—Chapter 12

Chapter 12 describes meiosis, the type of cell division used in sperm and egg cell production. In many ways, meiosis is similar to mitosis. However, the daughter cells produced by meiotic divisions contain only half of the parental cell's genetic information.

Chromosomes—Chapters 13 and 15

Chromosomes, cell structures made of DNA and protein, are responsible for carrying the cell's genetic information. Chapter 11 discusses how copies of each chromosome are provided to each daughter cell during cell division. In **Chapters 13** and **15**, you'll learn more about chromosomes, their importance, and how genes work.

DNA Synthesis—Chapter 14

Chapter 11 describes evidence that DNA synthesis occurs during S phase of the cell cycle. Details of DNA synthesis are given in **Chapter 14**.

Meiosis

The Cell Cycle—Chapter 11

While reading Chapter 12, you'll notice striking similarities between **mitosis** and **meiosis** and the basic cellular architecture enabling both processes. Examine **Chapter 11** and focus on cell components that are critical to **cellular division**. How might the **cytoskeleton** (filaments and microtubules), **cellular membranes**, and **chromosomes** be prepared for meiotic division? Would **DNA replication** precede this process?

KEY CONCEPTS

- **Mitosis** is the generation of two genetically identical cells from a primary cell.

- In **meiosis** or reduction division, a cell with two copies of each chromosome (2n) divides into four cells that each contain one copy of each chromosome (1n).

- Meiosis is used in sexual reproduction, where two **gamete** cells (each 1n) combine to form a **zygote**, the first cell of a new organism.

- Mitosis and meiosis processes proceed by similar, but not identical processes that include four steps: **prophase, metaphase, anaphase,** and **telophase** (PMAT).

- Meiosis generates genetic diversity in gametes through **genetic recombination** (DNA exchange between homologous chromosomes) and the **independent assortment** of these chromosomes.

- Asexual reproduction is displayed by **prokaryotes** and **protists**. Many plants and fungi also display asexual reproduction.

- A key research question in evolutionary biology focuses upon the potential benefits of sexual reproduction when compared with asexual reproduction. Why do many organisms invest so much energy and genetic risk in sexual reproduction?

A. CHAPTER OUTLINE

12.1 How Does Meiosis Occur?

An Overview of Meiosis

- Examine **Figure 12.1** and **12.2.** Consider meiosis from the perspective of a scientist trying to understand the nature of chromosome function in the **lubber grasshopper.** There is a clear logic to this process. To illustrate this logic, list the key outcomes of a complete cycle of meiosis. How are the products of meiosis (e.g., sperm/egg) different from the cells entering into meiosis? The biological implications of this process are clear when we begin to consider genetics and reproduction.

- **Table 12.2** offers a critical summation of terminology introduced in this section. Use these terms to revise your description of meiotic outcomes.

- As you begin studying the process of meiosis, it is important to realize that the germinal cell initiating this process contains **two homologs** of each chromosome (*2n* or diploid), and that the **gamete** cell receives only **one homolog** of each chromosome upon completion of meiosis (it will be *1n* or **haploid**). In 1902, **Walter Sutton** published his observations of homologous chromosomes in the lubber grasshopper (**Figure 12.1**—homologs identified by size and by pairing).

- In a diploid organism, what is the origin of each homologous chromosome? How might carrying this "extra" information be valuable? How might it be detrimental?

- Name a few haploid-only organisms. Do these organisms display meiosis or sexual reproduction? Compare with an organism that displays an alternating (haploid/diploid) life cycle, such as the *Ulva lactuca* species of green algae. What role do haploid cells play in the life cycle of this simple organism?

- The total number and characteristics of chromosomes in an organism is known as its **karyotype** (see **Box 12.1**). The karyotype of a human somatic (non-germ cell) cell is 46. The term "n" is the number of chromosome types, in humans $n = 23$, and $2n = 46$. What is the karyotype of a human with Down syndrome?

- Define the changes in DNA content, or ploidy, during the process of meiosis (see **Figure 12.4**). Is ploidy reduced in meiosis I or meiosis II?

- Just before meiosis begins, each of the chromosomes in the $2n$ parental cells is **replicated**. This copying of each chromosome doubles the cell's DNA content, but does not change the cell's ploidy. The newly synthesized chromosomal strands remain attached to the original chromosome from which they were copied until a later stage of meiosis.

- In general, each replicated chromosome looks like the letter *X*, with two linear **sister chromatids** being attached in the middle at a structure called the **centromere**.

- **Meiosis I** involves the complex mechanism of crossing over in prophase I (described in the next section), followed by the separation of homologous chromosomes into two separate cells (each are now $1n$).

- **Meiosis II** repeats the steps of meiosis I, the key difference being the physical separation of the sister chromatids (X → <>), and the migration of each into a gamete cell that will contain one copy of each chromosome ($1n$ or **haploid** gametes).

- **Key point**: Don't confuse the **homologous chromosomes**, which make the precursor cell $2n$, with the copies of each that are produced before meiosis begins. As long as the sister chromatids are attached to each other at the centromere, they are considered a single replicated chromosome.

✔ CHECK YOUR UNDERSTANDING

(a) On a sheet of scratch paper, make a diagram that compares the ploidy of diploid ($2n$) cells proceeding through mitosis and meiosis, and indicate where each change in ploidy takes place.

The Phases of Meiosis I

- Meiosis I contains two key events in the generation of genetic diversity from generation to generation. The first event is **crossing over** between homologous chromosomes in prophase I, and the second event is the **separation of homologous chromosomes** in anaphase I. These two steps generate incredible genetic diversity in the gametes, and recognizing how each contributes to this diversity should be a key focus in your study of meiosis I.

- Review **Figure 12.7** for a visual summary of the details of meiosis.

- **Prophase I**—This step of meiosis takes the most time, primarily due to the complexity of crossing over. As it progresses, the chromosomes **condense** into tightly wound balls of DNA (X-shaped, visible with microscopy). Chromosome condensation is mediated by small, positively charged **histone proteins** that bind to the negatively charged chromosomal DNA and wrap the DNA around their structure. Crossing over occurs in prophase I and involves the reciprocal exchange of long stretches of DNA between homologous chromosomes. This is a key step in recombination of DNA for the production of genetically diverse gametes.

- **Metaphase I**—The condensed chromosomes migrate to the midline of the cell and form a line. Homologous chromosomes are adjacent to each other at the midline. In mitotic and meiotic cells, tubulin proteins polymerize to form long ropelike chains known as microtubules. Migration of condensed chromosomes is facilitated by these microtubules, attached to each chromosome at the centromere. The spindle apparatus includes microtubules and some additional proteins.

- **Anaphase I**—This is the second key step in generation of genetic diversity in gametes. The paired homologues are separated from each other, migrating to opposite ends of the cell. Following migration, each pole of the cell contains one complete set of chromosomes.

- **Telophase I**—The chromosomes complete their migration and are enclosed in two newly formed daughter cells (each is $1n$; see the provided **Figure 12.7** and confirm by counting the centromere number in each cell).

The Phases of Meiosis II

- **Meiosis II** is really just a repeat of meiosis I, with a few important differences. At the start of meiosis II, each daughter cell contains haploid chromosome content ($1n$). Each replicated chromosome is

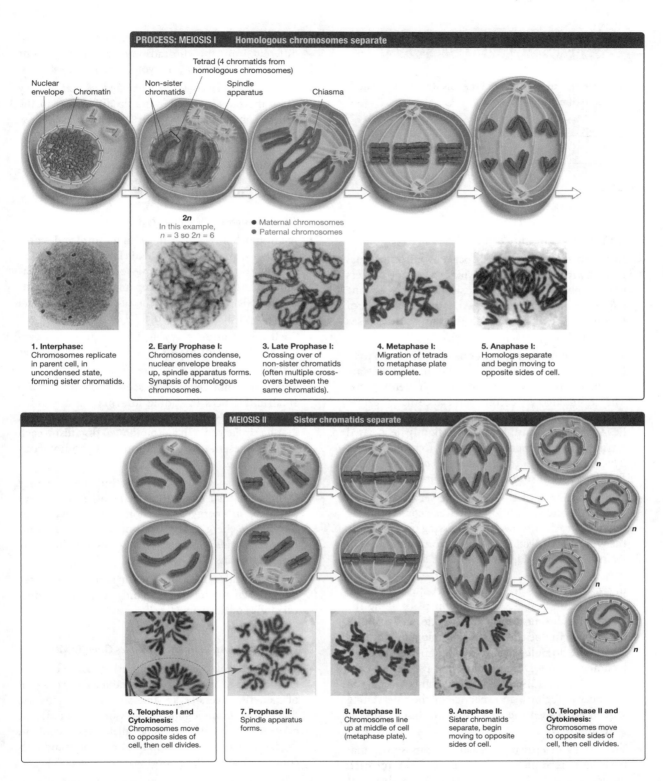

PROCESS: MEIOSIS I — Homologous chromosomes separate

Nuclear envelope · Chromatin · Non-sister chromatids · Tetrad (4 chromatids from homologous chromosomes) · Spindle apparatus · Chiasma

2n In this example, n = 3 so 2n = 6

● Maternal chromosomes
● Paternal chromosomes

1. Interphase: Chromosomes replicate in parent cell, in uncondensed state, forming sister chromatids.

2. Early Prophase I: Chromosomes condense, nuclear envelope breaks up, spindle apparatus forms. Synapsis of homologous chromosomes.

3. Late Prophase I: Crossing over of non-sister chromatids (often multiple crossovers between the same chromatids).

4. Metaphase I: Migration of tetrads to metaphase plate is complete.

5. Anaphase I: Homologs separate and begin moving to opposite sides of cell.

MEIOSIS II — Sister chromatids separate

6. Telophase I and Cytokinesis: Chromosomes move to opposite sides of cell, then cell divides.

7. Prophase II: Spindle apparatus forms.

8. Metaphase II: Chromosomes line up at middle of cell (metaphase plate).

9. Anaphase II: Sister chromatids separate, begin moving to opposite sides of cell.

10. Telophase II and Cytokinesis: Chromosomes move to opposite sides of cell, then cell divides.

composed of the two sister chromatids. These chromatids, attached to each other at the centromere, move into separate cells during anaphase II. Once the chromatids have segregated into gametes, they are called chromosomes. Because there is no crossing over in prophase II, this step is more rapid than prophase I. Anaphase II does not include the assortment of homologous chromosomes we noticed in anaphase I, only the separation of sister chromatids.

(b) Describe the two steps in meiosis where most genetic recombination occurs.

_____

_____

_____

A Closer Look at Prophase I—Crossing Over

- In prophase I, a complex process of **chromosomal recombination** is initiated. This is a unique process that is only seen during meiosis. Let's focus on one pair of chromosomes, knowing that each pair of homologues will go through the same process (see **Figure 12.9**).

- The homologous chromosomes each underwent DNA replication before meiosis began, and they are attached to their sister chromatid at the centromere. The homologous chromosomes come together (**synapsis**), recognizing each other by their matching DNA sequence and forming a tightly associated **tetrad** or **synaptonemal complex** of chromosomes. *This complex contains both homologs and their copies, for a total of four strands of chromosomal DNA.*

- The four strands of DNA associate very closely with each other, and association is between corresponding regions of each chromosome. Crossing over is simply the exchange of sections of DNA from one strand of DNA to another (**Figure 12.9**). The site of crossing over, the **chiasma**, can sometimes be visualized microscopically as an X-shaped structure along the paired chromosomes. At a random place along the length of a chromosome a cut is made, and a similar cut is made in one copy of the homologous chromosome. The homologues are then connected by reattachment of the DNA strands (red connected with blue in **Figure 12.9**). When reattachment occurs at two nearby regions, an exchange of the intervening sequence between the two cut/reattachment sites is generated.

- The results of crossing over are chromosomes that have been recombined. Regions of DNA on each homologue have been exchanged. Remember that your homologous chromosomes originated from your mother and father. When meiosis occurs in your body, this process of crossing over generated chromosomes that have some sequence from your mother, and some sequence from your father. These

are chromosomes with DNA content that is new, different from the chromosomes of either of your parents and different from your chromosomes.

- *The regions exchanged are homologous;* that is, they carry the same genes, but subtle differences in the genes contributed by each parent make these crossover products unique from the original chromosomes (see following "alleles" discussion).

12.2 The Consequences of Meiosis

Chromosomes and Heredity

- A chromosome is a long, linear **macromolecule** composed of deoxyribonucleic acid (DNA). As discussed in **Chapter 13** and **14**, chromosomes hold information in the form of distinct regions of DNA that contain a code, found within the ordering of chemical subunits of the chromosome.

- Each information-carrying DNA region that carries the blueprint for a single **protein** is known as a **gene**.

- Two genes that encode the same protein but have small differences in coding information are called **alleles** of each other. There can be hundreds of alleles of a single gene. In a diploid ($2n$) organism, each gene is present two times, on each homologous chromosome.

- Proteins carry out functions within the cell; they form the cellular infrastructure (cytoskeleton), carry out steps in energy production, break down sugar into usable energy, and carry out cellular communication. *Proteins do things*, while genes are blueprints for the proteins.

- In comparing sexual reproduction and asexual reproduction, which one produces only genetically identical **clones** of the original cell?

Independent Assortment Produces Genetic Variation

- The separation (or segregation) of homologous chromosomes in anaphase I generates a great deal of **genetic diversity** in the subsequent gametes. Remember: homologous chromosomes are different because of the unique gene alleles they contain. **Figure 12.10** uses an example of eye color and hair color to illustrate the importance of alleles.

- Diploid organisms, carrying two **homologs** of each chromosome, divide the diploid DNA content into two cells at the completion of meiosis I. This process is random, ensuring that each cell gets one replicated copy of each chromosome (a total of 23 in humans).

- In the production of gametes, the homologous chromosomes **segregate** into different cells at the completion of meiosis I. Notice that the chromosome number is now reduced to a haploid ($1n$) status.

- Because each gamete receives one of the two homologs of each chromosome, and there are 23 chromosomes in humans, we are capable of generating 8.4 million (2^{23}) genetically unique gametes. This process of distributing the homologous chromosomes in anaphase I is called **independent assortment**.

- When you add to that the additional diversity generated by crossing over in prophase I, the potential number of genetically unique gametes becomes much larger.

- **Sexual reproduction** offers incredible genetic diversity to offspring. Why is such effort expended to ensure genetic diversity from generation to generation? Doesn't this increase the likelihood of mistakes occurring during gamete formation?

12.3 Why Does Meiosis Exist?

- List the taxonomic kingdoms that display sexual reproduction, those that display asexual reproduction, and those that contain examples of both forms of reproduction.

The Paradox of Sex

- Many types of organisms thrive through the process of **asexual reproduction**. A fascinating question in biology focuses on why meiosis and sex is such a dominant reproductive strategy in multicellular organisms. Can you define the meaning of Maynard Smith's "two-fold cost of males" in a population?

- Does the answer lie in the genes? Are organisms that reproduce sexually able to produce more diversity and adaptability in their offspring?

- Maynard Smith predicted that asexually reproducing organisms should reproduce faster and outcompete similar organisms that invest in sexual reproduction. What assumptions of this model might be problematic when you consider the influence of sexual/asexual reproduction on inheritance? What environmental factors might explain the success of sexual reproduction?

- Two models that address the paradox of sex are (1) the **purifying selection hypothesis** and (2) the **changing environment hypothesis**.

- Individuals reproducing asexually will transmit deleterious mutations to all of their offspring, while individuals reproducing sexually will transmit deleterious mutations to only half of their offspring (on average). The purifying selection hypothesis predicts that over time, natural selection will reduce the number of deleterious alleles in a population that is reproducing sexually. Individuals with deleterious mutations will display diminished fitness, will be less likely to reproduce, and will be less likely to pass their genes on to the next generation.

- The changing environment hypothesis predicts that populations reproducing sexually (with genetically diverse offspring) will have an improved ability to adapt and survive when environmental conditions change (**Figure 12.13**, provided on the next page).

- Examine the examples offered in support of each of these hypotheses (parasitism and fitness, etc.). Can you list a few biological conditions in which asexual reproduction might be advantageous? List a few additional examples where sexual reproduction might be advantageous? Which type of reproduction would generate a population that adapts rapidly to a changing environment?

✓ CHECK YOUR UNDERSTANDING

(c) How might asexual reproduction offer an advantage in some environments? Describe the two steps in meiosis where most genetic recombination occurs.

_____

_____

_____

_____

(d) Why is the freshwater snail (*P. antipodarium*) a good choice for experiments comparing sexual and asexual reproduction?

_____

_____

_____

12.4 Mistakes in Meiosis

How Do Mistakes Occur?

- Meiotic mistakes are an inherent risk in the complex process of generating genetic diversity during gamete production.

- Some zygotes carry severe genetic abnormalities: abnormal numbers of chromosomes (**aneuploidy**) or

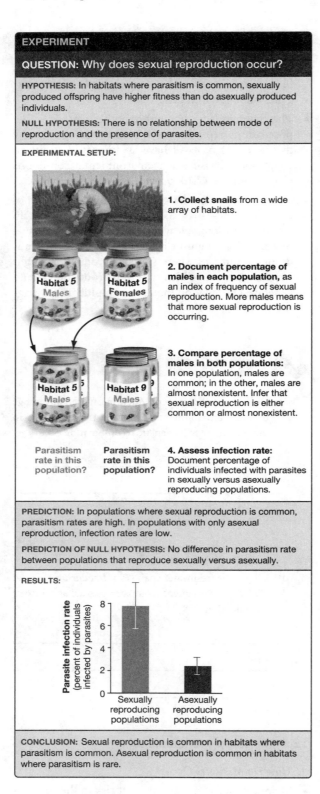

EXPERIMENT

QUESTION: Why does sexual reproduction occur?

HYPOTHESIS: In habitats where parasitism is common, sexually produced offspring have higher fitness than do asexually produced individuals.

NULL HYPOTHESIS: There is no relationship between mode of reproduction and the presence of parasites.

EXPERIMENTAL SETUP:

1. **Collect snails** from a wide array of habitats.

Habitat 5 Males   **Habitat 5 Females**

2. **Document percentage of males in each population,** as an index of frequency of sexual reproduction. More males means that more sexual reproduction is occurring.

Habitat 5 Males   **Habitat 9 Males**

3. **Compare percentage of males in both populations:** In one population, males are common; in the other, males are almost nonexistent. Infer that sexual reproduction is either common or almost nonexistent.

Parasitism rate in this population?   Parasitism rate in this population?   4. **Assess infection rate:** Document percentage of individuals infected with parasites in sexually versus asexually reproducing populations.

PREDICTION: In populations where sexual reproduction is common, parasitism rates are high. In populations with only asexual reproduction, infection rates are low.

PREDICTION OF NULL HYPOTHESIS: No difference in parasitism rate between populations that reproduce sexually versus asexually.

RESULTS:

Parasite infection rate (percent of individuals infected by parasites) — Sexually reproducing populations (~7.7); Asexually reproducing populations (~2.3)

CONCLUSION: Sexual reproduction is common in habitats where parasitism is common. Asexual reproduction is common in habitats where parasitism is rare.

fragmented chromosomes that were damaged at some point during meiosis. These zygotes typically do not survive to produce viable offspring.

- Down syndrome is a relatively common birth defect in humans (0.15 percent of live births), and is caused by an extra copy of chromosome 21, a trisomy-21 karyotype.

- Aneuploidy is caused by a mistake in meiosis, specifically a nondisjunction event (**Figure 12.14**). In nondisjunction, chromosomes that should segregate into separate cells are improperly segregated into the same cell. This generates one gamete that is lacking a chromosome and one gamete that has an extra chromosome. If either of these gamete types is involved in forming an embryo, the embryo will have an abnormal chromosome number (aneuploidy).

✔ **CHECK YOUR UNDERSTANDING**

(e) Explain nondisjunction as it relates to meiosis; where does this problem predominantly arise?

_____

_____

_____

_____

B. MASTERING CHALLENGING TOPICS

Mitosis versus Meiosis

Mitosis and meiosis share many characteristics—chromosome condensation, alignment of chromosomes at the cellular midline, migration of chromosomes to the poles, and formation of two new cells. What characteristics are observed only in mitosis or only in meiosis? How is meiosis I different from meiosis II?

Carefully examine these processes, focusing on the ploidy (or DNA content) at each stage and on how segregation at anaphase occurs. Some fundamental differences between these processes illustrate the purpose of each process. Mitosis occurs in all of our somatic cells, and it requires the careful transfer of all genetic information from mother cell to the daughter cells. Meiosis expends energy and engages in the dangerous process of rearranging our chromosomes to generate genetic diversity.

Genetic Recombination

Crossing over is a complex process that is often confusing at first. This is especially true when you consider the impact of crossing over on the independent assortment of genes. Diagram the formation of a tetrad and a crossover event between two homologs (use color to make the crossover more visible). Resolve the crossover by illustrating what would then happen to these homologs in anaphase I; notice how all four chromosomes are now unique, each carrying a slightly different DNA content.

Terminology

This chapter introduces many new terms and concepts that are crucial to your understanding of this, and future, chapters. Take time to learn these concepts, and you'll be prepared to move on to more complex concepts in genetics. Here are some terms to carefully examine and define as you study:

aneuploidy	chiasma	gene
diploid	gamete	life cycle
haploid	homologous	outcrossing
chromosomes	nondisjunction	zygote
karyotype	trisomy	
self-fertilization	sexual	
synapsed	reproduction	
chromosomes	chromatids	

C. ASSESSING WHAT YOU'VE LEARNED

1. Which of the following occurs during meiosis II, but does not occur as part of meiosis I?
   a. Crossing over between homologous chromosomes
   b. Separation of homologous chromosomes
   c. Separation of sister chromatids
   d. Migration to middle of cell

2. Crossing over:
   a. occurs between sister chromatids on the same chromosome.
   b. occurs between chromatids of homologous chromosomes.
   c. occurs very rarely.
   d. occurs during prophase II of the cell cycle.

3. An individual plant that exhibits self-fertilization can produce offspring different from itself primarily as the result of:
   a. mutations that occur during mitosis.
   b. independent assortment and crossing over during the formation of gametes.

   c. crossing over (only) during the formation of chromosomes.
   d. the fact that the number of chromosomes in pollen is always different from the number of chromosomes in egg cells.

4. **Figure 12.12** in your textbook demonstrates which of the following?
   a. Organisms produced by asexual reproduction are more variable than those produced by sexual reproduction.
   b. Sexual reproduction is always more successful than asexual reproduction.
   c. Asexual reproduction is always more successful than sexual reproduction.
   d. Asexual organisms generally produce more offspring than sexually reproducing organisms.

5. Robert inherited alleles A and B from his father, and alleles a and b from his mother (both genes are on one chromosome). If Robert marries a woman who has only a and b alleles on the corresponding chromosomes, how is it possible that their child has a and b alleles on one chromosome and A and b alleles on the other?
   a. Independent assortment
   b. Crossing over of Robert's wife's chromosomes
   c. Crossing over of Robert's chromosomes
   d. Crossing over of Robert's father's chromosomes

6. A **proximate explanation** of a biological mechanism would describe:
   a. why something happens.
   b. how something happens.

7. Curtis Lively generated evidence in support of the "changing-environment hypothesis" when he found:
   a. sexually reproducing *Daphnia* populations with high numbers of deleterious mutations.
   b. sexually reproducing *Daphnia* populations with high numbers of deleterious mutations.
   c. asexually reproducing populations of snails to be less susceptible to parasitism.
   d. sexually reproducing populations of snails to be less susceptible to parasitism.

8. Bacterial conjugation mediates a form of genetic recombination in these prokaryotic cells. Conjugation offers recombination via:
   a. the bidirectional exchange of chromosomal DNA.
   b. the transfer of portions of chromosomal DNA from a donor bacterium to a recipient bacterium.
   c. the transfer of an intact chromosome from one bacterium to another.
   d. a process almost identical to meiosis in sexually reproducing eukaryotic organisms.

9. Diagram meiotic crossing over between two adjacent genes on the same chromosome (*A* and *B*, shown in Chart 1). The parental cell entering into meiosis is genotypically *aB/Ab*, and will generate four haploid gametes at the completion of meiosis. Diagram the genetic makeup of the gametes produced by this meiotic process, and compare these gametes to the original parental contributions.

Chart 1

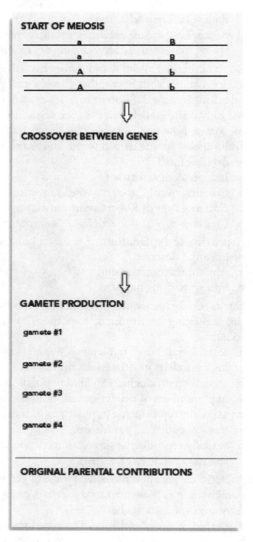

CHAPTER 12—ANSWER KEY

Check Your Understanding

(a) You can determine ploidy by counting the centromeres; a diploid cell with replicated chromosomes is still diploid, although it has doubled its DNA content.

   Mitosis—Before entering into mitosis, the 2*n* cell first replicates its genome. At anaphase, *sister chromatids* are separated and migrate to opposite poles of the cell. The cell then divides into two 2*n* daughter cells.

   Meiosis—Before entering into meiosis, the 2*n* cell first replicates its genome. At anaphase I, *homologous chromosomes* are separated and migrate to opposite poles of the cell. The cell then divides into two 1*n* cells (sister chromatids are still attached). At anaphase II, *sister chromatids* are separated and migrate to opposite poles of each cell. Then the cell divides into two 1*n* gametes.

(b) **Prophase I**—Homologous chromosomes (replicated) synapse into a tetrad and exchange homologous regions of DNA. This crossing over between chromosomes happens at least once for each tetrad.

   Anaphase I—The tetrad separates and homologous chromosomes migrate to opposite poles of the cell. Each centromere remains intact, with two sister chromatids attached, until anaphase II. The various combinations of homologs generated at the end of meiosis I offer dramatic diversity to the subsequent gamete.

(c) The most common site of nondisjunction is meiosis I, during which homologous chromosomes do not properly migrate to opposite poles of the cell. This can occur during meiosis II, but it is less common (see **Figure 12.14**).

(d) Reproduction in the absence of sex is faster, requires less energy, and does not require the fusion of germ cells from two sources. Asexually reproducing organisms can reproduce much more rapidly than sexually reproducing organisms. But, researchers have shown that asexual reproduction offers less adaptability (genetic variability) to the organism.

(e) This snail can reproduce both sexually and asexually. Populations of snails with high numbers of males tend to be reproducing sexually, while asexually reproducing populations have more female snails. The researcher (Curtis Lively) could thus analyze the ratio of males to females in populations of snails that are resistant or sensitive to disease (parasitism).

Assessing What You've Learned

1. c; 2. b; 3. b; 4. d; 5. c; 6. b; 7. d; 8. b;
9. Once the tetrad is formed, strands are exchanged between the homologous chromosomes. If this occurs between the two genes as diagrammed below, then half of the resulting gametes will have a unique combination of alleles relative to each parent.

Chart 1

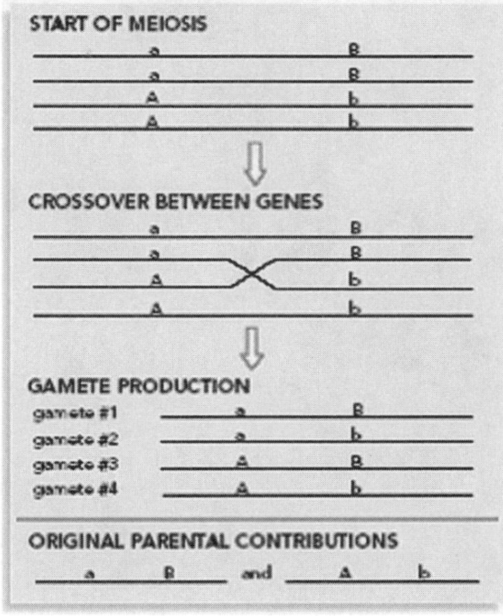

START OF MEIOSIS

CROSSOVER BETWEEN GENES

GAMETE PRODUCTION
gamete #1   a   B
gamete #2   a   b
gamete #3   A   B
gamete #4   A   b

ORIGINAL PARENTAL CONTRIBUTIONS
a   B   and   A   b

Looking Forward—Key Concepts in Later Chapters

Mendel and the Gene—Chapter 13

Mendel worked out much of the meiotic process with no knowledge of what a chromosome really was. The **principles of segregation and independent assortment** both describe what happens to chromosomes in meiosis.

DNA and the Gene—Chapter 14

We'll learn about DNA replication in **Chapter 14**, identifying how signals in the cell activate the deliberate process of copying the **genome** before division can occur.

Principles of Development—Chapter 21

Meiosis is a key process in **reproduction** and **fertilization**. In **Chapter 21**, we examine **gametogenesis** in more detail and follow the gamete through fertilization.

Chapter 13

Mendel and the Gene

Looking Back—Key Concepts from Earlier Chapters

Inside the Cell—Chapter 7

As you explore the work of Mendel, review the cell biology behind **phenotype** ("show-type" or form). For example, what aspects of cell biology might result in the loss of specific coloration in a flower? The genes that encode pigment can be **mutated**, and their loss or defect is a very obvious change in flower coloration (abnormal phenotype).

Meiosis—Chapter 12

Without knowledge of the **meiotic** process, it is difficult to follow the key conclusions of Mendel's work. Review meiosis, examining how gene segregation, independent assortment, and gene linkage fit into the production of gametes. All of Mendel's work was really a characterization of chromosomal behavior in meiosis, though meiosis was not discovered until many years later.

KEY CONCEPTS

- The **principle of segregation** states that during the production of gametes, pairs of homologous chromosomes separate in the production of gametes. This was observed in Mendel's monohybrid crosses with pea plants.

- The **principle of independent assortment** states that homologous chromosomes assort independently of one another in the production of gametes. This was observed in Mendel's dihybrid crosses with pea plants.

- The **chromosome theory** was proposed in response to observed similarities in chromosome behavior during meiosis and the behavior of genetic determinants (genes) described by Mendel. This theory predicts chromosomes to be composed of information-bearing material and to be responsible for inheritance.

- **Thomas Morgan** used fruit flies (*Drosophila melanogaster*) to study inheritance. His work supported the chromosome theory.

- Genes located on the same chromosome are **linked**; they do not assort independently. Maps of chromosomes can be developed by studying the degree of linkage between genes on a specific chromosome.

- There are exceptions to Mendel's observed patterns of inheritance, including the **codominance** of some alleles, **incomplete dominance** of some alleles, and traits that are influenced by multiple genes (**polygenic traits**).

A. CHAPTER OUTLINE

13.1 Mendel's Experimental System

- **Gregor Mendel** was an Augustinian priest and a scientist, studying inheritance using a simple plant found in his garden, a pea plant (*Pisum sativum*). He lived from 1822 to 1884.

- Mendel worked to describe **heredity**, the process in which **traits** (or characteristics) are passed from generation to generation.

- Mendel was an active member of an Agricultural Society in Brno, a town in what is now the Czech Republic. A key interest for this group was the process of selective breeding to improve traits in crops and farm animals.

What Questions Was Mendel Trying to Answer?

- **Blending versus acquired characters**: These competing hypotheses of Mendel's time were both severely damaged by data produced from his simple experimentation with pea plants. Review these hypotheses. As we examine Mendel's work, list the specific findings that contradicted these popular views of inheritance.

- **Incomplete dominance** is a form of blending, with the offspring displaying a trait that is intermediate of the parent's traits. For instance, a red flowered plant crossed by a white flowered plant might generate plants with pink flowers. This occurs with alleles of some genes, but did not occur in Mendel's analyses with peas. We'll talk more about incomplete dominance later in the chapter.

Garden Peas Serve as the First Model Organism in Genetics

- Consider all of the characteristics you'd want in a **model organism**. If you were setting up the experiments, what category of organism would you select— a plant, insect, worm, fungus, mammal, or primate? Your research focus would likely determine which organism would be the most appropriate. How would you select from among all the different possible representatives of each group?

- Mendel chose a common garden pea (*Pisum sativum*) and revolutionized our knowledge of genetics for all eukaryotic organisms. He used the garden pea because:

  1. Pea plants are easy to grow.
  2. Growing pea plants is inexpensive; there is little requirement for specialized equipment.
  3. Matings can be controlled completely (**Figure 13.1**).
  4. Pea plants display easily observed traits (phenotypes).
  5. Pea plants display a rapid life cycle (seedling to reproductively mature).

✔ **CHECK YOUR UNDERSTANDING**

(a) Mendel's choice of the pea plant (*Pisum sativum*) as his experimental organism was critical to his success. Describe how things might have turned out differently if he had chosen laboratory mice.

_____

_____

_____

_____

Here are some critical terms and concepts from Chapter 13:

- **Allele**—A single gene may have multiple alleles. Each allele is different from all other alleles, with some difference(s) in information content. Differences between alleles start at the DNA sequence. Any difference from the typical or **wild-type** DNA sequence for that gene identifies a unique allele of that gene.

- **Dominant/recessive traits**—Most typical or wild-type traits are dominant; mutant traits are often recessive. In other words, the mutation is hidden in organisms that carry both a wild-type and a mutant allele (different types of the same gene). Some mutations are dominant to the wild-type allele, but these are unusual.

- **F_1 generation**—The progeny of the parental organisms are called the "first filial" or F_1 generation. These progeny are genetically composed of one copy of each chromosome from the mother and one copy of each chromosome from the father. F_1 plants might also be referred to as **hybrids** of the two parental lines.

- **Genotype**—The collection of alleles comprising the genome (total DNA content) of an organism.

- **Heterozygous**—In somatic cells, individuals carrying two different alleles of the same gene are heterozygous for that gene.

- **Homozygous**—In somatic cells, individuals carrying two copies of the same allele are said to be homozygous for that gene.

- **Parental generation**—In a standard mating, these would be the original parental organisms that donate gametes. In Mendel's experiments, parental plants were **true-breeding** or **pure lines**, indicating several generations of self-fertilization. This increased the likelihood of homozygosity of alleles for each gene in each parental plant. When true-breeding plants are self-fertilized, they generate offspring that are phenotypically identical to themselves.

- **Phenotype**—The phenotype of an organism is its specific collection of detectable traits. Purple flowers, green eyes, and premature balding are all phenotypes.

- **Reciprocal crosses**—This refers to two separate matings (crosses) performed with reversed male and female contributions. For example, pollen taken from a purple plant fertilizing an oocyte from a white plant (cross 1); and, in a second cross, pollen from a white plant fertilizing an oocyte from a purple plant. These crosses are valuable in identifying genes that are located on the sex-determining chromosomes (see the examples offered in **Figure 13.3** and **Figure 13.11**).

13.2 Mendel's Experiments with a Single Trait

- Gregor Mendel performed many simple crosses, focusing on *one trait* (such as flower color in experiments known as **monohybrid crosses** [**Figure 13.4**]). Consider, for example, when a wild-type, purple-flowered plant (genotype *PP*) is crossed by a mutant, white-flowered plant (genotype *pp*). If the F_1 progeny (genotype *Pp*) are all purple, what does that indicate about the relationship between the two alleles?

- You will remember from **Chapter 12** that the product of gamete fusion is **diploid**. Thus, the F_1 plant is heterozygous for the flower color gene (*Pp*), and since the F_1 progeny are all purple, the allele generating purple flowers (*P*) is **dominant** to the allele generating white flowers (*p*). The dominant allele is usually indicated using capital letters, while the **recessive** is indicated using lowercase letters. *Mendel did not observe blending!*

- Mendel crossed the purple-flowered F_1 progeny (or allowed **self-fertilization**), resulting in an F_2 generation. These plants displayed a **3:1 ratio** of purple- to white-flowered plants (three-fourths of all progeny were purple, and one-fourth of all progeny were white). The recessive phenotype reappeared in the next generation, and thus must have been "hidden" in the F_1 plant.

- The segregation of alleles in gamete production is revealed in the ratios of F_2 progeny phenotypes. Review meiosis (**Chapter 12**) and the segregation of chromosomes in this process. Focus on one pair of homologous chromosomes in a diploid organism. One chromosome carries the "P" allele of a specific gene and the other carries the recessive "p" allele of that same gene. These alleles will segregate in meiosis, with half of the gametes receiving each allele (50 percent P, 50 percent p). The **Punnett square** is a visual way to display this segregation in a specific cross. Simply draw a box, and along the left side and the top side, list the haploid genotype of gametes produced by each parent (in this case, P and p along each line). Inside the box combine the gamete contributions from left and top to predict diploid progeny. In a monohybrid cross, you will generate four diploid genotypes inside the box (see **Figure 13.4**).

- In the case of a monohybrid cross, there are four combinations that will appear inside the box (each represent 25 percent of the progeny). However, three-fourths of these genotypes (75 percent) will be either *PP* or *Pp* and will display the purple phenotype, while one-fourth (25 percent) will be *pp* and will display the white phenotype.

- Mendel described gene segregation in meiosis long before scientists even began to look at chromosomes in the production of sperm or oocytes. Mendel developed a model in which a plant carrying a genotype of *Pp* generates 50 percent *P* gametes and 50 percent *p* gametes. Thus, the parental organism carries two determinants for each trait, and these segregate separately in the generation of gametes. This is referred to as Mendel's **principle of segregation**.

- Using the preceding example, what is the probability of producing a white-flowered (*pp*) plant if a heterozygous (*Pp*) plant is allowed to self-fertilize?

- Mendel knew that in order to make this probability calculation, he needed to include the individual probabilities for the contribution of the recessive allele in each gamete.

- The probability for any individual gamete to carry the recessive *p* allele is .50 (50 percent). To generate the white-flowered plant, a *p* oocyte must be fertilized by a *p* pollen grain. These two independent events must occur simultaneously to generate a homozygous recessive plant; thus, their probabilities are multiplied together.

- Thus, 0.25 or 25 percent of the progeny from a self-fertilized heterozygous plant will be white-flowered (*pp*).

- What is the probability of the same heterozygous F_1 plant generating a genotypically homozygous F_2 plant with respect to the *P* gene (i.e., either *PP* or *pp*)?

- This type of calculation, called an **either-or** situation, requires one additional step. You first use the product rule to determine the individual probabilities for the generation of a *PP* plant (.25) or a *pp* plant (.25).

- These independent probabilities are then added together to calculate the probability of generating either one or the other in any given fertilization event. Thus, .25 + .25 = .50, or 50 percent. Mendel used simple calculations of this type to make testable predictions about inheritance, and to analyze his data with statistical tests.

- **Table 13.2** summarizes fundamental observations of inheritance that are illustrated in simple monohybrid crosses. Diagram a simple monohybrid cross and identify where each of these five observations is supported by these simple data. The **principle of segregation** is a fundamental tenet of genetics, meiosis, and the study of inheritance.

13.3 Mendel's Experiments with Two Traits

The Dihybrid Cross

- Mendel did additional studies in which alleles of two different genes were studied at the same time. These analyses are call **dihybrid crosses**.

- Consider the miotic process in a diploid cell with four chromosomes ($2n = 4$). The cell contains two homologues of each of two chromosomes. Annotate those chromosomes and view their migration into a gamete during the steps of meiosis. Notice that the Punnett square of a dihybrid cross has 16 boxes; this is because there are four types of gametes that will be generated by each parent (4 × 4 = 16).

- One example given in the text (**Figure 13.5**, provided here) includes the cross of a true-breeding plant that displays round (*R*), yellow (*Y*) peas by a true-breeding plant that displays wrinkled (*r*), green peas (*y*). Thus, the parental cross is *RRYY* × *rryy*, and the F_1 progeny would be *RrYy*. The F_1 plants display the dominant phenotypes (round, yellow), but the F_2 plants display four combinations of phenotypes. Mendel found traits in these dihybrid crosses to segregate in a 9:3:3:1 pattern. Examine the Punnett square in the figure to confirm that this ratio makes sense.

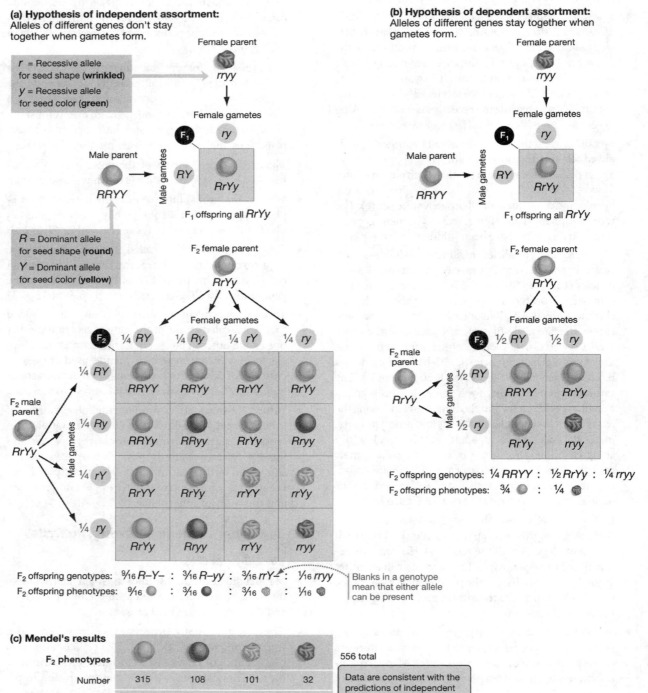

(a) Hypothesis of independent assortment:
Alleles of different genes don't stay together when gametes form.

r = Recessive allele for seed shape (**wrinkled**)
y = Recessive allele for seed color (**green**)

Female parent
rryy

Female gametes
F_1 *ry*

Male parent
RRYY

Male gametes
RY   *RrYy*

F_1 offspring all *RrYy*

R = Dominant allele for seed shape (**round**)
Y = Dominant allele for seed color (**yellow**)

F_2 female parent
RrYy

Female gametes
F_2 ¼ *RY*   ¼ *Ry*   ¼ *rY*   ¼ *ry*

F_2 male parent
RrYy

Male gametes
¼ *RY*   *RRYY*   *RRYy*   *RrYY*   *RrYy*
¼ *Ry*   *RRYy*   *RRyy*   *RrYy*   *Rryy*
¼ *rY*   *RrYY*   *RrYy*   *rrYY*   *rrYy*
¼ *ry*   *RrYy*   *Rryy*   *rrYy*   *rryy*

F_2 offspring genotypes: ⁹⁄₁₆ *R–Y–* : ³⁄₁₆ *R–yy* : ³⁄₁₆ *rrY–* : ¹⁄₁₆ *rryy*
F_2 offspring phenotypes: ⁹⁄₁₆ ● : ³⁄₁₆ ● : ³⁄₁₆ ◉ : ¹⁄₁₆ ◉

Blanks in a genotype mean that either allele can be present

(b) Hypothesis of dependent assortment:
Alleles of different genes stay together when gametes form.

Female parent
rryy

Female gametes
F_1 *ry*

Male parent
RRYY

Male gametes
RY   *RrYy*

F_1 offspring all *RrYy*

F_2 female parent
RrYy

Female gametes
F_2 ½ *RY*   ½ *ry*

F_2 male parent
RrYy

Male gametes
½ *RY*   *RRYY*   *RrYy*
½ *ry*   *RrYy*   *rryy*

F_2 offspring genotypes: ¼ *RRYY* : ½ *RrYy* : ¼ *rryy*
F_2 offspring phenotypes: ¾ ● : ¼ ◉

(c) Mendel's results

F_2 phenotypes	●	●	◉	◉	556 total
Number	315	108	101	32	
Fraction of offspring	⁹⁄₁₆	³⁄₁₆	³⁄₁₆	¹⁄₁₆	

Data are consistent with the predictions of independent assortment.

- By careful analysis of his results, Mendel was able to establish the **principle of independent assortment**. *This principle states that genes segregate independently of each other in the production of gametes* (**Figure 13.5a**). This is true for most genes; but some genes located on the same chromosome do not assort independently, and are described as linked (see **Section 13.4**).

- A simple way to "see" independent assortment is to diagram a simple meiotic division. Without knowing anything about DNA or chromosomes, Mendel was describing chromosomal assortment in meiosis.

  You can predict the genotypes and phenotypes of progeny from specific dihybrid crosses by using the product and sum rules.

CHECK YOUR UNDERSTANDING

(b) Using pea plants, you are analyzing two genes (*A* and *B*) using simple dihybrid crosses. You generate F_1 progeny that are heterozygous at both loci (*AaBb*), and interbreed these plants to generate F_2 progeny. What percentage of the F_2 progeny will carry the genotype *AABB* or *aabb*? Try to answer this question using the Punnett square, then use probability calculations to determine the answer. Do they agree with each other?

_____

_____

_____

_____

Using a Testcross to Confirm Predictions

- The **testcross** is a simple analysis used in controlled mating experiments. It allows the investigator to focus on the meiotic events in only one of the two parents.

- A testcross is accomplished by crossing the organism of interest with a true-breeding organism that typically contains only recessive alleles for the genes being studied. Thus, for the genes being studied, the true-breeding organism will produce only one type of gamete.

- Review the example given in the text (an *RrYy* plant crossed by an *rryy* plant; **Figure 13.6**). This cross isolates the meiosis of the *RrYy* plant, because only *ry* gametes will be produced by the testcross plant. Thus, the progeny genotype/phenotype reveals what happened during meiosis of the *RrYy* plant. This powerful strategy has many applications in gene analysis.

CHECK YOUR UNDERSTANDING

(c) Explain the value of a testcross. What advantage does it have for a geneticist? Compare a standard dihybrid cross of *AaBb* by *AaBb* to a testcross of *AaBb* by *aabb*.

_____

_____

_____

_____

13.4 The Chromosome Theory

- The **chromosome theory of inheritance** was born out of careful observations of meiosis and the movement of individual chromosomes during the stages of meiosis. In the early 1900s, Walter Sutton and Theodore Boveri realized that the reduction in chromosome number during **gametogenesis** (gamete formation) was consistent with Mendel's principle of segregation. They also realized that the movement of homologous chromosomes was consistent with Mendel's principle of independent assortment (**Figure 13.8**), and that the chromosomes likely contained heritable units of information that were passed from generation to generation.

- The scientific community did not immediately embrace the chromosome theory of inheritance; rather, scientists began to test this theory by searching for the information-bearing (heritable) material found within cells. By 1950, it was clear that in most cells the stable heritable material is made up of deoxyribonucleic acid (DNA). Chromosomes are primarily composed of DNA, consistent with the chromosome theory.

- **Sex chromosomes:** Nettie Stevens (Bryn Mawr College) examined the chromosomes in a beetle (*Tenebrio molitor*, $2n = 20$ chromosomes). She found that during meiosis, male beetles had an unusual pair of "homologous" chromosomes that paired in prophase I, but were dramatically different in size. The small chromosome in the pair was ultimately called the "Y" chromosome. Male beetles displayed a unique heteromorphic pair of chromosomes (XY), while females contained two copies of the larger chromosome (XX).

- It was observed that XY males generate X gametes and Y gametes, and that the composition of the sperm regarding X or Y content determined the sex of any subsequent fertilized embryo.

- Early support for the chromosome theory came from studies with the fruit fly, *Drosophila melanogaster*. These flies are simple to culture, have a rapid reproductive cycle, and display a number of easily observed traits (such as eye color). Recall our discussion of model organisms in Section 13.1.

- In 1904, **Thomas Hunt Morgan**, a biologist at Columbia University, first began to work with fruit flies to study inheritance. Morgan found inheritance patterns similar to those described by Mendel.

- Morgan's work supported the chromosome theory of inheritance, especially when you consider the discovery of **sex-linked inheritance**. The "w" mutation generates white-eyed flies (wild type is red eyes—"w^+"). Morgan noted that when a white-eyed female was crossed with a red-eyed male (both true-breeding), the male F_1 progeny displayed the white-eyed trait while the females were red-eyed. Morgan predicted that the W gene was on the X chromosome.

- Examine the reciprocal crosses shown in **Figure 13.11**. Notice that the F_1 and F_2 generations are different depending on which parental fly displays the white-eyed trait. *Note:* Whenever the results of reciprocal crosses generate different ratios of males/females that display the trait, the gene is located on a sex-determining chromosome (the X or Y).

- In pedigree analysis, whenever a phenotype within a pedigree is predominantly associated with males or females, one should suspect that the gene is located on a sex-determining chromosome (the X or Y).

- Review Morgan's data using a **Punnett square**; remember that if the gene is located on the X chromosome, the sex of parental cells used can dramatically influence the phenotypical ratios of the progeny.

✔ **CHECK YOUR UNDERSTANDING**

(d) A practice problem: In mice, gene "G" is located on the X chromosome. A recessive allele of this gene ("g") causes unusually small ears. The male genotype of a normal mouse may be represented as X^GY. In gamete production G will segregate with the X chromosome and will be present in 50 percent of the gametes (carrying the X), while the other 50 percent of the gametes receive the Y chromosome. Using a piece of scratch paper, diagram Punnett squares to analyze reciprocal monohybrid crosses for mice displaying the small-ear phenotype. Cross 1 is a small-eared male crossed by a wild-type female. Cross

2 is a small-eared female crossed by a wild-type male. Calculate the percentage of small-eared F_1 progeny for each cross. In each cross, do males and females display the trait at equal frequency? If you generated F_2 progeny for each cross, what percentage of males would display the trait?

_____

_____

_____

_____

13.5 Extending Mendel's Rules

What Happens When Genes Are Located on the Same Chromosome?

- When examining dihybrid crosses, all examples of independent assortment we've discussed to this point depend on genes being located on separate chromosomes. What happens when two genes are located near each other, on the same chromosome?

- First, their assortments are not independent of each other; they are physically attached (or **linked**) to each other on that chromosome. Linkage can be identified by segregation patterns that do not show independent assortment of the two genes (something other than a 9:3:3:1 for a dihybrid cross).

- This is a difficult concept to visualize; review the meiotic process again, taking into account two genes located on the same chromosome. The alleles of each of those two genes are physically attached on their respective homologous chromosomes and will segregate into the same gamete *every time*, unless a crossover event separates them onto homologous chromosomes. But any gametes that require a crossover event to be produced will be uncommon.

- **Figure 13.12** illustrates gene linkage using an example from fruit flies. Genes/alleles included in the example are *W*—red eyes, *w*—white eyes, *Y*—gray body, and *y*—yellow body. The parental fly in this example generates *Wy* gametes and *wY* gametes. *Alleles of these two genes are linked together.* The only way to generate a *WY* gamete or a *wy* gamete is for crossing over to occur during meiosis.

- If the female fly defined in Figure 13.12 were crossed by a male carrying only the recessive alleles (*wy*) on his X chromosome, what two categories of progeny will be observed? How would this result be different if the W and Y genes were on separate chromosomes and were assorting independently?

- Reexamine the dihybrid cross illustrated in **Figure 13.5**. If the two genes being studied (R and Y) had been located on the same chromosome, would the F_2 generation display a 9:3:3:1 segregation pattern? The *RY* and *ry* gametes will dominate each axis of the Punnett square as they are linked together. Only crossing over between these genes will generate *Ry* and *rY* gametes (**Figure 13.14**).

- Linked genes have also been shown to be valuable in generating genetic maps of chromosomes. The key to this strategy was the recognition that the frequency of a crossover event between two linked genes is dependent upon the distance between those two genes. The closer the genes are, the less frequent crossover events will be. Thomas Morgan and his research students were the first to develop this strategy, developing genetic maps of *Drosophila* chromosomes (see **Box 13.1**).

Do Heterozygotes Always Have a Dominant or Recessive Phenotype?

- **Mutant alleles** of a single gene are not always clearly dominant or recessive to the **wild-type alleles**. Sometimes the heterozygous organisms display an intermediate phenotype, between the wild-type and mutant phenotypes; this is described as **incomplete dominance**.

- A nice example of incomplete dominance is flower color (four o'clock). When breeding purple by white, the F_1 plants will have lavender flowers, an intermediate trait. Why? Remember that genes encode proteins that perform a cellular function. In this case, the gene involved in flower color encodes an enzyme that synthesizes flower pigment. Plants that lack this enzyme, "*rr*" genotype, make no pigment. Plants that are heterozygous at this gene, "*Rr*" genotype, have only one functional copy of the gene and make half the normal amount of enzyme. Thus, less pigment is made by less available enzyme, and the flowers display an intermediate phenotype—lavender. In many allelic relationships, a functional allele can completely compensate for a nonfunctional allele, thus generating the familiar **dominant/recessive allele** terminology.

- A heterozygous organism, displaying the phenotype of both alleles of a single gene, is said to display **codominance**. Neither allele is dominant or recessive to the other. Examples of codominance in humans include ABO blood typing (A and B are alleles of a single gene) and the L gene that encodes a membrane glycoprotein.

How Many Alleles and Phenotypes Exist?

- There can be many **alleles** of a single gene, since minor differences in gene sequence are common. Some genes in the human genome have many alleles,

CHECK YOUR UNDERSTANDING

(e) Allelic codominance is a genetic relationship that Mendel did not observe, but it was identified later as an exception and extension to common patterns of inheritance. Give an example of allelic codominance.

_____

_____

_____

_____

an example being the -globin gene. This gene encodes a component of hemoglobin (carries oxygen in red blood cells). Some of these allelic differences in gene sequence will affect the function of the encoded protein, generating a **trait** or **phenotype** specific for that allele. A polymorphic trait is any phenotype with more than two distinct forms.

Pleiotropy

- A pleiotropic gene influences many traits, rather than just a single trait. An example is the gene responsible for Marfan syndrome.

Genes Are Affected by the Physical Environment and Genetic Environment

- Phenotypes produced by most genes and alleles are strongly affected by the individual's physical environment. Thus, an animal's phenotype is a combination of its genotype and its physical environment.

Interactions with Other Genes and the Environment Effects Phenotypes

- Alleles of different genes affect each other in a way that can influence a single observable phenotype. **Figure 13.19** illustrates this using comb shapes in chickens as the example. This is essentially a dihybrid cross with a 9:3:3:1 pattern in the F_2 generation. However, in this example, a single trait is impacted by alleles of two different genes—resulting in four types of combs, depending upon genotype.

- Another excellent example includes the genes for coat color in mice. The *A* (agouti) gene controls the production of a yellow stripe down the length of each hair. The *B* gene controls the production of black pigment in hair. Both genes impact "coat color" in mice. An *A-B-* mouse is agouti, an *A-bb* mouse is cinnamon, an *aaB-* mouse is black, and an *aabb* mouse is brown.

- However, the recessive allele of a third gene (*C*) can completely turn off pigment synthesis. So a *BBcc* animal is white (as are *bbcc* and *Bbcc*), illustrating that *cc* is **epistatic** to *allele B* and eliminating its impact on phenotype.

✔ CHECK YOUR UNDERSTANDING

(f) A true-breeding agouti mouse is bred by a true-breeding brown mouse. All of the F_1 pups are agouti. If male F_1 mice are bred with female F_1 mice, what coat colors will be displayed within the F_2 generation? In what proportions (frequencies) will the coat colors appear in the F_2 generation?

_____

_____

_____

_____

What about Traits Such as Human Height and Intelligence?

- Some traits are influenced by the contribution of multiple genes and environment, and display wide variability (e.g., skin color, height, weight). These **quantitative traits** are observed as varying by degree. Rather than being either tall or short, heights vary across a range with the intermediate values being most common within a population. This distribution of values (illustrated in **Figure 13.20**) is called a **normal distribution**.

- In 1909, Herman Nilsson-Ehle (Lund University, Sweden) used wheat to show that quantitative traits are generated by the complex interactions of many genes. Alleles of each gene contribute to the trait in a small way (**Figure 13.21**). This type of inheritance is sometimes referred to as **polygenic inheritance**.

13.6 Applying Mendel's Rules to Humans

- In analyzing human inheritance, **pedigrees** are often a valuable tool in determining the mode of inheritance for disease-causing gene alleles.

- A **dominant mutation** will be passed from affected individuals to their offspring, which will also have the mutation.

- **Recessive mutations** display a more complex pattern of inheritance: affected individuals can be born of phenotypically normal parents who each carry a copy of the disease allele. If two unaffected parents have a child that is affected with a recessive trait, the parents must each be carriers of the recessive allele.

- If two parents are heterozygous for alleles of a gene (*Aa*), what is the likelihood of their having an affected (*aa*) child? Calculate the probability for each allele from the mother and the father. The probability for an "a" gamete is one-half from both the mother and the father. The probability of an "aa" embryo is the product of the individual parental probabilities: $(\frac{1}{2})(\frac{1}{2}) = \frac{1}{4}$.

- If a recessive mutation is found on the X chromosome, a pedigree will show that most of the affected individuals are male (e.g., hemophilia). This is because males that inherit an X chromosome with that mutation will display the trait. They have only a single X, and thus the recessive gene will generate the phenotype (**Figure 13.24**).

B. MASTERING CHALLENGING TOPICS

Linkage Analysis

A prominent experimental goal in genetic analysis is to determine the location of a specific gene—first identifying the chromosome and then the region of that chromosome where the gene is located. Disease-causing genes can sometimes be "mapped" to a specific place by using the medical records of families that carry the disease-causing allele. Finding the gene that causes a specific disease is a difficult task, but the rewards of finding the gene make the incredible effort worthwhile. There are many success stories of scientists identifying disease-causing genes (Huntington's disease, breast cancer). But how does this process work? Scientists use various strategies, one of which is fundamentally described in this chapter—linkage analysis (**Box 13.1**).

Determining genetic linkage involves looking for two genes (or markers) in the genome that do not assort independently of each other. The alleles being studied are often found to segregate together from the parental genome. The easiest way to see linkage is to examine the testcross example. Examine the cross of an *AaBb* organism by an *aabb* organism. What predictions would you make if the genes are unlinked and assort independently of each other? How would linkage between the two genes change the results?

Linkage between *A* and *B* would be revealed by the ratios of progeny produced by this cross. Using these data, scientists can actually determine the distance between two linked genes. As the geneticists gradually refine their assessment of where the gene is located, they can then begin the process of actually finding the gene itself.

C. ASSESSING WHAT YOU'VE LEARNED

1. Draw a diagram of a diploid cell with two sets of chromosomes, placing alleles of two genes on the chromosomes as follows: gene *A* on chromosome 1, and gene *a* on chromosome 1'; gene *B* on chromosome 2, and gene *b* on chromosome 2'. Now diagram the haploid ($1n$) products of meiosis in these cells. In what four possible ways will these alleles segregate? *Hint:* The key to this segregation is anaphase 1.

2. If an organism is heterozygous for two traits that are linked ("AB" and "ab" on homologous chromosomes), how many different genotypes are possible in the gametes? Assume that no crossing over occurs.
   a. 1
   b. 2
   c. 4
   d. 8

3. If an organism is heterozygous for two traits that are linked, how many genotypes are possible in the gametes produced from a single germ-line cell if a single crossover event occurs during meiosis I?
   a. 1
   b. 2
   c. 4
   d. 8

4. Pepper color is a trait that is controlled by several genes. The *Y* allele generates red peppers while the recessive *y* allele generates yellow peppers (*yy*). If a pepper plant that is heterozygous for these alleles (*Yy*) is mated by a genetically similar plant (*Yy*), what percentage of the progeny will be yellow?
   a. 50 percent
   b. 33 percent
   c. 25 percent
   d. 75 percent

5. A female housefly with red eyes was mated with a male housefly with red eyes. The gene allele for red eyes is dominant to the allele for white eyes. Their offspring were: all females displayed red eyes while 48 percent of males were red-eyed and 52 percent of males were white-eyed. Based on this information, which of the following statements is most accurate?
   a. The gene for red eyes must be autosomal.
   b. The gene for red eyes is Y-linked.

   c. The gene for red eyes must be on the X chromosome and the mother was homozygous for the dominant "red" allele.
   d. The gene for red eyes must be on the X chromosome and the mother was heterozygous for the dominant "red" allele.

6. As researchers study genes more closely, they are finding that many genes are pleotropic. This means that:
   a. the wild-type allele displays incomplete dominance over mutant alleles.
   b. the genes encode multiple unique proteins.
   c. the phenotype of mutant alleles depends upon a second gene.
   d. mutations in these genes generate more than one distinct phenotype (trait).

7. When flipping three coins simultaneously, what is the probability of getting all heads?
   a. One-half or 50 percent
   b. One-fourth or 25 percent
   c. One-eighth or 12.5 percent
   d. Three-fourths or 75 percent

8. A woman is about to deliver triplets. If each child is the product of an independent fertilization event, what is the probability that she will deliver EITHER all girls OR all boys?
   a. One-half or 50 percent
   b. One-fourth or 25 percent
   c. One-eighth or 12.5 percent
   d. Three-fourths or 75 percent

9. Using Punnett square analysis, examine reciprocal crosses of fruit flies carrying recessive copies of the g^+ gene (located on the X chromosome). First cross a mutant male fly (*gY*) with a wild-type female fly (g^+g^+). Next cross (cross 2) a wild-type male fly (g^+Y) with a mutant female fly (*gg*). Do these crosses generate different ratios of F_1 progeny? Calculate the ratio/phenotypes.

Cross 1

Cross 2

10. Walter Sutton and Theodore Boveri hypothesized that the chromosomes they visualized during meiosis were the hereditary determinants predicted by Gregor Mendel 30 years earlier. They utilized light microscopy to observe the behavior of chromosomes over time as meiosis proceeded. Define Mendel's principle of segregation and independent assortment, and the meiotic processes Sutton and Boveri would use to explain these principles (what did they see?).

11. Define a genetic map. How would a testcross be used to identify the genetic difference between two genes? What is the difference between physical distance and genetic distance?

CHAPTER 13—ANSWER KEY

Check Your Understanding

(a) Laboratory mice are much more difficult to study in large numbers; they reproduce in litters of 5 to 10. In studying pea plants, Mendel analyzed hundreds to thousands of progeny from each cross and then analyzed the data using statistical analysis. If he was examining only a few progeny, it would have been much harder to establish statistically relevant data.

Peas have many other advantages over mice for Mendel's analysis. Peas are easier to take care of, reproduce more rapidly, and display easily identified phenotypes; further, it is fairly simple to generate true-breeding lines, and they are edible. Mice are important experimental organisms in genetics; but for Mendel's experiments, peas were the better choice.

(b) Determine this answer by making a simple 4×4 Punnett square like the one in **Figure 13.5**. The genotypes AABB and aabb each appear in 1/16 of the progeny. For either AABB or aabb, the frequency is 1/16 + 1/16 = 2/16, or .125 (12.5 percent).

Also make this determination by calculating individual probabilities of each genotype. Remember that for unlinked genes in a dihybrid cross, an AaBb organism generates an equal distribution of the four possible gametes (25 percent AB, 25 percent ab, 25 percent Ab, and 25 percent aB). The probability of generating an AABB plant will require the fertilization of an AB gamete by another AB gamete. These two independent events must occur simultaneously; thus, you will use the **product rule** to calculate the probability of this fertilization event.

Probability for AABB: (.25)(.25) = .0625 (or 6.25 percent)
Probability for aabb: (.25)(.25) = .0625 (or 6.25 percent)

Now you can use the **sum rule** to calculate the probability of generating either AABB or aabb: (.0625) + (.0625) = .125 (or 12.5 percent).

(c) The testcross allows for careful examination of meiosis in one parent and controls the gametes introduced by the other parent. For example, when crossing an AaBb plant by an aabb plant, all of the gametes from the second plant will be ab. Thus, meiosis in the AaBb plant dictates ratios of genotypes and phenotypes in the progeny. The dihybrid cross AaBb by AaBb generates a 9:3:3:1 ratio of phenotypes. The testcross AaBb by aabb will generate a 1:1:1:1 ratio of phenotypes.

(d) Reciprocal crosses are two crosses analyzing the same traits, but in each cross the genotype of the parents is reversed. If reciprocal crosses generate different ratios of progeny genotypes, the gene involved is likely located on a sex chromosome. In this example of an X-linked gene (alleles G and g), Cross 1 would yield all normal-eared F_1 progeny. In Cross 2, all male F_1 progeny would be small-eared and all female F_1 progeny would be normal-eared. If the F_1 progeny from Cross 1 are interbred, the F_2 females will be normal and one-half of the male progeny will be small-eared. If the F_1 progeny from Cross 2 are interbred, one-half of the F_2 progeny will be small-eared, regardless of sex.

(e) The ABO blood groups in humans illustrate allelic codominance. A, B, and O refer to phenotypes possible from the allelic composition of a single gene (I). Three alleles of this gene—I^A, I^B, and I—control the composition of cell-surface molecules on red blood cells. If you are genotypically ii, there are none of these cell surface molecules; if you are $I^A I^A$, you have the A cell-surface marker; but if you are $I^A I^B$, you have both A and B on the cell surface. These two alleles are codominant relative to each other; the heterozygotes display both phenotypes.

(f) The F_2 generation will display the following traits in the following frequencies: agouti (A-B-, 9/16), cinnamon (A-bb, 3/16), black (aaB-, 3/16), and brown (aabb, 1/16).

Assessing What You've Learned

1. Four haploid gametes are possible (AB, Ab, aB, and ab). If the chromosomes assort independently, each of these gametes will be produced in equal proportions.
2. b; 3. c; 4. c; 5. d; 6. d; 7. c; 8. b

9.

Cross 1

	g	Y
g^+	g^+g (normal female)	g^+Y (normal male)
g^+	g^+g (normal female)	g^+Y (normal male)

Because g^+ is dominant to g, all of the flies will display a normal phenotype.

Cross 2

	g	Y
g^+	g^+g (normal female)	g^+Y (normal male)
g^+	g^+g (normal female)	g^+Y (normal male)

All male flies display the mutant phenotype. All female flies are wild type. The reciprocal crosses have different results! The mutant phenotype is associated with males only in the second cross. This gene must be located on the X chromosome.

10. The principle of segregation states that plants have two determinants (genes) that control each trait, and these determinants segregate into separate gametes (meiosis). The principle of independent assortment states that determinants controlling different traits (separate genes) are transmitted to gametes independently of one another, and they assort independently. Sutton and Boveri noticed darkly staining structures that appeared to be copied at the start of mitosis and meiosis. Then, during meiosis, they would pair with a similarly sized structure at the early stages of meiosis and ultimately segregate into separate gametes. The structures were chromosomes, and this behavior was consistent with the principle of segregation. In addition, they noticed multiple pairs of chromosomes, each segregating independently as predicted by the principle of independent assortment.

11. Genetic map is a linear diagram of genes and their ordering on a chromosome. This can be determined using genetic crosses. A testcross is especially helpful in genetic mapping as meiotic events in *one* parent are clearly displayed in the progeny generated from that cross. Physical distances on chromosomes are the actual distances between genes, determined from DNA sequencing analyses.

Looking Forward—Key Concepts in Later Chapters

How Genes Work—Chapter 15

Mendel established the foundations of gene inheritance without knowing what a gene really was! **Chapter 15** examines the basics of **gene structure and function**, revealing details of these heritable units. Each chapter in Unit 3 builds upon Mendel's work, moving toward the utilization of genes in powerful engineering strategies that have revolutionized biology.

Plant Form and Function—Chapter 35

Mendel was able to control mating in plants by manually fertilizing each cross to ensure that he could identify both parents of each cross. **Chapter 35** offers additional information on **plant structure** and **plant reproduction**. Mendel also examined traits that include flower color, seed shape and color, and even plant height.

DNA and the Gene: Synthesis and Repair

Looking Back—Key Concepts in Early Chapters

Nucleic Acids and the RNA World—Chapter 4

The **double helix**, **antiparallel strands**, the 5′ and 3′ **ends** of each strand, and the role of **nitrogenous bases** in DNA structure—these are key concepts in Chapter 14, and their introduction in **Chapter 4** should be reviewed carefully as you begin to study DNA synthesis.

The Cell Cycle—Chapter 11

Finally, in Chapter 14 we begin to truly define the process of **DNA replication** that precedes cell division. This chapter fills in some of the holes from **Chapter 11**. Replication of a cell's genome is no small matter, and performing this task carefully is crucial to the successful transfer of information to the daughter cells.

KEY CONCEPTS

- The heritable material is the deoxyribonucleic acid (DNA) found in chromosomal structures. Key evidence: bacterial viruses transmit DNA into host bacterial cells during the initial stages of infection.

- Chromosomal DNA is a **double helix**, with antiparallel strands composed of **ribose, phosphate**, and **nitrogenous bases** (adenine, guanine, cytosine, and thymine).

- **Semiconservative DNA replication** is accomplished by separating the strands of the double helix and using each of the original strands as a template to synthesize new DNA in the 5′ to 3′ direction.

- DNA replication involves a rapid but careful process that includes the coordination of many different enzymes. Defects in DNA replication are typically lethal, revealing the importance of this process.

- Damage to DNA can be detected and repaired by enzymes involved in **nucleotide excision repair**. The loss of DNA repair capability results in higher rates of mutation and diseases that include cancer.

A. CHAPTER OUTLINE

14.1 What Are Genes Made Of?

- A key to the study of chromosome theory was defining the biological molecule responsible for carrying information from generation to generation. Since chromosomes are made of DNA and protein, they were popular candidate molecules.

- If you were looking for a molecule that carried a complex "code of life" within its structure, what types of molecules might you focus upon? Protein is composed of 20 different amino acids in different orders. DNA is composed of four types of nucleotides that are different in minor ways. It is easy to understand why many scientists suspected protein was the heritable material.

- In the 1940s and 1950s, a series of critical experiments was performed that clearly identified DNA as the inheritable material. It is the accumulating impact of these experiments that firmly established DNA as the heritable material.

- Experiments that might reveal the heritable molecule were not easy to design and develop. Bacteriophage lambda (bacterial virus) offered an ingenious way to examine DNA and protein. During the infection of a bacterial cell by a virus, something is transferred into the host cell by the virus to initiate the infection. The transferred molecule contains the information necessary for building a new virus. Detecting whether protein or DNA is transferred into the host cell should offer valuable insight into the composition of the hereditary material.

The Hershey–Chase Experiment

- In 1952, Martha Hershey and Alfred Chase developed a bacteriophage-based experiment to examine the role of DNA and protein in an infectious cycle.

- A bacteriophage is made of a protein coat, with DNA inside the phage head structure. This virus attaches to the bacterium and *transfers material into the bacterial cell*, forcing the bacterium to synthesize hundreds of new viral particles (**Figure 14.1**). The material carries information, but is the transferred material made of protein or DNA?

- As illustrated in **Figure 14.2** (provided here), viral proteins were radioactively labeled sulfur (^{35}S) and viral DNA were labeled phosphorus (^{32}P). Proteins contain the amino acids methionine and cysteine, each of which has sulfur (S) in its structure. There is no sulfur in the structure of DNA. Similarly, phosphorus (P) is found in DNA but is not found in the basic structure of protein. Thus, growing a virus with radioactive phosphorus will selectively label the DNA molecule, and labeling a virus with radioactive sulfur will selectively label proteins. This differential labeling of biomolecules was a key component in the unique experimental design of Hershey and Chase.

- Separate bacterial cultures were infected with either type of radioactively labeled bacteriophage. The bacteria were then separated from any bacteriophage present in the culture medium using a simple kitchen blender to agitate and break up the virus/bacteria complexes. The amount of radioactivity of the bacterial cells was measured to determine if either the radioactive DNA or protein had been transferred into the infected bacterial cells. The scientists found that when bacteria were exposed to ^{32}P-labeled bacteriophage, the radioactivity was transferred into the bacterial cells. The same was not true when ^{35}S-labeled bacteriophage were used.

- This experiment strongly supported the hypothesis that DNA was the transforming factor and that DNA carried information on its structure. But questions remained. What type of code could be contained on such a simple molecular structure? How could this information-containing molecule be faithfully copied during cell division?

The Secondary Structure of DNA

- To better understand material in this chapter, review DNA structure, described in **Chapter 4**. The double

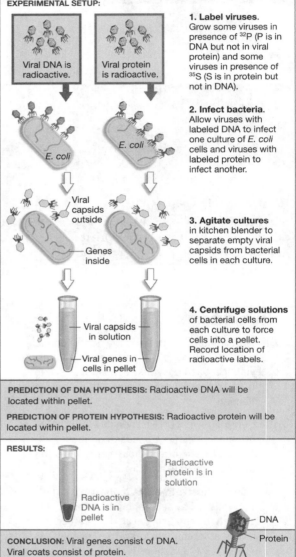

EXPERIMENT

QUESTION: Do viral genes consist of DNA or protein?

DNA HYPOTHESIS: Viral genes consist of DNA.

PROTEIN HYPOTHESIS: Viral genes consist of protein.

EXPERIMENTAL SETUP:

Viral DNA is radioactive.

Viral protein is radioactive.

E. coli

E. coli

Viral capsids outside

Genes inside

Viral capsids in solution

Viral genes in cells in pellet

1. Label viruses. Grow some viruses in presence of ^{32}P (P is in DNA but not in viral protein) and some viruses in presence of ^{35}S (S is in protein but not in DNA).

2. Infect bacteria. Allow viruses with labeled DNA to infect one culture of *E. coli* cells and viruses with labeled protein to infect another.

3. Agitate cultures in kitchen blender to separate empty viral capsids from bacterial cells in each culture.

4. Centrifuge solutions of bacterial cells from each culture to force cells into a pellet. Record location of radioactive labels.

PREDICTION OF DNA HYPOTHESIS: Radioactive DNA will be located within pellet.

PREDICTION OF PROTEIN HYPOTHESIS: Radioactive protein will be located within pellet.

RESULTS:

Radioactive protein is in solution

Radioactive DNA is in pellet

DNA

Protein

CONCLUSION: Viral genes consist of DNA. Viral coats consist of protein.

helix and the nature of this macromolecular structure are vital review topics.

- DNA is a polymer of nucleotides, each of which is composed of three key components: **ribose**, a **phosphate group** (PO_4), and a **nitrogenous base**. Each nucleotide contains a single nitrogenous base: **adenine (A)**, **guanine (G)**, **cytosine (C)**, or **thymine (T)**.

- The carbon atoms of ribose are numbered 1 to 5, with the first carbon being to the right of the oxygen molecule within the ring and numbering around the ring in a clockwise direction. The 3 carbon and the 5 carbon serve as attachment points for the formation of the DNA polymer (see **Figure 14.3**).

- The ribose and phosphate group form the **phosphodiester backbone** of a DNA strand. The DNA strand has directionality—a 5′ end and a 3′ end depending upon the orientation of the ribose groups. The nitrogenous bases are attached to each ribose group.

- Chromosomal DNA is double stranded, with opposing strands oriented in opposite or antiparallel directions. James Watson and Francis Crick published the correct structure of the double helix in 1953.

- The double helix is stabilized by hydrogen bonding between the nitrogenous bases in the DNA sequence. Only specific pairs of nitrogenous bases will bond with each other to hold the double helix together; A bonds with T and G bonds with C. This is called **complementary base pairing**.

- Watson and Crick proposed a mechanism of DNA replication that involved using existing strands as templates to synthesize new DNA strands. Subsequent work by others discovered the process of DNA replication in detail.

14.2 Testing Early Hypotheses about DNA Synthesis

The Meselson–Stahl Experiment

- Compare the three mechanisms of DNA replication that were proposed following Watson and Crick's discovery of the double-helix structure (**Figure 14.4**).

  1. **Semiconservative replication**—The double helix separates, and each old strand is copied to generate two new chromosomes. Thus, *each new chromosome is composed of one strand of old DNA and one strand of newly synthesized DNA.*

  2. **Conservative replication**—During replication, the original chromosome is copied but remains unchanged. Both of the original strands would remain at the end of this replication, and the *new chromosome will be completely new.*

  3. **Dispersive replication**—The replication process generates two new chromosomes, with *new and old sections of DNA mixed together randomly.* This model is complex and would require a great deal of cutting and splicing of DNA strands.

- *Escherichia coli* is a prokaryotic bacterium that divides rapidly, is easy to culture, and is readily available.

Many aspects of basic cellular function are shared among all living organisms; and though eukaryotic organisms are generally more complex, the underlying processes are similar. *E. coli* was an excellent **model organism** for early DNA replication studies.

- To allow for experimental separation of newly synthesized DNA from the original template DNA, Matthew Meselson and Frank Stahl grew *E. coli* bacteria in the presence of **"heavy" nitrogen** (^{15}N rather than the normal ^{14}N). Is nitrogen a part of the DNA structure? Where?

- After many generations of growing bacteria in the presence of ^{15}N-containing media, they changed the bacteria to normal ^{14}N media.

- How would this experimental design label DNA differently in each of the three models? Would each generate a unique result? How might you detect this difference? Sketch out DNA replication as it would proceed under each of the three proposed models. Would the DNA contributed to new cells after one generation of growing in ^{14}N-media be different in each of these models?

- Meselson and Stahl separated the DNA by *density* using high-speed centrifugation and found that the semiconservative mode of DNA replication was indicated by their data. Review the data, remembering that the more dense the DNA, the further the band will travel downward in the centrifuge tube (**Figure 14.5**). The semiconservative replication model predicted that after one cell division in ^{14}N-containing media, all of the DNA would be of intermediate density—having one complete strand of DNA with ^{15}N and a new strand with ^{14}N. After a second cell division, half of the DNA would be of intermediate density and the other half would be lower density (all ^{14}N).

CHECK YOUR UNDERSTANDING

(a) The Meselson and Stahl experiment focused on discerning between competing models of DNA replication. Describe or diagram the results the researchers would have generated if DNA replication proceeded by conservative replication.

_____

_____

_____

_____

14.3 A Comprehensive Model for DNA Synthesis

- DNA is synthesized from nucleotide triphosphates (dATP, dGTP, dCTP, and dTTP). When a nucleotide triphosphate is attached to the end of a DNA strand, the first phosphate is retained in the DNA structure while the other two phosphates are cut free from the structure.

- A key step in characterizing DNA replication was to isolate the enzyme that is primarily responsible for the process. Researchers were able to isolate **DNA polymerase I** from *E. coli*. The enzyme activity was detected in an assay using a **DNA template** strand, from which the new DNA was copied. Subsequent studies showed that another DNA polymerase, called **DNA polymerase III**, performed the bulk of DNA replication during bacterial cell division while DNA polymerase I performed a key role in the repair of mutation.

✔ CHECK YOUR UNDERSTANDING

(b) The synthesis of a DNA strand is an enzymatic process that proceeds quickly. Rapid enzymatic reactions are energetically favorable. Why is DNA synthesis an energetically favorable process?

_____

_____

_____

_____

How Does Replication Get Started?

- DNA polymerases catalyze DNA synthesis in the 5′ to 3′ direction. This makes more sense when you examine the structure of a nucleotide. That is to say, new nucleotides are added to the 3-carbon hydroxyl (–OH) on the ribose of an existing DNA strand. The phosphate attached to the 5′ end of the nucleotide triphosphate is attached to the hydroxyl (–OH) at the 3′ end of the DNA strand being extended (**Figure 14.6**).

- The origin of DNA replication is a site on the chromosome where the two strands of DNA are separated to form a "bubble" of single-stranded DNA. As replication of the DNA moves from the site of origination, new strands of DNA are synthesized using each single-stranded piece of chromosome as a template.

- The **replication fork** is a structure that moves forward from the site of origin; it is a Y-shaped structure moving along as DNA replication proceeds (**Figure 14.7**).

How Is the Helix Opened and Stabilized?

- The replication fork originates with the action of **DNA helicase**, an enzyme that opens the double helix to allow enzymes to attach to each strand. Small proteins known as **single-strand DNA-binding proteins** (**SSBPs**) bind the single-stranded DNA and keep it from re-forming double-stranded DNA.

- DNA polymerization requires a primer (a 3′ –OH) on which to add nucleotides. The **primase** enzyme makes short RNA primers to start the DNA synthesis. This RNA is later replaced with DNA.

- As the replication fork moves forward, **topoisomerase** enzymes release supercoiling that can occur as the double helix is pulled open.

- **Figure 14.8** offers a nice summary of the synthesis of the **leading strand** during DNA replication. The leading strand is the DNA synthesis process that polymerizes **toward** the replication fork.

How Is the Lagging Strand Synthesized?

- How can this replication work if the DNA polymerase works only in the 5′ to 3′ direction? Examine **Figure 14.9**; it displays the continuous **leading strand** and the discontinuous **lagging strand** found at a replication fork. The lagging strand is synthesized in a direction **away** from the overall movement of the replication fork. The lagging strand is composed of a series of small DNA fragments that are eventually connected together.

- Reiji Okazaki discovered the mechanism of lagging-strand synthesis. The short fragments generated at the site of lagging-strand synthesis are known as **Okazaki fragments**. These fragments are eventually connected to one another via the activity of DNA polymerase I and DNA ligase.

- Review the stepwise examination of DNA polymerization offered in **Figure 14.10**. Each of the key steps is presented in proper chronology and should make intuitive sense relative to the process we've introduced up to this point. Remember that this is a summary; DNA replication is a complex process requiring many proteins (some of which have yet to be identified). **Table 14.1** (provided here) offers a valuable summary of the proteins involved in the process of DNA replication. When you combine these enzymes and their activity, the large complex is known as the **replisome**.

(c) Explain the need for RNA in the process of DNA replication.

_____

_____

_____

_____

(d) Define the differences in functional roles between DNA polymerase I and DNA polymerase III during DNA replication.

_____

_____

_____

14.4 Replicating the Ends of Linear Chromosomes

- The ends of _linear_ chromosomes are called **telomeres**, and they contain many repeats of a six-base sequence repeated ~1000 times. In humans, the telomere sequence is 5′—TTAGGG.

- As the replication fork nears the end of a linear chromosome (like those found in human cells), the lagging strand makes an RNA primer for the end of the chromosome. DNA polymerase III uses the available 3-OH group on the RNA primer and polymerizes the last Okazaki fragment for that chromosome. But there is no way to replace this RNA with DNA, because there is no available primer for DNA synthesis (review **Figure 14.12**, step 4). The RNA primer is removed, leaving a section of single-stranded DNA at one end of each new chromosome.

- In most dividing cells, this residual single-stranded DNA is degraded and the telomere shortens by that length with each cell division. Telomere length then serves as a record of cell division and limits the number of times a somatic cell can divide. But some cells, like germ cells, replicate the telomere completely.

- **Telomerase** is an enzyme that mediates the replication of telomeres, solving the problem of the single-stranded DNA described earlier. As is illustrated in

SUMMARY TABLE 14.1 **Proteins Required for DNA Synthesis**

Name	Structure	Function
Opening the helix		
Helicase		Catalyzes the breaking of hydrogen bonds between base pairs to open the double helix
Single-strand DNA-binding proteins		Stabilizes single-stranded DNA
Topoisomerase		Breaks and rejoins the DNA double helix to relieve twisting forces caused by the opening of the helix
Leading strand synthesis		
Primase		Catalyzes the synthesis of the RNA primer
DNA polymerase III		Extends the leading strand
Sliding clamp		Holds DNA polymerase in place during strand extension
Lagging strand synthesis		
Primase		Catalyzes the synthesis of the RNA primer on an Okazaki fragment
DNA polymerase III		Extends an Okazaki fragment
Sliding clamp		Holds DNA polymerase in place during strand extension
DNA polymerase I		Removes the RNA primer and replaces it with DNA
DNA ligase		Catalyzes the joining of Okazaki fragments into a continuous strand

Figure 14.13, telomerase adds more six-base repeats to the end of the leading strand. Then primase makes an RNA primer, and DNA polymerase uses the primer to synthesize the lagging-strand sequence.

14.5 Repairing Mistakes and Damage

- The haploid human genome contains 3 billion nucleotides (3×10^9), and each diploid cell then contains 6 billion nucleotides. With each cell division, the entire genome must be copied completely and accurately. In the human body, there are trillions of cells, all derived from the original genome of that single embryonic cell. Thus, the process of DNA replication must be careful to avoid making mistakes and to repair sites of DNA damage.

Correcting Mistakes in DNA Synthesis

- During DNA synthesis, nucleotides are added to the new strand via their ability to form hydrogen bonds with the template nitrogenous base. Thus, using H-bonding interactions, an A in the template will direct the addition of a T to the new strand. Mistakes by DNA polymerase are rare, about 1 error for every 100,000 (10^5) nucleotides added to the newly synthesized strand. However, the overall rate of mutation for a typical cell is 1 mistake in 1 billion (10^9) nucleotides. Somehow, most mistakes made by the DNA polymerase must be corrected before DNA replication is completed.

- To examine how cells replicate their DNA so accurately, researchers mutated *Escherichia coli* cells and isolated those that now made frequent mistakes during DNA replication.

- *E. coli* bacteria carrying a mutation in a protein subunit of DNA polymerase III (the epsilon subunit) were found to make errors during DNA replication at a higher-than-normal rate. This subunit mediates **proofreading** during DNA replication by cutting out any nucleotides that were added incorrectly. The enzymatic removal of these nucleotides is termed **exonuclease** activity and acts in the 3′ to 5′ direction (**Figure 14.14**). After removal of the mismatched nucleotide, the correct DNA strand is then synthesized.

- DNA polymerase III makes one mistake every 100,000 nucleotides, and with proofreading activity, the error rate is reduced to 1 mistake every 10 million nucleotides (1 in 10^7).

- If the DNA polymerase leaves a mistake behind, the cell has other DNA repair enzymes available to it. *MutS* is one example of a DNA repair enzyme; it identifies **mismatched nitrogenous bases** and replaces the incorrect nucleotide.

Repairing Damaged DNA

- Mutagenic compounds generated by cellular metabolism (free radical compounds) can damage nitrogenous bases in DNA. Also, we are exposed to mutagenic compounds in our environment: ultraviolet light, poisons we inhale in smoke, and toxins found in mold-contaminated foods (aflatoxins). These compounds can change the chemical structure of nitrogenous bases, making it impossible to properly replicate this DNA.

- Ultraviolet light causes adjacent thymine bases to covalently bond with each other. These mutations are called **thymine dimers** or **cyclobutane pyrimidine dimers** (**Figure 14.15**).

- **Nucleotide excision repair** is one example of a DNA repair. The repair complex identifies sites of mismatch or thymine dimers by their unusual structure. The enzyme cuts the mutant DNA free (excision), allowing DNA polymerase and DNA ligase to fill in the excised region.

Xeroderma Pigmentosum: A Case Study

- **Xeroderma pigmentosum (XP)** is an inherited disease in humans; it results in sensitivity to UV light, X-rays, and other DNA-damaging treatments. People affected by this disease rapidly develop skin lesions following exposure to sunlight.

- This disease is caused by mutation of one of several **excision repair** enzymes utilized by humans. Diminished excision repair enzymes result in a deficient cellular capacity to repair damaged DNA.

- See **Figure 14.17** for an illustration of the structure of two key experiments characterizing DNA repair in cells isolated from XP patients. These studies simply examine the effect of UV light upon skin cells isolated from unaffected individuals and from XP sufferers. UV exposure was much more toxic to the XP cells.

B. MASTERING CHALLENGING TOPICS

One common theme in this chapter is the utilization of mutants to isolate and characterize the mutated gene. Scientists interested in a key cellular function can sometimes predict what a cell lacking that capacity will display as a mutant phenotype. One example from the chapter is the examination of DNA repair enzymes by isolating bacteria that made lots of mistakes during DNA replication. If the mutant phenotype is expected to be lethality, or a dramatic loss of cell division capacity, isolation of these mutants becomes difficult. Then temperature-sensitive mutations might be a better choice, because cells harboring this type of mutation may grow normally at the permissive temperature and display the mutant phenotype only at the restrictive temperature.

Humans afflicted with mutations that weaken the DNA repair mechanisms can display many phenotypes, including light sensitivity, brittle hair and nails, and skin that is easily sloughed off. Classic examples of this type of mutation are **xeroderma pigmentosum**, **Cockayne syndrome**, and **trichothiodystrophy (TTD)**. Recently,

Vermeulen and colleagues (2001) identified a group of people who suffer the symptoms of TTD whenever they run high fevers. Upon carefully examining a specific DNA repair gene in these people, the researchers discovered a mutation that results in a **temperature-sensitive** phenotype. This means the mutant enzyme works at normal temperatures, but at slightly elevated temperatures it is unable to function because of its mutation.* Wow! It is easy to think of bacteria or yeast as a cell type that might offer temperature-sensitive mutations to researchers, but this research actually identified them in humans!

C. ASSESSING WHAT YOU'VE LEARNED

1. Which of the following best describes the structure of a double-helical DNA molecule?
   a. A deoxyribonucleoside triphosphate with an adenine, thymine, guanine, or cytosine base attached to it
   b. Antiparallel ribose-phosphate backbones with adenine, thymine, guanine, or cytosine attached to each ribose and projecting toward the inside of the helix
   c. Antiparallel ribose-phosphate backbones with adenine, thymine, guanine, or cytosine attached to each ribose and projecting toward the outside of the helix
   d. Parallel ribose-phosphate backbones with adenine, uracil, guanine, or cytosine attached to each ribose and projecting toward the inside of the helix

2. Identify which of the following sequences would form double-stranded DNA with **5′–TTAACGTC-TAAGT–3′**.
   a. 3′–TTAACGTCTAAGT–5′
   b. 3′–AATTGCAGATTCA–5′
   c. 3′–UUAACGUCUAAGU–5′
   d. 5′–UUAACGUCUAAGU–3′

3. Which of the following represents the correct order of events as they occur in the process of DNA replication?
   a. Helicase opens helix, DNA polymerase III synthesizes RNA primer of leading strand, DNA polymerase I makes a new DNA strand.
   b. Primase synthesizes RNA primer, DNA ligase attaches primer to non-template strand, DNA polymerase III copies RNA to make new DNA strand.
   c. Helicase opens helix, DNA polymerase III elongates RNA primer and synthesizes leading

and lagging strands, DNA ligase links fragments of DNA.
   d. Primase synthesizes RNA primer, DNA polymerase III synthesizes lagging and leading strands, helicase opens helix.

4. The classic experiment performed by Alfred Hershey and Martha Chase revealed:
   a. RNA was not the hereditary material.
   b. Microbes could exchange genetic information.
   c. In a viral infection, protein was transferred into the infected cell.
   d. In a viral infection, DNA was transferred into the infected cell.

5. Semiconservative DNA replication involves:
   a. the separation and cutting of template strands; these sections are then copied and recombined to make the new DNA strands.
   b. the outward rotation of nitrogenous bases in a double helix, and the generation of completely new double-stranded DNA copy.
   c. synthesis of an RNA copy of the DNA, and the generation of a new DNA strand using reverse transcriptase.
   d. the separation of strands of DNA, and the synthesis of a new complementary strand on each old strand.

6. Cancer can occur when DNA is damaged. Which of the following statements is true regarding damage to DNA molecules?
   a. The structure of DNA can be damaged ONLY by ingested toxins such as cigarette smoke, aflatoxins, and other toxins in food.
   b. Excision repair enzymes are able to prevent all mutations in DNA.
   c. DNA damage generates mutations by causing errors in DNA replication.
   d. DNA molecules have a high rate of chemical instability, and commonly degrade in the absence of environmental toxins or radiation.

7. Imagine that you are a researcher working on developing a gene therapy for XP patients. Which of the following would be the best "gene" to incorporate into the skin cells of these patients, based on the data shown in **Figure 14.17**?
   a. A gene that stimulates DNA replication and cell division
   b. Genes that protect against telomere shortening or chromosome aging
   c. Genes encoding DNA excision/repair enzymes
   d. A gene that would increase the production of thymidine in damaged cells

---

* W. Vermeulen, S. Rademakers, N. G. Jaspers, E. Appeldoorn, A. Raames, B. Klein, et al. (2001). A temperature-sensitive disorder in basal transcription and DNA repair in humans. *Nature Genetics* 27(3), 299–303.

8. Once DNA strands are separated by DNA helicase, _____ block the re-formation of a double helix by the template strands.
   a. RNA primer strands
   b. topoisomerase enzymes
   c. single-stranded DNA-binding proteins
   d. RNA polymerase complexes

9. James Cleaver showed diminished DNA repair activity in XP cells by:
   a. isolating DNA polymerase I from these cells.
   b. culturing the damaged cells in radiolabeled deoxyribonucleotide.
   c. staining repaired DNA with a fluorescent dye.
   d. examining damaged DNA with a microscope.

10. Humans with a defect in the *mutS* gene display a defect in mismatch repair and develop:
    a. Alzheimer's disease.
    b. hereditary colorectal cancer.
    c. skin cancer.
    d. limb deformities.

CHAPTER 14—ANSWER KEY

Check Your Understanding

(a) Labeling over many generations with ^{15}N would generate heavy DNA throughout the bacterial culture. Transfer of these cells to ^{14}N media would result in light DNA with all new DNA synthesis. If conservative replication occurred, then density gradient centrifugation would generate only heavy and light DNA bands; no intermediate bands would be observed. Review **Figure 14.5** for an illustration of this process.

(b) DNA strands are synthesized from nucleotide triphosphate molecules (dATP, dCTP, dGTP, and dTTP). The triphosphate group on these molecules contains covalent bonds that "release" energy when they are broken. Thus, the breaking of the phosphoanhydride bond in a dNTP molecule during DNA synthesis makes this process energetically favorable.

(c) DNA polymerases are unable to synthesize DNA from a template strand unless there is an available primer on which to add new nucleotides. RNA polymerases are able to synthesize short RNA primers de novo, and they are utilized during DNA replication to prime new DNA synthesis. The RNA primers are later destroyed and replaced with the correct DNA sequence.

(d) DNA polymerase III performs the majority of template DNA copying during replication. It is a key enzymatic component of the replication fork. DNA polymerase I plays a role in filling in the gaps that are left between Okazaki fragments (and in DNA repair following mutation).

Assessing What You've Learned

1. b; 2. b; 3. c; 4. d; 5. d; 6. c; 7. c; 8. c; 9. b; 10. b

Looking Forward—Key Concepts in Later Chapters

Bacteria and Archaea—Chapter 28

Much of the early work in DNA synthesis and isolation of enzymes in chromosome replication was first identified in **bacteria**. These cells are simple compared to eukaryotic cells, but the underlying mechanisms of cell function are shared. **Chapter 28** offers a critical introduction to bacteria and prokaryotic cell biology.

How Genes Work

Looking Back—Key Concepts from Earlier Chapters

Nucleic Acids and the RNA World—Chapter 4

RNA is more than just a messenger. The capacity of RNA to aid in **protein synthesis**, to have **catalytic activity** (it can cut RNA), and to carry **complex messages** from the nucleus to the cytosol is truly remarkable. Review this introductory material on RNA to enhance your study of Chapter 15. Review the structure of **double-stranded DNA**; it will be important to your study of gene structure in this chapter.

Mendel and the Gene—Chapter 13

Phenotype and the influence of mutation on physical characteristics of organisms were crucial to Mendel's work. With an understanding of genes and the central dogma, we can now better understand what phenotype really entails. A **mutant gene** produces a **mutant protein** (or no protein at all), and the loss of this protein's function generates an abnormal phenotype. The mutant gene really is just an incorrect blueprint for a protein.

KEY CONCEPTS

- The heritable material is made up of DNA, and individual units of inheritance are called genes. *Each gene encodes a protein* that performs some kind of cellular function.

- The **central dogma** of gene expression holds that each gene is **transcribed** by RNA polymerase into an RNA strand (or mRNA), and that message is **translated** into protein by the ribosome.

- The basic unit of the genetic code is the **codon**. Three nitrogenous bases in the DNA (and the mRNA) define a codon, which specifies a single amino acid in the protein.

- The **genetic code** was deciphered through a series of ingenious experiments using synthetic RNA.

- Mistakes or **mutations** in DNA sequence can generate changes in the encoded protein that will affect its function. There are different types of mutations, from **silent** mutations that do not affect the protein, to **nonsense**, **missense**, and **frameshift** mutations that do alter the protein.

A. CHAPTER OUTLINE

15.1 What Do Genes Do?

- Archibald Garrod was a British physician who first began to think of inherited mutations and how they might influence cellular biochemistry (or **metabolic** processes). Garrod noticed a mutant phenotype that consistently appeared in the ancestry of several families, and he suspected that there was defect in the heritable material being passed from generation to generation.

- Garrod called the abnormality an **inborn error of metabolism**, and he suspected that the phenotype was the result of abnormal metabolic function. The disease, called **alkaptonuria**, caused affected individuals to display (among other things) black urine. Garrod published this work in 1902.

- How might one defective gene/protein generate abnormal metabolism? Think of metabolic processes as an assembly line, with stages along the line contributing consecutive pieces of the final product (**Figure 15.1**, **arginine biosynthesis**). Individual proteins (enzymes) perform a task at each stage, altering the compound to ultimately generate the final product. The loss of an enzyme results in the accumulation of an intermediate compound, stuck at a specific stage of processing. This buildup of an intermediate is exactly what happens in alkaptonuria; the accumulated intermediate (homogentisic acid) is disposed of in the urine, turning it black.

- Mutant alleles that generate a nonfunctional protein product are called **null mutations**, **knock-out mutations**, or **loss-of-function mutations**.

The One-Gene, One-Enzyme Hypothesis

- In 1941, George Beadle and Edward Tatum announced discoveries from their study of mutant cells of the bread mold, *Neurospora crassa*. They utilized biochemistry and genetic analysis to characterize the steps involved in biosynthesis of various cellular compounds, including the vitamin pyridoxine (B_6).

- If you recall the assembly-line example of metabolic pathways described earlier, you'll see that each step in the pathway modifies the substrate molecule, eventually generating the product of the pathway. These types of experiments led to the **one-gene, one-enzyme hypothesis**. A mutation in one gene generates one abnormal protein (or enzyme).

An Experimental Test of the Hypothesis

- Another team of researchers took these experiments further. Adrian Srb and Norman Horowitz used *N. crassa* to study the arginine biosynthetic pathway.

- Arginine is an amino acid that cells can make for themselves if it isn't in the growth medium. They isolated mutant cell lines that were unable to grow without arginine supplementation in the growth medium, indicating that the biosynthetic pathway was not functioning properly.

- To characterize *N. crassa* genes that are responsible for arginine biosynthesis, the researchers used radiation to mutate the DNA in a large population of cells. They then screened the cells for those that now lacked the ability to grow in arginine-depleted media. These arginine-dependent cell populations (or clonal cell lines) potentially carry mutations in arginine synthesizing enzymes.

- *The arginine-dependent cell lines likely carry mutations in genes that encode enzymes in arginine biosynthesis.*

- Srb and Horowitz knew some of the details about arginine synthesis, and they knew the identity of some of the intermediates (citrulline, ornithine). They added these intermediate forms to the growth medium.

- Why add an intermediate metabolite? If a single stage in the assembly line is blocked and you introduce an intermediate that occurs downstream from the blocked step, the intermediate will be successfully converted into product (arginine), allowing the cells to grow normally (review **Figure 15.1**).

- Using this strategy, Srb and Horowitz took individual mutant cell lines and identified where the mutation blocked the pathway. Did it work? Yes! Review **Figure 15.2** (provided on the next page) to examine their experiment more closely.

CHECK YOUR UNDERSTANDING

(a) How can a mutation in a gene destroy the activity of the encoded protein (or enzyme)?

_____

_____

_____

_____

CHECK YOUR UNDERSTANDING

(b) To initiate a "mutant hunt," the researcher must first isolate a large group of mutant populations of cells and then utilize them in further experimentation. Describe the process of isolating arg[2] populations of *Neurospora*.

_____

_____

_____

15.2 The Central Dogma of Molecular Biology

- If one gene = one enzyme, then how is the protein's structure encoded within the linear DNA code?

- DNA is double stranded, with antiparallel strands and the **nitrogenous bases** adenine (A), cytosine (C), thymine (T), and guanine (G). Francis Crick proposed that the sequence of nitrogenous bases in the DNA strand was a kind of molecular code, defining the protein structure through the linear grouping of the nitrogenous bases (i.e., AGT = serine).

- Francis Crick's hypothesis predicting a "genetic code" was fundamentally correct. The details of this code and the nature of how this code is translated into protein were worked by a collection of different scientists.

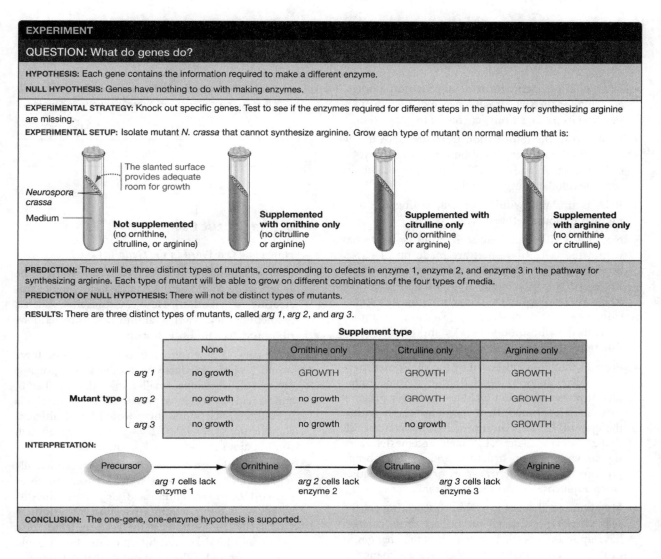

EXPERIMENT

QUESTION: What do genes do?

HYPOTHESIS: Each gene contains the information required to make a different enzyme.

NULL HYPOTHESIS: Genes have nothing to do with making enzymes.

EXPERIMENTAL STRATEGY: Knock out specific genes. Test to see if the enzymes required for different steps in the pathway for synthesizing arginine are missing.

EXPERIMENTAL SETUP: Isolate mutant *N. crassa* that cannot synthesize arginine. Grow each type of mutant on normal medium that is:

The slanted surface provides adequate room for growth

Neurospora crassa

Medium

Not supplemented (no ornithine, citrulline, or arginine)

Supplemented with ornithine only (no citrulline or arginine)

Supplemented with citrulline only (no ornithine or arginine)

Supplemented with arginine only (no ornithine or citrulline)

PREDICTION: There will be three distinct types of mutants, corresponding to defects in enzyme 1, enzyme 2, and enzyme 3 in the pathway for synthesizing arginine. Each type of mutant will be able to grow on different combinations of the four types of media.

PREDICTION OF NULL HYPOTHESIS: There will not be distinct types of mutants.

RESULTS: There are three distinct types of mutants, called *arg 1*, *arg 2*, and *arg 3*.

Supplement type

Mutant type	None	Ornithine only	Citrulline only	Arginine only
arg 1	no growth	GROWTH	GROWTH	GROWTH
arg 2	no growth	no growth	GROWTH	GROWTH
arg 3	no growth	no growth	no growth	GROWTH

INTERPRETATION:

Precursor → Ornithine → Citrulline → Arginine

arg 1 cells lack enzyme 1   *arg 2* cells lack enzyme 2   *arg 3* cells lack enzyme 3

CONCLUSION: The one-gene, one-enzyme hypothesis is supported.

RNA as the Intermediary between Genes and Proteins

- RNA is single stranded, and it is composed of A, U, G, and C (U stands for uracil; there is no thymine in RNA). In addition, the ribose found in RNA has a hydroxyl (–OH) on carbon 2 (DNA lacks this hydroxyl). Francois Jacob and Jacques Monod first proposed that RNA might serve as an intermediary (messenger) between DNA and protein synthesis machinery (**Figure 15.3**).

- **Ribosomes** are small structures found in the cytosol and associated with the **endoplasmic reticulum (ER)**. They synthesize proteins within the cell by reading the sequence of **messenger RNA (mRNA)** and **translating** that information into an amino acid sequence for a particular protein.

- Jerard Hurwitz and J. J. Furth isolated **RNA polymerase**, the enzyme that synthesizes mRNA, by reading

DNA sequence. The researchers examined cellular fractions for RNA synthesizing capacity. By adding the needed reagents for RNA polymerase activity (template DNA, ribonucleotides), they could measure RNA synthesis in each fraction. Newly synthesized RNA would be radioactive through incorporation of **radioactive ribonucleotide triphosphates** (not deoxyribonucleotides).

- How does RNA polymerase read the DNA to make the mRNA? In the coding sequence, the RNA polymerase will read a C and add the complementary G to the RNA strand (**Figure 15.5**). One unusual exception is that when the polymerase reads an A in the DNA, a U is added to the RNA (instead of thymine).

Dissecting the Central Dogma

- The **central dogma of molecular biology** defines information flow within the cell. DNA is **transcribed** into an

RNA message; the messenger RNA is **translated** into protein. This was first articulated by Francis Crick.

$$DNA \rightarrow RNA \rightarrow Protein$$

- DNA is a long-term form of information storage, allowing for the stable maintenance of the information and its passage from generation to generation.

- RNA carries the information from the DNA to the translation machinery (the ribosome). In eukaryotic cells, the DNA and the ribosomes are located in separate cellular compartments.

- RNA is unstable within the cell, lasting for only minutes to hours before it is destroyed.

- Not all RNA is translated; some RNA is functional within the cell. Ribosomes are made up of rRNA bound together with protein subunits, and transfer RNAs (tRNAs) facilitate translation. Functional RNAs are involved in (1) regulation of gene expression, (2) mRNA processing, (3) carrying amino acids to the ribosomes—tRNAs, and (4) ribosome function—rRNAs.

- Proteins that are synthesized by the ribosomes carry out cellular functions—they do things. Some examples of protein function include: (1) motor proteins that facilitate the movement of cellular "cargo" within the cytosol, (2) structural proteins that control cell shape (the cytoskeleton), (3) small proteins that carry signals within their structure, and (4) membrane proteins that facilitate transport of small molecules (such as glucose) across the membrane.

- Mutations or changes in DNA sequence impact the cells phenotype through the impact those changes have upon the encoded protein. It is the changes in protein function that typically generate a change in phenotype for the cell. Genes that control "coat color" in mice offer a wonderful example of this principle (**Figure 15.4**).

- One gene regulating coat color in mice is the melanocortin receptor (a type of membrane protein). This receptor protein helps to regulate the production of dark pigment in mouse hair cells, and it is regulated by melanocortin (a peptide hormone).

- A difference in the DNA sequence of the melanocortin receptor gene generates a change in the amino acid composition of the receptor protein. This can be illustrated in two mice with differences in coat-color phenotype, and corresponding differences in the melanocortin receptor gene. A small change in the gene generates a receptor that doesn't work properly, generating lighter-colored mice. *Thus, genotype typically influences phenotype via the functionality (or lack of functionality) of a protein.*

(c) Define one biological example in which the central dogma of molecular biology is violated or incorrect. Is this example a common phenomenon in cell biology?

_____

_____

_____

15.3 The Genetic Code

How Long Is a Word in the Genetic Code?

- George Gamow predicted that each code word in the genetic code would be three nucleotides long (**3 nucleotides = 1 codon**). How could he make such a prediction? Key: Twenty amino acids are used by ribosomes to synthesize proteins.

- If the genetic code were only two nucleotides, there would be only 16 different combinations possible ($4^2 = 16$), a number smaller than the number of amino acids used in most proteins. If the genetic code were three nucleotides, there would be 64 different combinations possible ($4^3 = 64$), easily enough combinations to code for all 20 amino acids.

- Francis Crick and Sydney Brenner experimentally revealed that the code is indeed based on three nucleotides for each amino acid. Crick made this discovery by making **deletion/insertion mutations** in viral DNA. He found that the **reading frame** of a gene could be destroyed by mutation, but it could then be restored if the total number of deletions or additions were multiples of three. Review *the fat cat ate the rat* as an illustration of reading frame. If one letter is inserted into the first word, all of the subsequent words are turned to gibberish. How is this outcome different if three letters are inserted into the first word? See **Figure 15.7** for additional visualization of this concept.

How Did Researchers Crack the Code?

- Identifying the three-nucleotide sequence coding for each amino acid was a difficult proposition. Eventually the genetic code was deciphered, and it is shown in **Figure 15.6**. Each of the 20 amino acids used by ribosomes to synthesize proteins has multiple codons that designate its addition to a protein.

- In the 1960s, Marshall Niernberg and Heinrich Matthaei were able to synthesize RNA and performed *in vitro* (in the test tube) translation experiments.

Having control over the content of the RNA, they could then identify which amino acids were attached to each other during translation.

- They found that a polymer of U residues in an RNA strand (5′–UUUUUUUUUUUUUU–3′) resulted in the translation of that sequence into a polymer of phenylalanine. Polymers of A (5′–AAAAAAAA AA–3′) were translated to polymers of lysine, polymers of C were translated to polymers of proline, and polymers of G were translated to polymers of glycine.

- Thus, if RNA is synthesized as the complementing sequence of the gene encoding DNA, then the codon for phenylalanine must be **AAA** (**UUU—phenylalanine, CCC—proline, GGG—glycine**). A piece of the code is revealed! This type of strategy revealed a great deal of the genetic code.

- Nirenberg worked with Phil Leder to characterize more complex codons. They synthesized RNAs with the three-letter codon of interest and mixed ribosomes and tRNAs with amino acids attached (charged tRNAs). They then determined which amino acid was bound to the ribosome in response to a particular codon.

- There is only one **start codon** (AUG), which signifies the start of the protein-encoding sequence in an mRNA.

- There are three **stop codons** in the genetic code (UGA, UAA, and UAG).

- Key findings regarding the genetic code: (1) it is **redundant**, containing multiple codons for each amino acid; (2) it **is unambiguous**, as each codon codes for a single amino acid; (3) it is **universal**, nearly identical among all living organisms; and (4) it is **conservative,** with similar codons usually coding for the same amino acids.

✔ CHECK YOUR UNDERSTANDING

(d) George Gamow predicted that the genetic code would contain "words" that were three nucleotides in length. His predictions were based on mathematical considerations rather than experimental data. If each "word" in the genetic code were four nucleotides in length, how many "words" would be mathematically possible?

_____

_____

_____

✔ CHECK YOUR UNDERSTANDING

(e) In cracking the genetic code, Marshall Nirenberg and Heinrich Matthaei performed an ingenious yet laborious series of experiments. Which two key experimental designs did they use to decipher individual codons?

_____

_____

_____

_____

15.4 What Is the Molecular Basis of Mutation?

Point Mutation

- Simple errors in DNA replication can generate point mutations, where one nitrogenous base is substituted for another (**Figure 15.8**). Once a point mutation is copied, it becomes incorporated into the DNA sequence permanently. A point mutation can cause a simple substitution of one amino acid for another, or can sometimes cause more serious problems.

- Point mutations can generate: (1) a **silent** mutation, where the nitrogenous base change doesn't change the amino acid being specified; (2) a **missense** mutation, where one amino acid is substituted for another; (3) a **nonsense** mutation, where a stop codon is generated at the site of the mutational change; or (4) a **frameshift** mutation can be generated when the point mutation involves the insertion or deletion of a nitrogenous base from the gene sequence. Review **Table 15.1** (provided on the next page) and the impact of each type of mutational change upon the synthesis of the encoded protein.

Chromosome-Level Mutations

- An **inversion** mutation is when a region of a chromosome becomes inverted within its location. This type of mutation is especially harmful to genes at the end of the inversion, as the rearrangement may break the gene apart, generating a null mutation.

- A **translocation** mutation is when a region of a chromosome is accidentally moved to another chromosome and inserted into the new chromosome. Translocation mutations can harm the DNA that has moved, but may also harm genes at the site of insertion.

SUMMARY TABLE 15.1 **Known Types of Point Mutations**

Name	Definition	Example	Consequence
		Original DNA sequence of non-template (coding) strand — TAT TGG CTA GTA CAT / Tyr – Trp – Leu – Val – His — Original polypeptide	
Silent	Change in nucleotide that does not change amino acid specified by codon	TAC TGG CTA GTA CAT / Tyr – Trp – Leu – Val – His	Change in genotype but no change in phenotype. Usually neutral with respect to fitness.
Missense (replacement)	Change in nucleotide that changes amino acid specified by codon	TAT TGT CTA GTA CAT / Tyr – **Cys** – Leu – Val – His	Change in primary structure of protein may be beneficial, neutral, or deleterious.
Nonsense	Change in nucleotide that results in early stop codon	TAT TGA CTA GTA CAT / Tyr **STOP**	Premature termination—polypeptide is truncated. Usually deleterious.
Frameshift	Addition or deletion of a nucleotide	TAT TCG GCT AGT ACA T / Tyr – Ser – Ala – Ser – Thr	Reading frame is shifted—massive missense. Usually deleterious.

- Chromosome-level mutations can be detected using **karyotyping** techniques that examine the cell's chromosomes. One new technology that assesses karyotype is called chromosome painting (or fluorescence in situ hybridization—FISH). Abnormal chromosomes are detected as abnormal using a fluorescence microscope (**Figure 15.9**).

B. MASTERING CHALLENGING TOPICS

This chapter contains some complex experimental data, generated by laboratory strategies that are new to the reader. Anticipate some confusion when first reading through these experiments; but realize that you can make sense of these experiments with a little extra review of the material. The characterization of the arginine biosynthetic pathway offers a good example. This experiment utilizes *Neurospora crassa* (a bread mold), a mutational screen, and finally a carefully designed series of phenotypic analyses. What is the relevance of each step in the experiment? Should you ignore these "minor details" and move on to memorizing the conclusions? No! The experimental strategy is critical to this discussion; work through the details one at a time to ensure that you understand why this experiment worked and what it was designed to accomplish. Read the text, the figure legends, and the study guide closely to ensure that you've thoroughly examined the experiment. Also review background material referenced in previous chapters. The relevance of each of the listed terms is described. *Scientists explore biology through careful experimentation. Recognizing the "how" of important discoveries is critical to your development as a scientist.* Don't give up on the experiments included in the text. They are included expressly to teach each student more than just a set of conclusions about important discoveries, and they offer details of the experiments themselves.

C. ASSESSING WHAT YOU'VE LEARNED

1. Adrian Srb and Norman Horwitz tested the one-gene, one-enzyme hypothesis via examination of the arginine biosynthetic pathway in *N. crassa*. In order to find cells with mutations in the arginine biosynthetic pathway, they exposed cells to mutagenic radiation and screened the cells for:
   a. cells that no longer properly make enzymes.
   b. cells that are poisoned by arginine.
   c. cells that cannot grow unless arginine is provided.
   d. cells with high levels of mutations in their DNA.

2. Which of the following serves as a template for making proteins from DNA?
   a. tRNA
   b. mRNA
   c. rRNA
   d. DNA polymerase

3. Which of these describes the function of RNA polymerase?
   a. Amplifies the "message" by making multiple copies of an mRNA molecule after it has been transcribed from DNA
   b. Converts a protein sequence to mRNA
   c. Transcribes DNA to RNA
   d. Translates DNA into protein

4. Which of the following is an *exception* to the central dogma of molecular biology?
   a. Production of RNA in the nucleus of a eukaryotic cell
   b. Double-stranded DNA from a virus inserts into the host genome
   c. Single-stranded DNA from a virus inserts into the host genome
   d. Single-stranded RNA from a virus is used as a template for DNA that inserts into the host genome

5. In a skin cell, which of the following changes may lead to skin cancer?
   a. A change in a single nucleotide in an mRNA molecule
   b. A change in a single nucleotide in a DNA molecule
   c. A change in multiple amino acids of a single protein molecule
   d. A change in the conformation of an enzyme molecule

6. Consider the following partial sequence from two mRNA molecules. The difference between the two molecules would most likely result in which type of change in the resulting protein molecule?

   Original RNA sequence: 5′–ACGUGUACCG

   Mutant RNA sequence: 5′–ACGUUGUACCG

   a. Frameshift mutation—all amino acids after the mutation are affected
   b. Silent mutation—no change in the amino acid sequence
   c. Ribosomal conversion mutation
   d. Missense mutation—one amino acid added improperly

7. George Beadle and Edward Tatum revealed that:
   a. one genetic mutation corresponded to one defective protein (enzyme).
   b. DNA is the heritable material.
   c. RNA is the heritable material.
   d. the arginine biosynthetic pathway is defective in bread molds.

8. Which of the following was considered important evidence that RNA was an intermediary between DNA and protein biosynthetic processes?
   a. RNA is located in the nucleus (with DNA) and the cytosol (with ribosomes).
   b. RNA is synthesized, but it is then fairly rapidly degraded.
   c. DNA does not co-localize with ribosomes in the cytosol.
   d. All of the above are correct.

9. If the genetic code was based on a two-digit system, how many different amino acids could be coded for by this system?
   a. 8
   b. 16
   c. 32
   d. 64

10. Which of the following statements distinguishes DNA and RNA?
    a. T is found only in DNA.
    b. DNA lacks a 2-hydroxyl on its ribose.
    c. DNA does not leave the nucleus.
    d. All of the above apply.

CHAPTER 15—ANSWER KEY

Check Your Understanding

(a) Enzymes catalyze each step in the biosynthetic pathway, modifying the intermediate forms to eventually generate the desired product. Each enzyme is a protein composed of a chain of amino acids. Changing the amino acid content of an enzyme, by altering the gene on which it is encoded, can profoundly influence the function of that protein.

(b) A culture of wild-type *N. crassa* would first be mutagenized using chemical treatment or exposure to radiation. The researchers would identify clonal populations of cells that grow in rich medium but cannot grow in medium that lacks the amino acid arginine. Cells unable to grow in the selective medium likely contain a mutation in one of the arginine biosynthesis genes. Review **Figure 15.2** to see how these isolated mutant cell lines can then be utilized to identify the genes involved in arginine biosynthesis.

(c) Human immunodeficiency virus (HIV) is a retrovirus whose genome is made up of RNA. After entering a host cell during an infection, the RNA is copied to DNA by a viral enzyme known as reverse transcriptase. This is counter to the central dogma of molecular biology.

(d) Gamow was correct. He knew that 20 amino acids were used by ribosomes to build cellular proteins. If the genetic code were only two nucleotides in length, then there would be only 4^2, or 16, codons available. But if they were three nucleotides in length, then the available number of distinct codons would be 4^3, or 64. If they were four nucleotides in length, there would be 256 possible codons.

(e) The first strategy was to synthesize polymers of RNA—poly(U), poly(A), poly(G), and poly(C). They added these RNA polymers to a mixture of reagents capable of translating RNA into protein. Uracil polymers (UUU-UUUUU) resulted in the synthesis of chains of phenylalanine, evidence that the codon for phenylalanine is UUU. The second strategy was to make specific codons and then measure which charged tRNA bound to the ribosome as it read that specific RNA sequence.

Assessing What You've Learned

1. c; 2. b; 3. c; 4. d; 5. b; 6. a; 7. a; 8. d; 9. b; 10. d

Looking Forward—Key Concepts in Later Chapters

Transcription and Translation—Chapter 16

In **Chapter 16**, we will examine the process of **protein synthesis** and learn how the information carried on a messenger RNA (mRNA) is **translated** into a functional protein.

Control of Gene Expression and Cell Differentiation—Chapters 17, 18, and 21

If the **human genome** is composed of 30,000 genes, and every cell in your body contains each of these genes, how are these utilized differently by different cells? Each cell type in the human body acts differently due to **differential gene expression**. The genes expressed in a cell dictate the shape, growth, and function of that cell. Regulating gene expression is central to how we utilize the information in the genome.

Viruses—Chapter 35

The life cycle of a virus can be quite complicated, sometimes including a period of integration into the host genome (**lysogeny**) before replicating itself and emerging from the host cell. Researchers have harnessed the lysogenic capacity of some viruses to carry genes into the human genome in **gene therapy** strategies.

The enzyme **reverse transcriptase** is produced by **retroviruses** to copy their RNA genome into DNA as part of the infectious process. Reverse transcriptase is an unusual enzyme that has become tremendously valuable in the production of **cDNA** for research.

Chapter 16

Transcription, RNA Processing, and Translation

Looking Back—Key Concepts from Earlier Chapters

Nucleic Acids and the RNA World—Chapter 4

Scientists are only beginning to understand the various roles for **RNA macromolecules** in cellular function. The discovery that RNA can be catalytic, functioning in the splicing of RNA strands, was amazing. Recent findings reveal that rRNA in the ribosome catalyzes the formation of **peptide bonds**. In this chapter we introduced several different examples of RNA function in the cell. RNA is an interesting molecule displaying an interesting diversity of activities.

Protein Folding and Processing—Chapter 3 and Chapter 7

Proteins are complex structurally; they must fold properly in order to function. The process of synthesizing new protein is much more complex than simply generating a linear array of amino acids attached via a peptide bond. The role of chaperones and post-translational modification should be reviewed. The endoplasmic reticulum and Golgi apparatus play a key role in cellular protein processing.

KEY CONCEPTS

- **RNA polymerase** reads DNA sequence to synthesize an RNA strand that complements the template strand of the gene. Transcription is in the 5′ to 3′ direction.

- In both eukaryotes and prokaryotes, the RNA polymerase is recruited to the start of a gene via a DNA sequence known as the **promoter**. The promoter binds to specific proteins critical for the recruitment of RNA polymerase, and the initiation of transcription.

- In eukaryotic cells, the protein-encoding portions of a gene (**exons**) may be interspersed with noncoding sequences (**introns**). After transcription to a primary RNA, the exons are spliced together via the activity of an enzyme known as the **splicesome**.

- After splicing, eukaryotic RNAs are processed further via the addition of a **7-methylguanosine cap (5′)** and a **polyA tail (3′)**. At the completion of these modifications, the RNA becomes a messenger RNA and is transported from the nucleus to the cytosol.

- Translation is mediated by the **ribosome**, an enzyme composed of both protein subunits and ribosomal RNAs. RNA mediates the catalytic activity of the ribosome.

- **Transfer RNAs** carry amino acids to the ribosome for protein synthesis or translation. The tRNA contains an anticodon that complements a specific codon in the genetic code. During translation, tRNAs serve as a key part of the translation of the genetic code into a protein.

- Translation is stopped when the ribosome encounters a **STOP codon**, terminating the translation process.

A. CHAPTER OUTLINE

16.1 An Overview of Transcription

- The expression of a gene begins with transcription, in which an **RNA polymerase** enzyme synthesizes an RNA copy of the expressed gene. In eukaryotic cells, the primary RNA product is processed in the nucleus to generate a mature messenger RNA (mRNA).

- Transcription requires a separation of the double helix to allow RNA polymerase to "read" one strand and synthesize a complementary, antiparallel, RNA strand.

- RNA polymerase reads the **template strand** of the DNA to synthesize RNA. This terminology is important to keep straight. The **non-template strand** of DNA matches the sequence of the synthesized RNA generated by transcription (except U is substituted for T). Because of this, the non-template strand is also called the **coding strand**.

Characteristics of RNA Polymerase

- An RNA polymerase enzyme attaches **ribonucleotide triphosphates** together, selecting the correct ribonucleotide by its hydrogen-bonding capacities and synthesizing the RNA strand in the 5' to 3' direction (**Figure 16.1**).

- How is a deoxyribonucleotide different from a ribonucleotide? Compare the ribose structures associated with DNA and RNA, as shown in **Figure 16.1** (five-membered ring structures).

- The presence or absence of an −OH (hydroxyl) functional group, on carbon 2 of the ribose, generates major differences in the characteristics of DNA and RNA chains.

- There is one RNA polymerase enzyme in prokaryotes.

- Eukaryotic cells contain three RNA polymerases, though mRNA synthesis is mediated by only RNA pol II. Thus, all protein synthesis in a eukaryotic cell depends on **RNA pol II**. RNA polymerase I makes ribosomal RNA (rRNA); RNA polymerase III is responsible for some rRNA and for transfer RNA (**Table 16.1**).

Initiation: How Does Transcription Begin?

- The prokaryotic RNA polymerase is composed of a core complex (holoenzyme or "whole" enzyme) that has the ability to synthesize RNA. The holoenzyme is made up of four globular proteins named alpha (α), alpha prime (α'), beta (β), and beta prime (β').

- The three-dimensional structure of RNA polymerase revealed grooves or channels through which the template strand of DNA passes during catalytic activity (**Figure 16.2**).

- Initiation of transcription requires an additional protein, called **sigma** (τ), that binds to holoenzyme not yet associated with DNA. Sigma can "read" DNA and identify the transcriptional start site of a gene. Thus, sigma is critical to the proper initiation of transcription at the correct place on the chromosome, recognizing the DNA sequence that precedes the coding region of a gene (the **promoter**).

- What sequences does the RNA polymerase/sigma complex commonly bind? **David Pribnow** examined promoter sequences in bacterial and viral genomes. He allowed the bacterial RNA polymerase/sigma complex to bind to DNA and then characterized the bound DNA. Pribnow identified two elements found within most bacterial/viral promoters: the **−10 box** (**Pribnow box**) and the **−35 box** (**Figure 16.2b**).

- The −10 box is located 10 bases upstream (in the 5' direction) from the transcription start site. It contains a 5'–TATAAT–3' consensus element.

- The −35 box is located 35 bases upstream from the transcription start site and contains a 5'–TTGACA–3' consensus element.

- Sigma identifies the −10 and −35 promoter sites, properly orienting the RNA polymerase core complex for transcription at the gene start site. The polymerase then opens the DNA double helix and begins to catalyze the polymerization of nucleotides, moving down the DNA template as RNA is synthesized.

- Sigma dissociates from the holoenzyme once transcription is successfully initiated.

- Because sigma has so much influence over which genes are expressed, it was anticipated that there would be multiple forms of sigma in microbial cells. This has been shown to be the case. There are sigma factors whose activity is induced by different environmental conditions (heat, starvation). Thus, cells adapt to these environmental changes by switching sigma factors, altering patterns of gene expression.

- *Escherichia coli* has seven different sigma factor genes, while *Streptomyces coelicolor* has more than 60 sigma factor genes. Different sigma factors recognize −10 and −35 boxes that are unique, and found in specific subsets of genes.

- The promoters of eukaryotic cells are highly variable, though many do contain a **TATA box** approximately 30 base pairs from the transcription start site.

- Eukaryotic cells display a more complex process of transcription. **Basal transcription factors** bind to the promoters of genes that are going to be expressed. These factors recruit the RNA polymerase to the transcriptional start site. Basal transcription factors are functionally analogous to sigma in prokaryotic cells.

- Basal transcription factors were discovered by *in vitro* (in the tube) transcriptional studies. Purified RNA polymerase was combined with template DNA and nucleotides, resulting in RNA synthesis. However, researchers discovered that by adding specific proteins to the test tube, the RNA polymerase activity became more specific for the promoters of genes. Eventually several proteins that regulate RNA polymerase activity were identified and classified as basal transcription factors.

CHECK YOUR UNDERSTANDING

(a) How did David Pribnow identify key characteristics of prokaryotic promoters?

_____

_____

_____

(b) What are the key characteristics of prokaryotic promoters?

_____

_____

_____

Elongation and Termination

- Once RNA polymerization is initiated, the enzyme **elongates** the RNA transcript by reading the template strand and synthesizing RNA in the 5′ to 3′ direction (moving down the template DNA strand in the 3′ to 5′ direction). This enzyme proceeds at a rate of ~50 nucleotides added/second.

- Notice there are grooves in the holoenzyme through which pass the template and non-template DNA strands, nucleotides, and the newly synthesized RNA strand (step 3 in **Figure 16.3**, provided here).

- Transcription is terminated when specific sequences, called **transcription termination signals**, are encountered in the DNA template. The process of RNA polymerase dissociating from the DNA can occur by several different mechanisms (**Figure 16.4**).

- One type of termination signal results in the RNA forming a hairpin structure (the RNA folds back on itself to make a short stretch of double-stranded RNA). Formation of the hairpin will "pull" the newly synthesized RNA out of the holoenzyme, terminating additional transcription. Another termination mechanism requires the action of a separate protein (rho factor) that binds the RNA at a termination sequence and initiates ribosomal disassembly.

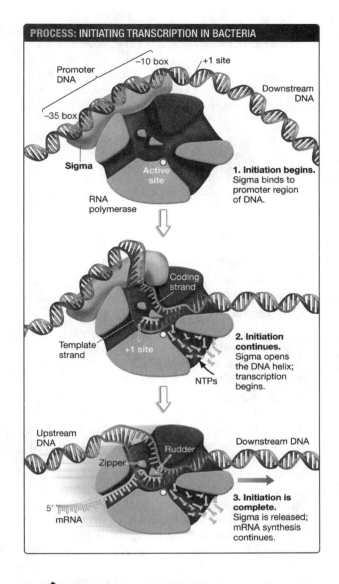

PROCESS: INITIATING TRANSCRIPTION IN BACTERIA

Promoter DNA
−10 box
+1 site
Downstream DNA
−35 box
Sigma
Active site
RNA polymerase

1. Initiation begins. Sigma binds to promoter region of DNA.

Coding strand
Template strand
+1 site
NTPs

2. Initiation continues. Sigma opens the DNA helix; transcription begins.

Upstream DNA
Rudder
Zipper
Downstream DNA
5′ mRNA

3. Initiation is complete. Sigma is released; mRNA synthesis continues.

CHECK YOUR UNDERSTANDING

(c) There are seven different sigma (σ) factors in most bacterial genomes. Each sigma factor plays a critical role in identifying promoters for the RNA polymerase to transcribe. Why would bacteria require seven different sigma factors?

_____

_____

_____

(d) Robert Roeder's research team characterized transcription in eukaryotic cells. They used an *in vitro* assay, adding purified RNA polymerase to DNA, but their early experiments resulted in transcription at random places on the DNA (not just at the promoter). They suspected that something was missing from the RNA polymerase. What might have been missing in these early experiments?

_____

_____

_____

16.2 RNA Processing in Eukaryotes

The Startling Discovery of Eukaryotic Genes in Pieces

- Prokaryotic genomes contain genes in one **continuous** DNA sequence.

- Characterization of eukaryotic genomes has revealed that genes are often **discontinuous**, having been broken into smaller fragments (**exons**) with noncoding sequences (**introns**) between each piece.

- In the 1970s, Philip Sharp and his colleagues developed and tested the **genes-in-pieces hypothesis**. This hypothesis was developed in response to an interesting experiment using DNA:RNA hybrids. His research team took chromosomal DNA for a specific gene, separated the strands of the DNA, and then combined the DNA strands with the corresponding mRNA gene sequence. The DNA and RNA strands were able to form a double-stranded structure due to their complementary sequences, just like double-stranded DNA:DNA form within the chromosome. When they examined this hybridized DNA:RNA structure, they noted large loops on the DNA side of the structure (**Figure 16.5**)! Why? The mRNA must be missing some of the sequence! They proposed that some sequence was removed as the primary RNA was processed to an mRNA.

- As more scientists began to examine this phenomenon, it became clear that eukaryotic genes are discontinuous in the genome, and that following transcription the gene-encoding fragments in the **primary RNA transcript** were spliced together to make the mRNA product.

- The coding fragments found in the chromosomal location of a gene were named **exons** (expressed), and the noncoding fragments interspersed between exons were named **introns** (intervening sequences). Introns are present in the chromosomal sequence of a gene, but are absent in the final mRNA.

Exons, Introns, and RNA Splicing

- How are the exons spliced together? This complex process is mediated by _small nuclear ribonucleoproteins_ (**snRNPs**, pronounced "snurps"). As their name suggests, these complex enzymes are composed of both protein and RNA. **Figure 16.6** illustrates how snRNPs identify introns by their sequence at each end (the splice junction site), catalyze the removal of the intron, and splice surrounding introns together.

- An interesting structure is formed during the splicing process. An adenosine (A) base inside the intron offers a site to which the 5′ end of the intron is attached (**Figure 16.6b**, step 2). This forms a looped structure, called the lariat, and it is quite unusual.

- The complex of snRNPs that mediates this process is called a **spliceosome**.

- A summary of the process: (1) A "GU" sequence at the 5′ end of the intron and an adenine base in the center of the intron are each recognized by snRNPs; (2) the large splicesome complex is recruited to the intron by the snRNPs; (3) the splicesome cuts the RNA at the 5′ splice site and attaches the free end to the internal adenine (forming a loop; and (4) the 3′ end of the upstream exon is attached to the downstream exon, releasing the intron as a lariat structure.

Adding Caps and Tails to Transcripts

- After splicing of the primary transcript, two additional steps of RNA processing are performed in eukaryotic cells—the addition of a **5′ cap** and the addition of a **poly(A) tail** (**Figure 16.7**).

- The cap is a modified nucleoside triphosphate (7-methyguanylate or m^7G) that is added to the front or 5′ end of the transcript.

- The 3′ poly(A) tail is another modification of the RNA transcript that aids mRNA stability in the cell. Approximately 100 to 250 consecutive adenosine nucleosides are added to the 3′ end of the RNA.

- mRNAs with caps and tails display improved levels of translation and increased overall stability in the cytosol. The cap is known to recruit key components of translation initiation to the mRNA. The cap and tail together improve the time that an mRNA exists in the cytosol, decreasing the impact of cellular RNA degrading enzymes (RNases).

16.3 An Introduction to Translation

- Once the mRNA moves to the **cytoplasm**, the information contained within its sequence can be translated by the **ribosome** to generate the functional protein.

- Proteins are synthesized from individual **amino acids**, and each amino acid in the protein is coded for in the mRNA. Each codon (three nucleotides in mRNA) directs the ribosome to add one amino acid to the protein.

Ribosomes Are the Site of Protein Synthesis

- Ribosomes are composed of both protein and RNA, called ribosomal RNA (rRNA).

- Roy Britton used **pulse-chase experiments** to examine whether ribosomes were indeed the site of protein synthesis. *Britton wanted to determine if protein synthesis was occurring at the site of ribosome localization.* He labeled proteins being synthesized for a short period of time by adding a 15-second pulse of radioactive sulfate to the culturing medium of *E. coli* (immediately chased with nonradioactive sulphur). This would result in the incorporation of the radioactive tag with amino acids containing sulphur (methionine and cysteine), and a short window of time in which all proteins being synthesized would be radioactive. But how did Britton detect these proteins? How did he determine that they were associated with ribosomes?

- Ribosomes can be isolated via **density-gradient centrifugation**, in which molecules are separated from each other according to their density. Britton found the pulse of radioactivity to localize to fractions containing ribosomes! As the radioactive pulse was chased with nonradioactive sulphur in the *E. coli* cultures, radioactivity disappeared from ribosomal fractions. This shows that the radioactive sulphur is associated with ribosome activity (the proteins being made), not with the ribosome structure itself.

- Because prokaryotic cells have no nucleus, translation occurs immediately following, or even during, transcription (**Figure 16.8**).

- Eukaryotic cells transcribe RNA in the nucleus, process the RNA to mRNA in the nucleus, and then export the mature mRNA to the cytosol for translation (**Figure 16.9**).

16.4 The Role of Transfer RNA

- **Adapter-molecule hypothesis**: In imagining how a codon of three nucleotides in an RNA molecule could somehow direct the recruitment of specific amino acids, **Francis Crick** (of Watson and Crick) developed a hypothesis. He hypothesized there was an as yet undiscovered adapter molecule that "read" the mRNA codon directly and carried with it an amino acid for protein synthesis. The predicted adapter molecule was experimentally identified and named **transfer RNA (tRNA)**.

- Researchers studied protein synthesis by identifying the essential ingredients of active protein synthesis. These ingredients are mRNA, ribosomes, amino acids, guanosine triphosphate, and adenosine triphosphate. One additional and indispensible ingredient came from cellular fractions and was eventually shown to be a small RNA. The molecules discovered by this research team were the transfer RNAs.

- A tRNA bound to its corresponding amino acid is called an **aminoacyl tRNA**; the addition of amino acids to tRNAs is mediated by a specific enzyme (**aminoacyl tRNA synthetase, Figure 16.11**). There are many types of tRNAs, each carrying a specific amino acid and recognizing a codon on the mRNA. They are the adapters between the amino acids and the mRNA that Crick had predicted.

- A key experiment is highlighted in **Figure 16.12**, revealing the transfer of an amino acid from a tRNA to a growing peptide chain. The ribosome mediates this transfer. How was the location of the amino acid determined in this experiment?

What Do tRNAs Look Like?

- It is hard to understand what a tRNA really does during protein synthesis without examining tRNA structure and putting the tRNA, ribosome, and mRNA together in a diagram. Carefully examine **Figure 16.13**. Notice the 3′ end of the tRNA to which amino acids are added; the 3′-ACC sequence is found at 3′ end of all tRNAs. Throughout the structure are a series of **stem-loop** structures that give shape to the tRNA. These structures are formed by complementary base pairing among adjacent, antiparallel strands. A third important characteristic is the **anticodon** region.

- The anticodon actually forms an **antiparallel association** (by hydrogen bonding) with the codon on the mRNA strand. If the aminoacyl tRNA matches the codon being presented, the amino acid on the 3′ end is presented to the ribosome for addition to the protein being synthesized.

How Many tRNAs Are There?

- Since the genetic code includes 61 amino acid codons, scientists originally thought there should be 61 distinct tRNAs that match each codon. However, most cell types contain only ~40 different tRNAs. How can this be?

- Francis Crick proposed the **wobble hypothesis**, predicting that the first two nucleotides in a codon are more important than the third (wobble) base for some amino acids. Recall that some amino acids are coded for by several different codons, but that all start with the same two bases (CAA and CAG both code for glutamine).

- It was experimentally shown that *some* tRNAs are capable of binding to codons that match at only two nitrogenous bases, while the third site contains a mismatch that is tolerated. This third position in the anticodon and the corresponding codon is called the "wobble" position. Thus, 40 tRNAs are capable of mediating the translation of all 61 codons.

16.5 The Structure and Function of Ribosomes

- Protein is synthesized by the ribosome as it reads the mRNA. This process occurs both in the **cytoplasm** and on the surface of the **endoplasmic reticulum**.

- The ribosome was first isolated using **density-gradient centrifugation**. A crude mixture of proteins can be divided into components, separating them by their relative density. Ribosomes are composed of protein and rRNA.

- Density-gradient centrifugation was used to isolate intact ribosomes and showed that the *E. coli* ribosome was composed of two major parts, now known as the **50S** (large subunit) and the **30S** subunits (small subunit). S is a unit of sedimentation density. Each subunit is composed of proteins and rRNA strands. The rRNA found in ribosomes is important to the function and structure of the ribosome; it is never translated.

- Ribosomes **self-assemble**, meaning that if the individual components are placed together, they immediately assemble into a ribosome.

- As shown in **Figure 16.14** (provided here), there are several key structural sites in the ribosome for tRNA

localization. These A, P, and E sites each will contain a tRNA during the process of translation.

- The A site is the ribosome's **acceptor** site, where the new tRNA enters the ribosome structure. If the anticodon on this tRNA complements the mRNA codon, the tRNA stays in the structure for involvement in translation.

- The P site is the ribosomal site where a peptide bond is formed between the amino acid attached to the tRNA in the P site and the existing peptide chain.

- The E site is the "exit" site, where a tRNA that has lost its amino acid is released from the ribosome.

- During translation the ribosome moves along the mRNA (**Figure 16.14**). A tRNA will stay associated with the same codon while in the ribosomal structure, though it moves from the ribosomal E site, to the P site, and then the E site.

Initiation

- Remember that mRNA contains sequences in addition to an open reading frame (coding for protein); there are noncoding 5′ and 3′ RNA sequences. Thus, the ribosome must locate the translation start site correctly to ensure that the correct information is translated into protein. How would you predict the process might occur? Think back to transcription; how was the correct starting site identified?

- Translation always begins with a **methionine** codon (**AUG** in the RNA sequence). The 30S subunit is recruited to a consensus sequence just 6 nucleotides upstream of the translation start site (**Figure 16.15**). Scientists have named it the **Shine-Dalgarno sequence** or the **ribosome binding site**. The consensus sequence at this site is 5′–AGGAGGU and is

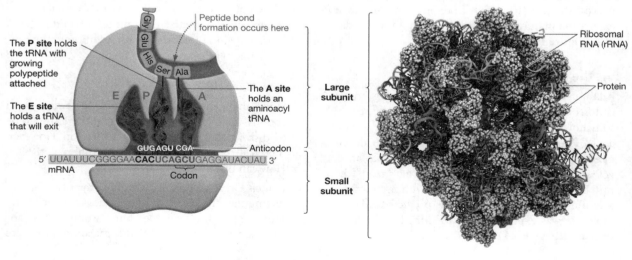

(a) Diagram of ribosome during translation (interior view)

The **P site** holds the tRNA with growing polypeptide attached

The **E site** holds a tRNA that will exit

Peptide bond formation occurs here

The **A site** holds an aminoacyl tRNA

Anticodon

5′ UUAUUUCGGGGAA**CAC**UCAGCUGAGGAUACUAU 3′
mRNA

Codon

(b) Model of ribosome during translation (exterior view)

Large subunit

Small subunit

Ribosomal RNA (rRNA)

Protein

recognized by a small rRNA in the 30S subunit via complementary strand hydrogen bonding.

- Once the 30S subunit is bound to the mRNA, the anticodon of an **initiator tRNA** (methionine) binds to the AUG. The initiator methionine is unique from other methionine tRNAs because the methionine carried on this tRNA is modified with a formyl group. The resulting methionine is called **N-formylmethionine**. Eukaryotes use a normal methionine to initiate translation. Initiator tRNA association is a complex process, mediated by a series of small proteins known as **initiation factors** (**IF**).

- The **50S subunit** binds next, associating with the 30S subunit that is already properly oriented to the mRNA translation start site.

✔ CHECK YOUR UNDERSTANDING

(e) Define the process of translation initiation in bacteria.

_____

_____

Elongation

- The ribosome-mediated process of protein synthesis can be summarized as regulated formation of peptide bonds between amino acids. *What is a peptide bond?* Review peptide bonds as discussed in **Chapter 3** to be sure you recognize the structure of amino acid and the reactive groups interacting to form a peptide bond (amine and carboxyl groups). *What characteristics of a specific amino acid make it unique from other types of amino acids?* Review the R-groups or side chains on amino acids; it is their makeup that generates the unique characteristics of each protein.

- Review **Figure 16.16** and the three steps of elongation in protein synthesis:

  1. **Filling the A site**—An aminoacyl tRNA moves into the A site of the assembled ribosome/mRNA complex. The anticodon of the rRNA forms hydrogen bonds with the codon in the A site. The adjacent P site contains the previously added tRNA with its amino acid bound by a peptide bond to the growing polypeptide chain.

  2. **Peptide bond formation**—The amino acid in the A site forms a peptide bond with the polypeptide chain, its amine group reacting with the carboxyl group of the P-site amino acid. As this bond is

formed, the P-site amino acid breaks its covalent attachment to the P-site tRNA.

  3. **Translocation**—The P-site tRNA diffuses out of the complex; the ribosome advances forward one codon (5′ to 3′), transforming the A site to the P site and making room for the next aminoacyl tRNA.

- Energy for elongation is derived from high-energy phosphate bonds in GTP molecules. Small proteins known as elongation factors also aid in the elongation process.

Termination

- When the A site encounters a stop codon, instead of a new aminoacyl tRNA, small proteins (**release factors**) enter the site and initiate release of the synthesized protein. The bond between the last amino acid in the protein and the tRNA in the P site is hydrolyzed. The ribosome then releases the protein, separates from the mRNA, and the 30S and 50S subunits dissociate from each other.

- Universal stop codons: **UAA**, **UAG**, **UGA**.

Post-Translational Modifications

- The completed protein product is often further modified to mediate proper folding and to effect protein function. These modifications can be complex, and they vary depending on the specific protein involved. As these modifications are completed after translation, they are called **post-translational modifications**.

- **Folding:** A class of proteins that mediates folding by interacting with newly synthesized proteins is called **molecular chaperones**. They are cellular proteins that function to bind to other proteins and enable their proper folding. Some chaperone proteins have been shown to associate with the ribosome tunnel from which the polypeptide chain emerges during translation. The expression of some chaperones is activated by cell stresses that might threaten protein folding. An excellent example is the **heat-shock-activated chaperone proteins**. High temperatures can dramatically influence protein folding; heat-shock proteins protect against this threat.

- **Chemical modifications:** Proteins can also be covalently modified as a part of their proper synthesis (see **Chapter 7** for details). These modifications occur in the rough endoplasmic reticulum and the Golgi apparatus. Modifications can include the addition of sugar molecules (**glycosylation**) or phosphate molecules (**phosphorylation**). Some proteins are processed by enzymatic **cleavage** of the final protein. Proteinases mediate this process and may be at the site of action for that protein, or they may be regulated by other signals.

B. MASTERING CHALLENGING TOPICS

X-Ray Crystallography and Molecular Structures (Figure 16.14)

As we examine ribosomes and protein synthesis, it is important to realize how researchers determine the three-dimensional shape of specific macromolecules. Chapter 4 introduced the process of solving the structure of a protein using X-ray diffraction analysis. Scientists have used this strategy to characterize DNA and protein interactions, protein and protein interactions, and even structural changes occurring with the activation of a specific protein (allosteric regulation).

You can find many crystallography images and resources on the Internet. Explore this technique and the data it generates with some simple searches. Here are two sites we recommend:

http://ncbi.nlm.nih.gov/Structure/MMDB/mmdb.shtml

http://boatman.med.wayne.edu/~xray/education.html

C. ASSESSING WHAT YOU'VE LEARNED

1. If translation were initiated at the AUG near the 5′ end of the RNA strand shown, what length of peptide product would be generated? (Figure 15.6 will help)
   5′–AUGAGACCGUCGAUUGAAUUAGCGUA
   ACGUAAA–3′
   a. 0
   b. 6
   c. 8
   d. 10

2. Which of the following was considered key evidence that amino acids are transferred from tRNAs to the newly formed protein during translation?
   a. Scientists used X-ray crystallography to show one amino acid being added to a new protein.
   b. When tRNAs were analyzed chemically, they typically had long polypeptide chains attached to their structure.
   c. A radioactive amino acid attached to a tRNA was found to be transferred to newly made protein in an "in vitro" translation experiment.
   d. When tRNAs were purified, they displayed high rates of protein synthesis.

3. Which of the following are required as an energy source for translation? (Hint: energy is required to add an amino acid to a tRNA, and energy is required for the ribosome to translocate down the mRNA.)
   a. Mitochondria
   b. Aminoacyl tRNA synthetase
   c. GTP and ATP
   d. GTP only

4. What is the best description of the "loops" in the tRNA molecule?
   a. Unpaired DNA bases
   b. Regions created as the result of hydrogen bonding between base pairs
   c. Regions of the molecules that remain as single-stranded RNA due to the absence of complementary base pairs
   d. Regions provided for binding to the mRNA codon

5. The tRNA shown in **Figure 16.13** in your textbook depicts the typical "L" shape of this molecule. Which of the following will most likely result if some hydrogen bonds in the region labeled "double-stranded stems" in the figure are broken using some type of chemical treatment?
   a. The tRNA will not be able to bind the amino acid properly.
   b. The tRNA will not have the correct anticodon.
   c. The anticodon will remain the same, but the amino acid bound to the tRNA will be different.
   d. The tRNA will not fit into the ribosome properly.

6. Refer to **Figure 16.15** in your textbook. Which of the following statements is *not* true regarding translation in bacteria?
   a. Initiation factors mediate the assembly of the large ribosome subunit and the mRNA.
   b. The peptide bonds between adjacent amino acids are formed within the large ribosomal subunit.
   c. The first amino acid in each protein is methionine (eukaryotes) or n-formylmethionine (prokaryotes).
   d. Translation proceeds from the 5′ end toward the 3′ end of the mRNA.

7. If there were a mutation in a ribosome that prevented tRNA molecules from moving from the E site, how many amino acids could be joined by this ribosome following translation initiation before the enzyme would arrest its activity?
   a. A maximum of 2
   b. A minimum of 2
   c. A maximum of 1
   d. None

8. Which of the following statements is correct?
   a. Protein enzymes catalyze the formation of peptide bonds between adjacent amino acids.
   b. Peptide bond formation occurs within the small subunit of the ribosome.
   c. Peptide bond formation occurs in the absence of any catalyst.
   d. RNA in the ribosome mediates the process of peptide bond formation. The ribosome is an RNA enzyme or "ribozyme."

9. Some proteins can act as infectious proteins, or prions. These proteins have an abnormal conformation and cause other "normal" proteins to change to the mutant form, resulting in a number of disorders, including mad cow disease. This process is similar to which of the following normal processes in the cell?
   a. Transcription
   b. Translation
   c. mRNA processing
   d. Post-translational modification

CHAPTER 16—ANSWER KEY

Check Your Understanding

(a) Pribnow examined the DNA sequence of a number of promoters for prokaryotic genes. He identified sequences that were always present in a promoter sequence, anticipating that this would reveal the RNA polymerase binding region. These sequence motifs are called consensus or conserved sequences.

(b) The promoters of prokaryotic genes have conserved sequences at the −10 and the −35 sites (distance to transcription start site). The −10 box contains a TATAAT sequence. The −35 box contains a TTGACA sequence. Both of these sites are critical for proper transcriptional activation.

(c) Bacteria have sigma factors present in the cell that activate the expression stress response genes. Upon heat shock, for example, a sigma factor that activates heat-shock protein expression is activated. Thus, genes that are important for stress response have unique promoters that are recognized only by the stress-activated sigma factors.

(d) Roeder and co-workers hypothesized that their RNA polymerase preparation was "too pure" and lacked a protein that normally helped to identify the proper promoter elements (a eukaryotic sigma factor?). When they repeated the experiments with crude cell extracts, the researchers found that transcription activity was promoter specific. Eventually they purified the components of the crude cell extracts that delivered this specificity—basal transcription factors.

(e) The ribosome must assemble around the mRNA to begin the process of translation. Upstream from the AUG is the Shine-Dalgarno sequence. Initiation factors bind to the Shine-Dalgarno sequence and enable the binding of a ribosomal small subunit (30S) to the region. Next, a charged tRNA (aminoacyl tRNA with *N*-formylmethionine) binds to the AUG with its anticodon sequence. Finally, the large subunit binds and completes the formation of the initiation complex.

Assessing What You've Learned

1. c; 2. c; 3. c; 4. c; 5. d; 6. a; 7. a; 8. d; 9. d

Looking Forward—Key Concepts in Later Chapters

Control of Gene Expression—Chapters 17 and 18

The next two chapters focus on the mechanisms controlling gene expression. Regulating gene expression at the chromosome is much more efficient than regulating RNA processing or ribosome activity. Much of the biology defining a cell type (e.g., a liver cell or a brain cell) is the direct result of **differential gene expression**.

Bacteria and Archaea—Chapter 28

Overuse of **antibiotics** in the United States and the resulting rise in **drug-resistant bacteria** are in the news today. Humans utilize molecules that attack critical cellular functions in bacteria (such as protein synthesis). Why don't these drugs harm our cells? Study the differences between bacteria and human cells; can you hypothesize how these drugs could be harmless to us?

Viruses—Chapter 35

Many early experiments in RNA polymerase activity were performed using viruses. **Bacteriophage viruses** infect prokaryotic cells, and **adenoviruses** infect eukaryotic cells. Experiments using each of these viruses to characterize RNA polymerase are shown in **Chapter 35**. Why was viral infection of a cell selected for these classical experiments? Upon entry into the cell, the virus takes over the native RNA polymerase for expression of the viral genes. Thus, scientists can examine the relatively few viral genes in their experiments, when compared to the large number of cellular genes. Examine this chapter for a detailed discussion of viral pathogenesis.

Chapter 17

Control of Gene Expression in Bacteria

Looking Back – Key Concepts in Early Chapters

How Genes Work—Chapter 15

With this chapter, you are adding to your knowledge of how genes work at the level of molecular biology. Regulating the expression of genes is a crucial aspect of their relevance to cell growth and adaptability. Regulation can be simplified to the interaction of proteins with specific sequences in the **promoter** region that precedes the **open reading frame (ORF)**. This interaction can increase or decrease the ability of **RNA polymerase** to transcribe that gene.

KEY CONCEPTS

☞ All organisms display a careful regulation of gene expression. A loss of regulated gene expression is extremely harmful or even lethal.

☞ During prokaryotic gene expression, regulation occurs at the level of transcription, translation, and post-translational protein processing.

☞ Glucose is a preferred catabolite in bacteria, and will be metabolically consumed before alterative catabolites, such as lactose, are consumed.

☞ The lactose operon is a classic example of regulated gene expression, and includes both negative and positive regulation of transcription. Lactose is an inducer of this operon, releasing negative regulation exerted by the lacI protein.

☞ Cyclic AMP (cAMP) is a general cellular signal for glucose depletion, activating the expression of alternative metabolic pathways such as the lactose operon.

A. CHAPTER OUTLINE

- *Escherichia coli* serves as an excellent model organism for the study of prokaryotic gene regulation. These bacterial cells adapt rapidly to **environmental changes** and can often continue to grow. The adaptation is usually based on altered gene expression, activating genes needed to respond to a specific change in conditions.

- Why not express all the genes all the time? Sets of genes are expressed for a specific cellular function, and most genes are expressed sporadically as they are needed. Continual expression of all genes would be devastating to cellular biochemistry, and would certainly cause the rapid destruction of the cell.

- Regulation of gene expression is fundamental to the identity of a cell, controlling cellular characteristics and physiology. When we discuss eukaryotic gene regulation, you will see how gene regulation controls cell identity in a multicellular organism.

17.1 Gene Regulation and Information Flow

- The central dogma of molecular biology involves several steps that can be regulated before an active protein is produced within the cell (**Figure 17.1**).

    1. **Transcriptional regulation**—Control of gene expression at the chromosome, allowing RNA polymerase binding and activation only under cellular conditions favoring expression of that gene.

    2. **Translational (post-transcriptional) regulation**—Regulatory mechanisms present in cells serve to regulate the stability of mRNA messages, to process a primary RNA transcript to an mRNA, and even to increase the efficiency of translation of that transcript.

    3. **Post-translational regulation**—The completed protein product sometimes undergoes additional processing steps to generate activity. These processing steps include phosphorylation, glycosylation, and/or enzymatic cleavage of the protein. See **Chapter 16** for more discussion of these mechanisms.

- These types of regulation occur in both prokaryotic and eukaryotic cells. Prominent within these mechanisms is the regulation of transcription.

(a) List all aspects of the central dogma of molecular biology that involve the regulation of gene expression.

_____

_____

_____

_____

Metabolizing Lactose—A Model System

- Sugars are metabolized in cells to generate energy in the form of ATP. The pathway that initiates the breakdown of glucose is called **glycolysis** (breaking down glucose and gathering energetically valuable electrons in the process).

- The fastest and easiest way to feed glycolysis is to import glucose molecules from outside the cell and utilize them in glycolysis. If glucose molecules are not available, the bacterial cell must convert available molecules to glucose or some other energetically useful molecule. To perform this conversion, enzymes are required. These enzymes are typically expressed only when glucose is unavailable but another convertible molecule is available.

- Lactose is a **disaccharide** (glucose and galactose molecules covalently attached to each other) that can be cleaved to the two constituent **monosaccharides** by an enzyme called **β-galactosidase**.

- In the absence of lactose, no β-galactosidase enzyme is expressed in *E. coli* cells. Lactose is an **inducer** of the β-galactosidase gene (and other genes in the lactose operon).

- What about cells having both lactose and glucose available at the same time? Is one sugar preferred over the other? **Jacques Monod** and colleagues tested the hypothesis that *E. coli* would prefer glucose over lactose (**Figure 17.2**). They measured the expression of β-galactosidase under three conditions: media with only glucose, media with only lactose, and media with both glucose and lactose. They found that β-galactosidase is expressed only when glucose is absent and lactose is present, consistent with Monod's hypothesis.

(b) In the presence of lactose and glucose, why do *E. coli* cells prefer glucose as a source of energy?

_____

_____

_____

17.2 Identifying Genes under Regulatory Control

Replica Plating to Find Mutant Genes

- Monod, working with **Francois Jacob**, initiated a hunt for mutant *E. coli* that are unable to metabolize lactose. This strategy allowed the scientists to screen for *E. coli* that carry mutations in the lactose utilization genes. A first step in characterizing a gene is to generate mutations in that gene and assess how those mutations impact cell function.

- Bacteria can be easily grown in **petri dishes**. Thousands of cells are spread onto each plate and allowed to grow into clonal groups of cells, or colonies. Each cell in a **colony** (a pile of cells) shares the same genetic makeup, because all have grown from a single cell. If the original cell carried a mutation in β-galactosidase, then all cells in the colony will contain this mutation.

- Examine replica plating, displayed in **Figure 17.3**. This technique is used to make "copies" of populations of cells. These copies can then be tested in lactose medium, and researchers will still have copies of each clonal population for analysis in later experiments. In the initial screen, normal (wildtype) cells were exposed to a mutation-inducing agent and then spread on glucose-containing media. The master plate was replica plated to plates that contained only lactose as a carbon source. Colonies that grew on the glucose media, but not the lactose media were retained for further study.

- **Indicator plates** were used in this mutant hunt. Monod and Jacob generated petri dishes containing the o-nitrophenyl-β-D-galactoside (**ONPG**) **indicator molecule** with a structure similar to that of lactose (two subunits bound by a β-1, 4 glycosidic bond). Thus, any enzyme that breaks down lactose would be expected to break down ONPG. Once broken down, the ONPG subunits generate a yellow coloration in the media, indicating enzyme activity. The researchers could thus examine thousands of *E. coli* colonies, identifying those that lacked the ability to break down ONPG by the absence of yellow coloration near the colony.

- Monod and Jacob were interested in these mutant cells; they hypothesized that the cells carried mutations in the genes that regulate or mediate lactose utilization.

✓ CHECK YOUR UNDERSTANDING

(c) Isolation of cells having mutations in specific types of genes is a common research strategy. Researchers studying utilization of lactose performed a screen for cells with a mutation in the *β*-galactosidase gene. How was this screen designed?

_____

_____

_____

_____

Different Classes of Lactose Metabolism Mutants

- Monod and Jacob recognized that several enzymes were likely important in the utilization of lactose. They isolated several classes of *E. coli* mutants, each containing a different defect in this process (**Table 17.1**).

  1. **β-glycosidic mutants**—Loss of this enzyme activity results in an inability to cleave lactose into glucose and galactose. These cells could not cleave ONPG either. The gene encoding *β*-galactosidase was named *lacZ*.

  2. **Galactoside permease mutants**—A protein located in the cell membrane mediates the transport of lactose into the *E. coli* cell. In the absence of this enzyme, cells cannot accumulate lactose for cellular metabolism. The gene encoding galactoside permease was named *lacY*.

  3. **Constitutive mutants**—This class of mutants has lost the ability to regulate expression of the genes needed for lactose utilization. Unlike wild-type cells, they display ONPG cleavage in the *absence* of lactose, indicating that *β*-galactosidase is being expressed continuously (constitutive expression). One mutation that resulted in

the constitutive expression of *β*-galactosidase was found in a gene named *lacI*. The "I" is for the phenotype where an inducer (lactose) is unnecessary for the expression of *β*-galactosidase. This makes perfect sense because the wild-type lacI⁺ protein **represses** gene expression, and the loss of this repression results in abnormal activation of gene expression. Many genes have been named after the phenotype observed when that protein is nonfunctional.

- In prokaryotic organisms, groups of genes with a related function are sometimes grouped together into an **operon**. The expression of an operon of genes is controlled by a single promoter element.

- Once scientists were able to generate a genetic map of the chromosomal region near the *lacZ* gene, they found that *lacY* **and a third gene**—*lacA* (an acetyltransferase)—were all located immediately adjacent to each other and under the control of a single promoter (**Figure 17.5**). The *lacI* gene was found nearby but under control of *its own promoter*. Ultimately it was shown that the *lacI* encoded protein (the *lac* repressor) regulated gene expression from the lactose operon (*lacZ, lacY, lacA*).

- A single promoter regulates the expression of all three *lac* operon genes, and that promoter is activated only under specific conditions (presence of lactose, absence of glucose).

17.3 Mechanisms of Negative Control: Discovery of the Repressor

- The lactose operon is regulated by **positive** and **negative control** mechanisms. Positive control is used when something must bind the promoter to activate expression; negative control is used when something bound to the promoter blocks expression of the operon (**Figure 17.6**).

- Leo Szilard (1950) suggested that Monod's data indicated the lacI⁺ protein was a negative regulator of *lacZ* and *lacY* expression. Thus lacI⁺ was a **repressor** of the lactose operon.

- Monod had shown that *lacI* expression was **constitutive**; in other words, it was always "on" regardless of the presence or absence of lactose in the medium. How then could lacI⁺ release its repression of the operon in the presence of lactose? Szilard proposed that the lacI⁺ protein interacted directly with lactose, and that when lacI⁺ was bound to lactose, it released the promoter element (**Figure 17.7**, provided on the next page).

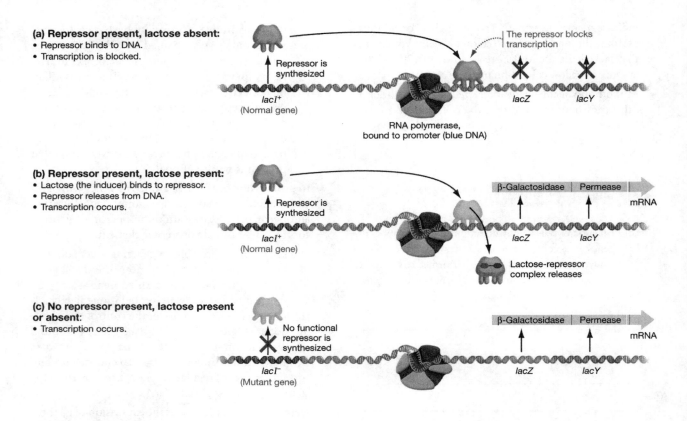

(a) Repressor present, lactose absent:
• Repressor binds to DNA.
• Transcription is blocked.

The repressor blocks transcription

Repressor is synthesized

lacI⁺ (Normal gene)

RNA polymerase, bound to promoter (blue DNA)

lacZ   *lacY*

(b) Repressor present, lactose present:
• Lactose (the inducer) binds to repressor.
• Repressor releases from DNA.
• Transcription occurs.

Repressor is synthesized

lacI⁺ (Normal gene)

β-Galactosidase | Permease
mRNA

lacZ   *lacY*

Lactose-repressor complex releases

(c) No repressor present, lactose present or absent:
• Transcription occurs.

No functional repressor is synthesized

lacI⁻ (Mutant gene)

β-Galactosidase | Permease
mRNA

lacZ   *lacY*

• To test this hypothesis, the researchers developed an experimental strategy for transferring DNA from one microbe to another (using the bacterial process of **conjugation**). A recipient group of cells carried the *lacI*- mutation; they expressed *lacY* and *lacZ* continually, regardless of whether lactose was present or absent. When DNA from a *lacI*⁺ *lacY*- *lacZ*- strain was introduced, the *lacI*⁺ protein from the introduced DNA restored negative control of the lactose operon in the recipient cells. This experiment and others showed Szilard's hypothesis was correct. The introduced *lacI*⁺ *was restoring negative regulation at the lactose operon.*

The lac Operon

• There are three central hypotheses to the Jacob and Monod model of *lac* operon regulation (1961):

   1. The *lacZ*, *lacY*, and *lacA* genes are adjacent to one another, and they are transcribed together as one polycistronic mRNA (a continuous molecule with all three encoded proteins). Expression of the three genes is coordinately regulated—the *lac* operon.

   2. The lacI protein represses transcription of the *lac* operon genes by binding to the operator motif within the promoter of the operon.

   3. Lactose is an inducer in this system, binding to the repressor protein and blocking its ability to bind the operator. The inducer activates transcription of the operon by releasing negative regulation.

17.4 Mechanisms of Positive Control: Catabolite Repression

• **Catabolism** is the metabolic breakdown of large molecules to smaller, energetically useful molecules or catabolites. If the cell has a ready supply of favored metabolites (such as glucose), catabolic pathways are typically stopped. This inhibition of a metabolic pathway is called **catabolite repression** (e.g., glucose inhibiting the lactose operon diagrammed in **Figure 17.9**).

The CAP Protein and Binding Site

• The absence of glucose actually helps to activate expression of the lactose operon via the action of another DNA-binding protein called the **catabolite activator protein (CAP)**.

- Inactivation of the *lacI*[+] repressor allows for some expression of the *lac* operon; but binding by the positive regulator protein, CAP, generates a massive induction of expression.

- CAP is **allosterically** regulated by binding to a molecule called cyclic adenosine monophosphate (cAMP), a modified nucleotide present in high concentrations only when glucose levels are depleted. CAP bound to cAMP induces expression of the lactose operon by binding the **CAP site** in the promoter region to recruit RNA polymerase to that site for transcription (**Figure 17.10**).

- The concentration of cAMP is directly controlled by an enzyme—**adenylyl cyclase**. This enzyme synthesizes cAMP from ATP and is inhibited by high concentrations of glucose (**Figure 17.11**).

- CAP protein is an example of positive control of an operon; when bound to the promoter region, CAP *increases* operon expression.

- **Figure 17.12** (provided here) is a good summary of positive and negative regulation of the *lac* operon.

CHECK YOUR UNDERSTANDING

(d) Define allosteric regulation and give an example.

_____

_____

_____

_____

(e) Catabolite repression is commonly seen in *E. coli*. It involves inhibition of some operons by the presence of a product of that operon's metabolic action. How does catabolite repression work in the lactose operon? Predict other cellular metabolic pathways that might experience catabolite repression.

_____

_____

_____

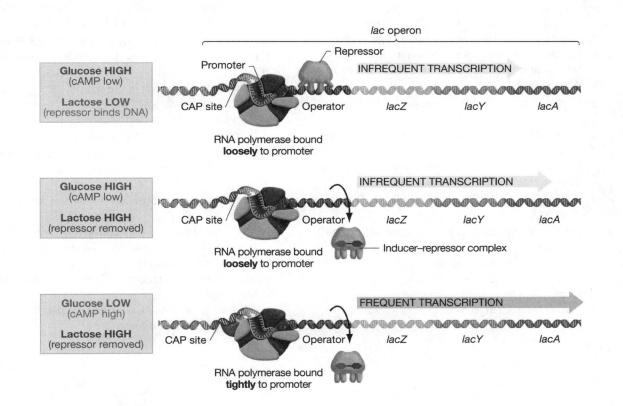

B. MASTERING CHALLENGING TOPICS

Scientists have been examining regulation of the lactose operon for decades. They have identified a complex but fascinating combination of regulatory proteins and DNA regulatory sequences. It is easy to follow how this regulation works when examining one component at a time; but when viewing both positive and negative regulation, things get complicated.

Examine **Figure 17.12** in your textbook, walking through the process that occurs for each of the conditions listed. Which conditions might result in weak expression of the operon? What interactions must occur to have full activation of the operon? This exercise is vital to study of this operon; ideally, it will resolve any confusion you may have, helping you see how positive and negative regulation of gene expression work together.

C. ASSESSING WHAT YOU'VE LEARNED

1. A hypothetical bacterium isolated from a Martian sea uses silicose as its main energy source. This silica-based sugar is one of many sugars present on Mars, and tests have shown that silicose is present everywhere on the planet. Which of the following would be the most efficient type of control for the production of silicase, the enzyme this bacterium uses to metabolize silicose?
   a. Constitutive transcription of silicase gene
   b. Negative control of transcription of silicase gene
   c. Catabolite repression transcription of silicase gene
   d. Inducible operon in control of transcription of silicase gene

2. The role of the operator sequence in the lactose operon is to:
   a. orient RNA polymerase at the transcriptional start site.
   b. bind a transcriptional repressor protein.
   c. bind an transcriptional inducer molecule.
   d. encode a transcriptional regulator (positive or negative).

3. The allosteric regulation of protein activity could be described as:
   a. regulation via allelic duplication.
   b. an increase in the expression of a protein.
   c. inactivation by proteolytic cleavage.
   d. a structural change in a protein, resulting in increased or decreased activity.

4. The pattern of β-galactosidase expression illustrated in **Figure 17.2** indicates that:
   a. lactose is the preferred source of energy in a bacterial cell.
   b. glucose is the preferred source of energy in a bacterial cell.
   c. glucose and lactose are utilized equally in bacteria.
   d. β-galactosidase is critical to glucose utilization.

5. An example of negative control of a promoter element would be:
   a. protein R binds to the promoter to induce transcription.
   b. protein R binds to RNA polymerase.
   c. protein R is allosterically regulated by cAMP binding.
   d. protein R binds to the promoter to inhibit transcription.

6. Which of the following statements is true regarding the reaction shown in **Figure 17.9b**?
   a. β-galactosidase tightly regulates cellular glucose concentration.
   b. Increasing the concentration of galactose will block β-galactosidase activity.
   c. Increasing the concentration of glucose will decrease the production of β-galactosidase.
   d. Increasing the concentration of glucose will induce production of β-galactosidase.

7. Which of these conditions would activate DNA binding by the catabolite-activated protein (CAP)?
   a. Low levels of glucose in the cell
   b. High levels of glucose in the cell
   c. Low levels of lactose in the cell
   d. High levels of lactose in the cell

8. Constitutive lactose operon mutants would be at a metabolic disadvantage compared to wild-type cells because:
   a. they would metabolize lactose more slowly than wild-type cells.
   b. they would exert energy converting a disaccharide (lactose) to monosaccharides when this process is unnecessary.
   c. constitutive mutations inhibit cell division processes.
   d. lactose metabolites would accumulate to toxic levels and harm the bacteria.

CHAPTER 17—ANSWER KEY

Check Your Understanding

(a) Gene expression is regulated at several places:

- transcription (regulation of promoter activity)
- post-transcriptional processing (RNA splicing, mRNA stability)
- translation initiation (availability of initiation factors)
- post-translational processing (modifications of proteins enhancing activity)

(b) Glucose can be acted on directly by glycolysis and other metabolic processes to produce energy. Lactose must be converted to glucose. Thus, the production of energy from glucose is more efficient and is favored by bacteria.

(c) Cultures of cells were exposed to a mutation-inducing compound and then screened for the ability to cleave a lactose analog (ONPG). Digestion of ONPG by β-galactosidase results in the release of a bright yellow molecule. Thus, cells that lack β-galactosidase activity are observed as white cells in the presence of ONPG, even under conditions that activate β-galactosidase expression.

(d) The allosteric regulation of a molecule involves alteration of its structure to regulate its activity. The binding of lactose to the lactose repressor, and cAMP to the CAP protein, are examples of structural changes that regulate the capacity of these proteins to bind DNA.

(e) Catabolite repression of the lactose operon involves binding of CAP to the promoter region. The presence of this protein dramatically enhances gene expression by helping to recruit the RNA polymerase to that site. CAP can bind to the promoter only when it is associated with cAMP, an allosteric regulator of its binding activity. The presence of high concentrations of glucose in *E. coli* cells generates a very low level of cAMP in the cytosol; the operon is thus repressed until the glucose is depleted.

There are many examples of catabolite repression in bacterial cells. One prominent example is a group of operons that mediate amino acid biosynthesis. In the absence of a specific amino acid, the operons are activated; but when the amino acid is readily available, the operon is repressed.

Assessing What You've Learned

1. a; 2. b; 3. d; 4. b; 5. d; 6. c; 7. a; 8. b

Looking Forward—Key Concepts in Later Chapters

Principles of Development—Chapter 21

Cellular differentiation in multicellular organisms is fundamentally due to the careful control of gene expression. Your introduction into regulated gene expression in this chapter builds a foundation for your future studies of **tissue differentiation**, **developmental biology**, and even the cause of diseases like **cancer**. Research into how genes are regulated has broad relevance to many areas of biology.

Animal Form and Function—Chapter 41

It is incredible to imagine a simple microbe "making careful decisions" about which genes to express at any given time. How did this come about? Varied environmental pressures exert a powerful selection on microbes, and those that survive from generation to generation must display the ability to adapt to change.

Control of Gene Expression in Eukaryotes

How Genes Work—Chapter 15

Structures of genes within chromosomes, and the nature of the **central dogma**, are complicated by the revelation of **exons and introns** in eukaryotic DNA. Each gene may produce a distinct protein, but the information may be broken up into several distinct units.

Control of Gene Expression in Bacteria—Chapter 17

Now that you've examined regulation of gene expression in eukaryotic cells, review **homologous processes of gene regulation** in prokaryotes. In what ways do regulatory mechanisms from these organisms parallel each other? What are key differences between these organisms?

KEY CONCEPTS

- The regulation of eukaryotic gene expression is significantly more complex than that of prokaryotic gene expression.

- In eukaryotic cells, **histone proteins** are closely associated with dormant genes and loosely associated with highly expressed genes.

- Eukaryotic gene expression often includes **alternative RNA splicing**. Because of this process, one gene transcript can be converted into diverse mRNAs via RNA splicing.

- Eukaryotic gene expression is regulated by short DNA sequences termed **enhancers** and **silencers**.

- One example of regulated gene expression in response to cellular stress (mutation) is both the *expression* of p53 and the *function* of p53 within the cell.

A. CHAPTER OUTLINE

18.1 Mechanisms of Gene Regulation—An Overview

- Regulation of gene expression in eukaryotic cells is more complex than that observed in prokaryotic cells. As is summarized in **Figure 18.1** (provided here), eukaryotic cells display regulation at six places within the gene expression pathway: chromatin remodeling, transcription initiation, RNA processing, mRNA stability, translation, and post-translational modification of proteins.

- Eukaryotic DNA is tightly associated with **histone** proteins, generating a DNA/protein complex known as **chromatin**. To express a specific gene, the **histones** at the site of that gene's promoter must be dislodged, a process known as **chromatin remodeling**.

- Eukaryotic gene expression involves **RNA processing**; introns are spliced out of the primary transcript, a 5′ cap is added, and a 3′ tail is added to the mRNA. The exons of some genes can be spliced together in different ways to generate unique protein products, a process known as **alternative splicing**. In addition, the stability of an mRNA in the cytosol can be regulated by cellular signals, sustaining or abbreviating the time that mRNA is available for translation.

18.2 Chromatin Remodeling

What Is Chromatin's Basic Structure?

- **Chromatin** is the molecular complex formed between DNA and proteins in the cell nucleus. Proteins in this complex are called histones, and are positively charged due to the high density of **lysine** and **arginine** amino acids in chromatin's

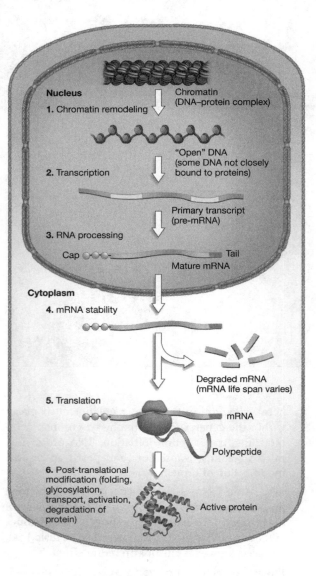

Nucleus
Chromatin
(DNA–protein complex)

1. Chromatin remodeling

"Open" DNA
(some DNA not closely
bound to proteins)

2. Transcription

Primary transcript
(pre-mRNA)

3. RNA processing

Cap ⬤○○▭ ———— Tail
Mature mRNA

Cytoplasm

4. mRNA stability

Degraded mRNA
(mRNA life span varies)

5. Translation

mRNA

Polypeptide

6. Post-translational
modification (folding,
glycosylation,
transport, activation,
degradation of
protein)

Active protein

structure. DNA is negatively charged due to the phosphate residues in the phosphodiester backbone; therefore, the DNA/histone complex forms spontaneously.

- **Histones** look like a round disk around which the DNA double helix can wrap, protecting DNA and compressing the long molecule. One chromatin complex with DNA wrapped around its structure is called a **nucleosome.** They are visible in preparations of chromatin and can be visualized using an electron microscope (**Figure 18.2**).

- **H1 histone** is important to chromatin structure; it is not part of the disk-shaped complex, but associates with the "linker" DNA between each nucleosome complex.

- The **30-nm fiber** forms as the nucleosomes form a helical structure, wrapping around an axis and

further condensing the chromosome. H1 histones interact with each other to mediate the formation of this structure.

Evidence That Chromatin Structure Is Altered in Active Genes

- **Key Research—Harold Weintraub** and **Mark Groudine** tested the hypothesis that transcription of a gene would not occur if it were tightly associated with histones in nucleosome complexes. They tested this by examining genes that are expressed differentially in varied tissues.

    1. **β-globin** is expressed at high levels in **reticulocyte** cells (a type of blood cell).

    2. **Ovalbumin** is expressed at high levels in the **female reproductive tract,** but it is not expressed in reticulocyte cells.

- To evaluate any histone association of these two genes in reticulocytes, the researchers employed an assay to degrade DNA that is *not* associated with protein (**Figure 18.3**). The DNase enzyme cannot attach and digest DNA if it is bound tightly to histone proteins.

- Chromatin (purified from reticulocyte cells) is incubated with the DNase I enzyme, which cuts any DNA not protected by protein association.

- But how can you identify the β-globin and ovalbumin chromosome fragments specifically? You must **probe** the DNA fragments produced for the intact genes (a Southern blot). The DNase fragments generated from reticulocyte cells contained intact ovalbumin genes and degraded β-globin genes. The β-globin gene was not associated with histones, consistent with their hypothesis.

- A second set of evidence was generated in studies with budding yeast (*Saccharomyces cerevisiae*). Mutations in histone genes resulted in abnormal patterns of gene expression, consistent with the hypothesis that chromatin has a key role in regulating gene expression.

How Is Chromatin Altered?

- If histone association with a chromosome occurs only at the site of inactive genes, activation of a gene must somehow be preceded by histone removal. How does this work?

- In the budding yeast (*Saccharomyces cerevisiae*), a group of proteins called the **chromatin-remodeling complex** is associated with the loss of histones from the chromosome. These complexes add acetyl ($-CH_2$ $COO-$) or methyl ($-CH_3$) groups to the histones. These additions disrupt histone association with the DNA and diminish the positive charge on the histone (**Figure 18.4**).

- **Histone acetyl transferases (HATs)** add acetyl groups to histones to allow gene expression; **histone deacetylases (HDACs)** remove acetyl groups to allow chromatin re-formation.

- The regulation of histone acetylation and deacetylation is a key area of research in gene regulation. There is much left to be discovered on this topic.

Chromatin Modifications Can Be Inherited

- In mitosis, the pattern of chromatin assembly on chromosomes is passed from mother cell to daughter cell. This serves an important role in maintaining cell identity and functionality by regulating which genes can be transcribed in that cell type. This is a form of epigenetic inheritance—where traits are influenced by factors independent of changes in DNA sequence.

✔ CHECK YOUR UNDERSTANDING

(a) Can you define the hypothesis that Weintraub and Groudine tested regarding histones and gene expression? Why did they choose to study the β-globin and ovalbumin genes?

_____

_____

_____

_____

18.3 Initiating Transcription: Regulatory Sequences and Regulatory Proteins

Some Regulatory Sequences Are Near the Promoter

- Most **regulatory sequences** are found 5' or upstream of the open reading frame. The **promoter** is the site where RNA polymerase binds to initiate transcription.

- In eukaryotic cells, the **TATA sequence** is found just upstream of the transcription start site. It is bound by the **TATA-binding protein** (or **TBP**). TBP acts in a similar manner to the sigma factor found in prokaryotic cells, locating the promoter for the RNA polymerase enzyme and aiding in the initiation of transcription. Once transcription is initiated, the TBP dissociates (like sigma factor).

- **Key research**—In the 1970s, **Yasuji Oshima** and co-workers used the budding yeast, S. *cerevisiae*, to first identify an additional **positive regulator** that aids in

regulating gene expression. They focused on genes required to utilize galactose; the genes are transcribed only when galactose is present in the growth medium.

- Yeast cell lines with a mutation in *GAL4* displayed inability to properly activate expression of galactose utilization genes. The protein encoded by this gene was shown to be a positive regulator of galactose utilization genes (similar to the CAP protein in E. *coli*). The GAL4 protein contains a helix-turn-helix motif, and a **zinc-finger domain** common in DNA-binding proteins. But what does the GAL4 protein bind?

- The GAL4 binding site was identified as a short DNA sequence 20 base pairs long and located 5' of galactose utilization genes (**Figure 18.5**). The binding site is outside of the RNA polymerase binding site (basal promoter) and called a **promoter-proximal element**. The GAL4 protein binds this site and increases transcriptional activity at the adjacent gene.

Some Regulatory Sequences Are Far from the Promoter

- **Enhancer sequences** are another type of regulatory sequence, but they are unique due to the ability to influence gene expression from a long distance (up to 100,000 base pairs from the gene). They can be located upstream or downstream of the gene they regulate, or in intron sequences within the gene.

- **Susumu Tonegawa** and co-workers selected an **immunoglobulin**-heavy chain gene for study. They recognized that regulation of this gene would be dynamic, and they focused attention on the regulatory elements they hypothesized to be present in the gene's introns.

- **Key research**—Review **Figure 18.6**, which summarizes the work of **Tonegawa** and co-workers as they characterized regulatory elements associated with this immunoglobulin gene. They suspected that a region within an intron was influencing the expression of this gene. They selectively removed regions of this intron using restriction enzymes (**see Chapter 19**), and then assayed the remaining gene locus for expression. This selective cutting and reconnecting (ligating) is a fundamental process in the manipulation and cloning of gene sequences.

- Gene expression was measured in this assay using a **Northern blot**. This type of analysis allows the researcher to measure the expression of specific genes by quantifying the amount of mRNA present in a cellular extract. RNA samples are separated on an agarose gel (similar to the Southern blot), transferred to a nylon membrane, then hybridized with a radioactive probe for a specific gene.

- The native and recombinant forms of the immunoglobulin gene were assayed for gene expression. They identified a region in the intron that was required for expression of the immunoglobulin gene; *when it was spliced out, no expression was observed.*

- This was the first demonstration of a regulatory element that regulated transcriptional activation from a considerable distance away. In addition, this element was *downstream* of the promoter.

- This element and others like it are called enhancers. They are grouped together due to the following characteristics:

  1. They can regulate a promoter from long distances upstream or downstream (up to 100,000 bases).

  2. Enhancers can regulate the promoter with similar efficiency when flipped backwards at their normal location.

  3. Enhancers can regulate the promoter with similar efficiency when moved to a new location near the gene.

  4. Enhancers increase transcriptional activity at the regulated promoter.

  5. Enhancers are specific to eukaryotic gene expression.

- **Silencers** are similar to enhancers, with varied location and distance to the regulated gene; but they repress rather than activate gene expression.

The Role of Regulatory Proteins in Differential Gene Expression

- **Key research—Julian Banerji** further characterized enhancer sequences, identifying their role in generating a **tissue-specific activation** of gene expression. They utilized the immunoglobulin gene discussed earlier, revealing that enhancer activity found in the intron was activated only in B cells (where immunoglobulins are made).

- The researchers then spliced the immunoglobulin enhancer sequence into the regulatory regions of the β-globin gene (normally not expressed in B cells). The β-globin transgene was now expressed in B cells, revealing the capacity for the enhancer element to activate a tissue-specific gene expression.

- Transcriptional **regulatory proteins** that are found in these differentiated tissues must bind enhancer, silencer, and regulatory elements found on the chromosome. Indeed researchers have identified a number of classes of regulatory proteins that are specific for different types of tissue. Muscle cells express muscle-specific genes because they express muscle-specific regulatory proteins that recognize the regulatory elements of muscle genes.

✔ CHECK YOUR UNDERSTANDING

(b) The GAL4 protein offers important clues into the mechanisms of gene regulation in eukaryotic cells (*S. cerevisiae*). What role does this protein play in activating gene expression under conditions where galactose is present? Define its similarities with the CAP protein in bacteria.

_____

_____

_____

_____

(c) Summarize what is known about enhancer sequences.

_____

_____

_____

_____

The Initiation Complex

- Many DNA-binding proteins that regulate the initiation of eukaryotic transcription have been identified.

- The two classes of regulatory proteins are: (1) **regulatory transcription factors** that bind to enhancers and silencers (promoter-proximal elements), and (2) **basal transcription factors** that associate with the core transcription complex and the promoter sequence.

- **Figure 18.8** summarizes the current model of transcription initiation in eukaryotic cells. This model summarizes the work of many researchers. Key to this model is the physical contact between regulatory proteins and the core transcription complex. For this to occur, the DNA must actually fold back on itself, allowing enhancer-binding proteins to come into contact with the core complex.

- Review **Figure 18.8**; initiation of transcription is summarized in the following steps:

  Step 1. Regulatory transcription factors recruit the chromatin-remodeling complex and histone acetyl transferases (HATs).

  Step 2. Acetylation of histones results in a weakening of histone-DNA interactions and generates "naked DNA" for TFIID binding and transcription initiation.

  Step 3. Additional regulatory transcription factors bind enhancers and promoter-proximal elements. These aid in recruiting basal transcription factors to assemble at the promoter, forming the **basal transcription complex** (~10 proteins). The TBP in the complex identifies the TATA box and orients the complex with respect to the transcription start site. Its binding is the first event in assembly at the promoter sequence.

  Step 4. The RNA polymerase enzyme and associated proteins join the basal transcription complex at the transcription start site to form the **core transcription complex**. Enhancer-binding proteins continue to interact with the large complex through the looping of DNA between their respective locations.

- A key question in the regulation of eukaryotic gene expression focuses on the regulatory mechanisms of these enhancer/silencer-binding proteins. How do they communicate with the core transcription complex?

18.4 Post-Transcriptional Control

- Gene expression is carefully regulated even after transcription to generate a primary RNA transcript. Additional regulatory steps include: (1) alternative splicing of exons to generate distinct mRNAs, (2) the regulated stability of mRNAs in the cytosol, and (3) regulated activation of the encoded protein.

Alternative Splicing of mRNAs

- Eukaryotic genes are often broken into exons that are separated by noncoding introns. The primary transcript contains both introns and exons; in the process of post-transcriptional processing, the introns are spliced out to yield the intact open reading frame for translation.

- Some genes exhibit **alternative splicing** patterns, depending on the tissue type or environmental conditions.

- Alternative splicing simply means that the splicesome attaches different exons together, depending on the cell type or condition, generating a unique protein

with each splicing pattern. The process of alternative splicing is regulated by cell-type-specific proteins that regulate the splicesome itself.

- One example from the text is the mammalian muscle-specific protein **tropomyosin**, encoded in a large gene with 14 exons. Exons that are spliced together depend on the identity of the cell in which the primary transcript is generated (**Figure 18.9**). Smooth muscle cells and striated muscle cells generate alternatively spliced forms of the tropomyosin protein, conferring one important difference in these two cell types. There are many examples of alternative splicing; each generates two or more unique proteins from a single gene.

- At least 35 percent of human genes undergo alternative splicing; though we have only around ~25,000 genes, it is anticipated that we express over 50,000 different protein products.

- The *Drosophila melanogaster* gene *Dsam* is an example of a single gene with a large number of known splice variants. This gene encodes proteins important in neuronal development in the fly embryo. This single gene can generate up to 38,000 different protein splice variants.

mRNA Stability and RNA Interference

- All mRNAs in the cytoplasm have a relatively short life span, though some mRNAs have a much shorter life span than others. Also, the life span of specific mRNAs is regulated by cellular signals. An example is the mRNA for casein—a prominent protein in milk generated during lactation. In non-lactating females the casein mRNA has an average life span of 1 hour, while in lactating females the life span is ~30 hours.

- One mechanism for the regulation of mRNA life span is known as RNA interference (**Figure 18.10**, provided here).

- **Key steps in RNA interference:**

  1. A small RNA with sequence complementarity to mRNA of interest is transcribed.

  2. In the nucleus, the small RNA spontaneously forms a hairpin structure with itself and is then trimmed to a smaller sized hairpin.

  3. In cytosol, small RNA is cut into smaller fragments (typically 22 bases in length). Now it is called a **mature micro RNA**.

  4. One strand of micro RNA associates with the enzyme complex called RISC (RNA-induced silencing complex).

  5. The RISC/micro RNA complex now targets mRNAs that have sequence complementary to

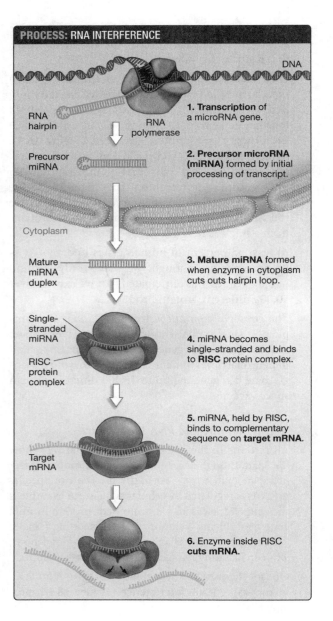

PROCESS: RNA INTERFERENCE

DNA

RNA hairpin

RNA polymerase

1. Transcription of a microRNA gene.

Precursor miRNA

2. Precursor microRNA (miRNA) formed by initial processing of transcript.

Cytoplasm

Mature miRNA duplex

3. Mature miRNA formed when enzyme in cytoplasm cuts outs hairpin loop.

Single-stranded miRNA

RISC protein complex

4. miRNA becomes single-stranded and binds to **RISC** protein complex.

5. miRNA, held by RISC, binds to complementary sequence on **target mRNA**.

Target mRNA

6. Enzyme inside RISC **cuts mRNA**.

micro RNA in enzyme complex. The matching mRNAs are destroyed by RNA-degrading activity of the RISC enzyme.

- RNA interference is a new discovery, but it is estimated that 20–30% of animal and plant genes are regulated by this process.

How Is Translation Controlled?

- In the cytosol, additional regulatory mechanisms control gene expression.

- Cytosolic factors can bind mRNAs to block their translation until needed. Oocytes (eggs) contain high concentrations of maternal mRNAs whose translation is induced following fertilization. Translation of the mRNAs is blocked by regulatory proteins that bind to them until the release of the mRNA is induced by cellular signals.

- The initiation of translation is mediated by translation initiation factors (TIFs). These can be inactivated by phosphorylation during times of cellular stress, blocking the translation of all but a few transcripts. Viruses can induce this type of response as they try to force the cell to synthesize huge amounts of viral proteins. If the endoplasmic reticulum becomes overwhelmed by proteins in need of processing, the **unfolded protein response** signal can be activated. This signal induces the phosphorylation of key initiation factors, blocking the initiation of most cellular translation.

Post-Translational Control

- The activity and stability of proteins is tightly regulated by post-translational events that can include protein glycosylation, cleavage, and phosphorylation (introduced in **Chapter 16**).

- Chaperone proteins bind to proteins being produced by the ribosome. Chaperones have the ability to mediate the process of protein folding, ensuring that hydrophobic domains in the protein are properly oriented in the center of the globular protein.

- Glycosylation (attachment of sugar) of proteins is critical to protein folding and structure. Phosphorylation or cleavage of a protein offers a rapid method for activating or inactivating protein activity.

- The signal transducers and activators of transcription (STATs) are an excellent example of post-translational regulation via protein phosphorylation.

- **STAT proteins** are **signal transducers and activators of transcription**. These proteins are present in cytosol in a dormant state, and they are activated rapidly following receipt of an extracellular signal (e.g., phosphorylation of STAT). Following activation, these proteins can migrate to the nucleus and activate the transcription of specific genes. Genes activated by STAT proteins are often involved in regulation of cell division.

- Mutant forms of some STAT proteins have been shown to bind DNA in the absence of any external signal, and to activate unregulated cell growth (cancer).

- The active form of STAT3 protein is a **dimer** of two activated STAT3 proteins, and formation of the dimer is induced by phosphorylation of a specific amino acid in the protein.

18.5 How Does Gene Expression in Bacteria Compare with That in Eukaryotes?

- **Table 18.1** compares the gene-expression regulatory mechanism displayed by eukaryotic and prokaryotic cells.

- The four primary differences between prokaryotic and eukaryotic gene expression are: (1) packaging of DNA in eukaryotes, (2) splicing of mRNA in eukaryotes, (3) complexity of transcriptional control in eukaryotes, and (4) operons in prokaryotes.

✔ CHECK YOUR UNDERSTANDING

(d) What enzyme is regulated to activate alterative splicing in eukaryotic cells? How is this regulatory mechanism activated?

_____

_____

_____

_____

18.6 Linking Cancer with Defects in Gene Regulation

- Humans are composed of trillions of cells, working together to produce a healthy organism. Each day many old, damaged, or infected cells are selectively removed from your body and replaced by the mitotic division of healthy neighboring cells. The maintenance of your size and shape requires a dynamic balance of cell death and cell division. **Cancer** is one disease that results when this balance is lost, and a group of cells grows uncontrollably, ignoring the signals sent to control its division.

- Cancer is frequently the result of mutation. Mutation-causing toxins (**mutagens**) often are shown to be **carcinogenic** (cancer causing) and are labeled as **carcinogens**.

- But how is a normal cell transformed into a cancer cell? As you may suspect, cancer often results from mutations in genes that (1) stop or slow the cell cycle or (2) trigger cell growth/division. Genes that stop/slow the cell cycle are known as **tumor suppressor genes**. When cell activity is lost due to mutation (loss-of-function mutations), cells are released from this negative control of the cell cycle. Genes that trigger cell division are called **proto-oncogenes**. Mutation can modify a proto-oncogene into an oncogene (a gain-of-function mutation), where the encoded mutant protein now induces uncontrolled cell division.

p53—Guardian of the Genome

- **p53** is a critical regulatory protein that has been shown to be abnormal in many types of cancer. Its normal cellular function is to regulate the progression of the cell cycle. Loss-of-function mutations in p53 are present in half of all human cancers.

- **Key research—Warren Maltzman** and **Linda Czyzyk** revealed that p53 expression is dramatically activated following exposure to a chemical mutagen.

- p53 was later shown to arrest cell division following DNA damage (**Figure 18.11**) and allowing the cell to properly repair any damaged genes.

- Active p53 has the ability to bind the enhancer sequences of specific genes, inducing their expression. The encoded proteins arrest the cell cycle.

- Mutations in p53 that are associated with cancer typically modify the DNA-binding domain of this protein.

- The loss of p53 function allows the cell to replicate DNA despite the presence of mutations, thus dramatically elevating the mutation rate and increasing the likelihood that a cancer-causing mutation may result.

- In cells that have experienced severe DNA damage, p53 has the ability to activate cell suicide (apoptosis). This removes severely damaged cells from the body, eliminating their potential threat to the entire organism.

✔ CHECK YOUR UNDERSTANDING

(e) Is p53 an oncogene or a tumor suppressor gene? Explain your answer.

_____

_____

_____

_____

B. MASTERING CHALLENGING TOPICS

Processing of mRNA as a form of gene regulation is a recent addition to our understanding of eukaryotic biology. The genes and enzymatic mechanisms that control alternative splicing are not yet well understood. The concept of differentially splicing a primary transcript depending upon cell type and environment is interesting, yet confusing at the same time. How did genes gain this adaptability, enabling a kind of "mix-and-match" strategy of exon combinations? This has become one of the most important areas of study in eukaryotic gene regulation. Scientists had anticipated that the genome project would reveal humans to have ~100,000 genes, basing their predictions on cellular complexity, mRNA diversity, and genome size. Completion of the genome in 2001 (see **Chapter 20**) revealed that humans have only ~20,000 genes. At first this did not seem to make sense, but subsequent research has shown that alternative splicing is very common in human cells. So, though we have only 20,000 genes, we may still be able to generate 50,000 different mRNAs by differentially splicing exons together. Alternative splicing is far from an interesting side note in eukaryotic gene expression; it is just as important as promoter-proximal elements. Certainly there will be many more surprises as scientists characterize the regulation of gene expression.

C. ASSESSING WHAT YOU'VE LEARNED

1. Histones bind tightly to DNA because of their:
   a. clamp structure located on each histone.
   b. high concentration of positively charged amino acids.
   c. DNA-specific groove structure on the histone periphery.
   d. ability to bind to specific nitrogenous bases.

2. Histone acetyl transferase (HAT) enzymes regulate gene expression by:
   a. adding acetyl groups to RNA polymerase, inactivating this enzyme.
   b. inactivating genes by modifying nitrogenous bases.
   c. strengthening the interaction of the histone with negatively charged DNA.
   d. adding acetate to histones, which recruits chromating remodeling enzymes to that site.

3. Which of the following is *not* true about eukaryotic regulatory sequences?
   a. Some increase transcription, and others suppress it.
   b. They can function if their 5' to 3' orientation is reversed.
   c. They always have the same function in different cell types.
   d. They can function if they are moved to a new location.

4. Which of the following regulatory elements is matched correctly to its chemical composition?
   a. Promoter-proximal element; DNA
   b. Enhancer; protein
   c. RNA polymerase; DNA
   d. Silencers; protein

5. Which of the following is (are) involved in transcription of prokaryotes, but *not* in eukaryotic transcription?
   a. Enhancers
   b. RNA polymerase
   c. Promoters
   d. Operators

6. In smooth and striated muscle cells, DNA for tropomyosin is the same, yet mRNA produced is different. Which of the following is the cause of this phenomenon?
   a. The presence of cell-type-specific protein influences the splicing patterns of the mRNA.
   b. The tropomyosin gene is very large and contains a large number of introns and exons.
   c. Distinct tropomyosin proteins are produced from identical mRNA sequences, depending upon the cell type.
   d. A shorter region of the tropomyosin DNA in striated muscle cells is transcribed into mRNA, resulting in a smaller mRNA transcript.

7. If a hormone bound to a cell in order to induce transcription of a given gene by using the STAT pathway, which of the following would most likely occur?
   a. An inactive STAT protein would bind to the enhancer.
   b. An activated STAT protein would bind to the gene, inhibiting transcription.
   c. The STAT proteins would become non-phosphorylated and remain in the cytoplasm.
   d. The STAT proteins would become phosphorylated and move into the nucleus.

8. Which of the following could be a mechanism to help prevent cancer?
   a. Increased expression of p53
   b. Decreased expression of p53
   c. Generation of a mutant p53 molecule that cannot bind DNA
   d. Exposing the cells to high levels of radiation

9. Tonegawa and colleagues discovered an enhancer element in a(n) _____ of an antibody-producing gene.
   a. exon
   b. poly A tail
   c. intron
   d. operator

10. The most startling discovery of Tonegawa's classic experiment was that:
    a. enhancers are large DNA sequences.
    b. enhancers encode a specific regulatory protein.
    c. enhancers can regulate a gene's expression even when located quite far from that gene.
    d. enhancers are identical in all genes they characterized.

CHAPTER 18—ANSWER KEY

Check Your Understanding

(a) Weintraub and Groudine hypothesized that an inactive gene would be more closely associated with histones than a gene being actively expressed. They used a DNase enzyme to analyze whether specific genes were associated with histones. If a gene is bound tightly to histones, the DNase is inhibited from cutting the gene into as many pieces as would be generated by a gene that is not bound to histones. They chose to study the DNA of reticulocyte cells. They examined the β-globin and ovalbumin genes because β-globin is highly expressed in these cells while ovalbumin is not expressed. They showed β-globin genes to display diminished histone association. A second experiment was performed in the yeast *Saccharomyces cerevisiae*. Yeast cells that carry mutations in histone genes display abnormal regulation of gene regulation. Thus, the histones must play an important role in regulating gene expression.

(b) GAL4 is a zinc-finger, DNA-binding protein that is analogous to the CAP-binding protein utilized in the lactose operon. In the presence of galactose, this protein binds upstream from the transcription start site of select genes and positively regulates activation of expression via interactions with RNA polymerase. The *E. coli* CAP (catabolite activator protein) binds promoter elements only when CAP is bound to cAMP (a molecular signal for glucose depletion).

(c) Enhancers elevate the level of transcription for a specific gene. Enhancers can generate a tissue or cell-type-specific regulation of gene expression.

Enhancers can work even if their normal $5' \rightarrow 3'$ orientation is flipped.

(d) Higher eukaryotic organisms divide their genes (open reading frames) into small fragments called exons, which are dispersed along the chromosome. The primary RNA transcript for the gene must then be spliced together to restore the continuous reading frame for the ribosome. There are now many examples of genes that splice the exons together differently, depending on cellular conditions. The process of splicing is mediated in the nucleus by the splicesome enzyme complex. Alternative splicing is induced when splicesome-regulating proteins are expressed in specific cell types. These proteins bind the splicesome and regulate its ability to select exons for splicing. Consequently, exons are added or deleted from the final product depending on the splicing decisions. The result is a dramatically altered protein product.

(e) The normal p53 protein functions to inhibit tumor formation in cells experiencing DNA damage. A loss of p53 elevates the liklihood of a cell becoming cancerous. Thus, p53 is a tumor suppressor gene.

Assessing What You've Learned

1. b; 2. d; 3. c; 4. a; 5. d; 6. a; 7. d; 8. a; 9. c; 10. c

Looking Forward—Key Concepts in Future Chapters

Principles of Development—Chapter 21

Cellular differentiation in multicellular organisms is fundamentally due to the careful control of gene expression. Your introduction into regulated gene expression in this chapter builds a foundation for your future studies of **tissue differentiation, developmental biology,** and even the cause of diseases like **cancer.** Research into how genes are regulated has broad relevance to many areas of biology.

The Immune System in Animals—Chapter 49

In this chapter, we've examined the regulated expression of **immunoglobulins** in **B cells.** The **immune system** is one of the most fascinating examples of cellular differentiation to generate cells with a specific function. Your body is constantly replenishing the immune system from undifferentiated **stem cells** in bone marrow, stimulating their maturation to an array of cells that fight infection.

Chapter 19

Analyzing and Engineering Genes

Mendel and the Gene—Chapter 13

Sex-linked traits were introduced in **Chapter 13** with a discussion of inheritance patterns displayed by a **white-eye trait** in *Drosophila melanogaster*. Thomas Morgan was able to make sense of the unusual segregation pattern of X-linked traits in reciprocal crosses. In humans, examining patterns of inheritance over several generations can reveal X-linked traits.

Control of Gene Expression in Eukaryotes—Chapter 18

Restriction enzymes are a key tool in the characterization and recombination of DNA sequences. Microbes make these enzymes as a defense against viral infection (they degrade the viral DNA during infection). The enzymes are key in analyses such as **Southern blotting, RFLP analysis**, and **gene cloning** into vectors.

KEY CONCEPTS

- This chapter introduces the strategies scientists use to study individual genes and to determine the function of their encoded proteins. The ability to manipulate genes has revolutionized the study of biology.

- **Gene cloning** involves the isolation of gene sequences from an organism of interest, transferring those sequences into a bacterial plasmid DNA, and amplifying the plasmid and clone within bacterial cells.

- **Genetic engineering** involves the manipulation of DNA sequences (cutting and recombining gene fragments) to enable experimental study of those sequences.

- The **polymerase chain reaction (PCR)** is an essential tool for the amplification of specific DNA sequences for cloning or genetic testing.

- **Genetic markers** on chromosomes have been used to identify the genes that cause some diseases.

- One potential application of gene cloning is **gene therapy**, where defective genes in the human body are suppressed by inserting a functional gene into affected cells.

A. CHAPTER OUTLINE

19.1 Case 1: The Effort to Cure Pituitary Dwarfism: Basic Recombinant DNA Technologies

- **Pituitary dwarfism** results from the abnormal production of **human growth hormone (hGH)**, encoded by the gene *GH1*. This disease can be inherited from generation to generation, or it can occur spontaneously.

- The mutant form of the *GH1* gene is **recessive** to the dominant allele; thus, humans can carry the mutant allele without displaying any outward abnormal phenotype.

- Humans affected by pituitary dwarfism grow slowly, reaching a maximum adult height of about 4 feet.

- hGH can be purified from pituitary glands dissected out of **human cadavers**. Isolating the hormone in this way is expensive, but it is effective in treating pituitary dwarfism.

- In the 1980s researchers discovered contamination of this hormone supply with the infectious agent that causes **neurodegenerative** disorders (**Kuru** and **Creutzfeld-Jacob disease**). Kuru was first observed in cannibalistic tribes in **New Guinea**, passing from person to person via consumption of human tissue. The discovery that some patients receiving hGH therapy had acquired the infectious agent caused this source of hGH to be banned from human treatment.

*Steps in Engineering a Safe Supply
of Growth Hormone*

- **Key research**—Recombinant DNA strategies for producing human growth hormone focused only on **cloning the human gene**, introducing the gene into the *Escherichia coli* bacterium, and having these microbes synthesize the hormone (a protein). This supply of hormone would be comparatively inexpensive and disease free.

- **Overview of genetic cloning (Figures 19.2 through 19.5)**—To isolate the *hGH* gene from human cells, researchers first isolated GH1 mRNAs from pituitary gland tissue, copied all of these mRNAs into **cDNA** using **reverse transcriptase**, and inserted these cDNA fragments into circular plasmid DNA.

- The process of copying an mRNA into a cDNA is summarized in **Figures 19.2 and 19.4**. The reverse transcriptase enzyme is unique to a type of RNA virus (retrovirus) that has the ability to convert its RNA genome into DNA inside an infected cell. This enzyme is now used to make single-stranded DNA copies of mRNAs, called cDNAs (complimentary DNA). To convert the single-stranded cDNA into double-stranded cDNA, a DNA polymerase is typically used.

- DNA fragments can be attached to each other using an enzyme called **DNA ligase**. The double-stranded DNA can be ligated into a circular plasmid, as is shown in **Figure 19.3**, generating a recombinant molecule.

- The plasmids are then used to carry the cDNA into living *E. coli* cells. Researchers can identify bacterial cells carrying the plasmid by their ability to grow in the presence of antibiotics (antibiotic resistance genes are engineered onto the plasmid).

- They then screen through all of the bacteria in search of those carrying the *hGH* cDNA (*GH1*) within a plasmid. The scientists must use a GH1-specific DNA probe to examine each colony of cells for the presence of this human gene. Following is a description of some of the key processes used in the cloning of a gene.

  1. **Using plasmids**—Plasmids are small, circular segments of DNA carried in prokaryotic cells (separate from the chromosome; see Chapter 7). Scientists use these plasmids as **vectors** (or vehicles) for DNA cloning and manipulation; they have engineered many new plasmids just for use in research, typically adding an **antibiotic resistance** gene and other components to the plasmid sequence. Thus, simply adding antibiotics to the media can determine the presence of plasmid within a bacterial cell; cells without the plasmid will die, while those with plasmid grow normally. Thus, scientists can amplify millions of copies (clones) of a piece of DNA simply by growing bacterial cells that carry that plasmid.

  2. **Using restriction endonucleases (Figure 19.3)**—Scientists can choose from a large group of enzymes that cut DNA at specific places. These restriction enzymes are made by bacteria to defend against viral infection. The enzymes cut the viral DNA but ignore similar sequences in the bacterial DNA as the bacteria block those sites by methylating nitrogenous bases at that site. Restriction enzymes are a key tool in molecular biology; they give researchers a method of cutting DNA at specific places. Thus, scientists can selectively isolate a specific piece of DNA from a plasmid or other DNA segment. Restriction endonucleases cut at specific **palindrome** sequences in the DNA (they read the same backward and forward). The process in **Figure 19.3** (steps 2–4) requires the utilization of a restriction enzyme to ensure that the cDNA is inserted into the correct site. The pieces of DNA being attached to each other must have complementary "sticky ends," which are the short, single-stranded sequences generated by a restriction enzyme digestion. After ligating the pituitary cDNAs into plasmids (now called a **library** of cDNAs), the scientists now needed to amplify these plasmids for further screening. Amplification is done easily in bacterial cells.

  3. **Transformation of bacteria**—Transferring a plasmid into a bacterial cell requires the DNA to cross the cell wall. To do this, scientists use **chemical treatment** or even a **pulse of electricity** to open holes in the cell wall. Once the DNA is inside, the cells are grown in the presence of antibiotics to select for cells carrying a plasmid and to kill cells that are not carrying plasmid. Typically a tube of cells is transformed with the collection of plasmids and then placed on a petri dish containing media plus antibiotic. Individual cells carrying the plasmid will grow into a pile of cells, known as a clonal group or colony of cells.

  4. **Screening the cDNA library for GH1**—To probe for the bacterial cells carrying the *GH1* gene, scientists had to generate a **probe** for the gene. Scientists use many types of probes to search DNA or RNA for specific sequences; but regardless of the type, each probe searches for its target by looking for nucleic acids to which it can **hybridize**. Double-stranded DNA is bound together through hydrogen bonding of its **nitrogenous bases**. DNA probes the hydrogen bond with **complementary sequence** in exactly the same way (**Figure 19.5**), and will search through long DNA and RNA sequences to

hybridize to its exact match. **Figure 19.6** (provided here) displays the screening of thousands of individual bacterial colonies with the *GH1* DNA probe. An imprint of each colony is generated on a positively charged filter paper (DNA is negatively charged). The DNA from each imprint is treated to make it single stranded, and it is then permanently attached to the filter. The filter is then probed with a radioactively labeled *GH1* probe to identify which colony on the original plate contains the *GH1* gene.

5. Once the correct plasmid-bearing bacteria are identified, the researchers can use the cloned gene to synthesize large quantities of human growth hormone (*hGH*). This is accomplished by placing the cloned gene under control of a strong bacterial promoter and putting a plasmid carrying this construct into bacteria. Bacterial cells then make a large quantity of the hGH protein that can be purified and used to treat patients.

Ethical Concerns over Recombinant Growth Hormone

- With human growth hormone being safe and readily available, concerns were raised about the appropriate use of this drug. Should doctors be allowed to prescribe this drug to any child who is merely short, or should it be restricted to individuals suffering from a more severe condition (such as hereditary dwarfism)? Currently, use of this drug is limited to children in the shortest 1.2 percent of the general population.

CHECK YOUR UNDERSTANDING

(a) Why is recombinant hGH better as a therapeutic agent than hGH isolated from human cadavers?

_____

_____

_____

(b) In the process of using a DNA probe to detect single-stranded DNA sequences on a filter paper, how does the probe "search" for the target sequences?

_____

_____

_____

19.2 Case 2—Amplification of Fossil DNA: The Polymerase Chain Reaction

- The **polymerase chain reaction (PCR)** is simply a methodology for amplifying short fragments of DNA from a few copies to billions of copies. Kary Mullis developed PCR in the early 1980s. The strategy exploits a heat-resistant DNA polymerase cloned from a bacterium that grows readily in hot springs (*Thermus aquaticus*).

- The technique is extraordinary due to its simplicity and its value to several areas of genetic analysis: gene cloning, forensic analysis, and rapid analysis of relatedness or ancestry, to name a few.

- The technique is limited only by the need for target DNA sequence, which is used to design primers for that specific target (**Figure 19.7**).

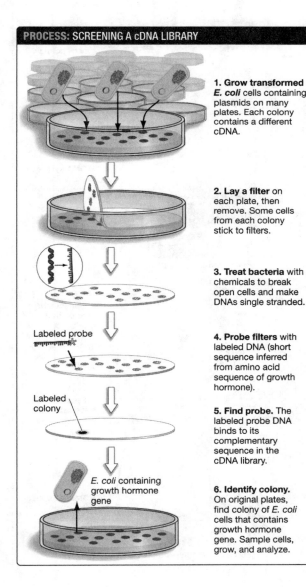

PROCESS: SCREENING A cDNA LIBRARY

1. Grow transformed *E. coli* cells containing plasmids on many plates. Each colony contains a different cDNA.

2. Lay a filter on each plate, then remove. Some cells from each colony stick to filters.

3. Treat bacteria with chemicals to break open cells and make DNAs single stranded.

Labeled probe

4. Probe filters with labeled DNA (short sequence inferred from amino acid sequence of growth hormone).

Labeled colony

5. Find probe. The labeled probe DNA binds to its complementary sequence in the cDNA library.

E. coli containing growth hormone gene

6. Identify colony. On original plates, find colony of *E. coli* cells that contains growth hormone gene. Sample cells, grow, and analyze.

- A PCR reaction contains just a few key ingredients:
  1. DNA template
  2. Two DNA primers matching the template target site
  3. dNTPs (nucleoside triphosphates)
  4. Buffer
  5. Heat-resistant DNA polymerase (known as *Taq* DNA polymerase)

- A PCR reaction requires about 30 cycles (**Figure 19.8**), with each cycle containing time points at three different temperatures:
  1. A denaturing step to separate the strands of the target sequence DNA (95°C)
  2. An annealing (primer binding) step to allow the primers to attach to the target site via hydrogen bonding of complementary sequence (45–70°C)
  3. An extension step to allow the thermostable DNA polymerase to extend the primer and synthesize new DNA in the 5′ to 3′ direction (72°C)

- How do these changes in temperature amplify DNA? Examine **Figure 19.7**, which illustrates several keys to understanding PCR.
  1. Denaturation of the template means to separate the double-stranded DNA to single strands (using temperature).
  2. PCR primers are designed for a DNA target sequence, annealing on each side of the target and offering a 3′-OH group for DNA polymerization.
  3. The primers anneal or bind to opposite strands of the DNA template. Remember that if DNA synthesis is always 5′ to 3′, this process will work only if the primers activate DNA synthesis toward each other (steps 3 and 4).
  4. The extension step allows the thermostable DNA polymerase (*Taq*) to synthesize a new strand of DNA, starting at the 3′ end of each primer.
  5. In one cycle, we've doubled the number of copies of the target sequence. Imagine repeating this for 30 cycles!

PCR in Action

- **Key research—Svante Paabo** and colleagues used PCR to compare DNA from *Homo neanderthalensis* and *Homo sapiens*, testing a hypothesis that these species may have interbred in the past. They first amplified small pieces of *H. neanderthalensis* mitochondrial DNA, isolated from ancient bones (estimated to be 30,000 years old). With the same DNA primers, they used PCR to amplify the same regions of *H. sapiens* mitochondrial DNA. They then used dideoxy sequencing on the amplified DNA fragments to compare them (see next section). They found that

H. sapiens and *H. neanderthalensis* mitochondrial DNA sequences were very different; thus, it is unlikely these two species ever interbred.

19.3 Case 3—Sanger's Breakthrough Innovation: Dideoxy DNA Sequencing

- Fredrick Sanger developed dideoxy sequencing, a key strategy for DNA analysis (summarized in **Figure 19.9**, provided here).

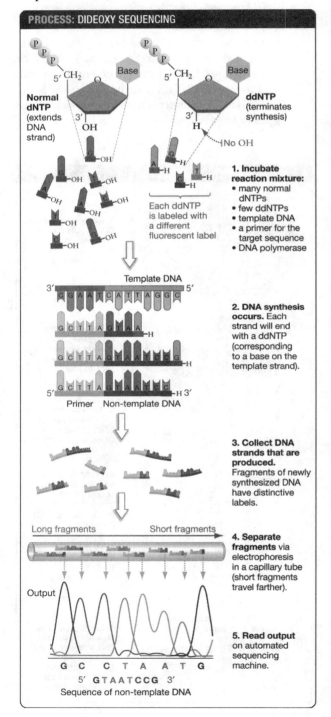

PROCESS: DIDEOXY SEQUENCING

Normal dNTP (extends DNA strand) — 3′ OH

ddNTP (terminates synthesis) — 3′ H — No OH

Each ddNTP is labeled with a different fluorescent label

1. Incubate reaction mixture:
- many normal dNTPs
- few ddNTPs
- template DNA
- a primer for the target sequence
- DNA polymerase

Template DNA

Primer   Non-template DNA

2. DNA synthesis occurs. Each strand will end with a ddNTP (corresponding to a base on the template strand).

3. Collect DNA strands that are produced. Fragments of newly synthesized DNA have distinctive labels.

Long fragments   Short fragments

4. Separate fragments via electrophoresis in a capillary tube (short fragments travel farther).

Output

G C C T A A T G
5′ GTAATCCG 3′
Sequence of non-template DNA

5. Read output on automated sequencing machine.

- Gene sequence information is of tremendous value to a scientist studying that gene. The DNA sequence can be examined to determine the amino acids in the encoded protein. In addition the DNA sequence can be used to identify other genes with related DNA sequence and presumably related protein function.

- The DNA sequencing procedure is based on dideoxynucleotides. Review the structure of a normal **deoxynucleoside** and a **dideoxynucleoside**. Why is the missing 3′–OH group so important in DNA polymerization? How might the addition of a dideoxynucleotide—to the 3′ end of the DNA strand being synthesized—affect the polymerization process?

- Once a dideoxynucleotide is added during DNA polymerization, the extension of that strand is stopped. Though the DNA polymerase is still present at the site, there is no available 3′–OH group to bond with an incoming nucleotide. Thus, the reaction is stopped precisely at that site.

- All polymerization reactions are initiated with a primer that anneals to the region of DNA to be sequenced. Each extension from this primer proceeds for some distance, unless a dideoxynucleotide is added to the chain. Thus, you can "see" where A, T, G, or C nucleotides are added by examining the fragments generated during a series of polymerization reactions containing the dideoxynucleotides.

- How is the termination of each reaction visualized experimentally? The dideoxynucleotides are each added to a sequencing reaction with the other components for the sequencing reaction. Each dideoxynucleotide is attached to a fluorescent dye (ddATP—green, ddGTP—purple, etc.). Thus, when each primed reaction stops due to the addition of a dideoxy nucleotide, the identity of that terminal nucleotide is revealed by the color of that strand of DNA. Following the reaction, the mixture of components is run into a capillary tube that denatures the DNA and separates DNA fragments by size. The color and size of each generated fragment is revealed as the DNA strands emerge from the tube. The ordering of the colored fragments emerging from the column reveals the DNA sequence of the template strand. This is automatically reported to a computer connected to the capillary tube electophoresis instrument.

19.4 Case 4—The Huntington's Disease Story: Finding Genes by Mapping

- **Key research—Nancy Wexler** and colleagues' search for the mutation behind **Huntington's disease** is an excellent example of how molecular biology has dramatically enhanced research strategies. This disease is an inherited **neurodegenerative disorder** resulting from a dominant mutation. It includes symptoms such as loss of motor function, changes in personality traits, and decreased intelligence. Onset of symptoms is usually at about 40 years of age.

- Once scientists were certain that this disease passed from generation to generation, they set out to use **pedigree analysis** to find the gene that is mutated in these families. But how does this work? How does a pedigree give information about the gene's identity?

- Researchers must first determine the mode of inheritance (dominant or recessive, X-linked or autosomal). By examining the pedigree data, scientists can generate reliable predictions of the mode of inheritance, though human pedigrees often offer a limited amount of information.

  1. In this example, *all affected individuals were born from affected parents* (no "skipping of generations"). There is no example of unaffected parents generating an affected child (which would be consistent with a recessive mutation).

  2. Also, consistent with autosomal mutations, there seems to be fairly equal distribution of affected males and females.

     The mutation responsible for Huntington's disease clearly displays an **autosomal dominant** mode of inheritance.

How Was the Huntington's Disease Gene Found?

- The DNA samples available from affected families can be analyzed with genetic markers to follow the inheritance of specific DNA regions from generation to generation.

- Researchers look for cosegregation (or linkage) of chromosomal markers with the disease phenotype, hoping to find which chromosome, and where on that chromosome, the disease-causing mutation is located (**Figure 19.10**). But to establish cosegregation of the markers and the mutation, researchers must analyze DNA from many members of affected families. Nancy Wexler and colleagues examined the DNA of people living near Lake Maracaibo, Venezuela. People in that region have a high incidence of early-onset Huntington's disease (symptoms appearing in early adulthood). By comparing patterns of inheritance in these families, and searching for regions of chromosomes that are passed from one affected individual to his or her children, the location of the mutation was eventually discovered.

- Genetic markers are simple differences in chromosomes, generally within noncoding regions. Genetic markers used include restriction enzyme cut sites (RFLP analysis), hypervariable repetitive sequences, and more recently the detection of single nucleotide polymorphisms (SNPs). Each individual inherits these polymorphic sites from his or her parents, and thus a specific DNA region inherited from a parent can be identified by the genetic markers **associated** with it (described as a haplotype). Through this type of genotyping, we can "see" a chromosomal region as it is passed on through several generations. We see that region by identifying its associated marker in individuals within the pedigree.

- **Restriction Fragment Length Polymorphisms (RFLPs)** are differences between chromosomes that are measured as the loss or gain of a restriction enzyme cut site. Digestion of the chromosome at this site generates a specific group of DNA fragments that can be detected by a probe for that DNA region. The presence or absence of restriction enzyme cut sites generates an altered pattern of bands on the resulting filter. Thus, individuals with slightly different DNA sequences are revealed through this RFLP analysis. Wexler and colleagues used RFLP analysis to examine the patterns of marker segregation through families carrying the Huntington's disease mutation. Eventually they identified a region on chromosome 4 that was always passed from one affected individual to the affected offspring.

- Imagine performing this analysis on DNA samples from a family affected by a disease-causing mutation. These analyses might find an RFLP that is *always* present in affected individuals and is *never* present in unaffected individuals! This result may suggest that the disease-causing gene is closely linked to the chromosomal location of that RFLP. Researchers use probability-based statistical analyses to establish linkage between a marker and the specific disease.

- Once the general location on chromosome 4 was established, the researchers had to examine all the genes at that site to determine which gene is mutated in Huntington's sufferers.

- Pinpointing the defect: Once the location of the Huntington's disease gene was narrowed to ~500,000 base pairs on chromosome 4, the next step was to isolate all the genes in this region and examine them for mutations.

- Each mRNA generated from this site was isolated, and the gene sequence was determined from its cDNA.

- **Key research**—Scientists found one gene within the isolated chromosomal region that was abnormal in people with Huntington's disease. It had an expanded **trinucleotide repeat** (CAG) at the 5′ (front) end of the coding region. Normal humans have about 11 to 25 of these repeats, and affected individuals have 40 or more. The codon codes for **glutamine**, and affected individuals have a protein that is inactivated due to the effect that all the extra glutamines have on its configuration.

- The protein encoded by this gene was called **huntingtin**, and the gene was named *IT15*.

What Are the Benefits of Finding a Disease Gene?

- Improved understanding of phenotype: Identifying the specific protein whose deficient function generates a disease phenotype provides researchers with an enhanced insight into that disease. Example: Huntington's disease is induced by the formation of aggregates of the huntingtin protein (IT15). These aggregates induce apoptosis in neurons. Thus, successful treatment of this disease will likely require inhibition of the formation of these aggregates.

- Therapy: A key route to developing effective therapies for a disease includes the development of an animal model. If the gene defect is known, animals can sometimes be engineered to display the disease phenotype. These model animals are an excellent starting point for the development of therapies.

- Genetic testing: Once the disease-causing gene is known, reliable tests for the defective gene can be developed and offered to the public. These tests can help to determine if one is a carrier for a disease before one chooses to have children. Also, prenatal testing can be made available.

✓ CHECK YOUR UNDERSTANDING

(c) What characteristics in a pedigree would suggest to you that a specific mutation was autosomal and recessive?

_____

_____

_____

_____

_____

19.5 Case 5—Severe Immune Disorders: The Potential of Gene Therapy

- Gene therapy involves the introduction of a gene that will replace or augment a mutant gene that is causing an abnormal phenotype. Three criteria must be met for this process to be successful:

  1. The wild-type allele for the gene in question must be sequenced and its regulatory sequences understood.

  2. There must be a method for introducing a copy of the normal allele into affected individuals.

3. The normal allele must dominate a recessive mutant allele, or replace a dominant mutation completely.

How Can Novel Alleles Be Introduced into Human Cells?

- The greatest difficulty in gene therapy strategies is stable introduction of the therapeutic gene into many cells in a human patient.

- The two prominent vectors for introducing these genes into human cells are both **viral vectors**. These viruses have been stripped of genes that might cause damage to the recipient cells, and serve only as a vehicle for delivery of the transgene.

- The gene is transferred into the viral sequence, and the virus is used to infect the patient's cells.

- The two most commonly used viruses are

  1. **Retroviruses**—These contain an RNA genome and convert themselves to DNA upon infection of the host cell. This type of virus **integrates** into the chromosomal DNA and remains there indefinitely while expressing the new gene (**Figure 19.10**).

  2. **Adenoviruses**—These viruses contain a DNA genome and are a common type of virus that can infect cells in the respiratory tract through inhalation. These viruses do not integrate into the chromosome, and thus are carried in infected cells for only a transient period.

Using Gene Therapy to Treat X-Linked Immune Deficiency

- The disease known as **severe combined immunodeficiency (SCID)** is devastating; it profoundly weakens the immune system. In this disease, white blood cells are unable to properly mature into **T-cells**.

- **Key research**—A research team led by **Marina Cavazzana-Calvo** developed a strategy to cure one category of SCID, in which a gene called *SCID-X1* (a gene on the X chromosome) is mutated. The gene encodes a **growth factor receptor** that receives a growth-activating signal from outside the cell.

- The team developed a retrovirus carrying the wild-type copy of this gene and tested the retrovirus on cells from dogs and mice.

- The first attempt at using this strategy on humans included two male infants, each affected with the mutant *SCID-X1* gene. No bone-marrow donors were available for either child.

- The research team removed **bone-marrow cells** from each of the boys, and infected the cells with the retrovirus they had constructed (**Figure 19.13**). The cells were then reintroduced into each patient, and gradually each child began to produce his own T cells! Researchers are now carefully watching T-cell production in these young boys to ensure that the transgene continues to provide a normally functioning γc receptor.

19.6 Case 6—The Development of Golden Rice: Biotechnology in Agriculture

- Most strategies for genetic engineering in agriculture focus on reducing herbivore damage, reducing competition between the crop plant and weeds (making the crop more herbicide resistant), and improving the quality of the food product.

Synthesizing Beta-Carotene in Rice

- Beta-carotene is a dietary component that can be rapidly converted to vitamin A. In countries with diets dominated by rice, people commonly suffer vitamin A deficiency. There is a small amount of beta-carotene in the rice husk, but it is removed in the polishing step of rice processing.

- **Key research—Ingo Potrykus** and colleagues set out to make a rice crop that was enriched with beta-carotene to relieve this dietary problem.

- As shown in **Figure 19.14**, the synthesis of beta-carotene can be accomplished through the conversion of a molecule called geranyl geranyl diphosphate (GGPP) in an enzyme-mediated process. Potrykus recognized that GGPP was present in the endosperm of rice seeds. His team set out to introduce into the rice endosperm the enzymes needed to convert GGPP to beta-carotene. The result would be a rice seed with an accumulation of this dietary precursor to vitamin A; but to accomplish this, he would need to make a transgenic plant.

The Agrobacterium Transformation System

- Researchers who do genetic engineering in plants often use a bacterium (*Agrobacterium tumefaciens*) to mediate the process of transferring genetic information into plant cells. This bacterium often infects plants, and as part of the infection process, the bacterium induces growth of a gall (a tumorlike mass) in the plant (**Figure 19.15**).

- The gall-inducing genes are found on a plasmid in the bacterium (Ti plasmid). The plasmid has several components, but the key part is a DNA sequence (T-DNA) that is transferred into the host plant's cells to integrate into the chromosome.

- Researchers recognized the value of DNA being transferred into the plant cell this way, and have exploited the process to carry other gene sequences along with the T-DNA sequence.

- Potrykus used *Agrobacterium* to carry the beta-carotene genes into cells from a rice plant, eventually generating a transgenic plant—now called golden rice (it is yellow from the high concentration of beta-carotene).

✔ CHECK YOUR UNDERSTANDING

(d) When compared to retroviral gene therapy strategies, what disadvantages are there for using adenoviruses instead? Advantages?

_____

_____

_____

_____

(e) Loss of the gene *GH1* results in the type I form of pituitary dwarfism, in which affected individuals are benefited by simply introducing an external source of human growth hormone (hGH). In SCID, another example from this chapter, T cells are unable to properly mature, even in the presence of the growth-stimulating signal. Why is the introduction of the external signal sufficient in one case but not the other? Explain.

_____

_____

_____

_____

B. MASTERING CHALLENGING TOPICS

Imagine how tedious the early research into Huntington's disease really was. Researchers worked together to find regions in the human genome that displayed variability regarding the presence or absence of a specific restriction enzyme cut site. The researchers then would develop a probe for this region that could be used to screen blots of digested DNA. They prepared DNA samples and screened them for each of these polymorphic regions to determine the individual's RFLP makeup. The data for this type of analysis would be complicated to look at on a gel, but eventually each individual would be typed for each locus. However, now what do the researchers do?

Somehow within the data there is linkage between a marker and the disease-causing allele. With Huntington's disease the mutation was dominant, and thus each person with the mutant allele would display the disease eventually (after age 35).

In the pedigree analysis for Huntington's disease, an RFLP marker for chromosome 4 cosegregated with the disease in Venezuelan families. It is this type of correlation that scientists hope for as they examine *thousands* of individual genetic loci. If a disease were caused by a recessive mutation, would the process of mapping the gene be significantly influenced? Explain.

Today, the methodology is faster, using PCR to amplify polymorphisms and enabling the simultaneous screening of many loci. But even with these improvements, the mapping of a disease gene is truly a daunting task, requiring a great amount of patience and endurance.

C. ASSESSING WHAT YOU'VE LEARNED

1. Genetic engineering became possible with the discovery of which of the following compounds?
   a. Restriction enzymes and DNA ligase
   b. ddNTP molecules
   c. DNA polymerase
   d. Reverse transcriptase

2. Which of the following is the main reason it was necessary to develop a method of using recombinant DNA technology to produce human growth hormone for treating individuals with pituitary dwarfism?
   a. There was no available source of functional growth hormone.
   b. The drug was extremely expensive.
   c. The growth hormone supply led to hereditary diseases.
   d. The growth hormone supply was contaminated with infectious agents.

3. When finding a colony of bacteria carrying a specific cloned gene, the probe detects these cells specifically by:
   a. binding to the protein structure of the encoded cloned protein.
   b. using dideoxy sequencing to determine each clone sequence.
   c. binding to DNA from the bacterial cells via complementary DNA sequence.
   d. identifying any plasmids with inserted cloned DNA fragments.

4. The plasmid used in the experiment illustrated in **Figure 19.3** of your text must include which of the following, before being inserted into *E. coli* cells?
   a. A marker gene
   b. Restriction recognition site
   c. cDNA synthesized from mRNA of gene being examined
   d. All of these

5. Refer to **Figure 19.6** to answer this question. Which of the following would be the most likely result if step 3 of the procedure were omitted?
   a. The X-ray film would not show any black spots.
   b. The X-ray film would show multiple black spots (one for each colony on the plate).
   c. Half of the colonies would demonstrate radioactivity on the X-ray film.
   d. The results would be identical to step 5 in the figure.

6. Restriction enzymes are produced by bacteria to break down viral DNA and protect bacterial cells from viral infection. When human DNA and viral DNA are cut with the same restriction enzyme, the "sticky ends" of the fragments can recombine and be attached to one another.

   Pick the statement that best describes the paragraph above.
   a. This is not true; viral DNA and human DNA cannot be ligated to each other due to molecular incompatability.
   b. This is not true; viral DNA does not contain restriction enzyme cut sites.
   c. Human DNA does not contain restriction enzyme cut sites like viral DNA.
   d. This is an example of the utilization of recombinant DNA technology and is used in gene therapy strategies.

7. How is the pedigree of an individual generally constructed?
   a. By sequencing the genome of the individual
   b. By interviewing the individual regarding the sex and physical traits of his or her family members
   c. By sequencing the genes of the individual's parents
   d. By analyzing the genetic fingerprint of the individual and of all of his or her family members

8. One example of a disease caused by the expansion of a repetitive DNA sequence (codon repeats) is:
   a. leukemia.
   b. pituitary dwarfism.
   c. fragile-X syndrome.
   d. Down syndrome.

9. Which of the following is a disadvantage of using biotechnological techniques in agriculture?
   a. Genetically modified corn grows poorly and is less nutritious.
   b. Transgenic herbicide-resistant plants may transfer their resistance transgene to closely related weeds (through pollination), making them resistant as well.
   c. Genetically modified crops are often slightly poisonous.
   d. It is expensive to produce seed from transgenic crop plants because the T-DNA impedes seed production.

10. Golden rice offers a potential dietary benefit because the rice grains are engineered to synthesize beta-carotene from:
    a. vitamin A.
    b. vitamin C.
    c. geranyl geranyl diphosphate.
    d. glucose.

CHAPTER 19—ANSWER KEY

Check Your Understanding

(a) Isolation of hGH from cadavers is very expensive and carries with it the risk of transferring disease (prion disease).

(b) The probes used in Southern blotting, Northern blotting, and the blots used in RFLP mapping are composed of single-stranded DNA that has been labeled in some way (typically with radioactivity). The single-stranded DNA will anneal tightly with any DNA sequences that are composed of the reverse, complementary sequence of the probe sequence. The probe and the target will anneal to one another in a manner similar to double-stranded DNA annealing to itself. Researchers can then localize the probe attached to the blot by exposing an X-ray film to the filter paper.

(c) Recessive mutations often appear in pedigrees, with affected children being born to unaffected parents. Also, recessive mutations typically skip generations within the pedigree, and sometimes may not reappear unless two related individuals have children together. If the sexes of affected individuals in the pedigree are equally distributed, the mutation is likely in an autosomal gene.

(d) Adenoviruses can be inhaled by the patient, and this offers an excellent delivery mechanism for gene therapy strategies involving the lungs and respiratory system. But these viruses don't integrate into the host-cell genome and don't offer long-term transgene expression.

(e) Type I pituitary dwarfism is caused by mutation in a gene responsible for the synthesis of hGH. SCID, in the gene therapy example, is caused by a deficient growth factor receptor. Thus, in SCID-X1, the disease results from cells being unable to *receive* the signal, whereas pituitary dwarfism is caused by an inability to *send* the signal.

Assessing What You've Learned

1. a; 2. d; 3. c; 4. d; 5. a; 6. d; 7. b; 8. c; 9. b; 10. c

Looking Forward—Key Concepts in Later Chapters

Viruses—Chapter 35

The life cycle of a virus can be quite complicated, sometimes including a period of integration into the host genome (**lysogeny**) before replicating itself and emerging from the host cell. Researchers have harnessed the lysogenic capacity of some viruses to carry genes into the human genome in **gene therapy** strategies.

The enzyme **reverse transcriptase** is produced by **retroviruses** to copy their RNA genome into DNA as part of the infectious process. Reverse transcriptase is an unusual enzyme that has become tremendously valuable in the production of **cDNA** for research.

Genomics

How Genes Work—Chapter 15

With information available from genome analysis, we can look back at our introductory examination of genes in a new light. Key findings include the discovery of **families of genes** clustered together in the genome, **pseudogenes** scattered on our chromosomes, and the amazing complexity of **gene structure**.

Analyzing and Engineering Genes—Chapter 19

Dideoxy sequencing is a critical concept to understand, because we build on that technology with an improved fluorescence labeling strategy. The chemistry of Sanger dideoxy DNA sequencing is still the basis of all sequencing used today, though some details have changed.

KEY CONCEPTS

- Genomics is the study of the entire DNA sequence found in an organism's genome, including the genes, the promoter elements that regulate gene expression, and even the noncoding DNA sequences.

- An entire genome is sequenced by: (1) digesting chromosomes into small fragments and cloning the fragments, (2) sequencing each clone individually, and (3) aligning the sequences using computer programs.

- Prokaryotic genomes (bacteria and archaeobacteria) are smaller than eukaryotic genomes. Microbes with diverse metabolic capacities have larger genomes than microbes with limited metabolic capacities.

- Eukaryotic genomes are large and complex with genes that encode proteins, genes that encode small RNAs, repetitive sequences that are not expressed, and large stretches of DNA sequence with no known function.

- Repetitive sequences can be highly polymorphic or variable. Determining an individual's alleles for a select group of repetitive sequences defines his or her genetic fingerprint.

- Functional genomics is a new application of genomics data that should benefit our ability to diagnose and treat disease.

A. CHAPTER OUTLINE

20.1 Whole-Genome Sequencing

- The first complete sequencing of the genome for a prokaryotic (*Haemophilus influenzae*) cell was completed in 1995, and the first eukaryotic genome (*Saccharomyces cerevisiae*) followed soon after in 1996.

- The Human Genome Project was essentially completed in 2001, and determined the sequence of the human genome (3.3 billion nucleotides). Genome sequencing data are publicly available through the GenBank website, which now contains more than 100 billion nucleotides of DNA sequence from thousands of species (**Figure 20.1**).

- The capacity for laboratories to generate sequence data has advanced rapidly in recent years. Review **dideoxy sequencing**, a key research tool introduced in **Chapter 19**.

- Fluorescent tags are used in DNA sequencing. These are attached to the *dideoxynucleotide triphosphate* molecules (**ddNTPs**), and they label each DNA fragment generated in the sequencing reaction *at the termination event*. Thus, if fluorescently labeled ddNTPs (ddATP, ddGTP, ddCTP, ddTTP) are added to a sequencing reaction that also contains dNTPs, the ddNTP nucleotides will occasionally be incorporated as the polymerase synthesizes the complementary strand. With the incorporation of a ddNTP, the polymerization is arrested, generating a DNA fragment that contains the fluorescent tag on one end.

- The ddNTPs are each labeled with a different-colored **fluorescent tag**, allowing researchers to determine which ddNTP addition terminated the reaction (e.g., ddATP = red, ddGTP = blue, etc.). A computer reads the color of each fragment generated in the sequencing reaction (analyzed by size) and automatically displays the DNA sequence found in the template strand.

How Are Complete Genomes Sequenced?

- The availability of automated sequencing facilities dramatically speeds up the process of genome sequencing, but researchers still require DNA templates and primers to analyze the sequence of specific regions of a chromosome. How does this process work, and what methods improve the speed of this process?

- **Shotgun sequencing** is a traditional method of genome sequencing. It involves breaking the genome into small pieces, and sequencing the small fragments. The data from the small sequencing reactions are eventually combined into distinct chromosomes. This process is summarized into seven steps in **Figure 20.2**.

    Step 1. Chromosomes are very large strands of DNA, and they must be cut into smaller fragments, about 160,000 base pairs in size. Sound waves can be used to generate these pieces from the larger chromosomes.

    Step 2. The fragments can be inserted (ligated) into a **bacterial artificial chromosome (BAC)**. A BAC is unusual in that it can carry very large fragments of DNA (up to 200,000 base pairs) within its circular structure.

- In genome analysis, the entire genome of an organism is cut into fragments and transferred into BACs for amplification in bacteria. This collection of BACs is called a **BAC library**.

- In cloning, scientists use *Escherichia coli* to make many copies of the DNA being studied. The vector with a DNA insert (a BAC or similar plasmid vector) is transferred into bacterial cells, and the cells are grown in media that contains an antibiotic. Only the cells carrying the DNA vector (with its antibiotic resistance gene) will grow. Scientists can then isolate

the vector DNA from the bacteria and have many copies to work with.

   Step 3. The BAC inserts are then analyzed individually. Each is purified, chopped into smaller 1,000-base-pair fragments using sound waves (sonication).

   Step 4. The 1000-base-pair fragments are inserted into another circular vector, called a plasmid. The resulting collection of plasmids is called a **shotgun library** because it contains small fragments of the original large piece of DNA.

   Step 5. Sequencing reactions are performed on each plasmid, and the continuous sequence of the entire chromosomal fragment carried on the BAC is determined by computer analysis.

- How does the computer *align* these sequences? The computer exploits one detail in this analysis, and it is well illustrated in steps 5 and 6 of **Figure 20.2**. The fragments that are generated **overlap** with each other and upon sequencing, these overlapping sequences are valuable in identifying the correct order of the sequenced fragments.

- The study of massive biological datasets, like DNA sequence data, is now referred to as **bioinformatics**.

Which Genomes Are Being Sequenced, and Why?

- The first genome to be completely sequenced was that of a bacterium, *Haemophilus influenzae* (1995). Since that time, many microbial genomes have been sequenced because they are relatively small genomes (the *H. influenzae* genome is 1.8 million base pairs).

- In 1996, the first eukaryotic genome sequencing project was completed. The *Saccharomyces cerevisiae* genome project sequenced all *16 chromosomes of this organism* (12 million base pairs).

- At the writing of this text, complete genome sequences are available for over 800 species.

- The selection of organisms for sequencing projects is a challenging process. Organisms are selected by the federally funded Large Scale Genome Sequencing Program based on the anticipated value of their genome sequence to the public and to the scientific community (see http://www.genome.gov/10001691).

Which Sequences Are Genes?

- Imagine the task of finding the genes in a genome sequence. What sequences identify the presence of a gene? Computers can search a sequence for reading frames with large regions that lack a stop codon. These regions are called open reading frames (ORFs) and are usually a gene (**Figure 20.3**).

- In bacteria and archaea there are no introns, and genes can be identified by highly conserved promoter

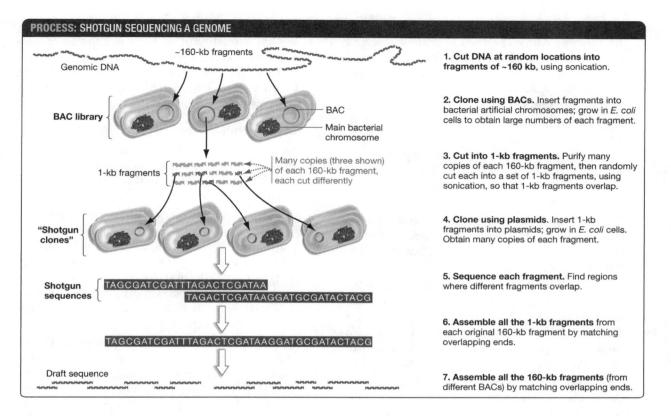

PROCESS: SHOTGUN SEQUENCING A GENOME

Genomic DNA

~160-kb fragments

BAC library

BAC

Main bacterial chromosome

1-kb fragments

Many copies (three shown) of each 160-kb fragment, each cut differently

"Shotgun clones"

Shotgun sequences

TAGCGATCGATTTAGACTCGATAA

TAGACTCGATAAGGATGCGATACTACG

TAGCGATCGATTTAGACTCGATAAGGATGCGATACTACG

Draft sequence

1. **Cut DNA at random locations into fragments of ~160 kb**, using sonication.

2. **Clone using BACs.** Insert fragments into bacterial artificial chromosomes; grow in *E. coli* cells to obtain large numbers of each fragment.

3. **Cut into 1-kb fragments.** Purify many copies of each 160-kb fragment, then randomly cut each into a set of 1-kb fragments, using sonication, so that 1-kb fragments overlap.

4. **Clone using plasmids.** Insert 1-kb fragments into plasmids; grow in *E. coli* cells. Obtain many copies of each fragment.

5. **Sequence each fragment.** Find regions where different fragments overlap.

6. **Assemble all the 1-kb fragments** from each original 160-kb fragment by matching overlapping ends.

7. **Assemble all the 160-kb fragments** (from different BACs) by matching overlapping ends.

sequences associated with a distinct open reading frame.

- In eukaryotic organisms, the process of finding genes is complicated by introns and diverse regulatory sequences. One strategy is first to identify mRNA sequences using a cDNA library and then to hunt for the genes within the genomic sequences. In the process of generating an mRNA, the introns have been spliced out; but the sequence of exons should match exactly with short sections of the genomic sequence.

- Also, researchers use known genes from other organisms to search for homologous (similar) genes in the organism being studied.

CHECK YOUR UNDERSTANDING

(a) In your own words, describe the process of aligning DNA sequences from a "shotgun cloning" strategy. Can you think of a good analogy?

_____

_____

_____

_____

CHECK YOUR UNDERSTANDING

(b) In higher eukaryotic cells, genes are often broken up into exons, with introns separating these coding fragments. How do researchers find these genes in the chromosomal sequence when performing genomic analyses?

_____

_____

_____

_____

20.2 Bacterial and Archaeal Genomes

- In this section, the discoveries generated from an analysis of prokaryotic genomes are highlighted and discussed.

The Natural History of Prokaryotic Genomes

- The size of the genome correlates with the complexity of microbial metabolic processes.

1. *Mycoplasma* bacteria have a very small genome and are unable to grow in the absence of a **host cell**. They parasitize the host cells, growing in

their cytosol and absorbing metabolites for their own use. With no host cell, *Mycoplasma* bacteria cannot survive (**obligate parasites**). They have a small, simple genome and display limited biochemical abilities as a result of having fewer available genes to utilize.

2. Prokaryotic cells that live in diverse environments must adapt to varying conditions, and as might be expected, they have large, complex genomes. They can survive in varying environments by generating enzymes that convert available nutrients to the required compounds.

- There is a great deal of genetic diversity among prokaryotes. In other words, microbes have many unique gene sequences (~15 percent) that are not encountered in other organisms.

- Some genes are found multiple times within the genome of a single species. The value of this redundancy is not known, but it may be due to slight differences in function or a requirement for a rapid induction of gene expression following activation. *Escherichia coli* displays one gene that appears 86 times on this microbe's chromosome.

- Genome sequencing revealed that there are a number of prokaryotic species with more than one chromosome.

Evidence for Lateral Gene Transfer

- Some genes or chromosome regions appear to have originated with other microbes. This transfer of large fragments of DNA from one microbe to another is called **lateral gene transfer**.

- Two key pieces of evidence for lateral gene transfer are (1) when a section of the chromosome in a species is more similar to genes in distantly related microbes than to those in closely related microbes, and (2) when sections of the chromosome are rich in either GC or AT nucleotides relative to the rest of the chromosome.

- *Thermotoga maritima* is a unique bacterial species thriving deep under the sea near hot-spring vents, where the temperature is very hot. In the same environment, some archaea species also grow well. Genome analysis revealed sections of the *T. maritima* genome that closely match the archaea genome. It appears that DNA was transferred from an archaeobacterial microbe to this eubacterial microbe, possibly improving this microbe's unusual ability to grow in a harsh environment.

- **Pathogenicity islands** have been discovered in some disease-causing strains of *E. coli* (called *E. coli* strain 0157:H7). These "islands" are chromosome regions that contain genes for aiding in infection; they also allow for localization of the bacterium to unique areas in the gut, and they generate devastating side effects such as severe diarrhea. Origins of these loci have been traced to other bacteria, such as *Shigella* spp., and even to the viruses that infect bacteria.

Environmental Sequencing

- **Metagenomics** (**environmental sequencing**) involves the simultaneous sequencing of a community of diverse microbial species that grow together in a specific environment.

- Researchers isolated a sample of microbes from the Sargasso Sea, a marine environment that is nutrient poor and metabolically challenging for microbial life. By rapidly sequencing DNA from all microbes in the sample, scientists learn about the metabolic strategies used by these microbes. By analyzing the sequence data, scientists estimate that 1800 species of microbes were present in their samples.

- Of the 1.2 million alleles of genes they identified, 780 encoded proteins that are structurally related to rhodopsin. In the human eye, rhodopsin is a component of a light-sensing biomolecule that facilitates characteristics of vision. In microbes, bacteriorhodopsin captures the energy in light to move protons across the microbial cell membrane. The proton pumping action of this molecule allows the bacteria to synthesize ATP from the proton gradient. Recall how a proton gradient is generated in the eukaryotic mitochondria to facilitate the synthesis of ATP. Notice the parallels between these microbes and the mitochondria.

- Though sequencing of individual genomes offers more explicit data about an individual organism, metagenomics offers a rapid method to characterize a community of organisms.

✔ CHECK YOUR UNDERSTANDING

(c) Using what has been learned from completed prokaryotic genome projects, what key points can now be made about genomes and prokaryotic natural histories?

_____

_____

_____

_____

20.3 Eukaryotic Genomes

Parasitic and Repeated Sequences

- **Exons** comprise less than 5 percent of the human genome, and repeated sequences (noncoding) comprise more than 50 percent of the genome. In prokaryotic cells, usually 90 percent of the genome is comprised of genes. Why is there such a dramatic difference?

- Repetitive **sequences** in the human genome are due to simple sequence repeats (noncoding), transposable elements, and families of genes.

- Sources of repetitive sequences include the **transposons**. These DNA sequences, found in our chromosomes, occasionally "jump" from one place to another (they are transposable). Our chromosomes have many transposon sequences. Transposons are sometimes referred to as "selfish genes," that function only to replicate themselves and to move about the genome.

- **Long interspersed nuclear elements (LINEs)** comprise one category of transposons. These elements have a promoter followed by two genes—**reverse transcriptase** and **integrase** (recall reverse transcriptase from our discussion of cDNA libraries).

- Expression of the LINE locus results in synthesis of both proteins; the reverse transcriptase acts on the mRNA to generate a cDNA of the LINE sequence. Integrase then nicks a new site in a chromosome and mediates integration of the cDNA fragment into that site. So the LINE transposon has moved, leaving a copy behind; a new copy is now present in the genome. See **Figure 20.5** in your text to review the mechanism of transposition.

Repeated Sequences and DNA Fingerprinting

- Short tandem repeats (STRs) are sequences that are repeated hundreds of times, nose to tail, along a portion of a chromosome. Microsatellites are small STR sequences that contain 1 to 5 bases per repeat. Minisatellites or variable number tandem repeats (VNTRs) are bigger STR sequences that contain 6 to 500 bases per repeat.

- Where do these repeats originate? One prominent hypothesis is that these repeats are produced by an abnormal process called slipped mispairing that can occur during DNA replication. Details of that model are beyond the scope of this text, but it can be thought of as a stuttering of the DNA polymerase, accidentally repeating small sequences.

- Repeating sequences are hypervariable, meaning that the number of repeats is highly variable from one person to the next and even between homologous chromosomes. Thus, each person has a unique set of repeat sequences—a genetic "fingerprint" that is unique to that one person.

- The highly polymorphic nature of these repetitive regions can be traced to crossing over during meiosis (**Figure 20.6**). Repetitive sequences would have the capacity to align improperly during prophase I, and if crossing over were to occur in this misaligned area, the chromosomes generated would have an expansion or a contraction of the size of the repetitive region. Examine the figure closely to understand this point.

- DNA fingerprinting is commonly used today in forensic science as authorities investigate and prosecute criminals. It is simply the determination of a person's haplotype at a number of highly polymorphic sites in the genome (**Figure 20.7**). There are typically 12 highly polymorphic sites analyzed for a DNA sample isolated from a crime scene. Except for identical twins, no two people will have the same haplotypes for all 12 of these sites.

- DNA fingerprinting has many other applications in the analysis of human ancestry, determining relatedness between people.

Gene Families

- Data from analysis of genomes has shown that duplicated genes are a key source of new genes. Evidence for duplication includes the discovery of groups of similar genes in one region of chromosome. Similarities include sequence, exon/intron organization, and regulatory elements. Over time, duplicated genes may change via mutation with no deleterious effect on the cell, because the original copy of the gene is still producing a functional protein. Sometimes the mutations in the duplicate gene may generate a protein with a new function.

- Genetic loci that are closely related to each other, and physically close to each other on the chromosome, are called gene families.

- One example is the globin gene family, which is a tandem array of related genes, with each member highly similar to other members of the family (**Figure 20.8**).

- Globin proteins are functionally important in the generation of hemoglobin, an oxygen-carrying molecule found in red blood cells. Some of the different globin genes contribute to generating unique forms of hemoglobin (such as fetal hemoglobin and maternal hemoglobin). In this hemoglobin example, members of this gene family enable the critical exchange of oxygen from maternal blood cells to fetal blood cells in the placenta.

- Some members of the gene family are nonfunctional pseudogenes. They are remnants of a functional copy of the gene; but due to mutation, they have lost the ability to be transcribed.

- Other examples of organisms with gene families that benefit their survival:

  1. *Anopheles gambiae* (a mosquito species) has 58 genes for fibrinogen-like proteins, used to prevent blood clotting at the site of feeding. Fruit flies eat rotting fruit (not blood) and have only 13 genes for fibrinogen-like proteins.

  2. *Plasmodium* spp. are single-celled, parasitic eukaryotes that are responsible for malaria in humans. In the blood, foreign cells are detected by our immune system cells through examination of cell-surface proteins and sugars. Plasmodium cells have between 600 and 800 copies of genes that code for cell-surface proteins. This is hypothesized to be a strategy for evading the immune system.

  3. *Prochlorococcus* spp. are prokaryotic bacteria commonly found in the ocean. These microbes are photosynthetic, gathering light to produce cellular energy. Researchers have found a correlation between the number of chlorophyll genes and the quantity of available light for each species' distinct habitat (depth in the ocean). Species living deeper in the ocean have more chlorophyll genes and a more sensitive light-catching antennae complex.

Insights from the Human Genome Project

- The analysis of genome data sets has revealed a great deal about the proportions of different categories of genes in each organism (**Figure 20.9**). Notice that most genes identified in each genome have no known function. There is a great deal more to learn.

- The human genome contains many genes for micro RNAs (miRNAs). These small RNAs are used in the regulation of the expression of other genes, as was discussed in **Chapter 18**.

- Transcripts of unknown function (TUF) are RNAs that are generated in the nucleus but never leave for the cytosol. The function of these transcripts is unknown, though scientists suspect they may play an important role for these RNAs in cell survival.

- **Table 20.1**—A major surprise from genome sequencing is that there is not a clear increase in gene number as cellular complexity increases. Humans have about the same number of genes as fruit flies and roundworms do (approximately 20,500).

- One explanation for this confusing result involves alternative splicing of RNA transcripts (discussed in **Chapter 18**). This mechanism allows for different proteins to be made from a single gene, depending upon how the RNA transcript is spliced together. Early attempts to quantify the number of alternative products for each human gene have generated a value of ~3 alternative products/gene.

- The human and chimp genomes are very similar (98.8 percent identical), with most proteins showing only one or two different amino acids. Yet, undoubtedly, humans and chimps display major differences in their biology. How can this be? One prominent hypothesis holds that the key differences between chimps and humans are in the regulation of gene expression. Small changes in regulatory elements can profoundly affect the timing and duration of gene expression.

✓ CHECK YOUR UNDERSTANDING

(d) What sequence-based evidence is there that a long interspersed element (LINE) originated as a transposon?

_____

_____

_____

_____

20.4 Functional Genomics and Proteomics

- With all the information contained in the genome, scientists must now begin to methodically analyze how these genes and their encoded proteins work together to generate a living cell—an area of study called **functional genomics**.

- Activation of gene expression is an important function of cell differentiation, stress response, and cell division. Scientists have used Northern blotting as a way to detect changes in gene expression, analyzing one gene at a time. DNA microarrays allow researchers to simultaneously measure the expression of every gene in the genome.

- Imagine comparing all of the genes expressed in normal liver cells, compared to cancerous liver cells. The differences in gene expression may offer important insight into the nature of mutations that generate a cancer cell.

- Researchers create microarrays by putting thousands of tiny spots onto a glass slide (or chip), with each spot containing short strands of single-stranded DNA (**Figure 20.10**). Then fluorescently labeled cDNA strands (copied from mRNA, introduced earlier in the chapter) can be used to probe the glass slide. The spots that bind to the cDNA probe reveal the expressed genes within that cell type. In a typical experiment, cDNAs from normal cells (control) are labeled one color, and cDNAs from abnormal cells are labeled another color. If the two cell types display the same level of expression for a gene, equal amounts of each color of cDNA will bind to that gene's spot on the slide (green = red). However, if the abnormal cells display an increase in that gene's expression, one color will increase relative to the other. Notice **Figure 20.11** (provided here), which illustrates this display of colored spots to reveal changes in gene expression.

- The microarray is an incredibly valuable tool for the simultaneous, quantitative analysis of all gene expression in a group of cells.

What Is Proteomics?

- All the proteins produced from an organism's genome are referred to as the proteome (Greek "ome" means *all*). Imagine being able to simultaneously analyze all of the proteins produced by a specific model cell rather than studying a single protein at a time. That type of analysis is a central goal of proteomics research.

One type of proteomic assay involves spotting proteins onto a solid matrix (such as glass) and then exposing these proteins to a pure preparation of a specific protein or molecule. Detection of protein-to-protein binding and even catalytic activity may then be measured, thus identifying which proteins in the array interact with the introduced protein/molecule.

Applied Genomics in Action: Understanding Cancer

- Genome projects offer huge databases of information for studying living organisms. One application of this knowledge is the pursuit of disease prevention and treatment strategies. For instance, one focus has been identifying the genes that contribute to the progression of cancer.

- Using microarrays, scientists identified genes that are expressed in cancer cells but are not expressed in noncancerous cells. In this study, data from 40 different comparisons of cancer to noncancer revealed 69 genes that were consistently mis-expressed in cancerous cells.

- This study and others like it will improve our understanding of the genetic changes that are responsible for the generation of cancerous tissue.

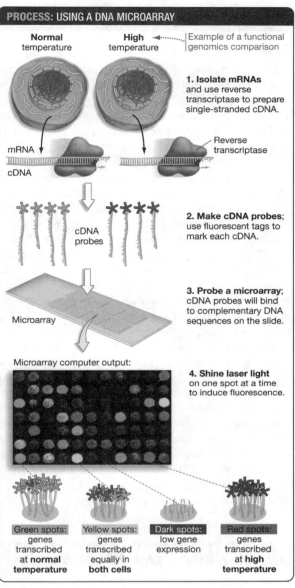

PROCESS: USING A DNA MICROARRAY

Normal temperature   High temperature   Example of a functional genomics comparison

1. Isolate mRNAs and use reverse transcriptase to prepare single-stranded cDNA.

mRNA   Reverse transcriptase
cDNA

cDNA probes

2. Make cDNA probes; use fluorescent tags to mark each cDNA.

Microarray

3. Probe a microarray; cDNA probes will bind to complementary DNA sequences on the slide.

Microarray computer output:

4. Shine laser light on one spot at a time to induce fluorescence.

Green spots: genes transcribed at **normal temperature**

Yellow spots: genes transcribed equally in **both cells**

Dark spots: low gene expression

Red spots: genes transcribed at **high temperature**

CHECK YOUR UNDERSTANDING

(e) When performing a microarray analysis, how is gene expression from the two categories of cells quantitatively examined? In other words, how do scientists generate a probe that reveals the level of gene expression for each of the genes being analyzed?

_____

_____

_____

_____

B. MASTERING CHALLENGING TOPICS

Rational Drug/Vaccine Design

Once scientists know all the genes of a particular organism, how can they use this information to develop better antibiotics, vaccines, or even medicines that prevent the progression of disease? Remember that the proteins within cells perform tasks and generate structure in a cell. Knowing a gene's DNA sequence allows researchers to predict the amino acid sequence in the protein that would be synthesized by that gene and offers clues to the gene's structure and function. Examining the genome offers vital insight into the "complete picture" of cellular function. Thus, strategies for treating a specific disease will now utilize genomics data sets to choose better targets for drug action.

One example is the production of a vaccine against a dangerous microbe (*Neisseria meningitidis*) that is one cause of infections in spinal fluid. A vaccine includes inactivated components of an infectious organism, and it is injected into a patient to reveal identifiable characteristics of the pathogen to the immune system. Thus, as soon as that organism enters the body, the immune system recognizes the danger and rapidly reacts. Genomic analysis of *N. meningitidis* has revealed all of the proteins encoded with that genome, so that researchers can now test all of these proteins for their ability to stimulate the immune system effectively. By understanding this genome sequence, scientists can make a better vaccine, one that will save many lives. The value of sequencing the human genome is undeniable; but certainly we are only beginning to grasp the best strategies for using this information. What applications do you see?

C. ASSESSING WHAT YOU'VE LEARNED

1. Using fluorescent tags in dideoxy sequencing is an improvement on using radioactive tags for the following reasons (choose all that apply):
   a. automated laser scanners can read results and compile data on a computer.
   b. radioactive sequencing is inaccurate.
   c. each ddNTP nucleotide is a different color, and thus the sequence can be determined from electrophoresis of a single sample (four are needed in radioactive labeling).
   d. fluorescent nucleotides are less expensive than radioactive nucleotides are.

2. Scientists studying the "normal" benign *E. coli* and the pathogenic *E. coli* strain O157:H7 have made which of the following determinations?
   a. The O157:H7 strain has fewer genes than the benign strain.
   b. The O157:H7 strain has more introns than the benign strain in key virulence-determining genes.
   c. The O157:H7 strain has the same genetic sequence as the benign strain, so differences in virulence must be the result of post-transcriptional modification of the mRNA of virulence determinants.
   d. The O157:H7 strain has gene sequences that are similar to *Shigella* bacteria that are not found in the benign strain.

3. Lateral gene transfer is indicated when (choose all that apply):
   a. a microbe has the genetic components for anaerobic growth.
   b. a microbe has an isolated region with a distinct guanine and cytosine (GC) content.
   c. a microbe has a genomic region containing genes that are very similar to genes found in the genome of a distantly related microbe.
   d. a microbe has distinct lateral gene transfer motifs in its DNA sequence.

4. *Prochlorotrichus spp.* offer an excellent example of:
   a. genomic adaptation.
   b. parasitism.
   c. microbial conjugation.
   d. microbial virulence.

5. Transposable elements:
   a. have the ability to replicate themselves and move from one location to another in a genome.
   b. occur very rarely in humans.
   c. are always active, if present.
   d. make up as much as one-half of a bacterium's genetic material.

6. Please refer to **Figure 20.5** in your textbook. If there were a mutation in the reverse transcriptase formed in this transposon-containing cell that prevented it from binding to the LINE mRNA, how would the resulting situation differ from that shown in the figure?
   a. The LINE mRNA would not be translated.
   b. A DNA copy of the LINE transposon would not be made.
   c. The reverse transcriptase would not be produced.
   d. The DNA would not be nicked.

7. Overall gene number and cellular/organismal complexity do not correlate as researchers had predicted. One explanation for this phenomenon is:
   a. our misunderstanding of biological complexity.
   b. our inability to identify genes in genome sequences.
   c. the importance of alternative splicing in multicellular organisms.
   d. the poor quality of genome sequence data.

8. The globin gene family shown in **Figure 20.8** in your textbook is thought to have arisen from which of the following processes?
   a. Polyploidy
   b. Transposable elements
   c. Unequal crossover during meiosis
   d. Alternative splicing

9. Shotgun sequencing involves:
   a. cutting genomic DNA into small overlapping fragments, sequencing each individually, and using computer software to align the sequences.
   b. isolating bacterial artificial chromosomes (BACs) containing huge genomic fragments, and sequencing each BAC using nonspecific "shotgun" primers.
   c. generating genomic DNA fragments using a high-pressure gas "shotgun" syringe, and sequencing each fragment generated.
   d. radioactive sequencing rather than fluorescent sequencing.

10. A "gene family" is:
   a. a group of functionally related genes that are located together and function together (i.e., lactose utilization).
   b. genes transferred together from one microbe to another, through conjugation.
   c. most often observed as a pathogenicity island.
   d. a group of genes that are similar in sequence, and are thought to be duplicated from a common ancestral gene.

CHAPTER 20—ANSWER KEY

Check Your Understanding

(a) Of course, this will be the student's answer and own analogy. Imagine having 100 freight trains, each 500 cars long, with exactly the same ordering of freight cars. If you were to break each train into fragments of 20 to 30 freight cars, each separated at different sites, could you examine the order of each of the sections and regenerate the original sequence of cars? Yes! You could read the sequence of each section and then put the sections together into the correct order by aligning sequences with information from overlapping sections. DNA sequencing with "shotgun cloning" uses exactly this strategy: the large BAC is broken up, sequenced in small fragments, and the original BAC sequence is determined by aligning the smaller sequences.

(b) This task is difficult, and several strategies are employed. The most productive is to search the DNA sequence for matches with known cDNA sequences isolated from cDNA libraries. The small matching regions will reveal the exons within the larger sequence.

(c) There is a general correlation between the size of a genome and the metabolic capacities of the organism.
   • Prokaryotic organisms display an incredible array of genetic diversity.
   • Prokaryotic organisms commonly display genetic redundancy (multiple copies of genes).
   • Prokaryotes display a high degree of lateral transfer of genetic information.

(d) LINEs contain a transcription start site, a reverse transcriptase gene, and an integrase gene. These two genes are known to mediate the transposition of a DNA sequence from one genomic site to another.

(e) In a microarray experiment, scientists isolate mRNA from control and treated cells. To generate a probe that quantitatively represents the mRNAs in each sample, cDNAs are synthesized from each mRNA using reverse transcriptase (see **Chapter 19**). For a microarray experiment, each cDNA is fluorescently labeled (perhaps green for control cDNAs and red for treated cDNAs). When these cDNAs are mixed, the ratio of green to red will represent the relative levels of expression, for each gene, between the control and treated cells.

Assessing What You've Learned

1. a, c; 2. d; 3. b, c; 4. a; 5. a; 6. b; 7. c; 8. c; 9. a; 10. d

Looking Forward—Key Concepts in Later Chapters

Bacteria and Archaea—Chapter 28

Bacteria display an incredible diversity of genetic sequence, and the evidence that genetic information is exchanged between even distantly related microbes is truly fascinating. Microbes appear to readily gather available genetic information via **lateral transfer of genes**, sometimes benefiting their ability to adapt and survive.

Principles of Development

Control of Gene Expression in Eukaryotes—Chapter 18

In Chapter 21 we have examined how vital the regulation of gene expression is in development. Many of the key regulatory molecules in early development are **transcription factors**, binding promoter elements to turn on genes at the proper place in the embryo and at the proper time (review the segmentation genes in *Drosophila*). Your understanding of **Chapter 18** content, transcription factors, and the regulation of gene expression is critical to your understanding of the homeobox and segmentation genes.

KEY CONCEPTS

- The process of development is complex and involves changes in groups of cells over time (temporal considerations) and changes in groups of cells in three-dimensional regions of the embryo (spatial consideration).

- To generate the complete body of an organism, embryonic development proceeds by cell division, cell death, cell movement, cell specialization or differentiation, and cell-to-cell communication.

- During development, cells become specialized or differentiated in response to the expression of specific sets of genes.

- Master regulator proteins generate the axes of the developing embryo. The axes are anterior/posterior, dorsal/ventral, and right/left. Genes that regulate early development are conserved even among distantly related organisms.

- Mutations in the genes that regulate development generate harmful defects in the embryo, but have the potential to occasionally generate new body shapes and structures that may benefit the organism.

A. CHAPTER OUTLINE

21.1 Shared Developmental Processes

- The development of a complex multicellular organism from a single cell is a fascinating process. Five categories of cellular processes drive development: cell proliferation, programmed cell death, cell movement or migration, cell differentiation, and cell-to-cell communication (**Table 21.1**, provided on the next page).

Cell Proliferation

- Cell division (mitosis) is regulated at several different checkpoints, affecting the activity of the central kinase in this process, called **mitosis-promoting factor** (MPF). Review **Chapter 11** for details on the cell cycle.

- Cells can stimulate or inhibit the growth of adjacent cells. Receptor proteins in the cell membrane typically mediate these **social controls** over the cell cycle.

- Plants and animals have a subset (small population) of cells that are not differentiated into a distinct tissue, and have the ability to rapidly divide when new cells are necessary. In plants, these cells comprise the **meristematic tissues** that give rise to branches, roots, leaves, flowers, and other structures. In animals, these

SUMMARY TABLE 21.1 **Five Essential Developmental Processes**

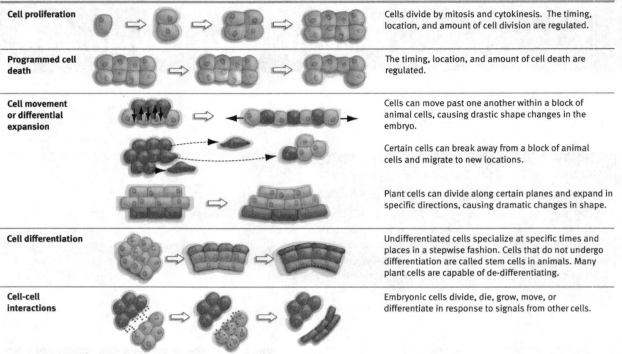

Cell proliferation	Cells divide by mitosis and cytokinesis. The timing, location, and amount of cell division are regulated.
Programmed cell death	The timing, location, and amount of cell death are regulated.
Cell movement or differential expansion	Cells can move past one another within a block of animal cells, causing drastic shape changes in the embryo.
	Certain cells can break away from a block of animal cells and migrate to new locations.
	Plant cells can divide along certain planes and expand in specific directions, causing dramatic changes in shape.
Cell differentiation	Undifferentiated cells specialize at specific times and places in a stepwise fashion. Cells that do not undergo differentiation are called stem cells in animals. Many plant cells are capable of de-differentiating.
Cell-cell interactions	Embryonic cells divide, die, grow, move, or differentiate in response to signals from other cells.

cells are called **stem cells**. In adult organisms these cells are located throughout the body, and divide to replace damaged tissue (skin, blood, liver).

Programmed Cell Death

- The preprogrammed loss of specific cells that need to be discarded during development is known as **programmed cell death** (PCD). A classic example is the development of the paw, hand, or foot. In early development a pancake-like structure forms at the end of each limb. The death of specific cells in this structure, and the retention of others, forms the digits of the hand or foot. PCD is an essential process in the development of a complex three-dimensional organ or tissue.

- A more general term that defines the regulated process of cell destruction is **apoptosis** ("falling away"). Apoptosis is essentially PCD, but occurring in different circumstances. Apoptosis is induced in cells that are functioning abnormally, typically due to mutation, viral infection, or some type of environmental insult. Apoptosis and PCD involve a protease and DNase-mediated process of destroying cellular protein and DNA, and breaking the cell into small vesicles that can be consumed by neighboring cells. The elimination of abnormal cells protects the multicellular system.

- The process of apoptosis was characterized in the roundworm *Caenorhabditis elegans*. As a worm develops from one cell to an adult, scientists recorded

each apoptotic event that occurred. They found that the same cells died in each developing roundworm embryo, and 131 total apoptotic events were observed each time a worm developed to adulthood.

- Eventually scientists identified a protease to be a key protein that regulated the apoptotic process in roundworms and mice. When the gene encoding this protease (called a caspase protease) was deleted in these organisms, apoptosis did not occur normally. **Figure 21.1** shows how the loss of apoptotic regulating genes affects the development of toes in chickens, and the brain in mice. In each case the embryo develops with an abnormal elevation in the number of surviving cells.

Cell Movement or Cell Growth

- The movement of cells from one region of an embryo to another is another key process in development. Often groups of similar-looking cells are generated and those cells subsequently migrate to distinct regions to build specific structures.

- In animal development, the process of **gastrulation** involves the organization of a mass of cells into three types of embryonic tissues. These three types of tissue are called the ectoderm (forms skin), mesoderm (forms muscle), and endoderm (forms gut). A key part of this organization of tissue is cell migration.

- Plant cells do not migrate like animal cells, but do display developmentally regulated cell division in which the plane (or direction) of cell division is carefully regulated.

Cell Differentiation

- The process of cellular differentiation is really a transformative process. The cell becomes something different structurally and functionally, though its DNA remains unchanged. How does this transformation occur? The fundamental mechanism of cell differentiation is altered gene expression. Inactivation of some genes and activation of others can transform the cell and generate differentiation.

- Cell differentiation is a progressive, sequential process as cells move from one cell type through a series of changes that culminate with fully differentiated sets of cells. Some cells have the capacity to de-differentiate (e.g., differentiated plant cells that have the potential to generate a new plant).

- **Stem cells** remain in an undifferentiated or partially differentiated state, and they are a reservoir for the production of differentiated cells in an adult organism.

Cell-Cell Interactions

- Cell-to-cell communication is critical to the process of organizing billions of cells into an organ or an organism. Review **Chapter 8** and the topic of signaling that is mediated by the extracellular matrix. Cells communicate with neighboring cells via contact points between the cells, and secrete small signaling molecules that bind to receptor proteins in the membranes of cells that are susceptible to that signal. Cellular communication is complex, yet is an area of research that has broad implications for the study of multicellular organisms.

✔ **CHECK YOUR UNDERSTANDING**

(a) What three cell types, from which are derived all adult tissues, are present in animal embryos? What tissues are generated by each type of cell?

_____

_____

_____

(b) What enzymes are important in the process of programmed cell death (or apoptosis)?

_____

_____

21.2 The Role of Differential Gene Expression in Development

- Patterns of gene expression define one differentiated cell type in comparison to another. That makes sense because ultimately proteins do things in cells, and gene expression controls which proteins are present in any given cell.

Evidence That Differentiated Plant Cells Are Genetically Equivalent

- A key question in the study of development and cellular differentiation focuses on the effect of differentiation on the genome (the complete DNA content of an organism). Is there an irreversible change in the DNA of a cell as it differentiates?

- Differentiated plant tissues are capable of de-differentiating when isolated from an intact adult plant. With proper culturing conditions the de-differentiated cells will organize into a new plant. This phenomenon is evidence that differentiated plant tissue retains all genetic information present in the embryo. This is because differentiated adult tissue retains all the genetic information necessary to generate a complete plant.

Evidence That Differentiated Animal Cells Are Genetically Equivalent

- Nuclear transfer experiments have been used in animal cells to examine if differentiation changes the DNA content. Nuclear transfer is when the nucleus of a differentiated cell is transplanted into an unfertilized egg. Researchers hypothesized that if there is no change to the DNA content of a differentiated cell, then the transferred nucleus should enable the normal development of a complete individual.

- In the 1950s, scientists were first able to isolate the nucleus of one frog cell and transfer it to an unfertilized frog egg by a process called **nuclear transfer**. Researchers found that the transfer of a nucleus from a blastula- or neurula-stage embryo into an egg was capable of generating a normal frog tadpole.

- However, isolation of nuclei from tadpoles for use in similar experiments was unsuccessful. Something was different about the DNA in these differentiated tissues.

- Ian Wilmut (1997) performed nuclear transfer experiments with sheep mammary gland cells and enucleated sheep eggs. Though the success rate was very low, this process was successful in generating cloned sheep (**Figure 21.2**).

- Tissue differentiation does not involve any irreversible change in DNA content, and it lies primarily within the differential regulation of gene expression in different cell types.

How Does Differential Gene Expression Occur?

- In **Chapter 18**, we learned that there are several places within the central dogma of gene expression where regulation occurs. In studies of development, it has become clear that transcriptional regulation is the key site of regulation used to drive cell differentiation.

- In eukaryotic cells, transcriptional control is accomplished through transcription factors that affect chromatin remodeling and bind to promoter-proximal elements.

21.3 Cell-Cell Signals Trigger Differential Gene Expression

Master Regulators Set Up the Major Body Axes

- **Pattern formation** in a developing embryo is the establishment of the three-dimensional, spatial organization that initiates many developmental processes. Cells receive molecular signals that indicate where they are within the embryo, and that information directs gene expression and cell-fate determination.

- The three body axes in an organism are

  1. the anterior (head) to posterior (tail) axis,

  2. the ventral (belly) to dorsal (back) axis, and

  3. the left to right axis (**Figure 21.3**).

- The fruit fly, *Drosophila melanogaster*, has been a key model organism for the study of pattern formation.

The Discovery of Bicoid

- Christiane Nüsslein-Volhard and Eric Wieschaus performed a large mutant hunt for genes that influence pattern formation in *Drosophila melanogaster*. They **mutagenized** flies and then examined the fly progeny for defects in development. They looked specifically for mutant flies that generated progeny with abnormal pattern formation.

- The *bicoid* mutation generated a dramatic phenotype, generating larvae that lacked structures on the anterior end (**Figure 21.4**). The normal head/thorax region develops into posterior structures, making the larvae *bicoid*, which means "two-tailed."

- The *bicoid* gene was shown to display **maternal effect inheritance**. This term means that the phenotype in the larvae is due only to the mother's genotype. If the mother lacks a functional copy of the *bicoid* gene, then the embryo will be two-tailed. The mother is responsible for providing functional *bicoid* mRNAs

to the egg, and these cytoplasmic determinants aid in larval development.

- The *bicoid* gene was isolated via **gene mapping** techniques. Review the examination of **polymorphisms** and **genetic linkage** discussed in **Chapter 13** to illustrate how this process works.

- Once the gene was isolated, researchers could generate probes for localizing *bicoid* mRNA (in situ hybridization) and protein (immunolocalization) within the embryo.

- A technique called in situ hybridization is used by researchers to localize specific mRNAs (**Figure 21.5**). If specific mRNAs are critical for the development of select cells during development, then that mRNA should be concentrated in those cells.

- In situ hybridization is a technique similar to Northern blotting, in which a labeled probe is used to screen unlabeled mRNAs for its complementary sequence. In this case the target mRNAs are located within existing tissue, thus they are said to be *in situ* (Latin for "in the normal or natural position").

- Using in situ hybridization analysis, *bicoid* mRNA was found in the anterior of the egg cytoplasm (**Figure 21.5**). This mRNA is from the mother and is present at fertilization as a crucial cytoplasmic determinant.

- Antibodies specific for the *bicoid* protein were used to determine when and where the protein was synthesized (**Figure 21.6**). Immediately following fertilization, *bicoid* protein is synthesized, and its concentration within the zygote is organized as a gradient.

- There are very high concentrations of the *bicoid* protein in the zygote's anterior and low concentrations at its posterior.

- The **gradient hypothesis** proposed by Nüsslein-Volhard stated that this gradient of *bicoid* protein was key in establishing anterior and posterior axes within the embryo. Thus, high concentrations of *bicoid* protein were hypothesized to signal anterior development, and low concentrations of *bicoid* protein were hypothesized to signal posterior development.

- To test the gradient hypothesis, Nüsslein-Volhard purified *bicoid* mRNA and injected it into different places in developing embryos. To determine if purified *bicoid* mRNA was functional when injected into an embryo, the researchers injected the mRNA into the anterior of embryos derived from mutant *bicoid* females. The injected embryos developed normally and did not display the two-tailed phenotype.

- When the researchers injected *bicoid* mRNA into wild-type (nonmutant) embryos, anterior structures always developed at the site of injection. These results strongly supported the gradient hypothesis.

- The *bicoid* protein was shown to act as a **transcription factor** within *Drosophila* embryos, binding DNA and regulating the expression of genes vital to development at the anterior of the embryo (**Figure 21.6**). Identifying what genes are regulated by *bicoid* was vital to unraveling some of the mysteries of early development.

Regulatory Genes Provide Increasingly Specific Positional Information

- Many other genes were identified via the mutant hunt conducted by Nüsslein-Volhard and Wieschaus, and each of these genes is vital to early developmental processes. Some of them can be grouped into one of three general categories of genes involved in embryo **segmentation** during *Drosophila* development (**Figure 21.7**).

- **Gap genes**—Loss-of-function mutation in gap genes results in a loss of several consecutive segments in the fly embryo. One key example of a gap gene is named *hunchback*. Mutation of *hunchback* results in embryonic phenotypes similar to those described for *bicoid* (lacking segments in the anterior of the embryo). However, *hunchback* functions via embryonic gene expression and is not a maternally derived cytoplasmic determinant like *bicoid*.

- **Pair-rule genes**—Loss of function in pair-rule genes results in the loss of **alternating** segments in the embryo, indicating that these genes are critical to the formation of every other segment formed during embryogenesis.

- **Segment polarity genes**—These genes are critical to formation of components of each segment; the loss of these genes generated embryos lacking portions of each segment.

- Segmentation genes are expressed in a precise order during embryogenesis:

  gap genes → pair-rule genes → segment polarity genes

- How does this pattern of consecutively activated genes get started? Do the genes activate each other? The maternal cytoplasmic determinant *bicoid* is known to activate expression of *hunchback* (a gap gene whose expression is located in the embryo anterior; similar to the *bicoid* expression pattern), so this is an example of a cytoplasmic determinant activating a member of the gap genes.

- But do the gap genes activate the pair-rule genes? Yes, many of the gap genes code for transcription factors that bind to regulatory sequences in pair-rule gene promoters, activating their expression at a precise time during development. As you might now suspect, members of the pair-rule gene family activate the segment polarity genes.

- This complex array of stepwise gene activation events is called a **regulatory cascade**, and it can be quite complex to consider in its entirety (**Figure 21.9**). These waves of gene expression are critical to establishing proper **pattern formation**. They continue through development, eventually being restricted to specific groups of cells as specific tissues begin to form. But how are distinct groups of cells given unique signals during development? In other words, what signals establish the individual segments?

- Edward Lewis (1940s) performed a mutant hunt with *Drosophila melanogaster*, isolating flies that displayed abnormal body patterns in the *adult fly*. Lewis isolated many mutant flies and recognized that some of the mutant phenotypes were due to abnormal development of individual segments. Some mutants had segments that developed into a phenotype that matched an adjoining segment, generating legs where antennae should be, an extra set of legs, or even an extra set of wings.

- This phenomenon of one segment developing incorrectly into the phenotype of another segment is called **homeosis** ("like condition").

- The genes responsible for proper identity in each segment were named **homeotic complex**, or *Hox* genes. Eight were identified by gene cloning and sequencing, each gene being responsible for the proper development of a specific segment or group of segments in the embryo (**Figure 21.10**, provided on the next page).

- All eight of these genes are located on one chromosome. The five genes expressed in the anterior of the embryo are called the *Antennapedia* complex; three that are expressed in the posterior of the embryo are called the *Bithorax* complex.

- The expression of the *Hox* genes is sequential, moving in order along the chromosome (*lab* to *AbdB*) and from the anterior to the posterior of the embryo (*lab* is expressed first and in the far anterior of the embryo; *AbdB* is expressed last and at the far posterior of the embryo). Examine **Figure 21.10a** closely to follow the parallels between chromosome location and the timing/distribution of gene expression.

- Each of the proteins encoded by the *Hox* genes contains a **DNA-binding region**; these regions are thought to act as transcription factors, activating genes critical for proper development of the relevant segments. The DNA sequence encoding these regions in each protein is called the **homeobox** (a 180-bp segment within each of the *Hox* genes).

- How are the *Hox* genes activated during development? There is clear evidence that the segmentation genes play a critical role in activating the proper set of

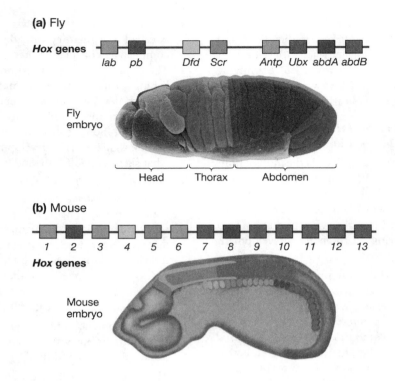

(a) Fly

Hox genes

lab pb Dfd Scr Antp Ubx abdA abdB

Fly embryo

Head Thorax Abdomen

(b) Mouse

1 2 3 4 5 6 7 8 9 10 11 12 13

Hox genes

Mouse embryo

Hox genes in each segment. Flies carrying mutant segmentation genes display abnormal activation of the *Hox* genes.

Cell-Cell Signals and Regulatory Genes Are Evolutionarily Conserved

- Homeobox genes, clustered into groups as in fruit flies, are found in the genomes of all animals examined to date. In these other organisms, they are expressed sequentially, and they appear to be transcriptional regulators with a function similar to that observed in fruit flies. Examine **Figure 21.10b**, revealing the *Hox* genes and their head-to-tail pattern of expression in a developing mouse embryo (same sequence as *Hox* genes in a fruit fly).

- The *Hox* genes play a key role in determining the head-to-tail axis of a developing embryo.

- The fruit fly *Antp* (*Antennaepedia*) gene and the mouse *Hox6b* are homologous to one another and affect the development of similar structures.

Common Signaling Pathways Are Active in Many Contexts

- Cellular signaling pathways and transcription factors regulate which genes to generate differentiation in specific tissues. Scientists have shown that specific signaling pathways and transcription factors can be used in more than one cell type, to activate unique target genes in unique groups of cells (different contexts).

- The *Wnt* (wingless) genes provide an example of signaling proteins that regulate different processes in different contexts. These genes establish the anterior-posterior axis in an embryo, but also play an important role in the development of muscle, the midbrain region, limbs, the gonads (testes and ovaries), hair follicles, parts of the intestine, and portions of the kidney. These are remarkable signaling proteins.

✔ CHECK YOUR UNDERSTANDING

(c) Within the *Drosophila* model of development, what early events establish the signals that initiate pattern formation?

_____

_____

_____

_____

(d) In *Drosophila*, how does the observed pattern of consecutively activated segmentation genes, as illustrated in **Figure 21.10a**, get started?

_____

_____

_____

21.4 Changes in Developmental Pathways Underlie Evolutionary Change

- Minor changes in the activity of key regulatory proteins (such as the homeobox proteins) can profoundly affect the development of an embryo. Mutations in these genes are thought to play a key role in change of form over time—the evolution of new body plans.

- **Tetrapods** (Latin, meaning "four-legged") include lizards (with four legs) as well as snakes. The genetic evidence is clear that these are closely related animals. Yet why do snakes lack functional legs? Some snakes have tiny pelvic bones, consistent with the belief that they are descendents of organisms that had functional legs.

- In tetrapods, *Hoxc6* and *Hoxc8* are expressed together in vertebrae that eventually develop ribs. In areas where *Hoxc6* is expressed without *Hoxc8*, forelimbs develop. In snakes, *Hoxc6* expression and *Hoxc8* expression co-occur, and functional forelimbs do not form. This indicates that the pattern of expression of these genes in snakes could have caused the loss of functional forelimbs.

- Sonic hedgehog is a transcription factor that controls hindlimb development in tetrapods. In snakes, this gene shows changes that have altered the functionality of the gene product.

- It is clear that changes in these transcription factors have played a primary role in the generation of snakes that lack functional limbs.

B. MASTERING CHALLENGING TOPICS

In Situ Hybridization Techniques (Figure 21.5)

A critical research tool introduced in this chapter is in situ hybridization, which enables scientists to detect the location of gene expression in an organism. The technique is similar to that used for a Northern blot, whereby mRNA transcripts are detected with a radioactive probe (DNA or RNA) that is complementary to a specific mRNA species (one gene). In Northern blotting, the pool of mRNAs is run on an agarose gel and transferred onto a membrane before the probe is used to screen through the transcripts. In situ hybridizations involve the addition of probe directly to the tissue.

Tissues to be examined using in situ hybridization are "fixed," freezing cellular macromolecules in place with a chemical treatment (formaldehyde). The probe enters into the cells of the tissue being tested and hybridizes to the mRNA present in the fixed cells. The site of probe hybridization is detected by placing an emulsion (a kind of liquid X-ray film) over the tissue, which is typically attached to a slide. The dots are generated as the emulsion is exposed to radioactivity and then developed in a manner similar to that used for photographic film. Researchers can then determine (with proper controls) the localization of gene expression within a tissue sample, or even within an entire organism, depending on the size of the organism. In **Figure 21.5**, the site of *bicoid* expression within a fly egg is revealed using in situ hybridization. How was this information valuable to scientists studying the process of embryo maturation in flies? Could this information have been gathered in another way?

C. ASSESSING WHAT YOU'VE LEARNED

1. List the advantages of using *Drosophila melanogaster* as a model organism for identifying the genes and processes responsible for cell-fate decisions in early development.
   a. _____
   b. _____
   c. _____
   d. _____

2. If the mother of larvae displaying the mutant *bicoid* phenotype carries only the mutant gene, how was the mother able to develop normally in the first place?
   a. The *bicoid* phenotype can be suppressed by culturing the embryos at cool temperatures.
   b. The *bicoid* mRNA is a maternal effect gene, donated to an egg by the mother regardless of the egg genotype.
   c. The mother likely also carried a *bicoid* suppressor mutation.
   d. All of the above apply.

3. What was the experimental focus of the screen performed by Nüsslein-Volhard and Wieschaus?
   a. To evaluate the mutagenic properties of varied treatments
   b. To discover the *bicoid* gene
   c. To generate two-tailed fruit-fly embryos
   d. To generate fruit flies with mutations in genes that control fruit-fly development

4. Which of the following is *not* an accurate description of the *bicoid* gene?
   a. Mutation generates larvae without anterior development.
   b. Encoded protein is a DNA-binding protein.
   c. The gene displays a maternal effect mode of inheritance.
   d. A male that is homozygous for this mutation cannot generate viable offspring.

5. Which of the following is true of maternal effect inheritance?
   a. A gene is transcribed in the offspring and affects the phenotype observed in the mother.
   b. A gene is transcribed in the mother and affects the phenotype observed in the offspring.

c. The genes of the mother negate the genes of the father.

d. The genes of the father negate the genes of the mother.

6. What do you call the 180-base-pair DNA sequence that was revealed in each gene in the *Antennapedia complex* and the *Bithorax complex*?
   a. Homeobox
   b. Homeotic complex
   c. Homeosis
   d. Homeotic genes

7. Defects in pair-rule genes cause:
   a. the loss of paired sets of structures, such as eyes.
   b. in a segmented developmental region, effects on alternating sections (every other section).
   c. defects in the anterior of the embryo.
   d. defects in the posterior of the embryo.

8. The *hunchback* gene is a type of:
   a. pair-rule gene.
   b. gap gene.
   c. segment polarity gene.
   d. *bicoid* family member.

9. When researchers injected *bicoid* mRNA at different sites of the early fruit-fly embryo, what happened?
   a. Development of legs at the site of injection
   b. Early lethality of embryo
   c. Development of anterior structures at site of injection
   d. Development of dorsal structures at site of injection

10. Stem cells are:
    a. cells in a mature organism that have the ability to develop into fully differentiated cells.
    b. found exclusively in the bone marrow.
    c. cells that occur only in early embryogenesis of an animal.
    d. a hypothetical cell type that scientists have not yet been able to isolate.

CHAPTER 21—ANSWER KEY

Check Your Understanding

(a) **ectoderm** (generates skin and nerve cells); **vegetal pole—endoderm** (generates gut and organs); **mesoderm** (generates muscle and organs).

(b) The process of programmed cell death involves the removal of unwanted cells as the developing embryo is sculpted into a functional organism. To remove a cell without damaging neighboring cells, the contents of the cell must be digested. Enzymes that destroy cellular macromolecules (DNA, RNA, protein, and lipid) mediate this process: DNases, RNases, proteases, lipases.

(c) The maternally derived cytoplasmic determinant, *bicoid*, is expressed in the anterior of the fertilized embryo and generates a gradient of *bicoid* protein (high concentration in the anterior to low concentration in the posterior). This gradient of concentration establishes a polarized pattern of gene expression.

(d) *Bicoid* is a transcription factor, activating the expression of some segmentation genes along its concentration gradient in the early embryo. The key example is the gap gene *hunchback*, whose expression is directly activated by *bicoid*. Then some of the gap genes activate the pair-rule genes, and some of the pair-rule genes activate the segment polarity genes, driving embryonic development forward.

Assessing What You've Learned

1.

(a) *Drosophila* generate large numbers of progeny very rapidly (the time from mating to reproductive maturity is about 3 weeks).

(b) *Drosophila* exhibit complex developmental processes (multicellular organism), serving as a better model than simpler multicellular organisms.

(c) *Drosophila* has well-characterized genetics (**Chapters 12** and **13**).

(d) Fruit flies are easily maintained and bred in a laboratory setting.

(e) Fruit flies can be mutagenized fairly easily (X-rays or chemical mutagens), and flies with mutant phenotype can be detected by visual inspection of progeny.

(f) Fruit-fly embryos are deposited into a rich nutrient source (in the laboratory, culturing medium), and are therefore easy to isolate for study.

2. b; 3. d; 4. d; 5. b; 6. a; 7. b; 8. b; 9. c; 10. a

Looking Forward—Key Concepts in Future Chapters

An Introduction to Animal Development—Chapter 22

Now that we have carefully developed a framework in which we can examine development, we will study animal development in more detail. **Chapter 22** introduces more detail into our understanding of early animal development, defining specific events that mold the embryo (**gastrulation**) and the genes that regulate this process.

An Introduction to Animal Development

Control of Gene Expression in Eukaryotes—Chapter 18

The regulation of **cadherin expression** is a critical component of development, enabling **selective adhesion** between specific groups of cells and disabling the adhesion between other cells. Beneath these complex developmental processes is the critical regulation of gene expression.

Analyzing and Engineering Genes—Chapter 19

Isolating the genes that regulate and mediate the developmental process enables scientists to perturb these processes in order to better characterize the molecular mechanisms behind development. Using genetic engineering, scientists can produce an abnormal pattern of gene expression, delete the expression of specific genes, and/or observe the influence of mutant proteins on the system. Examine the methodology used to clone the *GH1* gene, and then review the experiments used to clone and characterize *MyoD*.

KEY CONCEPTS

- The union of sperm and egg is dependent upon a complex array of protein interactions, physiological responses, and enzymatic reactions.

- Cell division in the early embryo is known as cleavage, and involves the rapid division of embryonic cells accompanied by little or no cell growth.

- The early embryo is organized into three layers of cell types: ectoderm (outer layer), mesoderm (middle layer), and endoderm (inner layer). These three tissues are responsible for all tissues, organs, and structures in the animal body plan.

- The process of differentiation involves the activation of specific genes within groups of cells, generating changes in cells that confer specialized function.

A. CHAPTER OUTLINE

22.1 Gamete Structure and Function

Sperm Structure and Function

- **Gametogenesis** is the formation of gametes (sperm and egg) in the reproductive organs. Fertilization is the union of one sperm with one egg to generate a diploid one-celled embryo.

- The mammalian sperm cell, which contains a haploid DNA genome, is composed of four major compartments (head, neck, midpiece, and tail) and several critical structural components (**Figure 22.2**):

  1. **Acrosome**—A lysosome-derived structure at the head of the sperm and containing a mixture of hydrolytic enzymes involved in degrading specific components of the egg cell wall. The acrosome is located between the nucleus and the anterior cell wall of the sperm head.

  2. **Centriole**—In the sperm's neck region, the centriole is a small, hollow, cylindrical structure comprised of microtubule fibrils.

  3. **Mitochondria**—Each sperm structure contains many tightly packed mitochondrial organelles within the midpiece region. The capacity for sperm motility depends on ATP production from these mitochondria.

  4. **Flagellum**—A proteinaceous "propeller" structure composed of microtubules, the flagellum generates the thrust necessary for sperm motility to the egg.

- **Pollen grains** serve as plant sperm and are small enough to be carried by the wind or to powder the extremities of insects. Once these grains come into contact with an ovule, a pollen tube grows from cells in the pollen grain, a single pollen cell then divides mitotically, and the resulting haploid sperm nuclei migrate down the pollen tube to the ovule (see **Chapter 23**).

Egg Structure and Function

- The mammalian egg cell is much larger than the sperm cell (**Figure 22.3**) and contains a cytosol enriched with nutrients as well as several structural components regulating the processes of fertilization and early development.

    1. **Yolk/yolk granules**—Structures in the egg cytosol, packed with proteins and lipids/fats for energy and macromolecular biosynthesis during early development. The yolk must provide all nutrition for the developing embryo until the **placenta** is formed.

    2. **Vitelline envelope**—A fibrous structure surrounding the cell membrane and lying beneath any associated gelatinous layer (egg jelly, zona pellucida, etc.).

    3. **Cytoplasmic determinants**—These protein and mRNA molecules are key regulators of early development in many animals (excluding mammals). Segregation of these molecules during cleavage controls key processes leading to cell differentiation.

    4. **Cortical granules**—These small vesicle-based organelles are localized to the periphery of the cytosol. They contain enzymes that are activated upon fertilization (see next section—blocking polyspermy).

22.2 Fertilization

- The sea urchin has proven to be an incredible research system for the characterization of fertilization. Researchers can isolate large quantities of sperm and eggs quite easily. Large numbers of sperm and eggs are released from adult sea urchins, and fertilization occurs externally in the surrounding seawater.

- The fertilization of a sea urchin egg involves a complex chain of events.

    1. The sperm is recruited to the egg by an attractant that is present in the **egg coat** and slowly diffuses out into the surrounding seawater. The **attractant** is a small protein (peptide) that is recognized by receptor proteins present in the sperm.

    2. Upon direct contact between the sperm head and the egg cell, the acrosome releases its contents from the sperm head.

    3. This mixture of hydrolytic enzymes degrades the gelatinous egg coat.

    4. At the same time the hydrolytic enzymes are released, a structure is built on the sperm head by the **polymerization of actin** proteins. This structure is called the **acrosomal process**, and it mediates the fusion of sperm-cell membranes with the egg-cell membranes.

    5. Once the membranes have fused, the sperm nucleus can enter the egg cytosol, eventually fusing with it to generate the zygote with a $2n$ chromosomal content.

How Do Gametes from the Same Species Recognize Each Other?

- When exposed to a mixture of sperm from different sea urchin species, the egg is fertilized only by sperm of its own species. If the mechanisms of fertilization are similar between these species, how is it possible for an egg to discriminate between different sperm so effectively?

- In 1910, Frank Lille identified a compound, present on the surface of sea urchin eggs, that appeared to bind sperm in a species-specific manner. The molecule was called **fertilizin**. His simple assay for fertilizin–sperm interaction was to mix the two together and measure any clumping of the sperm cells. Lille predicted there was a corresponding protein on the surface of the sperm cells and that it bound to fertilizin in a "lock-and-key" manner.

- In the 1970s Victor Vacquier and colleagues made another breakthrough in the study of species recognition. They isolated a protein on the surface of the sperm and named the protein **bindin**. Each species had a slightly unique form of this protein, which bound only to eggs from the same species.

- Kathleen Foltz and William Lennarz tested the hypothesis that fertilizin protein on the egg was the binding partner of bindin on the surface of the sperm. As summarized in **Figure 22.4** (provided on the next page), they used proteases to release cell-surface proteins from sea urchin eggs and then isolated any of these proteins that could bind sperm or bindin proteins of its own species. Key to this strategy was the prediction that in the collection of proteins released from the egg would be the fertilizin receptor proteins (attached to bindin proteins in a species-specific manner).

- The egg–sperm interaction was species specific due to the presence of the bindin proteins in sperm and the fertilizin receptors in eggs. These proteins bind tightly to enable fertilization but are very selective, binding to each other only if each protein originated from the same species (species lock and key).

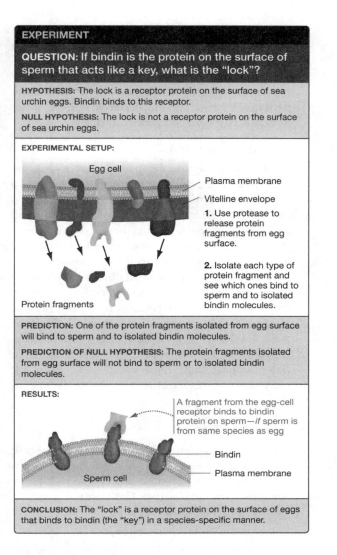

EXPERIMENT

QUESTION: If bindin is the protein on the surface of sperm that acts like a key, what is the "lock"?

HYPOTHESIS: The lock is a receptor protein on the surface of sea urchin eggs. Bindin binds to this receptor.

NULL HYPOTHESIS: The lock is not a receptor protein on the surface of sea urchin eggs.

EXPERIMENTAL SETUP:

Egg cell

Plasma membrane

Vitelline envelope

1. Use protease to release protein fragments from egg surface.

2. Isolate each type of protein fragment and see which ones bind to sperm and to isolated bindin molecules.

Protein fragments

PREDICTION: One of the protein fragments isolated from egg surface will bind to sperm and to isolated bindin molecules.

PREDICTION OF NULL HYPOTHESIS: The protein fragments isolated from egg surface will not bind to sperm or to isolated bindin molecules.

RESULTS:

A fragment from the egg-cell receptor binds to bindin protein on sperm—*if* sperm is from same species as egg

Bindin

Plasma membrane

Sperm cell

CONCLUSION: The "lock" is a receptor protein on the surface of eggs that binds to bindin (the "key") in a species-specific manner.

Why Does Only One Sperm Enter the Egg?

- One key question of fertilization is how only one sperm is successful in fertilizing the egg, despite the presence of hundreds of sperm nearby, all of them intent upon fertilizing the egg.

- A Ca^{2+}-based signal is induced rapidly after fertilization and inhibits fertilization by any additional sperm by generating a **fertilization envelope**. As summarized in **Figure 22.5a**, fertilization results in the release of Ca^{2+} from intracellular stores. The increase in cytosolic Ca^{2+} results in the fusion of cortical granules with the egg-cell membrane. These granules are small, membrane-bound vesicles located near the cell membrane; upon fusion with the membrane, they release their contents into the extracellular matrix.

- What do these cortical granules contain that could inhibit additional fertilization events?

1. **Proteases** that break down proteins associated with the egg-cell wall. These include the receptor protein (fertilizin), and the loss of these structures inhibits receptor binding to additional sperm.

2. A high concentration of **solutes**; these alter the concentration of solutes at the cell exterior and induce a sudden influx of water from outside the cell (**osmosis**).

- The influx of water generates a water-filled space between the vitelline envelope and the cell membrane, generating a barrier to further fertilization events. The thick matrix of cell-wall materials that have been pushed away from the cell membrane by water influx is called the **fertilization envelope** (**Figure 22.5b**).

- Mammalian fertilization is unique from that of invertebrates such as the sea urchin in several ways, including the absence of the species-specific binding between sperm and egg. Also, the sperm and egg meet within the female body, at the **oviduct (fallopian tube)**.

- Paul Wassarman examined the proteins that mediate binding between sperm and eggs during mammalian fertilization. He isolated **glycoproteins** (proteins with sugar additions to the structure) from the zona pellucida of mouse eggs, and identified one (**ZP3**) that binds to sperm heads. ZP3 did not display species-specific binding, as was observed with sea urchin proteins that shared a similar function.

- The interaction between the sperm and the glycoprotein in the egg's zona pellucida generates an acrosomal reaction. This reaction blocks polyspermy, but via a mechanism that is distinct from that observed in sea urchins.

- At fertilization, mammalian eggs do not produce a fertilization envelope, but they do release enzymes that modify the receptor glycoproteins. This causes the release of any additional sperm that are bound to the zona pellucida.

22.3 Cleavage

- **Cleavage** refers to the rapid cell divisions that follow fertilization (without growth) to generate a blastula. The cells generated by cleavage are called **blastomeres**. Cleavage is similar to normal mitotic cell division except it is more rapid, and it involves no increase in size. The two cells generated by cleavage will each be half the size of the original parental cell.

- Cleavage occurs in different patterns depending upon the species of the embryo (radial, spiral, discoidal are each patterns of embryo cleavage).

(a) What mechanisms are used by the sea urchin eggs to avoid polyspermy?

_____

_____

_____

_____

(b) Do mammals block polyspermy via a mechanism that is identical to the one used by sea urchins? Explain your answer.

_____

_____

_____

Partitioning Cytoplasmic Determinants

- Cytoplasmic determinants are found in specific locations within the egg cytoplasm, so they end up in specific populations of blastomeres (**Figure 22.6**).

- In 1894, H. E. Crampton established a link between the pattern of cleavage in early development and the establishment of shell-coiling direction later in development. Working with a species that produced a left-handed coil, Crampton was occasionally able to find mutant snails that display a right-handed coil. Surprisingly, when examining early development in these snails, Crampton discovered that the mutant cells underwent cleavage events with a switched orientation.

- An unusual phenomenon was observed with more extensive experiments. The coil direction of offspring was determined completely by the *mother's genotype*, not the offspring genotype. If the mother was homozygous for the recessive allele (*dd*), then all of the offspring displayed left-handed shells. If the mother carried the dominant allele (*DD* or *Dd*), then all of the offspring displayed right-handed shells.

- This **maternal effect** on embryonic development is due to mRNAs or proteins present in the egg before fertilization, and therefore it affects development independent of sperm genotype or zygote genotype.

- This is a key example of a **cytoplasmic determinant** regulating developmental processes.

- Another example of maternal effect inheritance is the *bicoid* mutation in the fruit fly. Review *bicoid* in

Chapter 21. This protein establishes the anterior-posterior axis of the embryo via its influence upon specific nuclei in the multinucleate fly embryo. Its expression establishes a concentration gradient of bicoid protein, the highest concentration being in the anterior portion of the embryo. Thus, the bicoid mRNA/protein is considered a cytoplasmic determinant.

Cleavage in Mammals

- Rotational cleavage occurs in human embryonic development. The oviduct or fallopian tube serves as a site for fertilization of the human oocyte and is the site of cleavage events. The oviduct is a small structure that connects the ovary to the uterus (**Figure 22.7**). Cleavage generates a blastocyst that is composed of a thin layer of **trophoblast** cells on the exterior surrounding a ball of cells known as the **inner cell mass (ICM)**. A water-filled cavity in the center of this structure is known as the **blastocoel**. The embryo hatches from the zona pellucida 3.5 days after fertilization and attaches (implants) to the wall of the uterus. The trophoblast cells combine with maternal cells to form the **placenta**, and the ICM continues to develop as the embryo.

22.4 Gastrulation

- During **gastrulation** a complex migration of cells occurs, organizing the blastula and further establishing the key embryonic cell types (endoderm, ectoderm, and mesoderm). Gastrulation begins as cleavage processes are completed.

- Within the blastula are the following:
  1. **Animal pole—ectoderm** (generating skin and nerve cells)
  2. **Central region—mesoderm** (muscle and organs)
  3. **Vegetal pole—endoderm** (generating gut and organs)

- Review **Figure 22.8 (provided here)** to examine the formation of distinct layers of cells during gastrulation in frog embryos. Early experiments characterizing these processes utilized simple staining of subsets of cells during cleavage and then followed the cells through gastrulation. This mapping of cell fate has proven incredibly useful in the study of early developmental processes.

- **Step 1:** The **blastocoel** is the water-filled hole in the center of the blastula.

- **Step 2:** Gastrulation begins with the formation of a **blastopore** (hole) on the surface of the blastomere. Cells migrate away from a specific site on the embryo

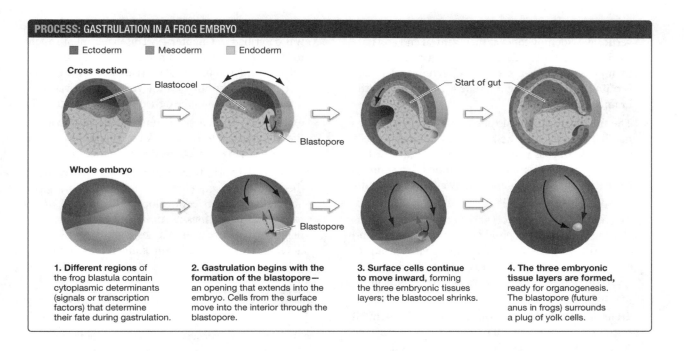

PROCESS: GASTRULATION IN A FROG EMBRYO

■ Ectoderm   ■ Mesoderm   ■ Endoderm

Cross section

Blastocoel

Start of gut

Blastopore

Whole embryo

Blastopore

1. Different regions of the frog blastula contain cytoplasmic determinants (signals or transcription factors) that determine their fate during gastrulation.

2. Gastrulation begins with the formation of the blastopore— an opening that extends into the embryo. Cells from the surface move into the interior through the blastopore.

3. Surface cells continue to move inward, forming the three embryonic tissues layers; the blastocoel shrinks.

4. The three embryonic tissue layers are formed, ready for organogenesis. The blastopore (future anus in frogs) surrounds a plug of yolk cells.

surface and form an **invagination**. This structure eventually develops into the anus.

- **Step 3:** Cells migrate inward, shrinking the blastocoel and gradually forming three distinct layers of cells.

- **Step 4:** The three layers of cells are clearly formed and ready for the formation of organs (organogenesis).

- The human tissues derived from each of the three embryonic cell types are:

  1. Ectoderm (outer layer)-derived: nervous system, epidermis of the gut, epithelial layers of the mouth and rectum, the lens of the eye, and the nasal cavities.

  2. Mesoderm (middle layer)-derived: skeletal system (bone), vascular system, muscular system, lining of body cavity, and reproductive system.

  3. Endoderm (inner layer)-derived: epithelial lining of digestive tract, respiratory tract, reproductive tract, urinary tract, and organs that include the thyroid, parathyroid, liver, pancreas, and thymus.

22.5 Organogenesis

- The process of forming tissue and organs is complex and involves cellular proliferation and differentiation. Differentiation can be defined as the generation of cellular diversity (structure, function), established through differential gene expression. Review **Chapter 21's** discussion on the signaling events that drive differentiation forward.

- **Morphogenesis**—The assembly of cells into recognizable tissues and organs.

- **Determination**—A cell's irreversible commitment to a particular cell type.

- **Differentiation**—The generation of cellular diversity (structure, function) that is established through differential gene expression.

- Chronology within development:

  pattern formation → morphogenesis →
  determination → differentiation

Organizing Mesoderm into Somites: Precursors of Muscle, Skeleton, and Skin

- At the completion of vertebrate gastrulation, the mesoderm is a thin layer of cells located between the endoderm and ectoderm. **Figure 22.10** offers an excellent summary of early organogenesis in chordates.

- As development proceeds, the endoderm forms the beginnings of the **digestive tract**.

- At the same time, the mesoderm begins to form the **notochord** on the dorsal side of the embryo. The notochord is maintained in some adult chordates, but in most it is a transient structure that is important in organizing the body plan during organogenesis and is destroyed by programmed cell death in later stages.

- Signals from the notochord stimulate the adjacent ectoderm along the dorsal side of the embryo to fold and form the **neural tube**. This process involves dramatic changes in the morphology of specific ectoderm cells.

The cytoskeletal proteins in these cells stretch the cells vertically, widening them at their ventral end and narrowing them at their dorsal end. This causes the cells to fold upward in the dorsal direction, eventually forming at tube. The neural tube is a precursor to the brain and spinal chord.

- The fully formed neural tube emits a molecular signal that organizes the neighboring mesodermal cells into blocks of cells known as **somites** (**Figure 22.11**). The neural tube is soon surrounded by somites, mediated by selective adhesion events (mediated by proteins called **cadherins**).

- The somites develop into a variety of adult tissues, including vertebrae, ribs, body wall, some muscle, and even skin (**Figure 22.12**). The fate of each group of somites is activated by cell-cell signals initiated in adjacent tissues (ectoderm, mesoderm, notochord, neural tube).

- Transplantation studies have shown that initially, all somites have the potential to become any of the somite-derived tissues. Transplanting somites from one site to another alters the differentiation fate of those cells. However, later in development, the somite's fate becomes irreversibly **determined**, and somatic cells that are transplanted to other sites in the embryo retain their initial differentiation fate.

✓ CHECK YOUR UNDERSTANDING

(c) Compare and contrast the terms *morphogenesis, determination*, and *differentiation* as they relate to early development.

_____

_____

_____

_____

Differentiation of Muscle Cells

- Cells in the somites receive molecular signals from several surrounding tissues (neural tube, ectoderm, and mesoderm). Cells located on the periphery of each somite are destined to become muscle cells.

- Researchers examined the molecular characteristics of myoblasts, somite-derived cells whose cell fate is determined to be muscle (**Figure 22.15**). The scientists isolated mRNA from myoblasts, converted the mRNA into cDNA through the process of **reverse transcription**, and reintroduced the **cDNAs** into fibroblast cells.

The generation of cDNAs and their value to research was described in **Chapter 19**. This experiment sought to identify genes whose expression in the fibroblasts results in the development of muscle-like characteristics. Scientists were hunting for the genes that activate cell-fate determination in the development of muscle cells.

- One cDNA (or gene) was found that could convert the fibroblasts into muscle-like cells, and they named the gene *MyoD* ("myoblast determination").

- *MyoD* codes for a transcription factor that binds to the regulatory sequences of muscle-specific genes and activates their expression. Activation of *MyoD* expression in somites is stimulated by signals from neighboring embryonic tissues, and it results in the development of muscle from some of the somites.

✓ CHECK YOUR UNDERSTANDING

(d) How can researchers establish when a somite cell's fate is determined? If the procedure shown in **Figure 22.13** were performed using somite cells from later in development, would the results be different?

_____

_____

_____

_____

(e) Describe the methodology used for the cloning of genes involved in muscle-cell differentiation. How were researchers able to assay each isolated gene for the capacity to activate muscle-specific differentiation?

_____

_____

_____

_____

_____

_____

B. MASTERING CHALLENGING TOPICS

Reverse Transcription and cDNA Libraries

Our discussion of the work of Harold Weintraub and colleagues (*MyoD* cloning) introduces an important research tool, the utilization of a **cDNA library** to hunt for genes. The researchers set out to find genes involved in muscle-cell determination, and they prepared mRNAs from a collection of myoblasts. Certainly, within the pool of mRNAs are some transcripts that signal and maintain the muscle-specific determination of these cells. But finding the relevant gene is not easy. It requires a functional screening of all the mRNAs to find those associated with cell determination.

To screen mRNAs, the researchers copied each single-stranded RNA molecule into a copied DNA or cDNA. An enzyme called **reverse transcriptase** is utilized for this purpose. Retroviruses (a category of RNA viruses) use reverse transcriptase to convert their RNA genome into DNA during the process of viral infection. The enzyme can be purified for research purposes, converting mRNAs into cDNAs.

Weintraub and colleagues generated cDNA for each of the mRNAs isolated from myoblast cells and then set out to screen each of these cDNAs for an ability to alter the determination status of fibroblast cells. Their goal was to find, amidst that collection of thousands of different cDNAs, those capable of activating "muscle-like" determination in the fibroblasts. To do this, each of the cDNAs would have to be expressed within a group of fibroblasts to identify any influence of that gene on fibroblast biology. The researchers used DNA plasmids, each carrying a strong promoter, and transferred the cDNAs into their structure. The result is a collection of plasmids with cDNA inserts in their structures, each of which expresses the gene encoded on the cDNA. The plasmids were then transferred into fibroblasts, and the plasmid-carrying fibroblasts were analyzed for changes that might indicate muscle determination.

In this instance, the research team was successful in identifying one plasmid carrying the *MyoD* cDNA, which dramatically altered the fibroblasts. The cDNA library had been successfully screened, and a critical insight in cell determination was achieved.

C. ASSESSING WHAT YOU'VE LEARNED

1. The *bindin* protein is best described in which of the following ways?
   a. A species-specific protein found in the egg outer membrane
   b. A species-specific protein found in the sperm outer membrane
   c. A protein that binds and inactivates competing sperm
   d. A protein that is located in the egg cytosol, and binds to the sperm nucleus upon fertilization of the egg

2. When is a cell considered differentiated?
   a. When a cell migrates to a new site in the body plan
   b. When a cell begins its pattern formation
   c. When a cell manufactures proteins that are specific to a particular cell type
   d. When a cell is part of recognizable tissues or organs

3. Which of the following is *not* necessary for fertilization to take place?
   a. The sperm and egg must be in the same place at the same time.
   b. The sperm and egg must fuse.
   c. The fusion of the two nuclei must trigger the onset of development.
   d. One sperm must release a toxin to inactivate competing sperm cells.

4. In sea urchins, the gelatinous coat surrounding the egg contains a molecule that attracts sperm. Which of the following is true regarding this phenomenon?
   a. The attractant is a small peptide that diffuses out of the jelly layer into the surrounding seawater.
   b. The attractant is not species specific, and this can result in sperm from one species of sea urchin trying to fertilize an egg from a different species.
   c. Sperm respond to the attractant by swimming toward any area that contains it, regardless of the concentration.
   d. The release of the attractant also triggers the polymerization of actin into microfilaments that form a protrusion, which extends until it makes contact with the egg-cell membrane.

5. Research over the past 70 years has revealed a wide array of mechanisms that block polyspermy in various animal species. In sea urchins, fertilization results in the erection of a physical barrier to sperm entry via the formation of a fertilization envelope. Which of the following are involved in the mechanism that creates the fertilization envelope?
   a. The entry of the sperm causes calcium ions to be released from stored areas inside the egg.
   b. Cortical granules located just inside the membrane fuse with the egg-cell membrane and release their contents to the exterior.
   c. An influx of water separates the envelope matrix from the cell.
   d. The contents of the cortical granules include proteases that digest the exterior-facing fragment of the egg-cell receptor for sperm.
   e. All of the above apply.

6. Which of the following is true of cleavage?
   a. Cleavage divisions partition the egg cytoplasm during additional cell growth.
   b. These divisions proceed slowly in the initial creation of a multicellular embryo.

c. When cleavage is complete, the embryo consists of a sphere of cells that is ready to undergo gastrulation.

d. Cleavage occurs exclusively in plant cells.

7. What were researchers able to determine by transferring cells from one somite to another?

a. A cell in a somite can become any of the somite-derived elements in the body.

b. Somites are responsible for controlling apoptosis.

c. Somites are transient structures that appear relatively briefly during development.

d. Mesodermal cells in a somite are destined for a variety of structures.

8. Weintraub and co-workers hypothesized that myoblasts contain at least one regulatory protein that commits them to their fate. Their research found the gene that encodes this protein, which they named the protein MyoD. All of the following are true of MyoD *except*:

a. MyoD encodes a transcription factor.

b. *MyoD* is the gene that regulates activities of *ced-3* and *ced-4* genes.

c. MyoD protein binds to enhancer elements located upstream of muscle-specific genes.

d. MyoD protein activates expression of the *MyoD* gene.

9. Cells located at the periphery of a somite are destined to become:

a. bone.   c. neurons.

b. muscle.   d. skin.

10. A classic example of a cytoplasmic determinant that is located in specific places in the egg cytoplasm and is located into specific cells during cleavage is:

a. bicoid.   c. actin.

b. MyoD.   d. antennaepedia.

CHAPTER 22—ANSWER KEY

Check Your Understanding

(a) A Ca^{2+}-based signal is induced rapidly after fertilization and inhibits fertilization by any additional sperm by generating a **fertilization envelope**. As is summarized in **Figure 22.5a**, fertilization results in the release of Ca^{2+} from intracellular stores. The increase in cytosolic Ca^{2+} results in the fusion of cortical granules with the egg-cell membrane. These granules release proteases that cleave egg-receptor proteins, and solutes that induce water flow into the space between the cell membrane and the cell envelope.

(b) Mammalian embryos do not form a fertilization envelope like sea urchins. However, at fertilization, they do release proteases that degrade receptor proteins in the zona pellucida.

(c) *Morphogenesis* is the migration and assembly of cells into distinct structures/tissues. *Determination* refers to the molecular "commitment" to a cell type, activating cellular changes that result in cellular differentiation. Activation of *MyoD* within myoblasts is an excellent example of cellular determination. *Differentiation* refers to the generation of diversity in cellular structure and function, and it is established through differential gene expression.

(d) As illustrated in **Figure 22.13**, somites may be removed from one site and transplanted to another to determine if the transplanted somite is capable of differentiating into other cell types. Early in somite maturation, each transplanted somite cell is capable of differentiating into any of the somite-derived cell fates. Later during somite development, this is no longer possible, indicating the somite fate is now irreversibly determined.

(e) Harold Weintraub and colleagues isolated mRNAs from determined muscle precursor cells called myoblasts. The mRNAs were reverse transcribed into cDNAs and cloned into expression plasmids. This procedure enables the expression of these cloned fragments in any cell type. The researchers screened through their cDNA clones, transferring each into fibroblast cells and examining the cells for muscle-like differentiation following expression of the cDNA clone. The muscle-like changes within these cells would be identified as changes in cellular morphology and the expression of genes that are active only in muscle.

Assessing What You've Learned

1. b; 2. c; 3. d; 4. a; 5. e; 6. c; 7. a; 8. b; 9. b; 10. c

Looking Forward—Key Concepts in Future Chapters

Animal Reproduction—Chapter 48

Key to this chapter are the processes of egg fertilization and embryo development. Examine **Chapter 48**, recognizing the complexity of these processes and the molecular mechanisms that drive development and differentiation forward.

An Introduction to Plant Development

Looking Back—Key Concepts from Earlier Chapters

Analyzing and Engineering Genes—Chapter 19

In this chapter you learned about the use of gene mapping, cloning, sequencing, and in situ hybridization to characterize key genes that regulate the development of flowers. These techniques allowed researchers to test very specific hypotheses that had been developed from their studies of mutant plants. Notice the application of molecular biology to a fundamental question in developmental biology. Another example is the production of transgenic plants, which carry engineered mutations in the *PHAN* gene. This work enabled an improved understanding of leaf development.

KEY CONCEPTS

- In plants, gamete production involves an initial meiotic process to generate haploid cells followed by mitotic cell divisions to amplify the number of gametes.

- A pollen grain contributes two sperm to an ovule. One sperm fertilizes the egg cell to generate an embryo, and the other sperm fuses with a diploid cell to generate triploid **endosperm** cells.

- As in animal development, plants develop by establishing axes around which development proceeds. The three key types of embryonic plant tissue are the **cotyledon** (leaf precursor), the **hypocotyl** (stem precursor), and the **root**.

- **Meristems** are plant structures that supply undifferentiated cells for growth and tissue production. The two types of meristems are the **shoot apical meristem (SAM)** and the **root apical meristem (RAM)**.

- Reproductive development is controlled by transcription-regulating proteins that bind DNA and regulate the expression of subsets of genes.

A. CHAPTER OUTLINE

- This chapter extends our exploration of development to that of plants. A focus of this chapter is upon *Arabidopsis thaliana*, a flowering mustard plant that has been studied extensively (**Figure 23.1**). This plant grows to reproductive maturity rapidly, and it can be readily cultured in a greenhouse. The genome of this plant has been completely sequenced, and the sequence data are readily available in a public database. Researchers can readily make transgenic *A. thaliana* plants using the techniques discussed in **Chapter 19** (*Agrobacterium*-based transfer). Information gained from the study of *A. thaliana* has been applied to many other plants. This organism thus serves as an excellent model organism for plants, in the same way that fruit flies, worms, and mice have served as model organisms for animals.

23.1 Gametogenesis, Pollination, and Fertilization

- Gamete formation and fertilization are more structurally complex in plants than they are in animals. In flowering plants, the egg is fertilized inside an **ovule**—a structure at the base of the flower (**Figure 23.2**). For fertilization to take place, the pollen grain must extend a long pollen tube that reaches inside the ovule. Sperm move down the pollen tube to the egg inside the ovule.

- Once the egg is fertilized, the ovule and embryo together develop into a **seed** through a process known as **embryogenesis**. Notice the haploid (1*n*) and diploid (2*n*) phases of this life cycle. In **Figure 23.2**, haploid and diploid cells are labeled. As in animal reproduction, sperm and egg cells are haploid.

- Haploid **pollen grains** (male gamete) are produced on the anther of a mature flower via meiosis followed by mitosis. The two cells generated by this mitosis have different roles in fertilization. One cell mediates formation of the pollen tube, while the other cell forms sperm and fertilizes the egg (**Figure 23.2a**). These cells are surrounded by a tough coat that protects the pollen grain from dehydration. Pollen is distributed by the wind or by birds and insects that come into contact with the flower.

- To fertilize an egg, the pollen must be deposited on the stigma of a mature flower. Plants can determine if a particular pollen grain is the appropriate species via specific protein-protein interactions between the pollen and the stigma (**Figure 23.3a**). This is very similar to bindin and fertilizing proteins in animal fertilization (Chapter 22).

- If the protein-protein interactions are specific, the pollen grain **germinates**. This involves the elongation of a pollen tube that reaches down the stigma and toward the egg; it also involves the generation of sperm that will be carried by the pollen tube to the egg (**Figure 23.3b**).

- The plant ovule is fertilized by two sperm nuclei carried by the pollen tube. This **double fertilization** is key to the development of both the embryo and the **endosperm** tissue that is important to providing nutrition for the young plant (**Figure 23.4**). The embryo has a $2n$ or *diploid* chromosome number (sperm plus egg), and the endosperm has a $3n$ or *triploid* genotype (sperm plus two maternal cells).

- The endosperm surrounds the embryo and provides nutrition that includes proteins, carbohydrates, and fats for embryonic developmental growth. Endosperm is analogous to the yolk found in animal eggs, providing nutrition to embryo cells that are rapidly growing and developing.

✔ CHECK YOUR UNDERSTANDING

(a) Describe the composition of a pollen grain.

_____

_____

_____

(b) Define the cellular products of double fertilization, and the ploidy of each.

_____

_____

_____

_____

23.2 Embryogenesis

- Embryogenesis occurs inside the ovule following fertilization. Compare with cleavage, gastrulation, and organogenesis in animal development (introduced in **Chapter 22**).

What Happens during Plant Embryogenesis?

- Following fertilization, an *Arabidopsis* embryo undergoes a series of unique developmental processes. These are summarized into six steps in **Figure 23.5** (provided here).

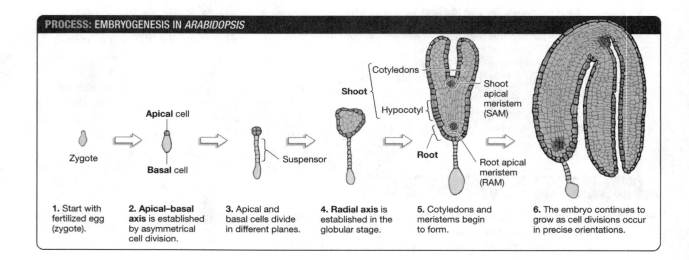

PROCESS: EMBRYOGENESIS IN *ARABIDOPSIS*

Zygote — **Apical** cell / **Basal** cell — Suspensor — Shoot — Cotyledons / Shoot / Hypocotyl / Root / Shoot apical meristem (SAM) / Root apical meristem (RAM)

1. Start with fertilized egg (zygote).

2. **Apical–basal axis** is established by asymmetrical cell division.

3. Apical and basal cells divide in different planes.

4. **Radial axis** is established in the globular stage.

5. Cotyledons and meristems begin to form.

6. The embryo continues to grow as cell divisions occur in precise orientations.

- **Step 1** is the fertilization of the egg by a sperm carried by the pollen tube. **Step 2** involves an asymmetric mitotic cell division, generating a **basal** cell (large) and an **apical** cell (small). In **Step 3**, the basal cell replicates several times in an axis that is perpendicular to the axis of the embryo, generating a column of cells called the **suspensor** at the base of the embryo. The apical cell ultimately develops into the mature embryo.

- The asymmetry of cell division in early development establishes the **apical-basal axis**. Apical refers to the tip of the embryo, and basal is the base of the embryo.

- In **Step 4 (globular stage)**, the apical cell divides to generate a **globule** (ball) of cells at the tip of the suspensor structure. The cells in the exterior layer on this globule are morphologically distinct from those in the center.

- In **Step 5, cotyledons** (precursors to leaves) begin to develop as two extensions at the apical tip of the embryo. The **hypocotyl** (precursor to stem) is attached to the base of the cotyledons. The cotlyedon and hypocotyl together are called the **shoot**. Basal to the hypocotyl is the **root** structure, which will become the belowground portion of the seedling. In **Stage 6** these basic tissue types grow through intense mitotic activity.

- The **meristem** cells of the embryo are like stem cells in an animal embryo. These are groups of cells that are undifferentiated, and they divide to generate a source of cells for building specific differentiated structures.

- The early embryo contains a **shoot apical meristem (SAM)** at the apical side of the embryo, and a **root apical meristem (RAM)** at the basal side of the embryo.

- A cross section of the embryonic stem reveals three tissues critical to the development of the adult plant (**Figure 23.6**). These tissues are the exterior layer of protective cells, or **epidermis**; the cells below the epidermis, or **ground tissue**; and the **vascular** tissue in the center of the tissue. These tissues could be compared to ectoderm, mesoderm, and endoderm in animals.

- The ground tissue can develop into several different cell types that will be involved in photosynthesis or other functions.

- The vascular tissue develops into a transport system for the movement of water and nutrients throughout the entire adult plant.

Which Genes and Proteins Set Up Body Axes?

- Gerd Jürgens and colleagues used a mutant hunt to isolate genes responsible for establishing the apical-basal axis (**Figure 23.7**). They predicted that plants carrying mutations in these genes would display abnormal development along this axis. They predicted that the gene products of these genes would be involved in regulating and signaling each cell's site along the apical-basal axis. The mutants they isolated were placed into three groups: **apical mutants** lacking cotyledons, **central mutants** lacking hypocotyl structures, and **basal mutants** lacking hypocotyl *and* root structures.

- The mutants that lacked hypocotyls and roots contained a mutation in a gene they named *MONOPTEROS*. The encoded protein appears to be a transcription factor. Analysis of the amino acid sequence of the protein shows that this protein is similar to DNA-binding proteins that regulate the transcription (expression) of specific genes.

- Auxin proteins regulate the expression of MONOPTEROS. Auxins are a class of plant hormones (phytohormones) that regulate plant cell growth and development.

- Researchers draw parallels between auxin and *bicoid* as molecules that establish polarity within the embryo through their differential concentration along the axis of the embryo.

23.3 Vegetative Development

- The adult plant continues to adapt and grow in response to environmental conditions. Trees extend branches in the direction of available sunlight. *Arabidopsis* plants that grow against a type of physical barrier will initiate growth away from that barrier. How does this work? There must be complex molecular signaling processes that regulate these processes.

Meristems Provide Lifelong Growth and Development

- In the adult plant, meristems (SAM, RAM) play a central role in mediating new growth and development. The adult meristems are found at the tip of each shoot and at the tip of each root (**Figure 23.8a**).

- Examine the labels in **Figure 23.8b**, which shows the groups of cells present in a SAM. Notice the cells emerging from the meristem and beginning to differentiate into epidermal, ground, and vascular tissue.

- Auxin has the ability to stimulate leaf growth from a SAM. This is a key cell-to-cell signal regulating vegetative growth within an adult plant.

- The new leaf, originating from the SAM, must develop all the tissues of a leaf in proper orientation. Therefore, the leaf must establish axes in all three dimensions (**Figure 23.9**).

- The *PHANTASTICA* (*PHAN*) gene has been shown to regulate establishment of the upper-lower axis in leaf development. Not surprisingly, the protein encoded by this gene encodes a transcription factor. This protein is expressed in the upper layer of the developing leaf. It induces the expression of genes that cause cells to form the upper layers of the leaf, and it suppresses genes that induce differentiation as lower leaf cells.

- To better characterize the function of *PHAN* in plants, scientists engineered tomato plants that carried a defect in this gene and then examined their leaf morphology. Review the methodology of generating transgenic plants discussed in **Chapter 19**. In the tomato experiments, transgenes were used to partially or totally block the effect of *PHAN* protein function. This allowed scientists to examine how different levels of *PHAN* function might affect leaf morphology.

- In **Figure 23.10**, some of their results are displayed. Diminished PHAN function can generate dramatic changes in leaf size and shape. Their results are consistent with *PHAN* playing an important role in tomato leaf growth, similar to what had been observed in *Arabidopsis* plants. PHAN is an example of a gene whose various gene alleles might confer dramatic variation in the form and function of an important organ (the leaf). This example and others like it have profound implications for our understanding of evolutionary biology.

CHECK YOUR UNDERSTANDING

(c) Define some of the effects of auxin on plant development.

_____

_____

_____

_____

23.4 Reproductive Development

- In animals, **germ cells** are set aside early in development, and somatic cells are involved in building most of the animal body plan. The gonads (testes, ovary) are sites of germ cell origination only and are established early in development. In plants, meristem cells generate flowers and their pollen and egg cells.

Meristem cells can go through many mitotic divisions during the normal life of the plant, with each division offering the risk of incorrect DNA replication (mutation). Thus, plants can pass on mutations as meristem cells with mistakes in their DNA are used to make new germ cells. Animal cells are thought to pass on fewer new mutations because they isolate and protect their germ cells.

The Floral Meristem and the Flower

- SAMs can convert from vegetative development to reproductive development, from making new leaves to making flowers.

- A SAM that has begun reproductive development is known as a **floral meristem**.

- The flower is composed of four layers of different cell types, and the pattern of layering is known as **whorls within whorls**. A whorl is a shape that coils or curls, like the swirls in icing on a cake. As shown in **Figure 23.11**, the four layers of whorls generate the flower structure. The layers are (1) a protective layer of **sepals** on the outside, (2) colorful **petals** inside of the sepals, (3) pollen-producing **stamen**, and (4) egg-producing structures, or **carpels**, in the center of the flower.

- In the 1800s, researchers identified mutant *Arabidopsis* plants that resulted in one category of reproductive organ replacing another (e.g., a flower that is missing the stamen layer and instead has two layers of petals). These mutants are called **homeotic** mutants. They can be compared to homeotic mutants in *Drosophila*, where a segment of cells develops abnormally due to defective signaling (**Chapter 21**).

- The powerful experimental strategies that are possible in *Arabidopsis* have enabled the discovery of mutations in specific genes as the source of the homeotic phenotype.

- Elliot Meyerowitz isolated a group of mutant *Arabidopsis* plants displaying abnormalities in flowering (homeotic mutants). He identified the genes responsible for each of the observed phenotypes. These genes can be divided into three categories, with each category of mutant phenotype displaying the absence of two adjacent whorls. The three phenotypes are carpel/stamen only, sepal/carpel only, and petals/sepals only (**Figure 23.12**).

- The research team predicted that each category of mutant phenotype was the result of a specific gene defect, and therefore three genes are the key regulators of flower development.

- Meyerowitz explained these three categories of genes/mutant phenotypes using the **ABC model** of

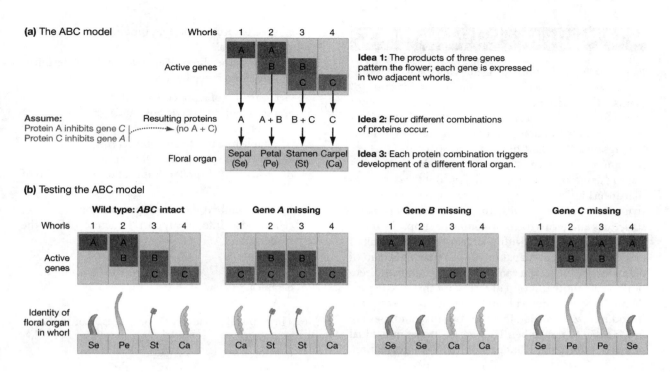

(a) The ABC model

Whorls 1 2 3 4

Active genes

Idea 1: The products of three genes pattern the flower; each gene is expressed in two adjacent whorls.

Assume:
Protein A inhibits gene C
Protein C inhibits gene A

Resulting proteins

A   A + B   B + C   C   (no A + C)

Idea 2: Four different combinations of proteins occur.

Floral organ

Sepal (Se)   Petal (Pe)   Stamen (St)   Carpel (Ca)

Idea 3: Each protein combination triggers development of a different floral organ.

(b) Testing the ABC model

| Wild type: *ABC* intact | Gene *A* missing | Gene *B* missing | Gene *C* missing |

Whorls 1 2 3 4

Active genes

Identity of floral organ in whorl

Wild type: Se Pe St Ca
Gene A missing: Ca St St Ca
Gene B missing: Se Se Ca Ca
Gene C missing: Se Pe Pe Se

---

flowering (**Figure 23.13**, provided here). This model includes three central components:

1. Each of the three genes regulating flowering (*A*, *B*, and *C*) is expressed only in two adjacent whorls.

2. If each gene is expressed in only two adjacent whorls, then four different combinations of gene products are possible (*A, AB, BC, C*). Each combination activates a unique molecular signal.

3. These unique signals are capable of activating different whorls of differentiated meristem tissue: sepal (*A*), petal (*AB*), stamen (*BC*), and carpel (*C*).

- An additional prediction regarding this model is that the presence of protein A inhibits the expression of protein C. Also, the expression of protein C inhibits the production of protein A. They are mutually inhibitory gene products—if one is expressed first, the other is suppressed. This relationship would explain the mutant phenotypes observed. Look at **Figure 23.13**. If a plant lacked a functional copy of gene *C*, what phenotype would be generated?

- To test their model, Meyerowitz and colleagues next had to clone the genes that encode each of these proteins. They mapped the chromosomal locus of each protein and cloned the associated gene using techniques described in **Chapter 19**. They then examined the pattern of expression of each of these genes using *in situ* hybridization, described in **Chapter 21**. Recall that this type of analysis uses a probe made of single-stranded DNA to infiltrate slices of tissue and show where the mRNA that complements that probe is located. The mRNA and the probe stick to each other because they have complementary sequences. This analysis supported Meyerowitz's model. Gene *A* was expressed in the outer two whorls, gene *B* in the middle two whorls, and gene *C* in the inner two whorls.

- Compare these results and the ABC model with the *Hox* genes and their pattern of expression during animal development (**Chapter 22**). The parallels between these developmental processes are striking.

- Each of the three genes was also shown to encode DNA-binding regions in its protein product, sharing functional homology with the homeobox sequences found in the *HOM-C* genes. The plant DNA-binding motifs are known as MADS-boxes.

CHECK YOUR UNDERSTANDING

(d) Using the ABC model, explain the mutant phenotype of a flower from an *Arabidopsis* plant that lacks the *C* gene.

_____

_____

_____

_____

B. MASTERING CHALLENGING TOPICS

Plant development presents a dizzying array of terminology used to describe the complex architecture of plants. It is challenging to a student learning about plants for the first time. As you work through this chapter, a helpful methodology for keeping track of terms is to diagram a flowchart of terms and structures that are relevant at each stage of development. Then move forward in developmental time and write out the structures that are critical at that stage. You can connect embryonic structures and terms with their corresponding adult structures and terms by color-coding or simply by interconnecting the terms with a drawn line. The result is similar to a concept map, but you are making a map of plant development. You can branch off in another direction to add another group of terms, for instance, reproductive organs. You will find that as new terms are introduced, you can add them to your map and they are much easier to remember. They are no longer individual points of information, but part of a web of information.

C. ASSESSING WHAT YOU'VE LEARNED

1. Compare and contrast an animal sperm cell and a plant pollen grain.

   _____

   _____

   _____

   _____

   _____

2. In flowering plants, an event known as double fertilization occurs when:
   a. two sperm enter the ovule; one fertilizes the egg, and the other fertilizes a $2n$ cell to generate $3n$ endosperm.
   b. two sperm cells fertilize one normal egg cell to make a $3n$ embryo; the embryo later becomes $2n$ during a unique mitotic division.
   c. two eggs are simultaneously fertilized by two sperm nuclei.
   d. None of the above apply.

3. Species-specific selection of pollen occurs via:
   a. allowing only species-specific pollen to come into contact with the stigma.
   b. toxin-based destruction of foreign pollen.

   c. specific protein-protein binding between pollen and stigma.
   d. none of the above: Plants do not block fertilization of their egg by foreign pollen.

4. Germination of a pollen seed refers to:
   a. rapid mitotic divisions and growth of the embryo.
   b. establishment of meristem cells.
   c. growth of roots and cotyledons.
   d. growth of a pollen tube and production of sperm.

5. In early embryonic development, the _____ cell divides into a support structure called the suspensor.
   a. basal
   b. ground
   c. apical
   d. vascular

6. The hypocotyl is an embryonic precursor to the:
   a. flower.
   b. root.
   c. stem.
   d. leaf.

7. Plant embryogenesis is comparable to which of the following developmental processes in animals?
   a. Segmentation and determination
   b. Fertilization
   c. Cleavage, gastrulation, and organogenesis
   d. Meiotic cell division

8. Using in situ hybridization experiments, the Meyerowitz research group examined the expression of flower development in gene *B* and found its expression to be limited to which of the following whorls in a normal developing flower?
   a. Sepal
   b. Petal
   c. Carpel
   d. Stamen

9. The PHAN gene is a transcription factor that regulates:
   a. axis development in the leaf.
   b. apical-basal axis development in the embryo.
   c. radial axis development in the flower.
   d. development of the suspensor structure.

10. Following fertilization of the egg, the plant ovule develops into:
    a. the exterior of the seed.
    b. the flower.
    c. a meristem.
    d. the shoot apical meristem (SAM).

CHAPTER 23—ANSWER KEY

Check Your Understanding

(a) Pollen grains (male gamete) are composed of two cells, essentially a cell within another cell. The outer cell mediates the formation of the pollen tube. The inner cell forms the sperm that fertilize the plant egg after moving down the extended pollen tube (**Figure 23.3a**). These cells are surrounded by a tough coat that protects the pollen grain from dehydration.

(b) Double fertilization is a unique fertilization process in plants. One sperm fertilizes the egg to generate the plant embryo (diploid or $2n$). A second sperm fertilizes a diploid ovule cell to generate endosperm (triploid or $3n$). The endosperm feeds the developing embryo by producing large amounts of carbohydrate, fat, and protein for the embryo. Seeds are primarily composed of endosperm and a small plant embryo.

(c) Auxin is a complex plant hormone that has wide-ranging effects on plant development. In this chapter, we have defined auxin as an inducer of MONOPTEROUS expression, a key protein in the establishment of the embryonic body axis. In addition, auxin stimulates the shoot apical meristem (SAM) to generate new leaves.

(d) In the ABC model, protein products from each gene work together to specify the type of tissue that develops: sepal (A), petal (AB), stamen (BC), or carpel (C). The C gene is normally expressed in the inner two whorls of the developing flower, and it normally suppresses expression of the A gene in those whorls. Thus, the whorls in a C-deficient plant would have the following protein products (from outer layer to inner layers): A, AB, AB, A. The flower of this plant would develop with sepals on the outer layer and the innermost layer, with petals in the middle layers. See **Figure 23.13**.

Assessing What You've Learned

1. The plant sperm cell is a pollen grain. This structure contains very little of the structural specialization displayed by the animal sperm cell. The pollen grain contains two haploid nuclei, whereas the animal sperm contains only one. The plant pollen grain relies on attachment to the fertile plant and growth of the pollen tube to the egg cell. Two nuclei are delivered. The relevance of these two nuclei involves the formation of endosperm ($3n$ in DNA content) along with the zygote ($2n$).

2. a; 3. c; 4. d; 5. a; 6. c; 7. c; 8. b, d; 9. a; 10. a

Looking Forward—Key Concepts in Later Chapters

Plant Sensory Systems, Signals and Responses—Chapter 39

In this chapter auxin was introduced as a vital, developmentally important hormone in plants. Auxin was the first hormone to be discovered. It is actually not one molecule, but a class of structurally similar hormones. It regulates root growth, shoot growth, and various other key developmental processes. Interestingly, most of these influences can be traced to the impact of auxin in regulating the expression of specific genes.

Evolution by Natural Selection

Evolution and Natural Selection—Chapters 1 and 4

The process of evolution by natural selection was introduced in **Chapter 1** along with some examples of how researchers study evolution. **Chapter 4** discussed the related process of chemical evolution and gave an example of experimental selection. Here in Chapter 24, natural selection and research on its action in modern organisms are described in greater detail.

Inheritance of Traits—Chapters 13 through 16

Chapter 24 emphasizes that only heritable traits can be acted on by natural selection to cause evolution in a population. The genetic basis of inheritance was explained in earlier chapters.

Mutation—Chapters 14 and 15

This chapter describes how a single-point mutation led to antibiotic resistance in *Mycobacterium tuber-culosis*. For additional information on the molecular basis of mutation, look back to **Chapters 14** and **15**.

Transcription—Chapter 16

Chapter 24 introduces you to an antibiotic, rifampin, that disrupts RNA polymerase's ability to transcribe DNA. See **Chapter 16** for the details of DNA transcription.

KEY CONCEPTS

- Species evolve over time due to allele-frequency change in populations.

- Heritable traits that increase an individual's ability to produce successful offspring are adaptations for that individual's environment. Adaptations are constrained by fitness trade-offs and by genetic and historical factors.

- Evolution by natural selection occurs when individuals with adaptive alleles contribute the most offspring to the population's next generation. Evolution changes the population's characteristics, not the individual's.

- Evolution does not do things for the good of the species and not all traits are adaptive.

A. CHAPTER OUTLINE

Chapter Introduction

- The theory of evolution by natural selection is one of the key theories in biology. This theory was controversial because it conflicts with the theory of special creation.

- Darwin's theory is radically different from special creation's claim that instantaneous creation of life by a supernatural being led to unchanging, independent species in the recent past.

24.1 The Evolution of Evolutionary Thought

- The theory of evolution by natural selection was revolutionary because it overturned one idea about how nature works, replacing it with a new and completely different idea.

Plato and Typological Thinking

- The theory that species are unchanging types created by God had dominated thinking about organisms since Plato articulated it more than 2000 years before.

- According to this typological thinking, variations within species are insignificant or even misleading.

Aristotle and the Great Chain of Being

- Another Greek philosopher, Aristotle, proposed the great chain of being that built on Plato's ideas by organizing species into a sequence from smallest and/or least complex to largest and/or most complex (**Figure 24.1**).

- Aristotle's ideas about species being fixed types where some species are "better" than others were still popular in the 1700s.

Lamarck and the Idea of Evolution as Change through Time

- Jean-Baptiste de Lamarck was the first to propose a theory of evolution when he argued that species had changed through time (1809). However, Lamarck thought organisms originated via spontaneous generation at the base of the great chain of being and then moved up the chain as they evolved over time.

- Lamarck proposed that organisms progressed from lower to higher forms by inheriting acquired characters—characters developed by individuals in response to environmental challenges.

Darwin and Wallace and Evolution by Natural Selection

- Darwin and Wallace argued that evolution occurs because individuals vary, and, based on their traits, some individuals contribute more offspring to the next generation than do others in the same population.

- Individuals of the same species living in the same area at the same time make up a **population**.

- Darwin and Wallace's theory requires **population thinking**; variation is necessary and essential rather than unimportant or misleading.

✔ CHECK YOUR UNDERSTANDING

(a) Why do traits have to be heritable in order for evolution by natural selection to occur?

_____

_____

_____

_____

- The theory of evolution is scientific. It proposes a mechanism that causes change in a species over time and makes predictions that can be tested.

24.2 The Pattern of Evolution: Have Species Changed through Time?

- Darwin described evolution as **descent with modification**—meaning that over time ancestral species change into somewhat different descendant species.

- The suggestion that species change through time and are related by common ancestry was a radical departure from the independently created and immutable species concept described by the theory of special creation.

Evidence for Change through Time

- Traces of organisms that lived in the past are called **fossils**. Those that have been found and described in the scientific literature make up the **fossil record**.

- Data from the fossil record and data from **extant species** (those living today) support Darwin's claim that extant species are modified forms of ancestral species.

The Vastness of Geologic Time

- The **geologic time scale** is a series of named intervals that represent major events in Earth history. These intervals were first associated with specific rock layers with certain types of fossils.

- Most fossils are found in **sedimentary rocks** that form in layers from deposits of sand, mud, or other materials at beaches or river mouths. Fossils from lower sedimentary layers are older than those from upper layers.

- Studies of modern rock formation suggested that it takes a very long time to make thick layers of sedimentary rock. Thus, Earth appeared to be much older than the 6000 or so years claimed by special creation.

- Researchers now use **radioactive isotopes** to date rocks. Dates are based on observed decay rates of unstable "parent" atoms and a comparison of parent/daughter atom ratios in newly formed rocks as compared with a sample rock.

- Earth is about 4.6 billion years old, and fossil data show the earliest signs of life in rocks that formed 3.4 billion years ago.

Extinction Changes the Species Present over Time

- Researchers have discovered many fossils unlike any known living organisms. These species are **extinct**; they no longer exist. More species have gone extinct than exist today, and species have gone extinct throughout Earth's history.

- Darwin interpreted extinction as evidence that species are dynamic and can change. He reasoned that if species have gone extinct, then the variety of species on Earth has changed through time.

Transitional Features Link Older and Younger Species

- Fossils often resemble living species found in the same region. This "law of succession" means that extinct fossil species are typically succeeded, in the same region, by similar species.

- Darwin interpreted this as evidence that extinct forms are the ancestors of modern (descendant) forms, and that species change over time.

- Improvements in the fossil record led to the discovery of many **transitional features** that are intermediate between older and younger species (e.g., **Figure 24.4** on the evolution of the tetrapod limb).

- Many transitions are well documented and provide strong evidence that species have evolved over time.

Vestigial Traits Are Evidence of Change through Time

- **Vestigial traits** are functionless structures similar to functioning structures in related species. Biologists interpret these vestigial traits as evidence that trait structure and function change over time (e.g., whale hip bones, human appendix, pseudogenes).

- Vestigial traits are strong evidence that species change over time. The existence of these traits is not adequately explained by the theory that a perfect supernatural being created unchanging species.

Current Examples of Change through Time

- Changes in modern species in response to environmental change are also well documented (e.g., evolution of antibiotic resistance in bacteria). Change in species over time continues today and can be measured directly.

✔ **CHECK YOUR UNDERSTANDING**

(b) Transitional forms and vestigial traits have both been used as evidence that species change over time. Comment on the strengths and weaknesses of these two types of evidence.

_____

_____

_____

_____

_____

- Evidence from both the fossil record and from living species supports the idea that life is ancient and that species have changed through time and continue to change today.

Evidence of Descent from a Common Ancestor

Similar Species Are Found in the Same Geographic Area

- Darwin, as ship naturalist, collected plants and animals on his *Beagle* voyage. The Galápagos mockingbirds he collected from different islands represented similar, yet distinct, species. Darwin proposed that the species were similar because they had descended with modification from a common ancestor (**Figure 24.6**).

- Molecular data support Darwin's hypothesis. These and other related species share a **phylogeny** (a family tree of evolutionary relationships). Species' ancestor–descendant relationships can be graphically depicted in a **phylogenetic tree** (see **BioSkills 3**).

Homology Is Evidence of Descent from a Common Ancestor

- **Homology**, in biology, refers to resemblance due to common descent (similar traits derived from a common ancestor).

- Genetic **homology** is similarity in DNA sequences (e.g., similarity in human and fruit-fly genes for eye formation, **Figure 24.7**).

- **Developmental homology** is similarity in embryo form and/or pattern of tissue differentiation. For example, biologists hypothesize that chick, human, and cat embryos all have a tail and gill pouches (a vestigial trait) because their common ancestor was a fishlike animal with gill pouches and a tail (**Figure 24.8**).

- Genetic homologies give rise to developmental homologies that in turn cause **structural homology**, meaning similarity in adult **morphology** (form). For example, all vertebrates share the same general limb structure (**Figure 24.9**).

- Evidence that all Earth's organisms descend from a single common ancestor include the most fundamental of all genetic homologies—the universal genetic code. Almost all organisms use the same rules for making proteins based on DNA's information.

- Some hypotheses about homology can be tested experimentally.

- For example, squid genes thought to be homologous to a fruit-fly gene that helps form eyes were inserted into fruit-fly embryos. When these genes were expressed in embryo parts that form appendages, eyes formed on legs and antennae.

- Thanks to homologies, the mutation-causing properties of potential cancer-causing chemicals can be tested in bacteria. Drugs for humans can be tested on mice and unknown sequences can often be identified by comparison with other species.

- The many examples of similarity in traits among different organisms suggest that species are related to each other by descent from a common ancestor. If species were created independently of one another, these similarities should not occur.

Current Examples

- Species continue to change and produce new species today. Biologists can thus directly observe the formation of descendant species.

✔ CHECK YOUR UNDERSTANDING

(c) Why is genetic homology "the most fundamental" level of homology?

_____

_____

_____

_____

Evolution's "Internal Consistency"— the Importance of Independent Data Sets

- Evidence that species change through time and are related is summarized in **Table 24.1**. Data from many independent sources support predictions that stem from the theory of evolution, as in the following cetacean evolution example.

- Cetacean fossils are easily identified by their unusual ear bones and include transitional forms. A phylogeny of fossil cetaceans indicates a gradual change from land-dwelling hippopotamus-like animals to modern whales (**Figure 24.11**).

- Relative dating based on the sedimentary rocks in which fossils were found agrees with the phylogeny conclusion of gradual change in the cetacean lineage. Absolute (radiometric) dating further confirms this conclusion.

- A phylogeny based on DNA sequences of living species provides evidence that hippos (semiaquatic terrestrial mammals) are the closest living relatives of cetaceans.

- Vestigial bones and embryonic hindlimb buds in certain cetaceans further supports the idea that cetaceans evolved from a terrestrial ancestor.

- In this and other examples, data from many different independent sources are more consistent with evolution than with special creation. Descent with modification is a powerful scientific theory because it explains so many different observations.

24.3 The Process of Evolution: How Does Natural Selection Work?

- The idea of evolution predates Darwin, but Darwin also described a process—**natural selection**—that satisfactorily explains how descent with modification occurs.

Darwin's Four Postulates

- A population experiences evolution by natural selection when (1) individuals vary in their traits; (2) some of these variations are heritable; (3) some individuals survive and reproduce better than other individuals; and (4) differential survival and reproduction are influenced by the heritable traits of individuals.

- In other words, some heritable traits help an individual to survive or reproduce better than other individuals. Such traits will, over time, increase in frequency in the population, causing evolution—change in a population's allele frequencies over time.

- Population thinking and an appreciation for the importance of variation among individuals are key to understanding evolution. Populations, not individuals, change over time when evolution occurs.

- Evolution by natural selection occurs when heritable variation leads to differential reproductive success.

The Biological Definitions of Fitness and Adaptation

- **Biological fitness** is the ability to produce more surviving offspring than other individuals in a population. It can be measured.

✔ CHECK YOUR UNDERSTANDING

(d) List some traits that might help some individuals survive or reproduce better than other individuals for each of these organisms: bacteria from a hot spring, tortoise in a woods, tree in a forest with tent caterpillars.

_____

_____

_____

_____

- **Adaptations** are heritable traits that increase an individual's fitness relative to individuals lacking that trait.

- Fitness and adaptation only have meaning in reference to a particular environment.

24.4 Evolution in Action: Recent Research on Natural Selection

- The theory of evolution by natural selection is testable. Here are two examples of how biologists test for evolution by natural selection in actual populations. Hundreds of other examples are available.

Case Study 1: How Did Mycobacterium tuberculosis Become Resistant to Antibiotics?

- *Mycobacterium tuberculosis* causes **tuberculosis (TB)**, a major public health threat in the nineteenth century.

- Better nutrition and antibiotics succeeded in greatly reducing deaths due to TB in industrialized countries. Unfortunately (but predictably), TB is on the rise again as antibiotic-resistant strains evolve and spread.

A Patient History

- Consider the evolution of antibiotic resistance in a young man diagnosed with TB. After more than 33 weeks of treatment with rifampin and other antibiotics, his lung fluid and chest X-rays were normal and he was considered cured.

- Two months later, the man was readmitted to the hospital with TB and treated with more antibiotics, including rifampin. Antibiotics did not kill these bacteria and the man died.

- Rifampin-resistant bacterial DNA was compared to DNA from the original antibiotic-sensitive bacteria cultured when the man was first admitted to the hospital with TB. The only difference between the two sequences was a mutation in the *rpoB* gene.

A Mutation in a Bacterial Gene Confers Resistance

- The *rpoB* gene codes for part of RNA polymerase—the enzyme that transcribes DNA to mRNA.

- The single point mutation (C → T) changed the normal codon TCG to TTG. This changes a serine to a leucine and prevents rifampin from efficiently binding to the polymerase (**Figure 24.12**).

- Researchers deduced that one or a few cells with this mutation were present in the original population.

- When antibiotics were added to the bacteria's environment, bacteria with the normal *rpoB* gene slowed growth or died; the bacterial population declined so much that the man appeared cured.

- However, cells with the mutant *rpoB* gene now had a selective advantage—they survived, reproduced, and eventually became so numerous that they killed their host (**Figure 24.13**).

Testing Darwin's Postulates

- Natural selection had caused evolution—a change in a population's allele frequencies over time.

  1. Variation existed in the initial bacterial population.

  2. The variation was heritable.

  3. Bacteria differed in their reproductive success.

  4. When rifampin was present, the rifampin-resistant bacteria produced more offspring than other bacteria; thus, selection occurred.

- Individual bacterial cells did not evolve; they merely survived or died and reproduced more or less. It was the population that evolved over time due to the effects of natural selection on individuals.

- Natural selection acts on the individual, but only the population evolves. Allele frequencies change in populations, not in individuals.

A Widespread Problem

- Evolution of antibiotic resistance in bacteria is increasingly common (**Figure 24.14**), as is the evolution of resistance to insecticides, fungicides, antiviral drugs, and herbicides in insects, fungi, viruses, and plants.

CHECK YOUR UNDERSTANDING

(e) How did the *M. tuberculosis* researchers establish heritability of rifampin resistance?

_____

_____

_____

_____

Case Study 2: Why Are Beak Size, Beak Shape, and Body Size Changing in Galápagos Finches?

- Biologists can study evolution in response to natural environmental change.

- Peter and Rosemary Grant led long-term research on changes in body size and beak size and shape in medium ground finches (**Figure 24.12**) on Isle Daphne Major of the Galápagos Islands.

- The medium ground finch lives on a small island, so researchers can catch, weigh, measure, and mark all individuals. Body size and beak size and shape varied among individuals and variation in these traits was heritable.

Selection during Drought Conditions

- In 1977, a severe drought eliminated 84 percent of the island's finch population. Most of the finches died of starvation since few plants were able to produce seeds.

- This sudden environmental change created a **natural experiment** that allowed researchers to compare groups created by a natural change in conditions (the before-drought and after-drought finch populations).

- Survivors had deeper beaks, on average, than did nonsurvivors (**Figure 24.16**). Grant's group hypothesized that individuals with large, deep beaks were more likely to survive because they could eat the tough fruits of *Tribulus cistoides*.

- Natural selection caused an increase in the finch population's average beak depth in just one generation. Alleles responsible for development of deep beaks increased in frequency as an adaptation for cracking large, tough fruits and seeds.

Continued Changes in the Environment, Continued Selection, Continued Evolution

- In 1983 there was much more rain than average and finches fed primarily on abundant small, soft seeds. Small finches with small, pointed beaks had particularly high reproductive success during this period, causing yet another change in the population's characteristics.

- The Grants continue documenting evolution in response to continuing environmental changes on Daphne Major (**Figure 24.17**). Long-term studies like these document how populations experience natural selection in response to environmental change.

✔ CHECK YOUR UNDERSTANDING

(f) A pattern of repeated speciation events is often found in island chains. What about islands promotes speciation?

_____

_____

_____

_____

Which Genes Are under Selection?

- Beak size, shape, and body size are influenced by many genes. To identify the genes involved, Clifford Tabin's lab looked at the pattern of expression of cell-cell signals known to be important in chicken beak development.

- In situ hybridizations showing where the cell-cell signal gene *Bmp4* is expressed revealed: (1) a strong correlation between *Bmp4* expression during beak development and the shape of adult beaks (**Figure 24.18**) and (2) that artificial increases in *Bmp4* expression caused wider and deeper beaks in chickens.

- Similar experiments show that variation in the calcium signaling molecule calmodulin affects beak length. Alleles for both the *Bmp4* gene and the calmodulin gene may be under selection in the Galápagos medium ground finch.

24.5 Common Misconceptions about Natural Selection and Adaptation

Selection Acts on Individuals, but Evolutionary Change Occurs in Populations

Individuals do not change during natural selection. Those that are selected simply produce more surviving offspring than other individuals do, leading to a change in the genetic makeup of the population in subsequent generations.

A Contrast with "Lamarckinan" Inheritance

- Lamarck suggested that individuals change in response to environmental challenges and pass these changes on to offspring.

- Under natural selection, individuals are sorted, not changed. Some individuals produce more or fewer offspring than others, but natural selection does not change alleles inside the selected individuals.

Acclimation Is Not Adaptation

- Sometimes individuals do change over their lifetime in response to environmental change (e.g., making more red blood cells at high altitudes). This is called **acclimation**.

- Acclimation does not cause evolution because these phenotypic changes are not heritable. They do not change an individual's alleles and thus cannot be passed on to offspring.

Evolution Is Not Goal-Directed

- Mutations, as in the tuberculosis example, occur randomly. Likewise, adaptations do not occur because organisms need them. They just happen, and when they do, natural selection may lead to their becoming more common.

Evolution Is Not Progressive

- Although groups that appear later in the fossil record are often more morphologically complex than related groups that appeared earlier, this tendency is not universal. Traits can be lost or gained due to evolution by natural selection.

- For example, populations that become parasitic tend to evolve simpler morphologies.

There Is No Such Thing as a Higher or Lower Organism

- Evidence does not support Lamarck's idea that evolution causes progress toward higher levels on a "ladder of life" (**Figure 24.19**).

- Different adaptations allow different groups to thrive in different environments, but all populations have evolved by natural selection based on their ability to gather resources and produce offspring.

Organisms Do Not Act for the Good of the Species

- Purely self-sacrificing behavior is not observed in nature. If a self-sacrificing behavior were to arise due to a mutation, that allele would be selected against as individuals with more selfish behaviors reproduced and left more offspring.

Limitations of Natural Selection

Nonadaptive Traits

- Adaptation is not a perfect process. For example, vestigial traits do not increase fitness; they exist simply because they were present in the ancestral population.

- Other nonadaptive traits include holdovers from structures present early in development and neutral mutations where a change in the DNA sequence does not change the amino acid sequence of the protein encoded by that gene.

Genetic Constraints

- Many alleles that affect body size also affect other aspects of size.

- In the finches studied by the Grants, beak depth was correlated with beak width, so deep, narrow beaks were not able to evolve even though birds with narrow beaks were better able to twist open the tough *Tribulus* fruits.

- When selection on alleles for one trait causes a correlated but suboptimal change in another trait, the possible evolutionary outcomes are limited. This type of constraint is called **genetic correlation**. It occurs when an allele affects multiple traits (pleiotropy).

- Lack of genetic variation can also constrain evolution because natural selection can work only on extant variation in a population.

Fitness Trade-offs

- A **fitness trade-off** is a compromise between traits. For example, there may be a trade-off between size of egg or seed and number of offspring.

- Fitness trade-offs occur because selection acts on many traits at once within a given individual.

Historical Constraints

- Because all traits evolve from previously existing traits, adaptations are constrained by history.

- For example, our middle ear bones derive from jaw and braincase bones of mammal ancestors. Natural selection had to work on the traits available—unlike an engineer, who is free to imagine the best possible way to solve each structural challenge.

✔ **CHECK YOUR UNDERSTANDING**

(g) Because most genes are inherited and used by both sexes, selection for traits that would be beneficial for males but harmful for females (or vice versa) is constrained. Is this a genetic constraint? A fitness trade-off? Both? Neither?

_____

_____

_____

_____

_____

B. MASTERING CHALLENGING TOPICS

Natural selection sounds relatively simple, yet it occurs in many different settings and with drastically different results. Don't lose sight of the basics when learning the details of individual instances. Instead, try to relate the details to the four steps leading to natural selection: variation, heritability, differential reproductive success, and reproductive success based on those variable heritable traits. Test your understanding by imagining how natural selection might have acted to produce one of your favorite species. What variable traits does that species have? Are those traits heritable? Would those traits affect survival and reproduction? How might that species have changed through time as individuals with the successful heritable traits had more offspring than those with less successful traits?

Remember that natural selection acts on the individual: an individual is either selected for or selected against. *On average*, some traits lead to greater success than do other traits (random exceptions will occur; they just won't matter in the long run if the population is large enough—more on this in the next chapter). This process makes the *population* evolve. Individuals do not change their heritable traits, but the frequency of those traits changes in a population over time in response to natural selection on individuals.

C. ASSESSING WHAT YOU'VE LEARNED

1. Which of the following are correct comparisons between Lamarckian and Darwinian evolution? (Choose all that are correct.)
   a. In Lamarckian evolution, individuals may acclimate, but these changes are not passed on to offspring, whereas in Darwinian evolution, individual change in response to environmental change is passed on to offspring.
   b. Lamarckian evolution relied on typological thinking, whereas Darwinian evolution requires population thinking.
   c. Only in Darwinian evolution do species change over time.
   d. Lamarckian evolution is progressive whereas Darwinian evolution is not.

2. Which of the following is a good restatement of Darwin's phrase "descent with modification"?
   a. Species change over time, and those that exist today descended from other, preexisting species.
   b. Humans are different from a preexisting ancestor species that lived in trees.
   c. Ancestral species are more primitive and lower down on the phylogenetic tree than descendant species alive today.
   d. Natural selection is survival of the fittest.

3. Which of the following is (are) supported by evidence from the fossil record? (Choose all that apply.)
   a. The concept that species can go extinct
   b. The hypothesis that individuals evolve by acquiring new traits
   c. The idea that the Earth is 4.6 billion years old
   d. The existence of transitional forms
   e. The existence of developmental homology
   f. The idea that species are typically succeeded, in the same region, by similar species
   g. The observation that species are only succeeded by larger and more complex descendants

4. An example of a vestigial trait in humans is:
   a. kidneys.
   b. cataracts.
   c. wisdom teeth.
   d. There are no vestigial traits in humans.

5. The use of RNA polymerase by all organisms is an example of:
   a. a vestigial trait.
   b. genetic homology.
   c. developmental homology.
   d. structural homology.

6. Which of the following is a good example of "internal consistency"?
   a. Beak depth is correlated with beak width in the medium ground finch.
   b. Special creation does not rely on observations from the natural world.
   c. Finch beak morphology changed during especially wet and especially dry years; *Bmp4* expression is correlated with the width and depth of adult beaks; and chickens with experimentally increased *Bmp4* expression develop wider and deeper beaks.
   d. An organism's internal organs are adapted to work together as an integrated unit so that the individual will be able to have more offspring.

7. Which of the following is considered Darwin's greatest contribution to science?
   a. Darwin recognized the pattern of evolution (descent with modification).
   b. Darwin recognized the process of evolution (natural selection).
   c. Darwin recognized "the original species."
   d. Darwin recognized the heritable basis of traits (genetics).

8. Which of the following is the best definition of Darwinian fitness?
   a. Darwinian fitness is the ability of an individual to survive in any environment.
   b. Darwinian fitness is the ability of a population of organisms to persist without changing for a long period of time.
   c. Darwinian fitness is the ability of an individual to reproduce more than others in the population.
   d. Darwinian fitness is the ability of an individual to stay healthy by eating well and exercising.

9. Which of the following is a modern definition of evolution?
   a. Evolution is a change in an individual's ability to survive and reproduce as it grows older.
   b. Evolution is a change in allele frequencies in a population over time.
   c. Evolution is an increase in the fitness of a population.
   d. Evolution is a change in an individual's fitness over time.

10. What are the four requirements for evolution by natural selection in a given population?
    a. (i) Population varies; (ii) all variation is heritable; (iii) some individuals produce more offspring than others do; (iv) populations differ in production of offspring.
    b. (i) Individuals in a population vary; (ii) the variation is neutral; (iii) offspring are produced that can survive; (iv) individuals produce offspring.
    c. (i) Individuals in a population vary; (ii) variation is heritable; (iii) more offspring are produced than can survive; (iv) all individuals have different alleles.
    d. (i) Individuals in a population vary; (ii) variation is heritable; (iii) some individuals produce more offspring than others do; (iv) certain traits lead to greater reproductive success than do others.

11. Which statement restates the fact that rifampin resistance in *Mycobacterium tuberculosis* is an adaptation?
    a. Rifampin resistance is a heritable trait that increases fitness relative to susceptible individuals in certain environments.
    b. Rifampin resistance is caused by a point mutation in the *rpoB* gene.
    c. A mutation occurred in the *rpoB* gene of one cell, and this mutant cell continued to grow and divide.
    d. Drug-resistant strains now account for about 10 percent of the *Mycobacterium tuberculosis*–caused infections throughout the world.

12. Resistance to a wide variety of insecticides, fungicides, antibiotics, antiviral drugs, and herbicides has recently evolved in hundreds of insects, fungi, bacteria, viruses, and plants. Why?
    a. Mutations are on the rise.
    b. Humans are altering the environments of these organisms, and the organisms are evolving by natural selection.
    c. No new species are evolving, just resistant strains or varieties. This is not evolution by natural selection.
    d. Humans have better health practices, so these organisms are trying to keep up.

13. Why did the Grants observe larger beak depths (in the medium ground finch), on average, following a drought?
    a. Deeper beaks help finches find water.
    b. Larger birds always survive better than smaller birds.
    c. During the drought, natural selection favored individuals with deeper beaks.
    d. The Grants did not observe deeper beaks after the 1977 drought.

14. Which of the following statements is true according to evolutionary theory?
    a. Acclimation can lead to evolution in a population.
    b. Flowering plants are more advanced or "higher" organisms than algae are.
    c. Natural selection always creates more complex species from simpler ancestor species.
    d. Individuals do not change during evolution by natural selection.

15. Which of the following are examples of evolutionary constraints?
    a. Trees that grow the tallest are more likely to fall over in a storm.
    b. Cheetahs have a small population with very little genetic variation.
    c. The mammal ancestor did not have an enzyme capable of breaking down cellulose.
    d. All of the above are examples of evolutionary constraints.

CHAPTER 24—ANSWER KEY

Check Your Understanding

(a) Traits that are not heritable may increase an individual's ability to survive or reproduce. But because the trait is not passed on to the offspring, it does not change the population frequency for that trait; thus evolution does not occur.

(b) Transitional form evidence typically comes in the form of intermediate fossils. Fossils are critical because they provide our most direct evidence of extinct life-forms. However, fossils are rare and the fossil record is biased toward organisms that are common, large, have hard parts, and so on. Vestigial traits however, especially ones like pseudogenes, can be found in a wide array of living organisms with less bias toward certain taxonomic groups.

(c) Although the underlying genetics is not always known or understood, developmental and structural homologies are based on genetic similarities in the genes controlling the structure or development. In Chapter 27 you learn that similarities in distantly related organisms with similar lifestyles are considered to be analogous rather than homologous if the evolutionary and genetic bases of the traits differ.

(d) *Bacteria*—ability to grow and divide quickly, heat tolerance of enzymes; *tortoise*—shell thick enough to protect against predators, but not too heavy to carry around; *tree*—leaves that taste bad (or are toxic) to tent caterpillars.

(e) Researchers identified the DNA mutation that caused the antibiotic-resistant trait. Any trait in a single-celled organism that is coded for by a specific DNA sequence is by definition heritable because DNA carries the genetic information. In multicellular organisms, mutations that occur only in nonreproductive cells (e.g., skin cancer) are not heritable.

(f) Islands promote speciation because the organisms on each island experience different environments and therefore different types of natural selection. In the next chapter you will learn some other reasons why islands can be sites of rapid speciation (genetic drift).

(g) The goal of this question is to provoke thought about genetic constraints versus fitness trade-offs.

This situation is a type of genetic correlation, but the correlation is between traits in different sexes rather than in the same individual. Thus, it can be evaluated only across a population or over many generations of male and female offspring. The fitness trade-off is also across generations rather than in an individual. For example, a trait that increases the success of male offspring but decreases the success of female offspring is selectively neutral for an individual that has equal numbers of male and female offspring.

Assessing What You've Learned

1. b, d; 2. a; 3. a, d, f; 4. c; 5. b; 6. c; 7. b; 8. c; 9. b; 10. d; 11. a; 12. b; 13. c; 14. d; 15. d

---

Looking Forward—Key Concepts in Later Chapters

Natural Selection and Mutation—Chapter 25

More information on natural selection and mutation can be found in **Chapter 25**, which also explores other possible causes of evolution.

Shared Ancestry—Chapters 26 and 27

Chapter 26 describes how two species can evolve from one ancestral population, and **Chapter 27** explains how relationships between species are studied and summarizes the history of the diversification of life on Earth.

Bacterial Diseases—Chapter 28

Mycobacterium tuberculosis is only one of many disease-causing bacteria. **Chapter 28** presents more information on bacterial diseases.

Chapter 25

Evolutionary Processes

Looking Back—Key Concepts from Earlier Chapters

Evolution—Chapters 1, 24

This chapter builds on discussions of evolution in earlier chapters. **Chapter 1** introduced evolution by natural selection as a key concept in biology and **Chapter 24** focused on the details of evolution by natural selection. You will now learn about additional evolutionary processes and how different types of natural selection have different effects on genetic variation in a population.

Mutation—Chapters 14, 15, 18

In **Chapters 14, 15,** and **18** you learned more about how mutations occur. In this chapter, the role of mutation as the ultimate source of genetic variation is highlighted. Mutation provides the variation that natural selection acts upon.

Genetics—Chapters 13, 15, 19

Chapters 13 and **15** provide background information critical to your understanding of how allele frequencies change in populations and how genes work to determine organism phenotype. The discussion in Chapter 25 of the Hardy-Weinberg principle shows how researchers study genetics at the population level. **Chapter 19** describes how researchers can artificially alter individual (and population) allele frequencies.

KEY CONCEPTS

- Researchers use the Hardy-Weinberg principle to generate a null hypothesis for tests of whether evolution or nonrandom mating is occurring for a particular gene.

- Mutation introduces new alleles, natural selection produces adaptation, genetic drift randomly changes allele frequencies, and gene flow reduces differences between populations.

- Inbreeding changes genotype frequencies but does not affect population allele frequencies.

- Sexual selection favors traits that help individuals attract mates and is typically stronger on males than on females.

A. CHAPTER OUTLINE

Chapter Introduction

- There are four processes that can change allele frequencies and cause evolution in a **population**—a group of individuals from the same species that live and breed together.

- Natural selection increases the frequency of alleles that improve reproductive success, genetic drift randomly changes allele frequencies, gene flow moves alleles from one population to another, and mutation introduces new alleles.

- Natural selection is the only process that consistently leads to adaptation.

- Effects of these evolutionary mechanisms can be compared to a null hypothesis that shows what happens to allele frequencies when no evolutionary processes are affecting a population.

25.1 Analyzing Change in Allele Frequencies: The Hardy-Weinberg Principle

- Evolutionary processes are often first analyzed using mathematical models. Model predictions are tested by collecting data. Results are often used to solve problems in conservation biology or human genetics.

- G. H. Hardy and Wilhelm Weinberg independently developed a mathematical model to calculate what happens to allele frequencies when no evolution is taking place.

- Instead of thinking about individual matings, Hardy and Weinberg considered allele frequencies in an entire population—they used population thinking.

The Gene Pool Concept

- Hardy and Weinberg considered all the gametes produced in one generation by individuals in a population. These gametes were placed in one single group called the **gene pool**.

- They then calculated what would happen if the next generation were made by randomly picking gametes from this gene pool. They predicted the genotypes of offspring produced and the frequency of each genotype.

Deriving the Hardy-Weinberg Principle

- Consider a population that has a gene with only two alleles (A_1 and A_2). The frequency of A_1 is designated p, and the frequency of A_2 is q. Since there are only two alleles, the two frequencies must add up to 1 (i.e., $p + q = 1$).

- Three genotypes are possible: A_1A_1, A_1A_2, and A_2A_2. When gametes are randomly picked from the gene pool, the resulting frequency of the A_1A_1 genotype in the new generation is p^2 (the probability of drawing two A_1 alleles $= p \times p$), and the frequency of A_2A_2 will be q^2. The frequency of A_1A_2 will be $2pq$ (see **Figures 25.1** and **25.2**).

- Because all individuals in the new generation must have one of the three genotypes, the sum of the three genotype frequencies must add up to one (100 percent of the population): $p^2 + 2pq + q^2 = 1$.

- When allele frequencies are calculated for this new generation, the frequency of A_1 is still p and the frequency of A_2 is still q. No evolution occurred.

- This result is called the **Hardy-Weinberg principle**: (1) If allele frequencies are p and q, genotype frequencies will be p^2, $2pq$, and q^2 for generation after generation; (2) meiosis and random combination of

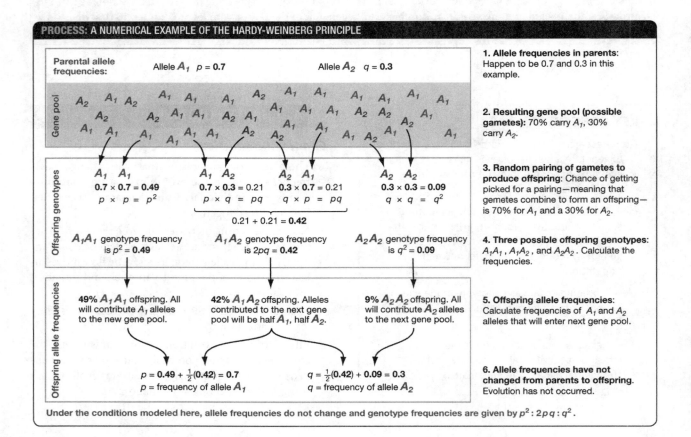

PROCESS: A NUMERICAL EXAMPLE OF THE HARDY-WEINBERG PRINCIPLE

Parental allele frequencies: Allele A_1 $p = 0.7$ Allele A_2 $q = 0.3$

1. Allele frequencies in parents: Happen to be 0.7 and 0.3 in this example.

Gene pool: A_2 A_1 A_2 A_1 A_1 A_1 A_2 A_1 A_1 A_1 A_1 A_1 A_1 A_2 A_2 A_1 A_1 A_1 A_1 A_2 A_1 A_1 A_1 A_2 A_2 A_1 A_1 A_1 A_1 A_1 A_1 A_2 A_1 A_2 A_1 A_1 A_2 A_2 A_1 A_2 A_1

2. Resulting gene pool (possible gametes): 70% carry A_1, 30% carry A_2.

Offspring genotypes:

A_1 A_1
$0.7 \times 0.7 = 0.49$
$p \times p = p^2$

A_1 A_2
$0.7 \times 0.3 = 0.21$
$p \times q = pq$

A_2 A_1
$0.3 \times 0.7 = 0.21$
$q \times p = pq$

A_2 A_2
$0.3 \times 0.3 = 0.09$
$q \times q = q^2$

$0.21 + 0.21 = 0.42$

3. Random pairing of gametes to produce offspring: Chance of getting picked for a pairing—meaning that gametes combine to form an offspring—is 70% for A_1 and a 30% for A_2.

A_1A_1 genotype frequency is $p^2 = 0.49$

A_1A_2 genotype frequency is $2pq = 0.42$

A_2A_2 genotype frequency is $q^2 = 0.09$

4. Three possible offspring genotypes: A_1A_1, A_1A_2, and A_2A_2. Calculate the frequencies.

Offspring allele frequencies:

49% A_1A_1 offspring. All will contribute A_1 alleles to the new gene pool.

42% A_1A_2 offspring. Alleles contributed to the next gene pool will be half A_1, half A_2.

9% A_2A_2 offspring. All will contribute A_2 alleles to the next gene pool.

5. Offspring allele frequencies: Calculate frequencies of A_1 and A_2 alleles that will enter next gene pool.

$p = 0.49 + \frac{1}{2}(0.42) = 0.7$
$p =$ frequency of allele A_1

$q = \frac{1}{2}(0.42) + 0.09 = 0.3$
$q =$ frequency of allele A_2

6. Allele frequencies have not changed from parents to offspring. Evolution has not occurred.

Under the conditions modeled here, allele frequencies do not change and genotype frequencies are given by $p^2 : 2pq : q^2$.

gametes do not, on their own, change a population's allele frequency over time.

✓ CHECK YOUR UNDERSTANDING

(a) It is easy to determine the frequency of homozygous recessive genes (q) because individuals with the homozygous recessive genotype show the recessive phenotype (whereas heterozygotes and homozygous dominants may have the same phenotype). There is some evidence that straight hairlines are recessive and widow's peaks are dominant in humans. If Hardy-Weinberg assumptions are met for this trait in a human population, and 81 percent of the population has straight hairlines (A_2A_2), then: q^2 must equal _____, q must equal _____, and p must equal _____. Thus, in this human population, the frequency of homozygous dominants (A_1A_1) must be _____, and the frequency of heterozygotes (A_1A_2) must be _____.

The Hardy-Weinberg Model Makes Important Assumptions

- The Hardy-Weinberg model assumes that mating is random with respect to the gene being examined and that none of the four evolutionary mechanisms is acting.
- The Hardy-Weinberg principle holds when there is (1) no natural selection; (2) no genetic drift; (3) no gene flow; (4) no mutation; and (5) random mating with respect to the gene in question.

How Does the Hardy-Weinberg Principle Serve as a Null Hypothesis?

- When biologists want to test whether natural selection, nonrandom mating, or some other evolutionary mechanism is acting on a particular gene, the Hardy-Weinberg principle serves as a null hypothesis.
- When biologists observe genotype frequencies that do not conform to Hardy-Weinberg proportions, it indicates that evolution or nonrandom mating is occurring in that population—at least one of the five Hardy-Weinberg conditions is being violated.

Case Study 1: Are MN Blood Types in Humans in Hardy-Weinberg Proportions?

- Most human populations have two alleles for the MN blood group. This gene codes for a protein on the surface of red blood cells, and human genotype (*MM*, *MN*, or *NN*) can be determined from blood samples.

- Using a large sample, geneticists can estimate genotype frequency for a population by dividing the number of individuals observed for each genotype by the total number in the sample. Allele frequencies can then be calculated from the genotype frequencies (**Table 25.1**).
- Allele frequencies can then be used to calculate the expected genotype distribution according to the Hardy-Weinberg principle; if mating is random and no evolution is occurring, genotype frequencies (*MM* : *MN* : *NN*) should equal $p^2 : 2pq : q^2$.
- Statistical tests (**BioSkills 5**) can determine whether any differences between observed and expected values are so small as to be due to chance or large enough to reject the Hardy-Weinberg null hypothesis for that gene.
- For MN blood groups, observed and expected genotype frequencies are almost identical. Because genotypes occur in Hardy-Weinberg proportions, evolutionary processes do not currently affect MN blood groups, and mating is random for this trait.

Case Study 2: Are HLA Genes in Humans in Hardy-Weinberg Equilibrium?

- The *HLA-A* and *HLA-B* genes code for proteins that help immune system cells recognize and destroy invading bacteria and viruses. Alleles for these genes are codominant, and they code for proteins that recognize slightly different disease-causing organisms.
- Researchers hypothesized that heterozygotes might be more fit than homozygotes because their immune systems would recognize more disease-causing organisms. They tested this hypothesis by determining the genotypes of Havasupai tribe members.
- Data on Havasupai tribe genotypes were used to calculate allele frequencies. These allele frequencies were then used to determine the expected genotype frequencies using the Hardy-Weinberg principle.
- **Table 25.2** shows observed and expected genotype frequencies. The Havasupai population had significantly more heterozygotes and fewer homozygotes than expected.
- At least one Hardy-Weinberg assumption must be violated for these alleles in this population. Mutation, migration, and genetic drift were negligible, so research focused on nonrandom mating and homozygote/heterozygote fitness differences.
- College students can distinguish genotypes at *HLA*-like loci by body odor and are more attracted to the smell of genotypes different from their own.
- Hutterite women with the same *HLA*-related alleles as those of their husbands had difficulty getting pregnant and had higher rates of spontaneous abortion.

- Research continues, but it appears that nonrandom mating, as well as low fitness of homozygous fetuses, may cause the higher-than-expected frequency of heterozygotes observed in the Havasupai tribe.

(b) Can you think of a situation in which the Hardy-Weinberg expected genotypes could be found even though the Hardy-Weinberg assumptions were not strictly met?

_____

_____

_____

_____

25.2 Types of Natural Selection

- Natural selection occurs when individuals with certain heritable phenotypes survive and reproduce better than others do. The alleles responsible for this increased reproduction then increase in frequency in succeeding generations, causing evolution.

- Biologists studying selection often focus on **genetic variation**—the number and relative frequency of alleles in a population.

- Low genetic variation can be dangerous for a population. If the environment changes and alleles that confer high fitness under the new conditions are not present, the average population fitness declines. Low genetic variation can even lead to extinction.

- How do different types of natural selection affect the genetic variation present in a population?

Directional Selection

- **Directional selection** is a type of natural selection that causes the average phenotype of a population to change in one direction.

Directional Selection Tends to Reduce Genetic Variation

- When many genes affect a trait, a graph of trait value (x-axis) versus number or frequency of individuals (y-axis) typically has a bell-shaped or normal distribution.

- Under directional selection, this distribution narrows and shifts toward the trait value that conferred higher fitness. The average value of the trait changes in the population (**Figure 25.3a**).

- If directional selection continues long enough, favored alleles approach a frequency of 1.0 (100 percent) and become "fixed," whereas alleles that are no longer found in the population are lost.

- Selection that causes a decrease in the frequency of disadvantageous alleles can also be called **purifying selection**.

Directional Selection on Body Size in Cliff Swallows

- Preferential survival of large cliff sparrows following a cold snap is an example of directional selection if body size differences are at least partly heritable (**Figure 25.3**).

- Directional selection is not necessarily constant throughout a species' range or over time. For example, severe cold snaps are rare events.

Countervailing Selection and Fitness Trade-offs

- Sometimes opposing patterns of directional selection on related traits maintain genetic variation in a population.

- For example, in cliff swallows, selection for large body size is counteracted by selection for maneuverability. This is known as a fitness trade-off.

Stabilizing Selection

- **Stabilizing selection** occurs when individuals with intermediate traits are more fit than either extreme (**Figure 25.4**).

- This type of selection maintains intermediate phenotypes in a population. Stabilizing selection decreases a population's genetic variation over time but does not change its average trait value.

- Human birth weight is under stabilizing selection because babies of average size are more likely to survive than are babies with low or high birth weights.

Disruptive Selection

- **Disruptive selection** is the opposite of stabilizing selection; it eliminates intermediate phenotypes and favors extreme phenotypes. Disruptive selection maintains genetic variation but does not change the mean value of a trait (**Figure 25.5**).

Disruptive Selection on Beak Size in Black-Bellied Seedcrackers

- For example, due to the types of food available, African seedcrackers with very short or very long beaks survive better than do those with intermediate phenotypes.

Disruptive Selection Can Lead to Formation of New Species

- Disruptive selection can cause speciation if individuals with one extreme of a trait start mating preferentially with individuals that have the same trait.

(c) You observe three lab populations of euglena (a photosynthetic protist that can engulf and digest bacteria) that differ in their environments. You determine the mean chloroplast number per euglena over several generations. Population A has a habitat that varies from dark with lots of bacteria to light with few bacteria. This population shows an increasingly bimodal distribution of chloroplast number per organism over time, but the mean number of chloroplasts per euglena does not change. Population B has a more constant habitat with both bacteria and light present throughout. The mean chloroplast number does not change, but you observe decreased variation in chloroplast number. Population C has a habitat with lots of light but no bacteria. The mean chloroplast number increases in this population. Which population(s) is/are experiencing selection, and what type of selection appears to be occurring?

_____

_____

_____

_____

Balancing Selection

- **Balancing selection** occurs when several alleles have similar fitnesses and frequencies. It maintains or increases variation in the population.

- Balancing selection occurs when heterozygotes have a higher fitness than homozygotes (as may be the case for HLA genes). This is called **heterozygote advantage**.

- Balancing selection can also happen when environmental or geographical variation favors different alleles at different times or in different places.

- **Frequency-dependent selection** is yet another type of balancing selection that occurs when rare alleles are favored over common alleles.

25.3 Genetic Drift

- **Genetic drift** is change in allele frequencies due to chance (**sampling error**). It causes allele frequencies to drift up and down randomly over time. Whereas natural selection increases fitness, genetic drift has a random effect on fitness.

Simulation Studies of Genetic Drift

- You can simulate genetic drift's effect on a gene with two alleles in a small population by flipping a coin to determine which of each parent's alleles is in each gamete used to form offspring.

- Due to random chance, the allele frequencies in the next generation are likely to be different from those in the initial population.

Computer Simulations

- Computer simulations have been used to model change in a population's allele frequencies due to random allele combination (**Figure 25.6**). These simulations show that genetic drift is more pronounced in small than in large populations.

- Over long periods of time, genetic drift can also affect large populations, especially for alleles that do not affect fitness and thus are not subject to natural selection (e.g., pseudogenes and silent mutations that do not affect the gene product).

Key Points about Genetic Drift

- Genetic drift is random with respect to fitness.

- Genetic drift is most pronounced in a small population.

- Over time, genetic drift can lead to the random loss or fixation of alleles.

Experimental Studies of Genetic Drift

- Warwick Kerr and Sewall Wright studied genetic drift in fruit flies using leg-bristle morphology (normal straight bristles versus "forked," bent-looking bristles) as a **genetic marker**—an easily identifiable trait with known alleles.

- Four females and four male flies were placed in each of 96 cages such that the initial frequencies of normal and forked-bristle alleles were 0.5 in each cage. These alleles do not affect fly fitness in the lab environment.

- In each succeeding generation, four male and four female offspring were randomly chosen to continue each population.

- Mutations for bristle morphology are rare and no gene flow between populations occurred. The only evolutionary process acting on the populations was genetic drift.

- After 16 generations, 29 of the 96 populations had only forked leg bristles, and 41 had only normal bristles. In these populations one allele had become fixed (**Figure 25.7**). Only 27 percent of the populations still had both alleles.

- Genetic drift decreased genetic variation within populations and increased genetic differences between populations.

✔ **CHECK YOUR UNDERSTANDING**

(d) Why did Kerr and Wright choose only four female and four male fruit flies from each population to breed in each generation?

_____

_____

_____

_____

What Causes Genetic Drift in Natural Populations?

- A sampling process occurs during fertilization in every population in each generation, but sampling error primarily affects population allele frequency in small populations. Other sampling events such as random accidents can also affect populations.

Founder Effects on the Green Iguanas of Anguilla

- A founder event occurs when a group leaves a population, immigrates to a new area, and establishes a new population (**Figure 25.8**).

- If the founding group is small, its allele frequencies probably differ from those of the source population. This sampling effect on the new population's allele frequencies is called a **founder effect**. It changes allele frequencies and is a type of genetic drift.

- Founder events and founder effects are especially common in the colonization of isolated habitats such as islands, mountains, caves, and ponds. For example, a raft of 15 or so green iguanas colonized the island of Anguilla.

Genetic Bottleneck on Pingelap Atoll

- A sudden decrease in population size, a **population bottleneck**, can cause a **genetic bottleneck**—a sudden reduction in the number of alleles in a population. Genetic drift occurs during the sudden reduction in population size.

- Genetic bottlenecks are commonly caused by disease outbreaks and natural catastrophes (e.g., the typhoon- and famine-induced bottleneck on Pingelap Atoll that led to increased frequency of the loss of function CNGB 3 allele for color vision).

- Genetic drift often continues in the resulting small population (thus the high percentage of the serious vision deficit achromatopsia on Pingelap today).

25.4 Gene Flow

- **Gene flow** is the movement of alleles from one population to another. It occurs any time individuals leave one population, join another, and reproduce.

- Gene flow reduces genetic differences between the source and recipient populations (**Figure 25.9**).

Gene Flow in Natural Populations

- Great tits breed in two woodlands on the island of Vlieland. Mated pairs are banded and nest in boxes provided by biologists.

- Tits hatched in the eastern woodland lay smaller clutches, on average, than tits hatched in the western woodlands, and females hatched in the eastern woodlands survive better than those hatched in the western habitats.

- The eastern bird population is 13 percent immigrants from the mainland and appears to be better adapted to the island environment. The western woodland population is 43 percent mainland immigrants: gene flow is three times higher in the western population.

- Females with more island-born grandparents have higher fitness (**Figure 25.10**).

How Does Gene Flow Affect Fitness?

- On Vlieland, immigrants bring in alleles that have relatively low fitness in the island environment. Gene flow can decrease the average fitness of individuals in a population where natural selection had produced adaptation to a specific habitat.

- Gene flow can increase the average fitness of individuals in a population where genetic diversity was lost due to a nonadaptive process such as genetic drift.

- Gene flow is random with respect to fitness; depending on the situation, it could increase or decrease average fitness. Either way, gene flow always decreases genetic differences between populations.

25.5 Mutation

- Most evolutionary mechanisms reduce genetic diversity. Gene flow removes alleles from the source population (but may add alleles to the recipient population) and genetic drift and selection usually decrease a population's genetic variation.

- **Mutation** restores genetic diversity and creates new alleles. A mutation occurs when DNA polymerase copy errors cause random change in a DNA sequence. Some changes alter protein amino acid sequence. Changes in chromosome makeup can also occur.

- Because errors are inevitable, mutation is always adding new alleles into populations at all gene loci.

- Mutation is a random process; because most organisms are well adapted to their current habitats, random changes usually result in **deleterious alleles**—alleles that lower fitness. These alleles tend to be eliminated by purifying selection.

- However, mutation does occasionally produce an advantageous allele. Beneficial alleles created by mutation should increase in frequency in a population over time due to the effects of natural selection.

Mutation as an Evolutionary Mechanism

- For a given gene (in humans), there is only about one mutation per 10,000 gametes. In eukaryotes, mutation rates are too low to significantly affect allele frequencies by themselves.

- Although mutation is extremely slow, organisms contain many genes (over 20,000 in humans), so almost every individual has a new allele at some gene due to mutation.

- Mutations are the only way genetic diversity can be increased. Mutation introduces new alleles into individuals in a population in every generation.

- If another evolutionary mechanism acts on alleles created by mutation, then significant evolutionary change may occur.

Experimental Studies of Mutation

Experimental Evolution

- Richard Lenski and colleagues tested the role mutation plays in evolutionary change by using single *Escherichia coli* cells to found many populations (**Figure 25.11**).

- Cells were grown under identical conditions. Each day some cells from each population were transferred to new growth media. At regular intervals, researchers froze sample cells from each population. The experiment continued for 10,000 generations.

- *E. coli* is asexual, so mutation was the only source of genetic variation. No gene flow was allowed, but natural selection and genetic drift operated in each population.

- At the end of the experiment, frozen "fossil" cells from previous generations were thawed and grown with cells from more recent generations in competition experiments. Cell populations were distinguished using a neutral color indicator.

- Populations of cells from newer generations grew faster than populations of cells from previous generations during these competition experiments. Thus, cells from more recent generations were better adapted to the experimental environment.

- Experimental outcomes were quantified by measuring relative fitness. The relative fitness of a descendant population is greater than one when recent-generation cells outnumbered older-generation cells at the end of the competition experiment.

Fitness Increased in Fits and Starts

- Relative fitness of the descendant populations increased by irregular intervals over the experiment, for an almost 30 percent total increase in fitness.

- Genetic drift probably had little effect because population sizes were large.

- Lenski's group proposed that each fitness increase was due to a new beneficial mutation that then rapidly increased in frequency through succeeding generations due to natural selection. Population fitness then stabilized until the next beneficial mutation occurred.

✔ **CHECK YOUR UNDERSTANDING**

(e) Could you do an experiment like Lenski's (**Figure 25.11**) on mice? In your answer, explain why or why not.

_____

_____

_____

_____

Take-Home Messages

- Mutation is the ultimate source of genetic variation. Without mutation, genetic variation would decrease and evolution would eventually stop.

- Mutation alone cannot normally change population allele frequencies. However, it becomes important when combined with natural selection as in Lenski's *E. coli* experiment.

- **Table 25.3** summarizes evolutionary mechanisms and their effects on fitness and genetic variation. Each of the four evolutionary forces has different consequences for allele frequencies.

- When one or more evolutionary processes affect a gene, population genotypes depart from Hardy-Weinberg proportions.

25.6 Nonrandom Mating

- The Hardy-Weinberg model assumes mating occurs at random, but in nature mating is rarely random. Inbreeding and sexual selection are two processes that violate the random-mating assumption.

Inbreeding

- Mating between relatives is known as **inbreeding**. Relatives share a recent common ancestor and are therefore likely to share alleles.

How Does Inbreeding Affect Allele Frequencies and Genotype Frequencies?

- In a self-fertilizing population, homozygous parents produce all homozygous offspring, but only half the offspring of heterozygotes are heterozygous.

- Thus, in each generation the proportion of homozygotes increases and the heterozygous proportion of the population is halved (**Figure 25.12**).

- Inbreeding decreases the frequency of heterozygotes and increases the frequency of homozygotes in each generation. This increase in homozygosity also occurs with less severe forms of inbreeding than self-fertilization.

- Although inbreeding affects genotype frequency, it does not change allele frequencies; thus it does not cause evolution.

Why Does Inbreeding Depression Occur?

- **Inbreeding depression** is the decreased fitness that occurs when population homozygosity increases.

- Many recessive alleles are loss-of-function mutations. These mutations are typically disadvantageous or even lethal in a homozygote but have little or no effect in heterozygotes.

- Homozygotes also have lower fitness for traits that show a heterozygote advantage (e.g., HLA-like genes that are involved in immune system function).

- In many cases, offspring of inbred matings have lower fitness than offspring of outcrossed matings (e.g., *Lobelia cardinalis*, **Figure 25.13**).

- Inbreeding depression has been shown in many human populations (**Table 25.4**). Many species, including humans, have mechanisms to avoid inbreeding.

- Inbreeding depression does not directly cause evolution since it does not change population allele frequencies. But it can increase the rate at which purifying selection removes disadvantageous recessive alleles by exposing these alleles in the homozygous phenotype.

Sexual Selection

- **Sexual selection** occurs when individuals in a population vary in heritable traits that affect their ability to attract mates. It is a type of natural selection that favors individuals that are more successful at obtaining mates.

CHECK YOUR UNDERSTANDING

(f) Sometimes mate choice is nonrandom, such that genetically different individuals are more likely to mate. This is called outbreeding. What would happen to allele frequencies and genotype frequencies in a population experiencing an outbreeding mating pattern?

_____

_____

_____

_____

Theory: The Fundamental Asymmetry of Sex

- The Bateman-Trivers theory states that sexual selection typically acts more strongly on males than on females. Thus, males tend to have more mate-attracting traits. The fact that making eggs requires more energy than making sperm explains this pattern.

- Females usually invest more in their offspring than do males. This is called the fundamental asymmetry of sex.

- Females typically produce relatively few offspring during their lifetime. Female fitness is thus primarily limited by ability to gain resources for producing and rearing young, not by ability to find a mate.

- Sperm are cheap to produce, so males can father many offspring. Male fitness is primarily limited by ability to acquire mates.

Predictions of the Bateman-Trivers Theory

- Females should be choosy about their mates, but males should be willing to mate with almost any female.

- Males should compete with each other for mates, especially if there is an approximately equal number of males and females in the population.

- Two types of sexual selection should occur: female choice and male-male competition.

- When number of mates limits male fitness, alleles that increase male attractiveness or success in male-male competition should be favored. Sexual selection should usually act more strongly on males than on females.

Female Choice for "Good Alleles"

- Female birds invest a great deal of time and energy in offspring, so they should be choosy about the males they mate with.
- Red and yellow carotenoid pigments cause bright feather and beak coloration in many bird species. Carotenoids also stimulate the immune system and protect tissues. Animals cannot make carotenoids and must obtain them in their diet.
- Birds with the brightest beaks and feathers are well fed and healthy, because carotenoids are used in beaks and feathers only when they are not needed by the immune system.
- Researchers fed zebra finches diets that differed in carotenoid content. Males eating a carotenoid-rich diet developed colorful beaks. Females preferred these bright-beaked males over duller-beaked males from the carotenoid-poor diet group (**Figure 25.14**).
- Other studies confirm that females prefer to mate with colorful, healthy, well-fed males. Presumably these males have alleles that will help their offspring fight disease and feed efficiently.

Female Choice for Paternal Care

- Males that provide resources for egg production or care of young are also preferred mates in many animal species.
- Females often choose mates based on (1) physical characteristics that signal good genes and/or (2) male resources or parental care. However, in some species, females have little choice in mating, and competition among males causes sexual selection.

CHECK YOUR UNDERSTANDING

(g) Male European blackbirds have a bright orange bill due to the presence of carotenoids. Researchers injected some males with sheep red blood cells, initiating a strong immune response. Based on your reading above, predict what happened to the beak color of these males.

_____

_____

_____

   If females were to choose between these blackbirds and a control group of blackbirds, which males do you think they would choose and why?

_____

_____

_____

Male-Male Competition

- Male elephant seals fight over beach patches where females come to bear their pups. Males that win battles obtain larger territories and father many offspring.
- A **territory** is an actively defended area where the owner has exclusive use—in this case, exclusive access to females inhabiting that part of the beach.
- Size may affect male ability to win fights, gain large territories, and produce offspring. Alleles of territory-owning males increase in frequency due to sexual selection. Sexual selection for large size in males may explain why male elephant seals are more than four times larger than females.
- A few male elephant seals father the vast majority of offspring produced each year, whereas females vary less in reproductive success (**Figure 25.16**). Sexual selection is intense and is driven by male-male competition in this species.

What Are the Consequences of Sexual Selection?

- Reproductive success usually varies more in males than in females, so sexual selection is typically more intense in males. Males, therefore, tend to have more traits that function primarily in courtship or male-male competition.
- **Sexual dimorphism**, or male-female differences, can be observed in many species. These sexually selected traits include weapons used in male-male competition as well as ornamentation and behavior used in courtship displays.
- Nonrandom mating due to sexual selection violates Hardy-Weinberg assumptions.
- Sexual selection, unlike inbreeding, can cause certain alleles to increase in frequency; sexual selection can cause evolution.

B. MASTERING CHALLENGING TOPICS

Understanding the Hardy-Weinberg equations can be challenging for those lacking an intuitive understanding of the math involved. It may help you to work through concrete examples like that given in the "Check Your Understanding" question (a). You might try mapping this example in a Punnett square figure like that shown in **Figure 25.2** in addition to doing the calculations. You could also use two candy colors or two different coins to go through the process of picking gamete alleles and calculating genotype frequencies, as shown in **Figure 25.1**.

Keep in mind that in a large population, if 70 percent of the alleles in the gene pool are A_1, then 70 percent of the gametes should be A_1 (under Hardy-Weinberg conditions). But if you try to replicate this process on your own, picking blindly, you may get a different percentage in the gametes due to random chance—in which case you have just shown the effects of genetic drift!

Now think about this process of picking alleles to go into the next generation when the other evolutionary processes are acting. How do the processes affect the allele frequencies? For example, if you are doing this exercise with orange and green candies, and you prefer the flavor of the green ones, you may find that you eat the green ones. The orange alleles survive better and are thus more likely to make it into the next generation. Presto—natural selection!

C. ASSESSING WHAT YOU'VE LEARNED

1. The Hardy-Weinberg principle acts as a null model because it describes the relationship between allele and genotypic frequencies under conditions where:
   a. none of the four evolutionary forces are acting, and mating is random.
   b. new species are differentiating.
   c. evolution by natural selection increases the fitness of individuals.
   d. individuals in a population are not mating randomly.

2. You survey a wild population of Mendel's garden peas and find that 36 percent have the homozygous recessive white-flowered phenotype. If the population shows Hardy-Weinberg proportions and there are only two alleles for this trait in the population (the dominant purple allele and the recessive white allele), what are the frequencies of the two alleles?
   a. The frequency of the dominant allele is .64 and the frequency of the recessive allele is .36.
   b. The frequency of the dominant allele is .8 and the frequency of the recessive allele is .2.
   c. The frequency of the dominant allele is .4 and the frequency of the recessive allele is .6.
   d. The frequency of the dominant allele is .75 and the frequency of the recessive allele is .25.
   e. The frequency of the dominant allele is .16 and the frequency of the recessive allele is .36.

3. You are studying the genetics of a large population of flowers. A gene for pigment production has two codominant alleles, X_1 and X_2. X_1 has a frequency of .3 in the population and X_2 has a frequency of .7. Use these measured allele frequencies to calculate expected genotype frequencies. Compare your expected genotype frequencies to the observed genotype frequencies shown in the following. What might you conclude?

Genotype:	X_1X_1 (white flowers)	X_1X_2 (pink flowers)	X_2X_2 (red flowers)
Observed	.15	.30	.55

   a. The difference between observed and expected frequencies is negligible, so this population does not appear to be evolving with respect to this trait.
   b. There appears to be a heterozygote advantage for this trait.
   c. Heterozygotes are disadvantaged; perhaps there is disruptive selection for this trait.
   d. This population appears to be experiencing directional selection for this trait.

4. Which of the following scenarios best illustrates heterozygote advantage?
   a. A population has more heterozygotes than homozygotes.
   b. For a given locus, heterozygous individuals are more fit than are homozygous individuals.
   c. Individuals in a population are nonrandomly mating without inbreeding depression.
   d. Parents with similar alleles produce more offspring than do parents with dissimilar alleles.

5. Which of the following examples of natural selection in action would tend to increase the genetic variation in the population?
   a. In a pond-dwelling fish species, higher average summer temperatures selects for fish with greater heat tolerance over time.
   b. A small population of butterflies is blown by a windstorm and colonizes a new mountaintop.
   c. Natural selection simultaneously selects against heavy field mice (they starve during winter) and light field mice (they freeze during winter).
   d. Males with bright red feet produce more offspring than do those with duller feet in one bird species.
   e. In alpine skypilots, large flowers set more seed above the timberline and small flowers set more seed below the timberline.

6. In mammals, young born from pregnancies that are much longer or shorter than the species average tend to have higher rates of infant mortality. This is an example of selection.
   a. disruptive
   b. directional
   c. sexual
   d. stabilizing
   e. frequency-dependent

7. Following the volcanic explosion of Mount St. Helens, prairie lupine seeds recolonized the ash plain around the mountain. Populations initially differed greatly from each other. These differences were most likely due to
   a. the founder effect.
   b. natural selection.
   c. gene flow.
   d. mutation.

8. You repeat Kerr and Wright's *Drosophila* experiment with 10 males and 10 females per cage instead of 4. Which of the following is the best prediction for how results of this new experiment will compare with Kerr and Wright's original results?
   a. Genetic drift should have a stronger effect in this new experiment because of the higher sample size.
   b. Gene flow should be more important because of the larger population size.
   c. More populations should lose the forked allele due to sexual selection.
   d. More populations should have both alleles present at the end of this new experiment.

9. Which of these statements most fully characterizes the fundamental asymmetry of sex?
   a. Female fitness is limited most by the ability to get resources for producing eggs and rearing young, whereas male fitness is limited by the ability to attract females.
   b. Female fitness is limited most by the ability to attract males, whereas male fitness is limited by the ability to get resources for provisioning the female.
   c. Female fitness is limited most by the ability to get resources for producing eggs and rearing young, whereas male fitness is limited by the ability to get resources for provisioning the female.
   d. Female fitness is limited most by the ability to attract males, whereas male fitness is limited by the ability to attract females.

10. Genetic diversity is required for natural selection to act, but natural selection can reduce or eliminate diversity. What process can restore genetic diversity to a population?
    a. Genetic drift
    b. Mutation
    c. Sexual selection
    d. Stabilizing selection

11. In Lenski's experiment with *E. coli*, if genetic drift had been the primary evolutionary mechanism acting on the bacterial populations, what different results would you expect?
    a. Most of the populations should have gone extinct after the fixation of deleterious alleles.
    b. Fitness should have decreased in the descendent populations more often than it increased.
    c. Relative fitness should have consistently increased or decreased throughout the experiment without the irregular steps found by Lenski.
    d. Beneficial mutations could not have occurred while genetic drift was acting.

12. What is an important consequence of gene flow in natural populations?
    a. Gene flow increases the mutation rate among sedentary organisms.
    b. Gene flow moves individuals from one habitat to another on a seasonal basis.
    c. Gene flow tends to separate allele frequencies among populations.
    d. Gene flow tends to reduce genetic differences among populations.

13. In what kind of population is random genetic drift most pronounced?
    a. Drift is greatest in large populations.
    b. Drift is greatest in small populations.
    c. Drift is greatest in migrating populations.
    d. Drift is greatest in fixed populations.

14. In some jacana bird species, males are solely responsible for incubating young and females defend territories of several males. Females are larger than males in this species. Based on this information, which of the following is most likely to be true for this species?
    a. Sexual selection does not occur.
    b. Male-male competition is more important than female choice.
    c. Males have greater variation in reproductive success than females.
    d. Females have greater variation in reproductive success than males.

15. Under extreme inbreeding where all population members self-fertilize, if genotype frequencies in one generation are 0.20 A_1A_1, 0.60 A_1A_2, and 0.20 A_2A_2, what would the genotype frequencies be in the next generation? (Assume no evolutionary mechanisms are acting.)
    a. 0.5 A_1A_1, 0.0 A_1A_2, and 0.5 A_2A_2
    b. 0.20 A_1A_1, 0.60 A_1A_2, and 0.20 A_2A_2
    c. 0.35 A_1A_1, 0.30 A_1A_2, and 0.35 A_2A_2
    d. 0.45 A_1A_1, 0.10 A_1A_2, and 0.45 A_2A_2
    e. None of the above

CHAPTER 25—ANSWER KEY

Check Your Understanding

(a) q^2 equals 0.81, q equals 0.9, and p equals 0.1. The frequency of homozygous dominants (A_1A_1) must be 0.01, and the frequency of heterozygotes (A_1A_2) must be 0.18.

(b) Theoretically this situation could occur if exactly the same allele proportions were migrating into a population as were migrating out. It could also occur if natural selection against some allele were exactly balanced by mutations and/or gene flow bringing that allele back into the population.

(c) All three populations appear to be experiencing selection. Populations A, B, and C appear to be experiencing disruptive, stabilizing, and directional selection respectively.

(d) They chose only four males and four females because they wanted to observe the effects of genetic drift, and genetic drift has a greater impact on allele frequencies in small populations.

(e) Lenski's experiment could not be performed on mice, because you cannot freeze mice and then use them to compete against later generations. Also, observing 2000 mice generations would take way too much time.

(f) Outbreeding would not change allele frequencies either, but it would increase the frequency of heterozygotes relative to homozygotes.

(g) Because carotenoids are used to stimulate the immune system, you should predict that the beak color would dim. This is, in fact, what occurred. Females probably would choose the brighter-beaked males (the control group) since, in nature, a dim beak might indicate either sickness or a low-quality diet. This prediction follows from the zebra finch results, but would have to be tested experimentally.

Assessing What You've Learned

1. a; 2. c; 3. c; 4. b; 5. e; 6. d; 7. a; 8. d; 9. a; 10. b; 11. b; 12. d; 13. b; 14. d; 15. c

---

Looking Forward—Key Concepts in Later Chapters

Evolution—Chapters 26, 27, 51, and others

Because evolution is one of the unifying themes of biology, it reappears throughout your textbook. Larger-scale evolutionary patterns are examined in **Chapters 26** and **27**. **Chapters 28** through **35** discuss the evolution of specific groups of organisms, and **Chapter 51** examines the evolution of behavior.

Adaptation—Chapters 27–35, 36, 41, and others

As with evolution, you can find mention of adaptations (traits that increase fitness) throughout this textbook. **Chapters 28** through **35** describe the major organismal groups—something that could hardly be accomplished without mentioning the salient characteristics, or adaptations, of each group. **Chapters 36** and **41** have more examples of plant and animal adaptations.

Populations—Chapter 52

Evolution occurs in populations. Although much of this textbook examines other biological levels (species, organism, cell), **Chapter 52** focuses on populations again, in a thorough discussion of their ecology.

Speciation

Evolution—Chapters 1, 24, and 25

Earlier chapters have focused on processes that cause evolution or change within a population. Chapter 26 focuses on how different evolution in different populations can cause speciation.

Meiosis—Chapter 12

Chapter 26's description of why triploid offspring would be unable to make healthy gametes relies on an understanding of the process of meiosis covered in **Chapter 12**. You may wish to review meiosis and think about what would happen to triploid chromosomes during synapsis.

Mitochondrial DNA—Chapter 20

Maternally inherited mitochondrial DNA, studied in this chapter's example of warbler hybrid zones, was first mentioned in **Chapter 20**.

KEY CONCEPTS

- Genetic isolation due to lack of gene flow leads to speciation if populations diverge due to selection, genetic drift, or mutation.

- Biologists use three criteria to evaluate whether distinct populations are different species: reproductive isolation, morphological features, and evolutionary history (phylogeny).

- Geographic separation can lead to genetic isolation. Populations living in the same area can also become genetically isolated if they cannot breed because one population is polyploid or if they use different habitats or resources.

- Diverged populations that come back into contact might fuse, continue to diverge or stay partially differentiated. Their offspring might form a new species.

A. CHAPTER OUTLINE

Chapter Introduction

- Populations that are no longer connected by gene flow may diverge genetically due to mutation, genetic drift, and/or natural selection.

- This genetic divergence may eventually lead to **speciation**, the creation of new species (**Figure 26.1**).

26.1 How Are Species Defined and Identified?

- A **species** is an evolutionarily independent population or group of populations.

- Identifying species can be more challenging than one might think. Biologists identify species using one or more of three different criteria: the biological species concept, the morphospecies concept, and the phylogenetic species concept.

The Biological Species Concept

- The **biological species concept** considers populations to be evolutionarily independent if they are reproductively isolated from each other (no gene flow).

- **Prezygotic isolation** occurs when individuals from different populations are unable to mate (e.g., breed at different times or have incompatible genitalia, **Table 26.1**).

- **Postzygotic isolation** occurs when individuals from different populations can interbreed, but the hybrid offspring die young or have low fitness.

- Applying the biological species concept of reproductive isolation is difficult for populations that are geographically separated (will the geographic separation continue?) and cannot be applied at all to asexual or fossil species.

The Morphospecies Concept

- Evolutionary independence can be estimated by looking at morphological differences using the **morphospecies concept**.

- Differences in size and form often evolve when populations are reproductively isolated. Different morphologies will only persist if gene flow is restricted.

- The morphospecies concept can easily be applied to most populations and species, including asexual and fossil species, but it is rather subjective and cannot identify **cryptic species** that differ in non-morphological traits (e.g., behavior).

The Phylogenetic Species Concept

- The **phylogenetic species concept** defines species as evolutionary distinct populations and is based on a reconstruction of evolutionary history. This approach is widely applicable.

- On phylogenetic trees, an ancestral population plus all its descendants (but no other groups) is called a **monophyletic group**—also known as a **clade** or **lineage**. (See **Bioskills 3.**)

- Monophyletic groups are distinguished by distinctive traits, called **synapomorphies**, which are only found in that group of organisms.

- A synapomorphy is found throughout a monophyletic group because it is inherited from the group's common ancestor (e.g., fur in mammals). These homologous traits can be genetic, developmental, or structural.

- A phylogenetic species is the smallest monophyletic group on a tree created from data on related populations (**Figure 26.2**).

- The phylogenetic species concept is applicable to any population and is logical—monophyletic groups, by definition, differ in some traits and are isolated from gene flow with other groups (they have an independent evolutionary history).

- Use of the phylogenetic species concept may lead to the recognition of more species. However, many groups do not yet have good phylogenetic trees available.

- In practice, biologists use all three species concepts to identify evolutionarily independent populations (**Table 26.2**). Sometimes the species they recognize differ, depending on the species concept used.

(a) Which species definition (i.e., biological, morphospecies, or phylogenetic) would you use for (1) a new fossil species, (2) a well-studied bacterial species, and (3) a well-studied mammalian species? Why?

_____

_____

_____

_____

Species Definitions in Action: The Case of the Dusky Seaside Sparrow

- A **subspecies** is a population that lives in a discrete geographical area that has certain distinguishing features but is not yet distinct enough to be its own species.

- Several seaside sparrow (*Ammodramus maritimus*) subspecies were named using the morphospecies concept.

- The seaside sparrow subspecies were thought to be genetically isolated because the populations are geographically isolated and young birds breed near their hatching ground (**Figure 26.3a**).

- Development of salt marshes where these birds live eliminates their natural habitat, causing population decreases. In 1980 the dusky seaside sparrow subspecies (*A. m. nigrescens*) had only six males left.

- In an effort to preserve this population, biologists bred these males with females from a nearby population (*A. m. peninsulae*) in hopes of reintroducing the hybrid offspring.

- Other biologists compared gene sequences from different seaside sparrow populations and determined that seaside sparrows belong to two monophyletic groups: one on the Atlantic Coast and one on the Gulf Coast (**Figure 26.3b**).

- *A. m. nigrescens* is not distinct from other Atlantic seaside sparrows and did not need to be individually preserved according to the phylogenetic species concept. Instead, the Atlantic Coast and Gulf Coast groups should be separately preserved.

(b) Why does the fact that the seaside sparrow phylogeny shows only two distinct groups suggest there may be more gene flow within those two groups than previously thought?

_____

_____

_____

_____

26.2 Isolation and Divergence in Allopatry

- Genetic isolation due to physical separation can occur when a group colonizes a new habitat (dispersal) or when a new physical barrier divides a population (**vicariance**).

- Speciation that stems from physical isolation is called **allopatric speciation** (**Figure 26.4**) and populations living in different areas are said to be in **allopatry**.

- The study of the geographic distribution of species and populations is called **biogeography**.

Dispersal and Colonization Isolate Populations

- Peter and Rosemary Grant observed five large ground finches from a nearby island colonizing Daphne Major. Because finches normally stay on the same island, the colonists were allopatric to their original or source population.

- The Grants caught, weighed, and measured large ground finches on Daphne Major over the next 12 years and found the average bill size of the new Daphne Major population was larger than that of the original population.

- Differences between the new population and its source could be due to genetic drift and/or natural selection for large beaks.

- Due to chance, a colonizing population may differ from its source population. Genetic drift often continues in small colonizing populations. Population divergence may then be hastened by natural selection in the new environment.

- Colonization followed by genetic drift and natural selection appears to have caused repeated speciation in many groups of island organisms.

(c) Colonization followed by genetic drift and natural selection has also led to speciation in aquatic organisms and organisms that live at the top of some mountains. Explain how this could happen.

_____

_____

_____

_____

_____

_____

Vicariance Isolates Populations

- Sometimes an existing population is separated by a new physical barrier. For example, uplift of a new mountain range or a change in a river course may separate populations. This separation is known as a vicariance event.

- For example, the Isthmus of Panama formed a land bridge between North and South America about 3 million years ago, separating Caribbean and Pacific marine populations—including the snapping shrimp (**Figure 26.5**).

- Phylogenetic analysis shows that, in snapping shrimp, many sister species are from opposite sides of the Isthmus. This pattern supports the idea that these populations diverged following physical isolation by the isthmus.

- Physical isolation by dispersal or vicariance can genetically isolate populations. Genetic isolation followed by genetic divergence due to mutation, selection, and genetic drift can cause speciation.

(d) Why is genetic drift more likely to be an important evolutionary factor in colonization events than in vicariance events?

_____

_____

_____

26.3 Isolation and Divergence in Sympatry

- Populations or species that live in the same geographic region (close enough to mate) live in **sympatry**.

- Researchers used to think that speciation could not occur among sympatric populations, because gene flow would tend to eliminate any evolutionary differences.

Can Natural Selection Cause Speciation Even When Gene Flow Is Possible?

- In some situations, natural selection *can* overcome limited gene flow to cause **sympatric speciation**. Often this happens when populations become isolated by habitat preference.

- Habitat preference appears to have facilitated divergence between the apple maggot fly and the hawthorn fly.

- Apple maggot flies court and mate on apples. Apple trees were introduced to North America less than 300 years ago. Thus, the apple maggot fly appears to have recently evolved from the closely related hawthorn fly.

- Hawthorns and apples often grow together, but biologists followed marked individuals and found that apple and hawthorn flies rarely mate with each other.

- Apple flies are attracted to apple scent and avoid hawthorn scent. Hawthorn flies are attracted to hawthorn scent and avoid apple scent. Flies do not differ in their response to no scent or mixed scents.

- Further experiments determined that the fly's ability to distinguish scents has a genetic basis, identified the odor cells responsible for response to these two smells, and showed that hybrids cannot discriminate between apple and hawthorn fruit.

- Disruptive selection is acting on these populations to genetically isolate apple flies from hawthorn flies. These two populations are in the process of becoming distinct species.

- Many insects are associated with specific hosts. Individuals that switch hosts often experience different selection pressures and reduced gene flow with the source population. Host switching may have been an important trigger for speciation among insects.

How Can Polyploidy Lead to Speciation?

- Mutation alone does not usually cause evolutionary change, because it is an inefficient evolutionary

(e) Would divergence occur if hawthorn flies preferred hawthorn scent, but apple flies had no preference? Why or why not?

_____

_____

_____

_____

_____

_____

mechanism. But one type of mutation does cause speciation in sympatric plant populations.

- **Polyploidy**—the presence of more than two sets of chromosomes—occurs when an error in cell division causes chromosome number to double. This is a massive mutation that affects entire chromosomes.

- Polyploid mutations can cause sympatric speciation by reducing or eliminating gene flow between mutant and normal (wild-type) plants.

- If a diploid ($2n$) parent with haploid (n) gametes mates with a mutant **tetraploid** ($4n$) parent with diploid gametes, the offspring are **triploid** ($3n$). Even if this triploid survives, it cannot reproduce, because the three chromosome copies will be unable to undergo meiosis correctly (**Figure 26.8**).

- Tetraploid and diploid populations are reproductively isolated because if they mate, their offspring are almost always sterile.

- Mutations that result in a doubling of chromosome number produce **autopolyploid** individuals. **Allopolyploid** individuals result from a mating between parents of two different species.

Autopolyploidy

- Biologists found several polyploid maidenhair ferns. In the normal fern life cycle, individuals alternate between haploid (n) and diploid ($2n$) life stages. But the polyploid fern instead alternated between diploid and tetraploid ($4n$) life stages.

- The parent of the polyploid plants had a defect in meiosis and produced diploid gametes. Maidenhair ferns can self-fertilize, so these gametes combined to form tetraploid offspring that could self-fertilize or mate with the parent plant.

- Polyploid individuals are an evolutionarily independent population because they are genetically isolated from the source diploid population. They may form a new species if genetic drift and selection cause the two populations to diverge.

Allopolyploidy

- Many hybrid offspring are sterile because their chromosomes do not pair normally during meiosis.

- However, if a mutation doubles the chromosome number, then homologs can synapse normally and gametes can be produced. These gametes may be able to self-fertilize, producing a tetraploid offspring (**Figure 26.9**).

- The sequence of events just described seems to have caused speciation in *Trogopogon*, a weedy European plant introduced to North America in the early 1900s. Two new tetraploid species appear to have originated following allopolyploid events.

Why Is Speciation by Polyploidy So Common in Plants?

- Many diploid plant species have closely related polyploid species, indicating that this type of instantaneous sympatric speciation by autopolyploid and allopolyploid mutation has been relatively common in plants.

- Plant reproductive cells don't separate from somatic cells until late in development. If sister chromatids don't separate during an early mitotic division, a tetraploid cell is produced that may later undergo meiosis to form diploid gametes.

- Many plants self-fertilize, so diploid gametes can fuse, creating a new, genetically isolated tetraploid population.

- Cross-species hybridization is also fairly common in plants and increases the rate of allopolyploidy in this group.

- Speciation by polyploidy is virtually instantaneous as compared to gradual speciation by geographic isolation or disruptive selection in sympatry (**Table 26.3**).

CHECK YOUR UNDERSTANDING

(f) Why aren't polyploid mutations an important cause of speciation in animals?

_____

_____

_____

_____

_____

26.4 What Happens When Isolated Populations Come into Contact?

- Populations that have been isolated may evolve differences in mating mechanics, time, place, or behavior. These prezygotic isolating mechanisms may prevent gene flow even if the populations regain contact, allowing divergence of the populations to continue.

- Alternatively, populations that regain contact may successfully interbreed. Gene flow may eventually eliminate differences that evolved during their separation.

- Other possible outcomes are reinforcement, hybrid zones, and speciation by hybridization.

Reinforcement

- Sometimes individuals from populations that had been separated can mate with each other, but the hybrid offspring have low fitness.

- Hybrids may be poorly adapted to both parental habitats or may even fail to develop normally. This postzygotic isolation typically causes strong selection against interbreeding.

- Individuals mating within their own population will be more successful at passing their genes on to future generations, so natural selection should favor any traits that reproductively isolate the populations. This process is called **reinforcement**.

- Researchers found that sympatric *Drosophila* species typically will not mate in a laboratory environment, whereas allopatric species often will. This pattern suggests reinforcement occurred among sympatric *Drosophila* species.

- Reinforcement can only occur when species live in the same geographic area; if two species never naturally interbreed, natural selection cannot act to reduce interbreeding.

Hybrid Zones

- Sometimes hybrid offspring are healthy, with traits that are intermediate between the two parental populations. A geographic area where interbreeding and hybrid offspring between two populations is common is called a **hybrid zone**.

- Hybrid zones can be narrow or wide, long- or short-lived depending on the extent of interbreeding and the fitness of the hybrid offspring.

- Townsend's warblers and hermit warblers hybridize in North America's Pacific Northwest. The species' ranges overlap in Washington State, and hybrid offspring with intermediate characteristics are common (**Figure 26.10**).

- Gene sequencing of the maternally inherited mitochondrial DNA (mtDNA) showed that most hybrids are the offspring of Townsend's males and hermit females.

- Townsend's males attack hermit males, but hermit males do not challenge Townsend's males. Thus, Townsend's males may invade hermit territories, chase off the males, and mate with the females.

- Further investigation showed that many other northern Townsend's warblers have hermit mtDNA. On some islands all the warblers had hermit mtDNA, even though they looked like typical Townsend's warblers.

- The Townsend's warblers may have gradually expanded their range into hermit territory, hybridizing with female hermit warblers along the way. As each new generation mates with Townsend's males, the offspring look more and more like Townsend's warblers but retain the original hermit mtDNA.

(b) Hybrids inherit species-specific mtDNA sequences from their mothers.

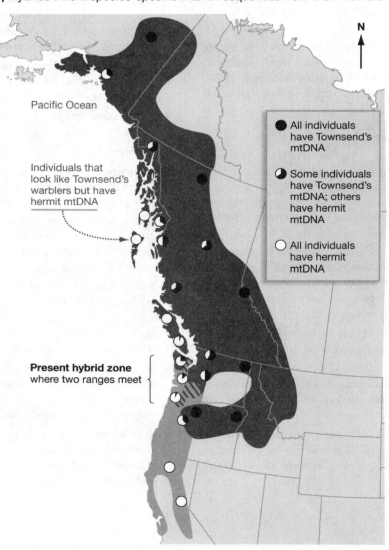

✓ CHECK YOUR UNDERSTANDING

(g) Why isn't there any Townsend's warbler mtDNA on some of the northern islands within the current Townsend's warbler range?

_____

_____

_____

_____

New Species through Hybridization

- The sunflower, *Helianthus anomalus*, resembles hybrids of *H. annus* and *H. petiolaris*. *H. anomalus* also has some gene sequences similar to those in *H. annus* and others similar to those in *H. petiolaris*.

- All three species have the same number of chromosomes. Perhaps hybridization between *H. annus* and *H. petiolaris* formed a new species with its own distinct traits. In fact, *H. anomalus* has a distinct appearance and grows in drier habitats.

- Biologists experimentally hybridized *H. annus* with *H. petiolaris* for several generations. These hybrids looked like *H. anomalus* (**Figure 26.11**).

- Genetic maps showed that the experimental hybrids had combinations of *H. annus* and *H. petiolaris* genetic sequences similar to those found in *H. anomalus*.

- These results support the hypothesis that *H. anomalus* originated from hybridization of *H. annus* and *H. petiolaris*.

- When two populations regain contact, many outcomes are possible: fusion of the populations, reinforcement of divergence, founding of stable hybrid zones, extinction of one population, or creation of a new species (**Table 26.4**).

B. MASTERING CHALLENGING TOPICS

In describing the process of speciation, this chapter shows that speciation is usually a gradual process. Because it is gradual, there is no obvious signpost announcing to researchers the point at which two populations have sufficiently diverged to be considered different species. (This determination is similar to the judgment call that is sometimes necessary when deciding whether a cell is in a particular phase of mitosis). Worse, questions that are relevant to these decisions are often unanswerable; for example, if two species are allopatric now, how long will they remain isolated? What will happen if contact is regained?

This ambiguity can be frustrating for beginning biology students, but it is an important reality in many areas of research. Biologists work with extremely complex systems, and sometimes there are no hard-and-fast rules for solving certain scientific questions. Researchers must then just do the best they can with the information they have.

C. ASSESSING WHAT YOU'VE LEARNED

1. The two key factors responsible for speciation among populations are:
   a. lack of gene flow and mutation.
   b. mutation and genetic drift.
   c. genetic isolation and genetic divergence.
   d. postzygotic isolation and morphological change.
   e. hybrid vigor and heterozygote advantage.

2. Which of the following are examples of postzygotic reproductive isolation? (Choose all that apply.)
   a. Two sympatric bird species are similar in plumage, but engage in dramatically different courtship dances.
   b. Two species of yucca plants can pollinate each others' seeds, but the seeds are then aborted.
   c. Two plant species have wind-dispersed pollen that lands on the styles and grows a pollen tube through the ovary of either species; however, in hybrid matings, the sperm cannot fertilize the ovum.
   d. Two frog species meet and can mate with each other, but the hybrid offspring are infertile.
   e. Two beetle species are superficially similar in appearance, but the structure of the male penis and the female genitalia prevent males of one species from copulating with females of the other.

3. A monophyletic group would be best described as:
   a. a grouping of all species descended from a common ancestor, including that ancestor.
   b. a grouping of all species descended from a common ancestor, excluding that ancestor.
   c. a grouping of all species that share a similar set of traits, regardless of whether the traits are homologous or analogous.
   d. a grouping of most, but not all, species descended from a common ancestor.

4. Paleontologists studying fossilized therapsids (mammal-like reptiles that are now extinct) would probably be using which of the following species concepts?
   a. The biological species concept
   b. The morphospecies concept
   c. The phylogenetic species concept
   d. None of the above; fossil species do not have synapomorphies.

5. To which of the following situations would it be difficult to apply the biological species concept? (You may wish to choose more than one situation.)
   a. Two populations of morphologically similar organisms that differ in breeding habits and physiology
   b. Two populations of asexual clonal organisms that are morphologically similar
   c. Two morphologically similar populations, one in Africa and one in South America, that exhibit no apparent prezygotic or postzygotic isolation
   d. Lions and tigers do not interbreed in nature, but can sometimes hybridize in zoos
   e. Two bird populations that live next to each other and look similar, but have different songs that attract only females that grew up in a nest with a father that sang that song

6. Why shouldn't the hybrid offspring of the male *A. m. nigrescens* and female *A. m. peninsulae* be released to replenish the dusky seaside sparrow's genetic diversity?
   a. There is already enough genetic diversity in the *A. m. nigrescens* population.
   b. Releasing these offspring would artificially cause gene flow between two genetically distinct populations.
   c. Due to hybrid sterility, these hybrids cannot reproduce.
   d. The morphospecies concept does not recognize subspecies.

7. Vicariance refers to:
   a. a pattern of speciation in which a population is subdivided by a geographic barrier.
   b. a kind of speciation where a small group of colonists founds a population on an island habitat.
   c. a pattern of speciation whereby new species evolve reproductive isolation in sympatry.
   d. a technique for constructing phylogenetic trees based on vicariant traits shared by all members of a taxon.

8. Peter and Rosemary Grant observed a small group of the large ground finch colonizing Daphne Major in the Galápagos Islands. In a few years, the descendants of the colonists had evolved beaks that were much larger than those in the original source population. What factors likely influenced this evolution? (Choose all that might apply.)
   a. Artificial selection due to the measurements researchers made
   b. Continued gene flow with the source population

   c. The original small size of the colonist pool resulting in divergence due to genetic drift
   d. Reinforcement due to postzygotic isolation on secondary contact
   e. Natural selection imposed by larger seeds found on Daphne Major than on the ancestral island

9. Which of the following is the best example of isolation and divergence in allopatry?
   a. Two similar yet distinct ground-squirrel species are separated by the Grand Canyon.
   b. A bacteria that infects birds switches hosts and starts infecting humans.
   c. Horticulturalists have created tetraploid flower species.
   d. The hybrid *H. anomalus* has characteristics that allow it to produce more offspring in dry microhabitats within the *H. petiolaris/H. annus* hybrid zone.

10. Which of the following would best be described as a case of speciation in sympatry?
    a. A population of lizards is subdivided by a natural barrier and subsequently diverges to form two species that cannot interbreed.
    b. A new, isolated population of fruit flies is founded by a small group of colonists, which then diverges from the ancestral source population.
    c. An individual plant undergoes meiotic failure, producing diploid pollen and ovules; these self-fertilize, germinate, and grow into fully fertile tetraploid plants.
    d. Speciation cannot take place in sympatry, only in allopatry, where geography poses a barrier to gene flow.

11. Soapberry bugs feed on the fruits of sapindaceous plants in North America. Following the introduction of three Asian species with smaller fruits, some soapberry bugs started feeding on these new species. Researchers measured beak length and found that bugs feeding on the Asian species have shorter beaks on average than bugs on North American species. What does this suggest?
    a. Postzygotic isolation is probably preventing gene flow between these two populations.
    b. This is probably an example of reproductive isolation due to vicariance.
    c. The bugs have different morphologies, so the population feeding on the Asian species is a new species.
    d. Disruptive selection may be driving morphological change in this population.

12. Parents from two different species mate and produce a fertile offspring with a polyploid chromosome number. This is called:
    a. triploidy.
    b. mutation selection.
    c. autopolyploidy.
    d. allopatric speciation.
    e. allopolyploidy.

13. Reinforcement of speciation refers to:
    a. selection for increased prezygotic isolation stemming from the reduced fitness of hybrid offspring.
    b. natural selection for increased postzygotic isolation due to differences in courtship rituals or pollination strategies.
    c. genetic drift that causes two subsets of a population to become more differentiated.
    d. the evolution of compatibility between hybrids, causing the fusion of two diverging populations.

14. Researchers used mitochondrial DNA to study the hybrid zone between Townsend's warblers and hermit warblers. Why was mitochondrial DNA so useful in this study?
    a. The mtDNA is maternally inherited, so researchers were able to discover that most hybrids resulted from Townsend's males mating with hermit females.
    b. The mtDNA is easy to extract, so the researchers could obtain lots of genetic information about the hybrids.
    c. The mtDNA shows that reinforcement has been occurring in this hybrid zone.
    d. The mtDNA is biparentally inherited, like most of the nuclear genetic material. Any gene could have been used in this study.

15. In this chapter's sunflower hybridization example, you are told that *Helianthus anomalus* grows in drier habitats than either of the parental species. If, instead, this species successfully grew in the same habitats as the parental species and freely interbred with both parental species producing successful offspring, what would be the likely outcome?
    a. Sympatric speciation through reinforcement
    b. Fusion of the parental species through increased gene flow
    c. Postzygotic isolation and reduction of gene flow
    d. Speciation through allopolyploidy

CHAPTER 26—ANSWER KEY

Check Your Understanding

(a) You would probably have to use the morphospecies concept for a new fossil species because phylogenetic trees with the species would not yet be available, and there is no way to determine with which other populations a fossil was able to mate. You would probably use the phylogenetic species concept for a well-studied bacterial group, although you might be able to use the morphospecies concept. Bacteria are asexual, so the biological species concept could not be used in this case. Finally, you could really use any of the species definitions for the well-studied mammal species; ideally, you would use all concepts and compare your conclusions using each method.

(b) Greater gene flow within the two monophyletic groups (perhaps by occasional dispersal of the young to a different population) would tend to eliminate any genetic differences.

(c) Movement of an aquatic organism from one lake to another (for example, during an unusual wet period that temporarily joins two lakes) also causes physical isolation and may involve a small colonist population. Similarly, mountaintops in the Southwest United States are separated by desert areas that are difficult for many mountaintop organisms to cross (these mountains are sometimes referred to as the sky islands). Colonization by a small population that then experiences genetic drift and natural selection can also occur in these habitats.

(d) Colonization events are more likely to produce small population sizes (at least initially).

(e) If apple flies had no preference, they would probably mate on hawthorn fruits as often as on apples. This would permit significant gene flow between the two populations, so unless some other evolutionary mechanism was acting, the populations would be unlikely to diverge.

(f) Animals typically cannot self-fertilize, so even if diploid gametes were formed, they would be unlikely to join with other diploid gametes to produce viable offspring. Polyploids produced by chromosome doubling in the zygote might survive but would probably be unable to reproduce.

(g) Presumably there is no Townsend's warbler mtDNA on those islands because no female Townsend's have made it to the islands (perhaps the male Townsend's warblers disperse farther than do the females). Thus, all the island birds are descended from the original female hermit warblers that first mated with the Townsend's males.

Assessing What You've Learned

1. c; 2. b, d; 3. a; 4. b; 5. b, c; 6. b; 7. a; 8. c, e; 9. a; 10. c; 11. d; 12. e; 13. a; 14. a; 15. b

---

Looking Forward—Key Concepts in Later Chapters

Speciation Patterns—Chapter 27

Now that you understand how speciation occurs, you are ready to consider large-scale speciation patterns (e.g., adaptive radiations), discussed in the next chapter.

Conservation—Chapter 55

Understanding of speciation is critical to conservation efforts, as demonstrated by the dusky seaside sparrow example from this chapter. Biodiversity and conservation are further discussed in the final chapter of the text.

Phylogenies and the History of Life

Homeotic Genes—Chapter 21

Clearly, homeotic genes (e.g., *Hox* loci) have played a major role in evolution because they affect the overall morphology of organisms.

Evolution—Chapters 1, 4, 24–26

Because evolution is one of the unifying themes of biology, many chapters have addressed this topic. Chapter 27 is the last to concentrate specifically on evolution. Nonetheless, watch out for continued references to evolution in future chapters.

Phylogenies—Chapters 1 and 26

Chapter 1 depicted a phylogeny for the tree of life, and **Chapter 26** described the phylogenetic species concept and what a monophyletic group is. Now you learn about large-scale patterns in the tree of life, and how phylogenetic information can be compared with information on specific genes to infer how different groups and morphologies evolved.

KEY CONCEPTS

- Biologists estimate evolutionary history from data and document their hypotheses in phylogenetic trees.

- The fossil record provides physical evidence of organisms that lived in the past.

- Adaptive radiations are rapid diversifications associated with new ecological opportunities and/or new morphological innovations.

- Environmental catastrophes that rapidly eliminate most species happen periodically and are called mass extinctions.

A. CHAPTER OUTLINE

27.1 Tools for Studying History: Phylogenetic Trees

- The evolutionary history of an organismal group is called a **phylogeny**. A **phylogenetic tree** shows ancestor-descendant relationships among populations or species, where each **branch** represents a population or species through time.

- The point where two branches diverge, a **node,** represents the splitting of an ancestral group into two or more descendant groups; and each branch tip represents a group that is living today or one that ended in extinction (**BioSkills 3**).

How Do Researchers Estimate Phylogenies?

- Phylogenetic trees summarize data on the evolutionary history of a group of organisms.

- Both morphological and genetic characteristics are used to estimate phylogenetic relationships among species. Closely related species should share many traits, while distantly related species should share fewer traits.

- The **phenetic approach** evaluates relatedness by computing overall similarity. For example, one can assess similarity using *genetic distance*, a measure of the average percentage difference in DNA bases between two populations.

- The **cladistic approach** argues that some characteristics are more informative than others in determining relatedness. This approach focuses on synapomorphies.

- Biologists use synapomorphies to identify monophyletic groups because each group inherits its own set of shared derived traits from a common ancestor (**Figure 27.1**).

- A characteristic that is found in an ancestor is an **ancestral trait**. A **derived trait** is a modified form of the ancestral trait found in a descendant. These terms are relative—which traits are ancestral and which are derived depends on which groups you are considering.

How Can Biologists Distinguish Homology from Homoplasy?

- Difficulties determining the true phylogeny arise when similar traits evolve independently in two species.

What Is Homoplasy?

- **Homology**, in biology, refers to similarities due to shared ancestry. **Homoplasy** occurs when traits are similar but are not inherited from a common ancestor.

- For example, icthyosaurs and dolphins have morphological similarities not found in their reptilian and mammalian ancestors. These traits evolved in both groups because streamlined bodies are advantageous in fish-eating marine animals (**Figure 27.2**).

- Species inhabiting similar environments may have similar traits due to **convergent evolution** when natural selection favors similar solutions to the challenges of a particular habitat. These convergent traits are not found in the species' common ancestor.

Evidence for Homology

- The key to distinguishing homology from homoplasy is determining whether the similar structures or genes derive from a common ancestor. This is often difficult, but can be done.

- For example, the *Hox* genes of insects and vertebrates (1) show similar chromosomal organization, (2) share the *homeobox* sequence that binds DNA and regulates gene expression, and (3) code for products with similar functions and patterns of expression.

- Many animals from lineages that branch off between insects and vertebrates have similar genes (**Figure 27.3**).

- When distantly related groups have traits that appear to be similar due to common ancestry, one expects many other groups found in between the two on the tree of life to have the same trait (because all inherited it from the same common ancestor).

Using Parsimony to Minimize the Impact of "Noise" in the Data

- Researchers often use **parsimony** to try to identify the phylogenetic tree that requires the least amount of change (**Figure 27.4**). This approach assumes that

convergent evolution occurs less frequently than similarity due to shared descent.

CHECK YOUR UNDERSTANDING

(a) Are bat wings and pterosaur wings an example of homology or homoplasy? Why?

_____

_____

_____

Whale Evolution: A Case History
A Phylogeny Based on Morphological Traits

- Phylogenetic trees based on morphological data place hippos, cows, and other animals with hooves and an even number of toes together in a group called the artiodactyls. Presence of a pulley-shaped astralagus ankle bone is a synapomorphy for this group.

- Whales do not have an astralagus and are shown as an **outgroup** for artiodactyles—a group that is closely related, but not part of the monophyletic group in question (**Figure 27.5a**).

A Phylogeny Based on DNA Sequence Data

- DNA sequence data suggest a close relationship between whales and hippos. A phylogenetic tree showing closely related whales and hippos is less parsimonious for morphological data because it requires evolution and then loss of the pulley-shaped astralagus in whales (**Figure 27.5b**).

A Phylogeny Based on Transposable Elements

- Data on parasitic gene sequences called **short interspersed nuclear elements** (**SINEs**) show that whales and hippos share several SINE genes that are absent in other artiodactyl groups (**Figure 27.5c**).

- Whales and hippos share four SINEs. These SINEs are synapomorphies supporting the hypothesis that whales and hippos are closely related. Other SINEs are present in some artiodactyls but not in others, and camels have no SINE genes.

- Biologists hypothesize that SINEs were not present in the artiodactyl ancestor and became inserted into the genomes of descendant populations after the camel group branched off. This interpretation supports the phylogeny that places whales and hippos as sister groups.

- Recent fossil finds of whale-like artiodactyls with pulley-shaped astralagus bones lend further support to the hypothesis that whales and hippos are closely related and that whales are, in fact, artiodactyls.

27.2 Tools for Studying History: The Fossil Record

- **Fossils** are physical evidence left by organisms from the past. They provide the only direct evidence of what ancient organisms looked like and when they existed.
- The **fossil record** includes all fossils that have been found and recorded.

How Do Fossils Form?

- Most fossils form when an organism is buried in sediment (ash, sand, mud, etc.) before decomposition occurs (**Figure 27.6**).

Preservation after Burial

- Fossils are formed in several ways (**Figure 27.7**).
- Organic remains may resist decomposition and be preserved intact.
- An organism may be cemented into rock when sediments compress the organic material into a thin film.
- A **cast** may form when remains decompose after burial and the remaining hole is filled with dissolved minerals.
- Permineralized fossils may form if dissolved minerals enter cells.
- Centuries later, fossils exposed by erosion or human action can become part of the fossil record if they are carefully separated from the surrounding rock, named, dated, described in a scientific publication, and added to a collection.

Fossilization Is a Rare Event

- Most organisms decompose rapidly and are buried slowly, if at all. Thus, fossilization is an extremely rare event.

Limitations of the Fossil Record
Habitat Bias

- Organisms living where sedimentation occurs (beaches, swamps, etc.) are more likely to become fossilized than are organisms living in other habitats. Burrowing organisms are more likely to fossilize because they are already buried at death.

Taxonomic and Tissue Bias

- Organisms with hard parts (bones, shells) are more likely to resist decay long enough to be fossilized than are organisms with only soft parts. Similarly, within an organism the hard parts (pollen, teeth) are more likely to fossilize than are the soft parts.

Temporal Bias

- Rocks in Earth's crust may be worn down by erosion or melted as one tectonic plate is pushed underneath another one. These processes destroy rocks and the fossils in them; recent fossils are thus more common than ancient ones.

Abundance Bias

- Because fossilization of any organism is a rare event, organisms that are common, widespread, and extant as a species for long periods are more likely to become fossilized and be found by researchers.
- The fossil record is highly biased; but it gives the *only* direct evidence about the morphology and ecology of extinct organisms. Study of the fossil record by **paleontologists** helps us understand the history of life.

CHECK YOUR UNDERSTANDING

(b) There are no known fossils of the very first living organisms. Give at least two reasons to explain why this lack of fossil evidence for these first organisms is not surprising.

_____

_____

_____

_____

Life's Timeline

- Earth formed about 4.6 billion years ago and life began around 3.4 billion years ago.
- Geologists initially identified the boundaries between named intervals based on distinctive rock formations or the presence of certain fossilized organisms.
- Researchers now use radiometric dating to assign absolute dates to events in the fossil record. However, these dates may underestimate the actual timing of events since a species could exist for millions of years before leaving fossil evidence.

Precambrian

- Major events in the history of life from its origination to the appearance of most animal groups about 542 million years ago are shown in **Figure 27.8**. This interval is called the **Precambrian** and includes the Hadean, Archaean, and Proterozoic eons.

- During the Precambrian (most of Earth's history), almost all life was unicellular and hardly any oxygen was present in Earth's oceans or atmosphere. Oxygen increases during the Proterozoic eon. The ocean is oxygenated by the end of the Precambrian.

Phanerozoic Eon

- The Phanerozoic eon includes the interval between 542 million years ago (mya) and the present and is divided into three eras that are further divided into periods (**Figure 27.9**).

- The appearance of many animal groups marks the beginning of the **Paleozoic era**. Land animals, land plants, and fungi appear and diversify during this period. This era ends with a mass extinction at the end of the Permian.

- The end-Permian extinction is followed by the **Mesozoic era**, during which gymnosperms were the dominant land plants and dinosaurs were the most important vertebrates. This era ends with a mass extinction at the end of the Cretaceous period.

- The Mesozoic is followed by the **Cenozoic era**. Mammals became the most important vertebrates and angiosperms became the dominant plants. The Cenozoic era continues today.

Changes in the Oceans and Continents

- Earth's crust is broken into enormous plates that move around the earth, changing the extent and location of Earth's oceans and continents (**Figure 27.10**). Earth's climate has also changed radically over time.

27.3 Adaptive Radiations

- The history of life is punctuated with (1) periods when species originate and diversify rapidly and (2) periods when species go extinct rapidly.

- An **adaptive radiation** occurs when a single lineage produces many descendant groups in a short period of time. Groups descended from an adaptive radiation are typically found in a wide diversity of habitats and/or use a wide array of resources.

- For example, Hawaiian silverswords rapidly diversified following colonization of the Hawaiian Islands by a Californian tarweed about 5 million years ago (**Figure 27.11**). Today's silverswords show amazing diversity in form and habitat.

- The Hawaiian silverswords have three key characteristics found in adaptive radiations: they are a monophyletic group, they speciated rapidly, and they diversified ecologically.

- A **niche** describes the range of conditions a species tolerates and the range of resources it uses. Silverswords diversified ecologically because they occupy many different niches.

Why Do Adaptive Radiations Occur?

- New resources and new ways to exploit resources can spark adaptive radiations.

Ecological Opportunity

- Adaptive radiations often follow periods of ecological opportunity when new or novel resources become available. This can occur when habitats are unoccupied by competitors.

- For example, tarweeds arrived on the Hawaiian Islands when few other flowering plants were present.

- The adaptive radiation of *Anolis* lizards on islands in the Caribbean is also thought to have occurred due to ecological opportunity in the absence of competitors.

- Among *Anolis* lizards, the size and shape of a species closely correlate with its habitat. Species living on the ground or on tree trunks have longer legs and tails than do those that live on twigs. Each island has several species, each occupying a different habitat.

- Biologists hypothesized that there had been a mini-radiation on each island. They tested this hypothesis by constructing an *Anolis* phylogeny based on DNA sequence data. This phylogeny revealed that each island had its own monophyletic *Anolis* group.

- On Hispaniola and Jamaica, the founding lizard colonists belonged to different ecological types; but over time, their descendents evolved to occupy all island habitats, forming four different species with four different morphologies (**Figure 27.12c**).

- Apparently each island did have its own mini-adaptive radiation. As predicted by the ecological opportunity hypothesis, colonial *Anolis* populations rapidly diversified to fill empty niches in the absence of competitors.

Morphological Innovation

- In the history of life, new morphologies that allow a group to live in new areas, exploit new resources, or move in new ways have also often triggered adaptive radiations.

- Examples of traits that allowed exploitation of new habitats, causing adaptive radiations, are (1) the evolution of wings, three pairs of legs, and a protective external skeleton in insects; (2) the evolution of flowers in angiosperms; (3) the evolution of a second set of jaws in cichlid fish; and (4) the evolution of feathers and wings in dinosaurs/birds.

CHECK YOUR UNDERSTANDING

(c) A biologist studies sunflowers on two adjacent mountains. Each mountain has four species—one that is tree-like, one that is bushy, one that is herbaceous, and one that has a rosette form. The biologist constructs a phylogeny based on DNA sequences and finds that each morphological type is a sister species to the same morphological type on the other mountain. Should this biologist conclude that these mountains have each had a mini-radiation as in the *Anolis* example? Why or why not?

_____

_____

_____

_____

_____

_____

_____

The Cambrian Explosion

- Animals first originated around 565 million years ago. Soon after that, at the start of the Cambrian period, animals diversified into almost all the major groups extant today. This is known as the **Cambrian explosion**.
- Life had been predominantly unicellular for almost 3 billion years when this spectacular period of evolutionary change occurred. Shells, exoskeletons, and limbs are only a few of the new traits that appeared during the Cambrian's adaptive radiation.

The Doushantuo, Ediacaran, and Cambrian Fossils

- Three major fossil assemblages record this diversification of animal life, each bed documenting its own distinctive **fauna** (collection of animal species).

- China's Doushantuo fossils record the first animal life 570 mya; Australia's Ediacaran fossils record life from 565 to 542 mya; and Canada's Burgess Shale fossils date from 525 to 515 mya (**Figure 27.14**).
- The *Doushantuo microfossils* include tiny sponges and what seem to be animal embryos undergoing cleavage. These first animals probably filtered organic debris from the water for food.
- *Ediacaran faunas* include sponges, jellyfish, comb jellies, and traces of other animals. No animals with shells, heads, mouths, or feeding appendages are present during this time period, so these animals also probably filtered or absorbed organic material.
- *Burgess Shale faunas*, in contrast, show incredible morphological and ecological diversity. Just about every major animal group can be found among the Burgess Shale fossils: arthropods, mollusks, echinoderms, worms, and vertebrates.
- The increased size and morphological complexity of the Burgess Shale fossils is associated with amazing ecological diversification. Besides filter feeders, there are predators, scavengers, and grazers. These animals had eyes, mouths, limbs, and shells; they swam, burrowed, walked, ran, slithered, clung, and/or floated.

What Triggered the Cambrian Explosion?

- Several hypotheses have been proposed to explain the Cambrian explosion. Most or all of these could be correct—they are not mutually exclusive.
- *Higher oxygen levels* may have increased the efficiency of aerobic respiration, making possible larger bodies and more active lifestyles.
- *The evolution of predation* may have caused natural selection for predation-defense and predation-avoidance structures and strategies (shells, exoskeletons, speed, etc.).
- *New niches beget more new niches.* As grazers and scavengers diversified and moved to new ocean depths and into the open ocean, predators diversified as they followed their prey into their new niches.
- *New genes, new bodies*—mutations may have increased the number of *Hox* genes in animals, making the evolution of larger, more complex bodies possible (**Figure 27.15**). These genes signal cell position in the embryo and were not present in the earliest animals.
- About 100 million years after the Cambrian explosion, a similar adaptive radiation occurred after the first plants adapted to life on land.

(d) How might new data contribute to our understanding of the Cambrian explosion? What evidence might convince researchers to favor one hypothesis over another in terms of importance? (Give a specific example.)

_____

_____

_____

_____

_____

_____

27.4 Mass Extinctions

- **Mass extinctions** are the rapid extinction of many lineages throughout the tree of life (loss of at least 60 percent of all species within 1 million years). These extinction events are caused by catastrophic episodes.

How Do Mass Extinctions Differ from Background Extinctions?

- Traditionally, five mass extinction events are distinguished from the lower **background extinction** rate that represents the normal loss of some species that always occurs (**Figure 27.16**).

- Background extinctions typically occur when normal environmental change or competition reduces a population to the point that it dies out. Mass extinctions occur when unusual large-scale environmental change causes the extinction of many normally well-adapted species.

- Natural selection causes most background extinctions, whereas random chance plays a large role in mass extinctions. In this way, mass extinctions function more like genetic drift.

The End-Permian Extinction

- The end-Permian extinction is the largest of the mass extinctions. Most families of organisms and 90 percent of all species went extinct during this period, yet researchers are not yet agreed about its cause. Likely contributing factors are listed below.

- Flood basalts called the Siberian traps occurred during the end-Permian. These outpourings of molten rock added heat, CO_2, and sulfur dioxide to the atmosphere. The CO_2 led to intense global warming and the sulfur dioxide caused acid rain.

- Rocks formed during this period show that the oceans lacked oxygen during this period. This would have killed organisms that rely on aerobic respiration.

- The sea level appears to have dropped, reducing the total volume of marine habitat.

- Terrestrial animals may have been restricted to low-elevation habitats due to the low oxygen and high CO_2 levels in the atmosphere.

- While it is unclear exactly why all these changes occurred (the "world went to hell hypothesis"), the resulting climate was a disaster for aerobic organisms.

What Killed the Dinosaurs?

- The well-supported **impact hypothesis** proposes that the end-Cretaceous extinction occurred when an asteroid struck the Earth 65 mya, killing 60–80% of multicellular species.

Evidence for the Impact Hypothesis

- High levels of iridium, common in asteroids but rare on Earth, are found worldwide in sedimentary rocks from the Cretaceous-Tertiary (K-P) boundary.

- Minerals found only at meteorite impact sites (e.g., shocked quartz and microtektites) are common in Haitian and other rocks dated to 65 mya.

- A huge crater the size of Sicily off the northwest coast of the Yucatán peninsula dates to the K-P boundary. Microtektites are abundant in the crater's walls (see **Figure 27.17**).

Nature of the Impact

- The asteroid was about 10 km across and hit the Earth at an angle, splashing material over much of southeastern North America.

- Computer and geological data indicate the following devastating consequences followed the asteroid impact.

- The impact caused a fireball of hot gas that spread out from the impact site, causing worldwide wildfires (evidenced by soot and ash deposits dated to 65 mya).

- The largest tsunami in the last 3.5 billion years disrupted ocean sediments and circulation patterns.

- Rock at the impact site contained sulfate, so SO_4^{2-} was released by the impact and reacted with atmospheric water to form sulfuric acid and cause extensive acid rain.

- Dust, soot, and ash due to the impact and subsequent fires blocked the Sun, causing rapid global cooling and decreased plant productivity.

Selectivity of the Extinctions

- Certain evolutionary lineages withstood the post-asteroid impact environmental change better than others did. Dinosaurs, pterosaurs, and large marine reptiles went extinct; but most mammals, crocodiles, amphibians, and turtles survived.

- One hypothesis is that extended darkness and cold affected large organisms more than small ones because they require more food. However, size selectivity is not found among marine clams and snails.

- Another hypothesis is that organisms capable of long periods of inactivity (by hibernating or as seeds or spores) were better able to survive the catastrophe.

Recovery from the Extinction

- Ferns replaced diverse woody and flowering plants in many habitats following the K-P extinction, marine environments show low diversity for 4 to 8 million years, and terrestrial organisms in the Paleogene are very different from those of the Cretaceous.

- Mammals diversified to fill niches left empty following the extinction of their dinosaur competitors. Within 10 to 15 million years, all major mammalian orders had appeared. This change in the terrestrial vertebrate fauna was basically due to a chance event.

✔ **CHECK YOUR UNDERSTANDING**

(e) What might happen if Earth were to be hit by another asteroid the size of the one that struck 65 mya?

_____

_____

_____

_____

B. MASTERING CHALLENGING TOPICS

A business analogy might help you remember why adaptive radiations occur and the difference between the ecological innovation and the morphological innovation mechanisms. Companies making some product often look for new business niches. Two ways to rapidly expand a product line are (1) to find a new emerging market in which to sell a product, preferably one with no competitors (analogous to the ecological innovation mechanism), or (2) to create a new type of product (analogous to the morphological innovation mechanism).

For example, suppose a developing country suddenly opens its formerly closed market to one company that sells refrigerator magnets. This company might become very successful as more people buy refrigerators and magnets for them. Suppose this company follows up on its initial success by starting a line of sticky notes, bumper stickers, and key chains. This diversification in a new environment with no competitors would be similar to adaptive radiation via ecological innovation (although the process is different—business diversification is intentional whereas adaptive radiations occur due to natural selection in a population that varies in heritable traits).

Next consider how creation of a new type of product is analogous to morphological innovation. For example, the development of the personal computer led to a rapid proliferation of software, hardware, and customer support services. Can you improve this analogy? How is it similar and how does it differ from the morphological innovation mechanism that causes adaptive radiations in organism?

C. ASSESSING WHAT YOU'VE LEARNED

1. A monophyletic group is:
   a. a taxonomic group known to have diverged prior to the rest of the taxa in the study.
   b. an ancestor and all its descendants.
   c. the branch tips of a phylogenetic tree.
   d. any two taxa that show convergent evolution.

2. Which of the following correctly identifies the key difference between the phenetic and the cladistic approach to estimating phylogenies?
   a. The phenetic approach uses homoplasies to identify monophyletic groups whereas the cladistic approach uses homologies.
   b. The phenetic approach uses derived traits to construct phylogenetic trees whereas the cladistic approach uses ancestral traits.
   c. The phenetic approach uses morphological data whereas the cladistic approach relies on DNA sequence data.
   d. The phenetic approach relies on overall similarity whereas the cladistic approach focuses on identifying synapomorphies.

3. Bird feathers in warblers and ostriches are:
   a. an example of homology.
   b. an example of homoplasy.
   c. an example of convergent evolution.
   d. an example of parsimony.
   e. all of the above

4. Researchers conclude that whales are most closely related to hippos because (choose all that apply):
   a. Whales and hippos share several SINEs that are not present in other artiodactyls.
   b. Modern whales have tiny pulley-shaped astralagus bones.
   c. Non-SINE whale and hippo DNA sequences are similar.
   d. Whales look more like hippos than they look like cows.
   e. all of the above

5. The fossil record is a biased sample of organisms that are now extinct. Which of the following is *not* a reason for that bias?
   a. Organisms with hard parts, such as teeth, shells, and skeletons, are more likely to be preserved.
   b. Organisms living in certain habitats were more likely to be buried in sediments and preserved than were organisms in other habitats.
   c. Organisms with broad geographic distributions and large populations are more likely to be preserved than are rare endemic species.
   d. Exposed fossils are often subject to weathering, whereas those beneath the surface are protected.

6. Which of the following correctly lists the geological eras from most ancient to most recent?
   a. Paleozoic, Mesozoic, Precambrian, Cenozoic
   b. Paleozoic, Mesozoic, Cenozoic, Precambrian
   c. Precambrian, Paleozoic, Mesozoic, Cenozoic
   d. Precambrian, Cenozoic, Paleozoic, Mesozoic
   e. Cenozoic, Paleozoic, Mesozoic, Precambrian

7. Which of the following correctly matches a geological eon or period with both its approximate start date and a first appearance that occurred during that geological eon or period?
   a. Precambrian—4.6 mya: first appearance of photosynthetic cells
   b. Mesozoic—251 mya: first mammals
   c. Cenozoic—65 mya: first seed plants
   d. Paleozoic—2500 mya: first multicellular organisms

8. Which of the following cases best illustrates an adaptive radiation?
   a. A fishless lake is colonized by a single fish species, which over a few thousand years gives rise to several species, each with a series of unique feeding adaptations.
   b. A small group of tree-dwelling lizards of a single species migrates to an uninhabited, treeless island and adapts to use the open grassland habitat.
   c. A population of a cricket species takes up residence in a cave. Over many generations, the cave crickets eventually lose their eyes, as have many other cave-dwelling animals.
   d. A colonizing species of fruit fly takes up residence on a new continent and displaces two closely related native species.

9. Why are morphological innovations often associated with adaptive radiations?
   a. Morphological innovations often allow a species to replace competitors.
   b. Morphological innovations can open up new adaptive options that may allow a species to colonize an underutilized resource.
   c. Morphological innovations disrupt natural selection, forcing species to find new adaptations.
   d. Morphological innovations prevent adaptive radiations most of the time.

10. In the study of *Anolis* lizards colonizing various Caribbean Islands, researchers found similar habitats and similar ecological types on Hispaniola and Jamaica. They conducted a phylogenetic analysis of the species on each island. Which observations suggested that adaptive radiation occurred on these two islands in response to habitat availability and the absence of competitors?
    a. Different islands were colonized by different species that differed in habitat preference. Subsequent speciation produced, on each island, a range of ecological specialists occupying four different habitat niches. Each island had its own monophyletic lizard group.
    b. The first colonists on every island were always the same species, a twig-dwelling specialist. But in every case, that species gave rise to the same set of other species that specialized on different habitats. The phylogeny for each island had the same root and the same pattern of speciation.
    c. Each island was initially colonized by a different species that occupied its preferred habitat. Subsequent colonizing species were successful only if they could use an unoccupied habitat type. The phylogeny for each island was the same throughout the region because no evolution occurred, only colonization.
    d. Each island was colonized by a different species initially, and that species underwent a radiation, giving rise to new species that competed with each other until all but one species went extinct.

11. Fossils of organisms from the Precambrian are extraordinarily rare, but many species are found in the Burgess Shale deposits. Which of the following statements best describes the Burgess Shale organisms?
    a. Mostly bacteria and simple one-celled plants and animals
    b. Primarily sponges, algae, and seaweeds
    c. Arthropods, mollusks, vertebrates, and a variety of soft-bodied organisms
    d. Jellyfish and sponges, along with some tracks of various other organisms

12. What is one genetic hypothesis proposed for the rapid diversification of body plans during the Cambrian period?
    a. Many mutations accumulated due to high exposure to UV radiation, creating many new kinds of organisms.
    b. Reproductive isolation was nearly nonexistent; many "species" interbred to form hybrids that went on to become new species.
    c. New genes appeared in many different organisms through a mechanism that has not been identified; these are associated with new body plans.
    d. Duplications of the *Hox* genes that organize body development facilitated the evolution of larger, more complex bodies.

13. Which of the following statements highlights an important contrast between background extinctions and mass extinctions?
    a. Background extinctions occur only sporadically; mass extinctions happen on a regular, periodic basis.
    b. Background extinctions are caused by rapid changes in the environment; mass extinctions occur primarily in response to biological competition.
    c. During mass extinctions, adaptations for survival and competition do not much affect the likelihood of extinction; the opposite is true for background extinctions.
    d. Mass extinctions, such as the K-P event, tend to take out large-bodied organisms; smaller-bodied organisms tend to be removed by background extinction.

14. Which of the following provide evidence that an asteroid impact caused the end-Cretaceous extinction? (Choose all that apply.)
    a. Flood basalts in Siberia date to this period.
    b. Sedimentary rocks at the K-P boundary have iridium in them worldwide.
    c. The Tunguska event increased dust levels across the Northern Hemisphere.
    d. Microtektites are found in sediments in the wall of a large Yucatán crater.

15. Which of the following statements best represents current thinking about the rise of the mammals after the K-P extinction?
    a. Mammalian diversity increased dramatically prior to the impact, but the population sizes of every species remained small until after the dinosaurs were extinct.
    b. Mammals were inherently better competitors than the dinosaurs; the K-P impact simply speeded up the replacement, which was already well under way.
    c. By chance, the K-P impact caused extinction of the dinosaurs but did not have as great an effect on small, nocturnal, scavenging mammals. The release from competition allowed the mammals to diversify.
    d. Both mammals and dinosaurs survived the asteroid, but because mammals were more advanced, they rapidly underwent an adaptive radiation and outcompeted the dinosaurs.

CHAPTER 27—ANSWER KEY

Check Your Understanding

(a) Bat wings and pterosaur wings are an example of homoplasy, because the early mammal ancestor of bats did not have wings and neither did the reptile ancestor of pterosaurs. This is a case of convergent evolution.

(b) It is no surprise that there are no known fossils from the very first living organisms, because (1) most of the sedimentary rocks that were formed then have since been eroded or otherwise destroyed; and (2) those organisms would have had no hard parts, and they would thus have been less likely to be fossilized in the first place.

(c) No, the biologist should not conclude that each mountain had its own mini-radiation. In the *Anolis* example, the lizards of each island formed a monophyletic group, whereas in this example, each sunflower species is more closely related to the species with the same form on the other mountain than to the sunflower species with different forms on the same mountain. This pattern of paired sister species instead suggests vicariance (see **Chapter 26**, perhaps gene flow between sister populations on the two mountains halted following local climate change). No information is given about whether the initial diversification of the sunflowers into different morphological types occurred rapidly in a new environment without competitors. Thus, it is not possible to determine whether that initial diversification might qualify as a mini-radiation.

(d) New fossil finds might contribute more information about feeding habits and niche partitioning among animals during the Cambrian explosion. For example, discovery of fossils showing adaptations for predation or predator defense might favor the evolution of predation hypothesis, whereas fossils from additional Cambrian habitats showing feeding adaptations not related to predation might favor the new niche hypothesis. On the other hand, additional information about *Hox* genes in sponges and other descendants of early animals might favor the new genes, new bodies hypothesis, especially if combined with experimental results demonstrating how *Hox* gene duplication might allow larger, more complex bodies.

(e) The impact could cause worldwide wildfires and global cooling from all the dust and ash in the air. However, acid rain would occur again only if the impact site contained the sulfate-rich anhydrite. It is difficult to predict which organisms would survive and which would become extinct since chance might play a role.

Assessing What You've Learned

1. b; 2. d; 3. a; 4. a, c; 5. d; 6. c; 7. b; 8. a; 9. b; 10. a; 11. c; 12. d; 13. c; 14. b, d; 15. c

---

Looking Forward—Key Concepts in Later Chapters

Diversity of Life—Chapters 28–35

Now that you know a bit about how evolution occurs and about evolutionary patterns in the tree of life, you are ready to learn specifics about the major groups of organisms comprising the tree of life. **Chapter 30** features details on another adaptive radiation—the "Devonian explosion" associated with the colonization of land by plants.

Competition—Chapter 53

Chapter 27 mentioned how adaptive radiations tend to occur when niches are unoccupied by competitors. **Chapter 53** discusses what happens when two or more species do compete within the same habitat.

Mass Extinction—Chapter 55

The last chapter of this text, on biodiversity and conservation biology, considers evidence that a mass extinction is occurring right now.

Chapter 28

Bacteria and Archaea

Respiration—Chapter 9

The diverse world of bacteria and archaea depends on the presence of electron donors and acceptors. There are many variations on the basic theme in bacteria and archaea; they can employ not only glucose, oxygen, and standard fermentation but also an entire suite of new substrates and reactions.

Photosynthesis—Chapter 10

Plants use water as their source of electrons, and bacteria can use many other molecules. Chlorophylls are also used in cyanobacteria in the same manner as in plants.

Carbon Fixing—Chapter 10

Plants use the Calvin cycle to get carbon from CO_2 and while some bacteria do this, others may fix carbon from other inorganic sources of carbon.

KEY CONCEPTS

- Bacteria and archaea are ancient, diverse, abundant, and ubiquitous.

- A few bacteria cause infectious disease and some are effective at cleaning up pollution.

- Bacteria and archaea live in virtually every habitat known and use diverse types of compounds in cellular respiration and fermentation.

- Bacteria and archaea play a key role in ecosystems. They cycle nutrients through terrestrial and aquatic ecosystems, and photosynthetic bacteria were responsible for the evolution of the oxygen atmosphere.

A. CHAPTER OUTLINE

28.1 Why Do Biologists Study Bacteria and Archaea?

Biological Impact

- **Bacteria** have a unique compound called peptidoglycan in their cell walls.

- **Archaea** have unique phospholipids in their plasma membranes.

- Bacteria are fascinating—some bacteria live over 1 km underground or in 95°C hot springs.

- The lineages in the domains Bacteria and Archaea are ancient, diverse, abundant, and ubiquitous.

- In terms of total volume of living material on our planet, bacteria and archaea are dominant life-forms.

Medical Importance

- Bacteria that cause disease are said to be **pathogenic**.

- **Key research**—Robert Koch 1880) established that bacteria can cause disease. He tested his hypothesis using **anthrax**.

- Koch's postulates:
  1. The microbe must be present in individuals suffering from the disease and absent from healthy individuals.
  2. The organism must be isolated and grown in a pure culture away from the host organism.
  3. If organisms from the pure culture are injected into a healthy experimental animal, the disease symptoms should appear.
  4. The organism should be isolated from the diseased experimental animal, again grown in pure culture, and demonstrated by its size, shape, and color to be the same as the original organism.
- Koch's experiments also became the basis for the **germ theory of disease**.
- **Antibiotics** are molecules that kill bacteria. Widespread use of antibiotics has led to drug-resistant strains of bacteria.

Role in Bioremediation

- **Bioremediation** is the use of bacteria and archaea to degrade pollutants:
  1. Fertilize contaminated sites to encourage the growth of existing bacteria that degrade toxic compounds.
  2. "Seed" or add specific species of bacteria to contaminated sites.

Extremophiles

- Bacteria or archaea that live in high-salt, high- or low-temperature, or high-pressure habitats are **extremophiles**.
- Extremophiles have become a hot area of research:
  1. Understanding extremophiles may help explain how life on Earth began.
  2. Astrobiologists are using extremophiles as their model organisms in the search for extraterrestrial life.
  3. Enzymes from extremophiles are useful in commercial applications.

✔ **CHECK YOUR UNDERSTANDING**

(a) Bacteria and archaea appear to be very similar. Are they closely related?

_____

_____

_____

_____

28.2 How Do Biologists Study Bacteria and Archaea?

Using Enrichment Cultures

- Biologists rely heavily on the ability to culture organisms in the lab.
- **Enrichment cultures** are based on establishing a specific set of growing conditions.
- Cells that thrive under the specified conditions increase in numbers enough to be isolated and studied in detail.

Using Direct Sequencing

- **Direct sequencing** is a strategy for documenting the presence of bacteria and archaea that cannot be grown in culture and studied in the laboratory.
- This technique is vitally important in understanding diversity since only a tiny fraction of the bacteria and archaea that exist can be grown in the lab.
- Direct sequencing has been used to discover new lineages of archaea found all over the world and even on the human body.

Evaluating Molecular Phylogenies

- **Ribosome complexes** are found in all organisms. Ribosomes are the site of protein synthesis and contain a **small subunit** (SSU) RNA.
- **Key research**—In 1960 Carl Woese described base sequences from SSU in many bacterial species. He drew a diagram now known as the **universal tree**, or **tree of life** (**Figure 28.1**; **BioSkills 3**).

✔ **CHECK YOUR UNDERSTANDING**

(b) What did Woese's follow-up work determine?

_____

_____

_____

_____

28.3 What Themes Occur in the Diversification of Bacteria and Archaea?

Morphological Diversity

- *Size*—Bacteria and archaea range from the smallest of all free-living cells—mycoplasmas have a volume of 0.15 μm^3—to the largest bacterium known,

Thiomargarita namibiensis, with volumes as large as $200 \times 106 \ \mu m^3$.

- *Shape*—Bacterial shapes include **filaments, spheres, rods, chains,** and **spirals** (**Figure 28.7b**).
- *Motility*— Some bacteria are **sedentary** and some are **motile**—using **flagella** or gliding.
- Within bacteria, two types of cell walls exist and are distinguished by a **Gram stain:**
  1. **Gram-positive** cells have a plasma membrane surrounded by a cell wall with extensive peptidoglycan.
  2. **Gram-negative** cells have a thin gelatinous layer containing peptidoglycan and an outer phospholipid bilayer.

Metabolic Diversity

- Bacteria and archaea are incredibly diverse in what compounds they can use for food.
- Bacteria produce ATP in three ways:
  1. **Phototrophs**—use light energy to promote electrons to the top of electron transport chains. ATP is produced by cellular respiration.
  2. **Chemoorganotrophs**—oxidize reduced organic molecules with high potential energy. ATP is produced by cellular respiration or fermentation.
  3. **Chemolithotrophs**—oxidize inorganic molecules with high potential energy. ATP is produced by cellular respiration with the inorganic compound serving as the electron donor.
- **Autotrophs** manufacture their own carbon-containing compounds, and **heterotrophs** must acquire them.

Producing ATP via Cellular Respiration: Variation in Electron Donors and Acceptors

- **Cellular respiration**—A molecule with high potential energy serves as an electron donor and is oxidized, while a molecule with low potential energy serves as a final electron acceptor and becomes reduced. The potential energy difference is converted into ATP.
- A common **electron donor** is glucose, and a **common electron acceptor** is oxygen.
- When cellular respiration is complete, glucose is completely oxidized to CO_2.
- When oxygen acts as the final electron acceptor, the by-product is water.
- Other electron donors range from hydrogen molecules and hydrogen sulfide to ammonia and methane.
- Other electron acceptors can be sulfate, nitrate, carbon dioxide, or ferric ions.

Producing ATP via Fermentation: Variation in Substrates

- **Fermentation** is a strategy for making ATP without using electron transport chains.
- No electron acceptor is used—redox reactions are internally balanced.
- Fermentation is less efficient compared to respiration and is often used as an alternative.
- Glucose is fermented into lactic acid or ethanol.
- *Clostridium aceticum* can ferment ethanol, fatty acids, and acetate as well.
- Species that ferment amino acids produce end-products such as cadaverine and putrescine—responsible for odors of rotting flesh.
- Other bacteria can ferment lactose, and in some species, propionic acid is an end-product—responsible for the taste of Swiss cheese.

Producing ATP via Photosynthesis: Variation in Electron Sources and Pigments

- Photosynthesis can happen in one of three ways:
  1. Light activates a pigment called bacteriorhodopsin, which uses the absorbed energy to transport protons across a membrane. The flow of protons drives ATP synthesis.
  2. Photosynthesis can be performed by absorbing geothermal radiation.
  3. Pigments can be used to absorb light and raise electrons to a higher energy state. Dropping electrons to a lower energy state releases energy, which is used to generate ATP.
- This process requires a source of electrons—cyanobacteria and plants use water. Splitting water to obtain electrons produces oxygen.
- Other bacteria split hydrogen sulfide or ferrous iron.
- It is important to recall the light-absorbing properties of chlorophylls *a* and *b*. Cyanobacteria have the same two pigments.

Obtaining Building-Block Compounds: Variation in Pathways for Fixing Carbon

- Autotrophs make their own building-block compounds; heterotrophs do not.
- In cyanobacteria and plants, enzymes from the **Calvin cycle** transform CO_2 into organic molecules. Other bacteria, animals, and fungi must obtain carbon by eating plants or animals or by absorbing organic compounds released in dead tissues.
- **Methanotrophs** use methane as their primary electron donor and carbon source.
- Some bacteria use carbon monoxide or methanol as starting material.

(c) Explain the difference between a chemoorganotroph and a chemolithotroph.

_____

_____

_____

_____

_____

_____

Ecological Diversity and Global Change

The Oxygen Revolution

- No free molecular oxygen existed for the first 2.3 billion years because:

    1. There was no plausible source of oxygen at the time the planet cooled to a solid.

    2. The oldest Earth rocks indicate any oxygen that formed immediately reacted with iron atoms to produce iron oxides.

- Even in the ocean, oxygen was produced by **cyanobacteria** through **oxygenic photosynthesis**.

- Quantities of oxygen did not begin to build up until 2.0 billion years ago. At 1.8 billion years ago, iron oxides in terrestrial environments appeared.

- Once oxygen was common in the oceans, it could be used as an electron acceptor; thus, **aerobic respiration** was now a possibility.

- The rate of energy production and metabolism could now rise dramatically. Coincidentally, 2.1 billion years ago is when the first macroscopic algae first appear in the fossil record.

Nitrogen Fixation and the Nitrogen Cycle

- Plant growth and overall productivity are often limited by the availability of nitrogen.

- Redox reactions that result in the production of ammonia (NH_3) from nitrogen (N_2) are referred to as **nitrogen fixation**.

- Certain species of cyanobacteria are capable of fixing nitrogen.

- On land, nitrogen-fixing bacteria live in close association with plants—often taking up residence in **nodules**.

- Plants need this nitrogen to make up large quantities of proteins.

Nitrate Pollution

- Widespread use of ammonia fertilizers (NH_3) on many crops not associated with N_2-fixing bacteria had an unforeseen consequence.

- Nitrate (NO_3) produced as a by-product of bacterial ammonia metabolism dissolves readily in water; eventually it enters aquatic ecosystems.

(d) Why is high nitrate detrimental to aquatic ecosystems?

_____

_____

_____

_____

_____

28.4 Key Lineages of Bacteria and Archaea

Bacteria

Firmicutes

- They are also called "low-GC Gram positives" and are rod shaped or spherical.

- They exhibit large metabolic diversity and include anthrax, botulism, tetanus, and strep throat.

Spirochaeles (Spirochetes)

- They are distinguished by their corkscrew shape and unusual flagella.

- Syphilis is caused by a spirochete, and so is Lyme disease.

Actinobacteria

- They are called "high-GC Gram positives" and vary in shape from rods to filaments.

- Over 500 distinct antibiotics have been isolated within the genus.

- Tuberculosis and leprosy are caused by members of this genus.

Chlamydiae

- They are spherical and very tiny.
- All known species live as parasites inside host cells (**endosymbionts**).
- *Chlamydia trachomatis* is a very common sexually transmitted disease.

Cyanobacteria

- They produce much of the oxygen and nitrogen and many of the organic compounds that feed other organisms in freshwater and marine environments.
- Cyanobacteria were responsible for the origin of atmospheric oxygen.

Proteobacteria

- Species in this group cause Legionnaire's disease, cholera, dysentery, and gonorrhea. Certain species can produce vinegars.
- *E. coli* is among the best studied organisms of all species.
- *Rhizobium* can fix nitrogen.

Archaea
Crenarchaeota

- None of these species have yet to cause disease in humans. They can inhabit very extreme environments.

Euryarchaeota

- None of these species have yet to cause disease in humans. *Ferroplasma* live in piles of waste rock near abandoned mines. Methanogens live in the soils of swamps and guts of animals.

✔ **CHECK YOUR UNDERSTANDING**

(e)   From which genus have researchers obtained over 500 distinct antibiotics? How do you think antibiotics are made from bacteria?

_____

_____

_____

_____

B.   MASTERING CHALLENGING TOPICS

In your first-year biology course, you are being exposed to many new terms in a short time. Often, these terms are introduced with a new branch of biology. In this chapter, the terms describing the metabolic diversity of bacteria can be confusing (phototroph, chemoorganotroph, chemolithotroph, autotroph, and heterotroph). Take special care in learning what each term means, and be able to verbally explain the differences between them to a classmate, laboratory partner, or friend. "Compare and contrast" questions are often used on tests, so be able to create a list of similarities and differences between all of these terms. To simplify, categorize the groups into terms explaining how bacteria make ATP (phototroph, chemoorganotroph, or chemolithotroph) and how the bacteria obtain carbon (autotroph or heterotroph). After that, the prefixes help a lot in defining the terms.

C.   ASSESSING WHAT YOU'VE LEARNED

1. The key substance making up bacterial cell walls is:
   a. polysaccharides.
   b. peptidoglycan.
   c. ribosomes.
   d. polymerases.

2. Any bacteria that can cause disease are said to be:
   a. pathogenic.
   b. contagious.
   c. resistant.
   d. prolific.

3. The use of bacteria and archaea to degrade pollutants is called:
   a. biocontamination.
   b. seeding.
   c. biodetoxification.
   d. bioremediation.

4. The oxygen revolution was so important because available oxygen in the oceans could now be used:
   a. as an electron donor.
   b. as an electron acceptor.
   c. for oxygen fixation.
   d. as a reducer.

5. Plant growth and overall productivity is often limited by the availability of:
   a. oxygen.
   b. nitrogen.
   c. carbon.
   d. xenon.

6. Direct sequencing is used to document the presence of bacteria and archaea that
   a. cannot be found in animals.
   b. cannot be detected.
   c. cannot be grown in culture.
   d. cannot infect humans.

7. Researchers generally evaluate bacterial molecular phylogenies by using sequences from:
   a. DNA.
   b. cDNA.
   c. membrane proteins.
   d. ribosomes.

8. Bacteria that use light to promote electrons to the top of electron transport chains and produce ATP using cellular respiration are called:
   a. phototrophs.
   b. autotrophs.
   c. heterotrophs.
   d. organotrophs.

9. The strategy for making ATP without using electron transport chains is called:
   a. aerobic respiration.
   b. fermentation.
   c. electron transport.
   d. carbon fixation.

10. The lineage of bacteria characterized by causing diseases such as tuberculosis and leprosy are the
    a. high-GC gram positives.
    b. low-GC gram positives.
    c. cyanobacteria.
    d. spirochaeles.

CHAPTER 28—ANSWER KEY

Check Your Understanding

(a) Although their relatively simple morphology makes bacteria and archaea appear similar, they are strikingly different at the molecular level.

(b) Follow-up work documented that bacteria were the first of the three lineages to diverge from the common ancestor of all living organisms.

(c) Chemoorganotrophs oxidize reduced organic molecules with high potential energy, whereas chemolithotrophs oxidize inorganic molecules with high potential energy.

(d) Cyanobacteria and algae use nitrates and grow and die in huge numbers. Decomposers such as heterotrophic bacteria and archaea then grow in huge numbers. These organisms use up all the available oxygen and produce "dead zones."

(e) Over 500 distinct antibiotics have been isolated within the genus *Streptomyces*. Originally, the metabolic products each species generated were used as antibiotics. Now, these products can be synthesized without the use of the bacteria.

Assessing What You've Learned

1. b; 2. a; 3. d; 4. b; 5. b; 6. c; 7. d; 8. a; 9. b; 10. a

Looking Forward—Key Concepts in Later Chapters

Biogeochemical Cycling—Chapter 54

Chapter 28 just touches on the importance of the nitrogen cycle in the global ecosystem, which is discussed in greater detail in **Chapter 54**. Many other interrelated cycles exist and function to maintain the Earth's balance.

Animal Behavior—Chapter 51

Animal behavior is intimately related to the process of disease transmission by bacteria. Many diseases travel to several animal species before ultimately settling on a host.

Looking Back—Key Concepts from Earlier Chapters

Cell Organelles—Chapter 7

All of the organelles of the cell, their structure, and function are described in detail in **Chapter 7**. Chapter 29 explains some of the diversity in function of the same organelles.

Diversity—Chapter 28

The variation in photosynthetic pigments observed in bacteria echoes the theme in diversification of protists. Photosynthetic species harvest different wavelengths of light to avoid competition.

Genetic Recombination—Chapter 18

Chapter 18 introduces the mechanisms that bacteria use to transfer genes and undergo genetic recombination. These cells exchange short stretches of DNA between individuals, usually via plasmids. In protists, the entire genome is involved.

Meiosis—Chapter 11

The reduction division called meiosis is explored in **Chapter 11**. The process is now applied and explained in terms of the first organisms to adopt this process.

KEY CONCEPTS

- Protists are a paraphyletic group that includes all eukaryotes except the land plants, fungi, and animals.

- Key morphological innovations occurred as protists diversified: the nuclear envelope, multicellularity, and the mitochondrion and chloroplast via endosymbiosis.

- Many protists are photosynthetic while others obtain carbon compounds by ingesting their food or parasitizing other organisms.

- Sexual reproduction evolved in protists although many species can reproduce both sexually and asexually.

A. CHAPTER OUTLINE

29.1 Why Do Biologists Study Protists?

- Protists do not make up a **monophyletic** group. Instead, they are a **paraphyletic** group—they represent some, but not all, of the descendants of a single common ancestor.

Impacts on Human Health and Welfare

- The most spectacular crop failure in history, the Irish potato famine, was caused by a protist: *Phytophthora infestans*.

Malaria

- *Plasmodium* is the causative agent of malaria and is transmitted to humans by mosquitoes. At least 300 million people worldwide are sickened by it each year, and over 1 million people die from the disease annually.

- Because each *Plasmodium* species spends part of its life cycle inside mosquitoes (**Figure 29.3**), most antimalaria campaigns have focused on controlling these insects.

Harmful Algal Blooms

- Harmful blooms occur when toxin-producing protists called dinoflagellates reach high densities in a particular area. Shellfish eat these protists, and people can become sick from eating the contaminated shellfish.

Ecological Importance of Protists

- Although protists represent 10 percent of the total number of named species, 1 milliliter of pond water can contain over 500 flagellated protists. Dinoflagellates can reach 60 million cells per liter.

Protists Play a Key Role in Aquatic Food Chains

- Photosynthetic species are major players in the global carbon cycle. Primary productivity by the world's oceans, in turn, is responsible for almost half of the total carbon that is fixed on the planet.
- Photosynthetic protists (**plankton**) are the basis of marine and freshwater **food chains**.
- Without protists, most food chains in freshwater and marine habitats would collapse.

Could Protists Help Reduce Global Warming?

- Protists could help reduce global warming by becoming carbon sinks.
- Several lineages of protists have shells made of calcium carbonate. When these protists die, they sink to the ocean floor and form sedimentary rock.
- Petroleum formation begins with accumulations of dead bacteria, archaea, and protists at the bottom of the ocean.

✔ CHECK YOUR UNDERSTANDING

(a) How is stable carbon deposited in the ocean?

_____

_____

_____

_____

_____

29.2 How Do Biologists Study Protists?

Microscopy: Studying Cell Structure

- Detailed studies of cell structure revealed that many protists have a characteristic overall form (**BioSkills 10**).

- Researches interpret these distinct morphological features as **synapomorphies**—shared, derived traits that distinguish major monophyletic groups (**Table 29.2**).
- Seven major groups have been identified on the basis of structures that protect or support the cell or that influence the organism's ability to move or feed.

Evaluating Molecular Phylogenies

- Using the gene that codes for the RNA molecule in the small subunit of ribosomes, eight monophyletic groups from the Eukarya were found.
- The current tree (**Figure 29.7**) shows that Amoebozoa and Opisthokonta form a monophyletic group recently named the Unikonta, and the Alveolata and Stramenopila form a monophyletic group called the Chromalveolata.

Discovering New Lineages via Direct Sequencing

- Direct sequencing has found a wide array of distinct ribosomal RNA sequences and thus a vast amount of new species.
- Eukaryotic cells are much more variable in size than previously imagined; many very tiny species were just discovered.

✔ CHECK YOUR UNDERSTANDING

(b) What do the synapomorphies listed in **Table 29.2** represent?

_____

_____

_____

_____

_____

29.3 What Themes Occur in the Diversification of Protists?

What Morphological Innovations Evolved in Protists?

- The earliest eukaryotes were probably single-celled organisms with a nucleus and endomembrane system, mitochondria, and a cytoskeleton, but no cell wall.
- It is also likely that these cells swam using a novel type of flagellum.

The Nuclear Envelope

- Although researchers are still gathering evidence on how the nuclear envelope originated, the leading hypothesis is that it derived from infoldings of the plasma membrane.

- The nuclear envelope is advantageous because it separates RNA transcription inside the nucleus and RNA translation outside. This allows for further control of gene expression.

Endosymbiosis and the Origin of the Mitochondrion

- Mitochondria are organelles that generate ATP using pyruvate as an electron donor and oxygen as the ultimate electron acceptor.

- The endosymbiosis theory proposes that mitochondria originated when a bacterial cell took up residence inside a eukaryote about 2 billion years ago.

- The **host** supplied the bacterial **symbiont** with protection and chemically reduced carbon compounds from its other prey, while the symbiont supplied the host with ATP.

- **Key research**—Lynn Margulis expanded on the original hypothesis of the origin of mitochondria.

- Observations consistent with the theory:

  1. Mitochondria are about the size of an average bacterium and replicate by fission, as do bacterial cells.

  2. Mitochondria have their own ribosomes and manufacture their own proteins. The ribosomes are similar to bacterial ribosomes.

  3. Mitochondria have double membranes, consistent with the engulfing mechanism (**Figure 29.9**).

  4. Mitochondria have their own genomes, which are organized as circular molecules, like bacteria.

- Researchers compared gene sequences isolated from the nuclear DNA of eukaryotes, mitochondrial DNA from the same eukaryote, and DNA from several species of bacteria. Mitochondrial gene sequences were found to be much more closely related to the sequences from the bacteria.

- Once the mitochondrion was present in the ancestor of today's protists, it underwent diversification.

Structures for Support and Protection

- Many protists have a rigid internal skeleton or a hard external structure called a **test** or a **shell** (**Figure 29.11**).

- Diatoms are surrounded by a glass-like, silicone-oxide shell.

- Dinoflagellates have a cell wall made up of cellulose plates.

- Foraminifera can use calcium carbonate or cover themselves with tiny pebbles.

- Parabasalids use an internal support of cross-linked microtubules.

- Euglenids have a collection of protein strips just under the plasma membrane.

Multicellularity

- Differentiation of cell types is a crucial criterion for defining multicellularity.

- In simple multicellular species, some cells are for producing or obtaining food, while others are specialized for reproduction.

- Not all cells express the same genes.

- Multicellularity is a synapomorphy shared by all of the brown algae and all of the plasmodial and cellular slime molds. It also arose in some lineages of red algae.

- The array of aforementioned novel morphological traits played a key role as protists diversified.

How Do Protists Obtain Food?

- Protists feed by ingesting packets of food, absorbing organic molecules, or performing photosynthesis.

Ingestive Feeding

- Predation and scavenging: Protists can eat bacteria, archaea, and other protists—dead or alive.

- The engulfing process is possible because protists lack a cell wall, and because their flexible membrane and cytoskeleton move as shown in **Figure 29.12a**.

Absorptive Feeding

- Some absorptive groups are decomposers, like the oomycetes.

- Many others live inside organisms as parasites, although in many cases the relationships are beneficial or mutualistic.

Photosynthesis—Endosymbiosis and the Origin of the Chloroplast

- The endosymbiosis theory contends that the organelle where photosynthesis takes place in eukaryotes originated when a protist engulfed a cyanobacterium (**primary endosymbiosis**).

- The evidence for endosymbiotic origin for the chloroplast:

  1. Chloroplasts have the same bacteria-like characteristics presented earlier for mitochondria.

2. There are many examples of endosymbiotic cyanobacteria living inside protists today.

3. The DNA sequences of chloroplasts are extremely similar to those of cyanobacteria.

4. They contain the same pigment molecules and similar cell wall structures.

Photosynthesis—Primary versus Secondary Endosymbiosis

- **Secondary endosymbiosis** occurs when an organism engulfs an already photosynthetic organism and retains the acquired chloroplasts as intracellular symbionts.

Photosynthesis—Diversification in Pigments

- The presence of different pigments means different species absorb different wavelengths of light.

How Do Protists Move?

- **Amoeboid motion** is a gliding motion. Fingerlike projections called pseudopodia stream forward over a substrate. This motion involves interactions between actin, myosin, and ATP.

- The other major mode of locomotion involves flagella or cilia. Flagella and cilia have identical structures and are actually extensions of the cell. Both consist of nine sets of doublet microtubules around two central, single microtubules.

How Do Protists Reproduce?

Sexual versus Asexual Reproduction

- **Asexual reproduction** produces offspring that are genetically identical to the parent.

- **Sexual reproduction** produces offspring that are genetically different from their parents.

- Importance of sexual reproduction:
  1. Production of better-quality offspring.
  2. Genetically variable offspring have a better chance of surviving during environmental changes.

Life Cycles—Haploid- versus Diploid-Dominated

Life Cycles—Alternation of Generations

- Every aspect of a life cycle is variable among protists.

- **Alternation of generations** is a phenomenon in which the haploid and diploid phases of the lifecycle are multicellular.

- Multicellular haploid and diploid forms of the same individual can be identical in morphology or radically different.

✔ **CHECK YOUR UNDERSTANDING**

(c) In alternation of generations, which life stage produces spores and which life stage produces gametes? Which life stage is haploid and which life stage is diploid?

_____

_____

_____

_____

_____

_____

29.4 Key Lineages of Protists

- Each of the seven major Eukarya lineages has at least one distinctive morphological characteristic. But once an ancestor evolved a distinctive cell structure, its descendants diversified into a wide array of lifestyles.

Amoebozoa

Myxogastrida (Plasmodial Slime Molds)

- Have a huge supercell with many nuclei.

- Move by amoeboid motion and are important decomposers in forest ecosystems.

Excavata

Parabasalida

- Cells lack both a cell wall and mitochondria, and have a single nucleus. A distinctive rod of cross-linked microtubules runs the length of the cell. They feed by engulfing.

- Several species live inside the guts of termites and break down the cellulose of wood.

Diplomonadida

- Each cell has two nuclei, each associated with four flagella, and all lack a cell wall.

- *Giardia* are common intestinal parasites in mammals.

Euglenida

- Cells lack a cell wall but have a system of interlocking protein strips.

- One-third of species perform photosynthesis; euglenids are important components of freshwater plankton and food chains.

Plantae

Rhodophyta (Red Algae)

- Have cell walls composed of cellulose and other polymers and have no flagella. Species contribute to reef building.

Rhizaria

Foraminifera

- Cells have multiple nuclei, and their tests are usually made of organic material. Dead forams often form extensive calcium carbonate deposits.

Alveolata

Ciliata

- Cells have a micronucleus and a macronucleus, can reproduce by conjugation, and use cilia for locomotion.

Dinoflagellata

- Cells from sexual reproduction may form cysts to remain dormant. Dinoflagellata are second only to diatoms in amount of carbon fixed.

Apicomplexa

- Have system of organelles at one end called the apical complex.
- All are parasitic. Malaria is caused by *Plasmodium.*

Stramenopila

Oomycota (Water Molds)

- Fungus-like and extremely important decomposers in aquatic ecosystems. An oomycete caused the Irish potato famine.

Diatoms

- Cells are supported by silicon-rich, glassy shells. They are photosynthetic and therefore the most important producer of carbon compounds in the water.

Phaeophyta (Brown Algae)

- Cell walls contain cellulose, and all species are multicellular.
- They are photosynthetic and sessile, though their reproductive cells may be motile. They form forests that are important habitats.

✓ CHECK YOUR UNDERSTANDING

(d) Which of the groups just described is likely to demonstrate alternation of generations?

_____

_____

_____

_____

_____

_____

B. MASTERING CHALLENGING TOPICS

In this chapter, the theory of endosymbiosis is used to explain the origins of the first eukaryotes. While the theory itself may be easy to understand, explaining it and the techniques used to support it is a different matter. Try starting with the basics. Outline the theory and note what it tries to explain. Then outline each experiment in the chapter that supports the theory and note why it supports the theory. As the text now has separated mitochondrion endosymbiosis from chloroplast endosymbiosis, so should you. If you do not understand the techniques used in a particular experiment, go back in the text to previous chapters that outline those techniques in more detail. Theories are modified constantly as new evidence (for or against) is published.

C. ASSESSING WHAT YOU'VE LEARNED

1. The following ranks as one of the most devastating diseases and is caused by a protist.
   a. Red tides
   b. AIDS
   c. Malaria
   d. Plasmodium
2. The theory of endosymbiosis proposes that mitochondria originated
   a. when a virus invaded and stayed in a bacterial cell.
   b. when a bacterial cell took up residence inside a eukaryote.
   c. when a virus took up residence inside a eukaryote.
   d. when a bacterial cell took up residence inside a prokaryote.

3. Which one of the following statements *does not* support endosymbiosis?
   a. Cyanobacteria look nothing like chloroplasts.
   b. Mitochondria and chloroplasts are about the same size as an average bacterium.
   c. Both mitochondria and chloroplasts have a double membrane.
   d. Both mitochondria and chloroplasts reproduce by fission.

4. The crucial criterion for defining multicellularity is:
   a. differentiation.
   b. colonial growth.
   c. lysosomes.
   d. cell walls.

5. Which of the following is *not* a mode of locomotion in protists?
   a. Ameboid movement
   b. Using flagella
   c. Using cilia
   d. Using mitochondria

6. The phenomenon in which the haploid and diploid phases of a life cycle are multicellular is called:
   a. double diploid.
   b. alternation of generations.
   c. haplodiplo generation.
   d. multigenerational.

7. Which type of reproduction produces offspring that are genetically different from their parents?
   a. Sexual reproduction
   b. Asexual reproduction
   c. Multisexual reproduction
   d. Unisexual reproduction

8. This group of protists includes the species responsible for malaria.
   a. Diatoms
   b. Red algae
   c. Apicomplexans
   d. Dinoflagellates

9. This group was responsible for the Irish potato famine.
   a. Diatoms
   b. Apicomplexans
   c. Ciliates
   d. Oomycetes

10. The key characteristic of the plasmodial slime molds is:
    a. their single nuclei per cell.
    b. their ability to parasitize.
    c. their huge supercell with many nuclei.
    d. their carbon fixing in aquatic habitats.

CHAPTER 29—ANSWER KEY

Check Your Understanding

(a) Carbon turnover in the ocean is significantly more rapid than in the terrestrial environment. However, limestone is a stable repository for carbon in the oceans. The sugars that protists produce are the basis for food chains in both freshwater and marine environments.

(b) The synapomorphies listed in the table represent changes in structures that protect or support the cell or that influence the organism's ability to move or feed.

(c) In alternation of generations, the diploid sporophytes produce spores and the haploid gametophytes produce gametes.

(d) Alternation of generations evolved independently in the brown algae and the red algae.

Assessing What You've Learned

1. c; 2. b; 3. a; 4. a; 5. d; 6. b; 7. a; 8. c; 9. d; 10. c

Looking Forward—Key Concepts in Later Chapters

Muscle Contraction—Chapter 47

The mechanism of amoeboid movement is closely related to muscle movement in animals, which is detailed in **Chapter 47.**

Chapter 30

Green Algae and Land Plants

Green Algae and Land Plants

Looking Back—Key Concepts from Earlier Chapters

Evolutionary Relationships—Chapter 24

Chapter 24 introduces us to how molecular and morphological data are used to infer evolutionary relationships. This investigation of evolutionary relationships is continued in Chapter 30 with the land plants.

Alternation of Generations—Chapter 29

The phenomenon of alternation of generations in the protists was introduced in **Chapter 29**. This phenomenon persists with further variation in the land plants, as discussed in Chapter 30.

Genetic Engineering—Chapter 19

The mechanisms and procedures for genetic engineering are explained in **Chapter 19** and applied to the production of land plants in Chapter 30.

KEY CONCEPTS

- Green algae are an important source of oxygen and provide food for aquatic organisms.

- Land plants release oxygen, hold soil and water in place, build soil, moderate extreme temperatures and winds, and provide food for terrestrial organisms.

- Land plants were the first multicellular organisms that could live with most of their tissues exposed to air. Plants dominate today's terrestrial environments.

- Important evolutionary changes made it possible for plants to reproduce in dry environments.

A. CHAPTER OUTLINE

30.1 Why Do Biologists Study the Green Algae and Land Plants?

Plants Provide Ecosystem Services

- Green algae and land plants enhance the life-supporting attributes of the atmosphere, surface water, soil, and other physical components of the ecosystem.

- Plants produce oxygen via oxygenic photosynthesis.

- Plants build soil by providing food for decomposers.

- Plants hold soil and prevent nutrients from being washed or blown away.

- Plants hold water and increase water-holding capacity of an area.

- Plants moderate the local climate.

- Plants are the dominant primary producers in terrestrial ecosystems.

Plants Provide Humans with Food, Fuel, Fiber, Building Materials, and Medicines

- The grains that form the basis of our current food supply were derived from wild species about 10,000 to 2,000 years ago. Humans were actively selecting seeds to plant the next generation of crops—a process called **artificial selection**.

- Sugars synthesized during the Carboniferous period laid the groundwork for the industrial revolution. There will probably not be another generation of plant-based fuels.

- Cotton and other plant fibers are still important sources of raw materials for clothing, rope, and other household articles.

- Primary interest in woody plants is for building materials and fibers used in papermaking. Relative to its density, wood is a stiffer and stronger building material than concrete, cast iron, aluminum alloys, or steel.

- In both traditional and modern medicine, plants are a key source of drugs.

✔ CHECK YOUR UNDERSTANDING

(a) Describe the process of artificial selection in producing corn with more oil.

_____

_____

_____

30.2 How Do Biologists Study Green Algae and Land Plants?

Analyzing Morphological Traits

Similarities between Green Algae and Land Plants

- Their chloroplasts synthesize starch and contain chlorophyll *a* and *b*.
- They have similar arrangements of thylakoids.
- Their cell walls, sperm, and peroxisomes are similar in structure and composition.

Major Morphological Differences among Land Plants

- Land plants were traditionally grouped into three categories:
  1. **Non-vascular plants** lack vascular tissue. (Bryophyta—mosses)
  2. **Seedless vascular plants** do not make seeds. (Pterophyta—ferns)
  3. **Seed plants** also have vascular tissue. (Flowering plants—angiosperms)

Using the Fossil Record

- The fossil record for land plants is massive; it began 475 million years ago (**Figure 30.6**) and is divided into five segments:
  1. **475 million years ago**—duration 59 million years

Most fossils are microscopic and consist of reproductive cells called **spores** and sheets of a waxy coating called **cuticle**.

  2. **"Silurian-Devonian Explosion"**—416 to 359 million years ago

Macroscopic fossils from most major plant lineages are found. Virtually all adaptations that allow plants to occupy dry, terrestrial habitats are present, including water-conducting cells, tissues, and roots.

3. **Carboniferous Period**—359 to 299 million years ago

Extensive deposits of coal are found; coal is a carbon-rich rock packed with fossil spores, branches, leaves, and tree trunks. Usually derived from lycophytes, horsetails, and ferns and in the presence of water, the makeup of coal indicates the presence of extensive forested swamps.

4. **Gymnosperms**—299 to 145 million years ago

These lineages include the cycads, conifers and gingkoes, gnetophytes, and pines. Because gymnosperms grow readily in dry habitats, biologists infer that both wet and dry environments on the continents became blanketed with green plants for the first time during this interval.

5. **Angiosperms**—145 million years ago to present

The woody plants that produced the first flowers are the ancestors of today's grasses, orchids, daisies, oaks, maples, and roses.

Evaluating Molecular Phylogenies

- The phylogenetic tree in **Figure 30.7** is a recent version from laboratories studying plant phylogenies. Note the following:
  1. The green plants are monophyletic.
  2. Land plants evolved from green algae.
  3. Green algae are paraphyletic.
  4. Charophyceae are the closest living relatives to land plants.
  5. The land plants are monophyletic.
  6. The bryophytes or non-vascular plants are the earliest-branching groups among land plants.
  7. The non-vascular plants are paraphyletic.
  8. The seedless vascular plants are paraphyletic but vascular plants as a whole are monophyletic.
  9. The seed plants are monophyletic.
  10. The gymnosperms and angiosperms are each a monophyletic group.

✔ CHECK YOUR UNDERSTANDING

(b) Explain the evolutionary relationship between green algae and angiosperms.

_____

_____

_____

_____

30.3 What Themes Occur in the Diversification of Land Plants?

Transition to Land, I: How Did Plants Adapt to Dry Conditions?

- Plants had to adapt to situations where only portions of their tissues were bathed in fluid.
- Solving the water problem was a breakthrough in evolution by natural selection, achieved in two steps:
    1. preventing water loss from cells
    2. transporting water from tissues with access to water to tissues without access

Preventing Water Loss: Cuticle and Stomata

- **Cuticle** is a waxy, watertight sealant that gives plants the ability to survive in dry environments. Gas exchange is accomplished on wax-covered plants by the presence of a **stoma**, which consists of an opening covered by **guard cells**.
- The **pore** opens or closes as the guard cells change shape. The presence of guard cells in later groups implies that many of the early land plants had the ability to regulate gas exchange and control water loss by opening and closing their pores.

The Importance of Upright Growth

- In a terrestrial environment, individuals that can grow erect have much better access to sunlight than individuals that are incapable of growing erect.

The Origin of Vascular Tissue

- **Key research**—Paul Kenrick and Peter Crane examined extraordinary fossils found in Scotland in a formation called the Rhynie Chert. This 400-million-year-old formation contains many plants fossilized in an upright position.
- Some of the fossilized water-conducting cells had simple cellulose-containing cell walls like today's mosses (**Figure 30.9a**).
- Some of the fossilized water-conducting cells had thickened rings containing lignin.
- **Lignin** is a polymer built from six-carbon rings and is extraordinarily strong for its weight.

Elaboration of Vascular Tissue: Tracheids and Vessels

- **Tracheids** are elongated cells that die after maturing, do not contain cytoplasm, and are the first true vascular tissues that made possible the efficient transport of water to aboveground tissues.
- **Vessel elements** are shorter and wider than tracheids and form a continuous pipelike structure.
- Tracheids and/or vessel elements can form **wood**.

Mapping Evolutionary Changes on the Phylogenetic Tree

- Cuticles, stomata, and vascular tissue were key adaptations that allowed early plants to colonize land.
- Water-conducting cells evolved independently in mosses and in the vascular plants.
- Vessels evolved independently in gnetophytes and angiosperms.

Transition to Land, II: How Do Plants Reproduce in Dry Conditions?

- Spores resisted drying because they were encased in a tough coat of sporopollenin.
- Gametes were produced in complex multicellular structures.
- Embryo was retained on the parent and nourished by it.

Producing Gametes in Protected Structures

- The gametophytes of early land plants contained reproductive organs called **gametangia**.
- The sperm-producing structure is called the **antheridium**.
- The egg-producing structure is called the **archegonium**.

Retaining and Nourishing Offspring: Land Plants as Embryophytes

- Instead of shedding eggs, land plants retain them.
- The zygote is retained on the gametophyte after fertilization; therefore, land plants do not have to make their own food early in life.

Alternation of Generations

- Individuals have a multicellular haploid phase (**gametophyte**) and a multicellular diploid phase (**sporophyte**).
- The two phases are connected by the reproductive cells: gametes and spores.
- This life cycle always follows the same key events:
    1. The sporophyte produces spores by meiosis. Spores are haploid.
    2. Spores divide by mitosis and develop into a haploid gametophyte.
    3. Gametophytes produce gametes by meiosis. Both the gametophyte and gametes are haploid.
    4. Two gametes unite during fertilization to form a diploid zygote.
    5. The zygote divides by mitosis and develops into a multicellular, diploid sporophyte.

The Gametophyte-Dominant to Sporophyte-Dominant Trend in Life Cycles

- In non-vascular plants, the sporophyte is small, short lived, and dependent on the gametophyte for nutrition.
- In ferns and other vascular plants, the sporophyte is much larger and longer lived.
- The trend to sporophyte-dominated life cycles is one of the most striking trends in land plant evolution.

Heterospory

- **Heterospory** is the production of two distinct types of spore-producing structures and thus of spores.
- **Microsporangia** produce **microspores**, which develop into the male gametophyte.
- **Megasporangia** produce **megaspores**, which develop into the female gametophyte.

Pollen

- Non-vascular plants and ferns need water for gametes to swim and perform fertilization.
- In heterosporous seed plants, the microspore germinates to form a tiny male gametophyte surrounded by a tough, dessication-resistant coat of sporopollenin—a **pollen grain**.

Seeds

- A **seed** is a structure that includes a developing embryo and a food supply surrounded by a tough coat. Seeds are often attached to a structure that aids in dispersal by wind, water, or animals.

Flowers

- Flowers contain stamens (where microsporangia develop) and carpels (where ovaries are found).
- When a pollen grain lands on a carpel and produces sperm, fertilization takes place.
- **Double fertilization** involves one sperm fusing with the egg to form the zygote and the other sperm fusing with two nuclei to form triploid nutritive tissue called **endosperm**.

Pollination by Insects and Other Animals

- Flowers are diverse in size, shape, and coloration. They also produce a wide range of scents. Scientists have hypothesized that flowers are adaptations that increase the probability that **pollination** will occur. Flowers provide a food reward to the pollinator in the form of sugar-rich nectar or protein-rich pollen grains—a mutually beneficial relationship.

Fruits

- A **fruit** is a structure that is derived from the ovary and encloses one or more seeds.

- Tissues derived from the **ovary** are often nutritious and brightly colored. Animals often eat fruits and disperse seeds.
- Once land plants had vascular tissue and could grow efficiently in dry habitats, their diversification revolved around sperm reaching eggs and seed dispersal.

The Angiosperm Radiation

- Angiosperms represent one of the great **adaptive radiations** in the history of life.
- The diversification of angiosperms is associated with three key adaptations: (1) vessels, (2) flowers, and (3) fruits.
- **Monocots** are monophyletic, but **dicots** are paraphyletic. Thus, a new term for the monophyletic group of dicots is the **eudicots**.

✔ **CHECK YOUR UNDERSTANDING**

(c) Why is a flower a key innovation in plants?

_____

_____

_____

_____

30.4 Key Lineages of Green Algae and Land Plants

Green Algae

Ulvophyceae (Ulvophytes)

- Many large green algae such as *Ulva*, the sea lettuce, belong to this group.
- Unicellular and small multicellular species inhabit the plankton of freshwater systems.

Coleochaetophyceae (Coleochaetes)

- Most grow as flat sheets of cells (e.g., water lilies), and the multicellular individuals are haploid.

Charaphyceae (Stoneworts)

- They commonly accumulate crusts of calcium carbonate over their surfaces.
- They can form extensive beds on lake bottoms.

Non-vascular Plants ("Bryophytes")

Hepaticophyta (Liverworts)

- They have liver-shaped leaves and can grow on bare rock or tree bark, which helps in soil formation.

Bryophyta (Mosses)

- Can be abundant in extreme environments and can become dormant.
- *Sphagnum* species are among the most abundant.

Anthocerophyta (Hornworts)

- Sporophytes have a horn-like appearance and have stomata.

Seedless Vascular Plants

Lycophyta (Lycophytes or Club Mosses)

- The most ancient plant lineage with roots.
- Tree-sized lycophytes dominated the coal-forming forests of the Carboniferous.

Psilotophyta (Whisk Ferns)

- They are restricted to tropical regions and have no fossil record.

Equisetophyta (or Sphenophyta) (Horsetails)

- They can flourish in waterlogged soils by allowing oxygen to diffuse down their hollow stem.

Pteridophyta (Ferns)

- The only seedless vascular plants to have large, well-developed leaves.

Seed Plants

Gymnosperms > Cycadophyta (Cycads)

- They harbor large numbers of symbiotic, nitrogen-fixing cyanobacteria, which are important sources of nutrients.

Gymnosperms > Ginkgophyta (Ginkoes)

- One species is alive today. It is deciduous, and individual trees are either male or female.

Gymnosperms > Redwood Group (Redwoods, Junipers, Yews)

- This group is named for its reproductive structure, the cone. This group dominates all high-latitude and high-altitude forests.

Gymnosperms > Gnetophyta (Gnetophytes)

- Three living genera; the drug ephedrine has been isolated from this group.

Gymnosperms > Pinophyta (Pines, Spruces, Firs)

- Also producing cones, pines are a mainstay in the forest products industry—wood, paper and turpentine.

Anthophyta (Angiosperms)

- The defining adaptation is the flower. They supply the food that supports virtually every other species.

✓ CHECK YOUR UNDERSTANDING

(d) What do the gymnosperms and angiosperms produce during reproduction that is different from all other groups?

_____

_____

_____

B. MASTERING CHALLENGING TOPICS

By far, understanding how alternation of generations works, and then changes during the evolution of land plants, is the toughest concept to master in this chapter. First, think about the definition of alternation of generations and look at the differences between sporophytes and gametophytes. Next, slowly go through **Figures 30.13** through **30.20** and really take in the general ideas first, such as how much time the groups spend in haploid versus diploid stages. Last, try to describe to your classmates these major changes in the different plant groups.

C. ASSESSING WHAT YOU'VE LEARNED

1. The process of actively choosing seeds to plant the next generation of crops is called:
   a. natural selection.
   b. artificial selection.
   c. general selection.
   d. hydroponics.

2. The Carboniferous period is most noted for:
   a. its extensive deposits of coal.
   b. its extensive angiosperm radiation.
   c. its extensive radiation of aquatic plants.
   d. an explosion of macroscopic fossils.

3. Which of the following is *not* a major grouping of land plants?
   a. Seedless vascular plants
   b. Non-vascular plants
   c. Cone-bearing plants
   d. Seed plants

4. The latest and most impressive plant radiation was that of the
   a. algae.
   b. angiosperms.
   c. gymnosperms.
   d. club mosses.

5. Which of these statements about plant molecular phylogenies is *true*?
   a. The group "green algae" is paraphyletic.
   b. The Charales is not the sister group to green algae.
   c. The land plants are paraphyletic.
   d. The non-vascular plants are the most recent group.

6. Plants can prevent water loss from cells by:
   a. secreting a waxy cuticle.
   b. exchanging respiratory gases through a pore using guard cells.
   c. regulating the opening and closing of pores.
   d. all of the above

7. Upright growth in nonwoody plants is attained by:
   a. extensive root systems for support.
   b. the production of thicker leaves.
   c. the absence of leaves.
   d. pumping water into cells to increase their rigidity.

8. The structure that encloses and protects the developing plant embryo is the:
   a. dicot.
   b. seed.
   c. pollen grain.
   d. ovule.

9. Diversification of the angiosperms is associated with the following adaptation:
   a. vessels.
   b. flowers.
   c. fruits.
   d. all of the above

10. The only major group of seed plants with only one species alive today is:
    a. Ginkophyta.
    b. Cycadophyta.
    c. Coniferophyta.
    d. Gnetophyta.

CHAPTER 30—ANSWER KEY

Check Your Understanding

(a) By selecting and planting seeds coming from corn plants with the highest oil content, eventually corn plants will yield more oil per plant.

(b) Angiosperms and green algae share a common ancestor—the common ancestor to all green plants.

(c) Flowers are diverse in size, shape, and coloration. They also produce a wide range of scents. Scientists have hypothesized that flowers are adaptations to increase the probability that pollination will occur. Flowers provide a food reward to the pollinator in the form of sugar-rich nectar or protein-rich pollen grains—a mutually beneficial relationship.

(d) Angiosperms and gymnosperms produce seeds instead of spores.

Assessing What You've Learned

1. b; 2. a; 3. c; 4. b; 5. a; 6. d; 7. d; 8. b; 9. d; 10. a

Looking Forward—Key Concepts in Later Chapters

Water Conduction—Chapter 37

Water conduction is mentioned as only one explanation for the massive aboveground growth of land plants. The mechanisms for water conduction are explored in detail in **Chapter 37**.

The Seed—Chapter 40

The structure of the seed is introduced in Chapter 30, but how it protects the developing embryo and the changes that occur during growth and development are discussed in detail in **Chapter 40**.

Fungi

KEY CONCEPTS

🔑 Fungi supply plants with key nutrients and decompose dead wood.

🔑 Fungi are the master recyclers of nutrients in terrestrial environments.

🔑 All fungi survive by absorbing nutrients from living or dead organisms. Fungi secrete enzymes so that digestion takes place outside their cells. Their morphology provides a large surface area for efficient absorption.

🔑 Many fungi have unusual life cycles. Fungal cells can contain haploid nuclei from two different individuals. Although most species reproduce sexually, a few species produce gametes.

A. CHAPTER OUTLINE

31.1 Why Do Biologists Study Fungi?

Fungi Provide Nutrients for Land Plants

- **Mycorrhizal** (fungal-root) associations between fungi and the roots of land plants are extremely common.

- Tree species can grow three to four times faster with these fungal associations.

Fungi Speed the Carbon Cycle on Land

- **Saprophytes** are fungi that make their living by digesting dead plant material. They did not grow in the Carboniferous period because the pH was too low.

- At the end of the great Permian extinction, researchers documented a huge fungal spike.

- Fungi make the carbon cycle (the fixation of carbon by plants and the release of CO_2 via respiration) turn more rapidly.

Fungi Have Important Economic Impacts

- Incidence of fungal infections in humans is low. The major destructive impact is on crops.

- Fungi have important positive impacts on the human food supply.

- In nature, fungi have killed 4 billion chestnut trees and tens of millions of American elm trees. The fungi were accidentally imported.

- The yeast *Saccharomyces cerevisiae* has been incredibly important in basic research on cell biology and molecular genetics.

- It has become the species of choice for studies on control of the cell cycle and regulation of gene expression in eukaryotes.

✔ **CHECK YOUR UNDERSTANDING**

(a) Describe some positive impacts fungi have on the human food supply.

_____

_____

_____

_____

_____

31.2 How Do Biologists Study Fungi?

Analyzing Morphological Traits

- Only two growth forms exist: (1) single-celled forms called **yeasts** and (2) multicellular filamentous forms called **mycelia**.

The Nature of the Fungal Mycelium

- Mycelia constantly grow in the direction of food and die back in areas that lack food.
- The body shape of a fungus can change almost continuously throughout its life.

The Nature of the Hyphae

- The filaments that make up a mycelium are called **hyphae**, and each filament is separated by cross-walls called **septa**.
- Most hyphae are haploid or **heterokaryotic**, meaning each cell contains several haploid nuclei from different parents.
- Most heterokaryotic hyphae are **dikaryotic**.

Mycelia Have a Large Surface Area

- Mycelia are incredibly thin and have an extraordinarily high surface area.
- This adapts fungi well to an absorptive lifestyle. The only thick, fleshy structures are the reproductive structures.

Reproductive Structures

- By analyzing reproductive structures, biologists have identified four major groups of fungi:

    1. **Swimming gametes and spores**
        - These are the only motile cells known in the fungi.
        - Species with swimming gametes are traditionally known as chytrids.
        - Chytrids reproduce sexually and asexually.

    2. **Zygosporangia**
        - Haploid hyphae come in several different mating types or "sexes." If chemicals released by two hyphae indicate that they are from different mating types, the individuals can become yoked together and form a spore-producing structure—a **zygosporangium**.
        - Species with zygosporangia are traditionally called the zygomycetes.

    3. **Basidia**
        - Mushrooms, bracket fungi, and puffballs are reproductive structures produced by members of the Basidiomycota. Their size, shape, and color vary enormously, but they all originate from hyphae of mated individuals.
        - These fungi form specialized structures called **basidia**, which produce spores.

    4. **Asci**
        - Inside complex cup-shaped reproductive structures are hyphae that have distinctive sac-like cells called **asci**.
        - Species with asci are traditionally called ascomycetes.

Evaluating Molecular Phylogenies

Fungi Are Closely Related to Animals

- In addition to DNA sequence data, three key morphological features link animals and fungi:

    1. Most animals and fungi synthesize tough structural material called chitin.
    2. The flagella that develop in chytrid gametes are similar to those in animals.
    3. Both animals and fungi store food by synthesizing glycogen.

What Is the Relationship among the Major Fungal Groups?

- To understand the relationships among the groups of fungi, biologists have sequenced a series of genes from an array of fungal species and used the data to estimate the phylogeny of the group. The data support several important conclusions:

    1. The single-celled eukaryote microsporidians are actually fungi.
    2. The chytrids and zygomycetes are paraphyletic. This means that either swimming gametes or yolked hyphae evolved more than once, or both were present in a common ancestor but then were lost in certain lineages.
    3. The Glomeromycota are monophyletic.

4. The Basidiomycota and Ascomycota are each monophyletic and together a monophyletic group.

5. The sister group to fungi consists of protists called choanoflagellates and the animals.

Experimental Studies of Mutualism

- Fungi and land plants often have a **symbiotic** relationship.

- **Mutualism** is a relationship that provides benefits to the host in return for the food the fungi absorbs.

- Researchers have begun to categorize these relationships as mutualistic, **parasitic**, or **commensal** (only the fungi benefit).

- By using radioactive isotopes of CO_2, phosphorus, and nitrogen, researchers have documented that up to 20 percent of sugars produced by a plant are transferred to fungi, and fungi facilitate the transfer of phosphorus and nitrogen to the plant.

CHECK YOUR UNDERSTANDING

(b) What is the likely ancestor of fungi, and what three things are similar between fungi and animals?

_____

_____

_____

31.3 What Themes Occur in the Diversification of Fungi?

Fungi Participate in Several Types of Mutualisms

Ectomycorrhizal Fungi (EMF)

- EMF form a dense network of hyphae around roots and are found on virtually all of the tree species that grow in temperate and boreal forests.

- EMF hyphae penetrate between cells in the outer layer of the root, but they do not enter root cells. Most are basidiomycetes, and a few are ascomycetes.

- Where the growing season is short, the decomposition of needles, leaves, twigs, and trunks is often sluggish. As a result, nitrogen tends to remain tied up in dead tissues.

- EMF however, release peptidases that cleave the peptide bonds between amino acids. The nitrogen released by this cleavage is absorbed by the hyphae and transported close to the tree roots, where it can be absorbed by the plant. In return, the fungi receive sugars and other reduced-carbon compounds from the plant.

Arbuscular Mycorrhizal Fungi (AMF)

- AMF grow *into* the cells of root tissue, and most are members of the Glomeromycota. They are found in 80 percent of all land plant species and are especially common in grasslands and in the forests of warm tropical habitats.

- In these habitats, phosphorus is the limiting nutrient, so fungi supply plants with phosphorus in exchange for carbon.

Are Endophytes Mutualists?

- Endophytes do not cause disease in the tissues they invade. Most are mutualistic, and some are commensal.

Mutualisms with Other Species

- **Lichens** are a mutualistic partnership between a species of ascomycete and either a cyanobacteria or an alga.

- Some ant species actively farm fungi inside their colonies.

What Adaptations Make Fungi Such Effective Decomposers?

Extracellular Digestion

- Fungi secrete enzymes outside their hyphae into their food. Digestion takes place outside the fungus, or **extracellularly**.

- Only a handful of organisms on Earth can digest **cellulose**, and members of the basidiomycetes are the only organisms that can degrade lignin completely to CO_2.

Lignin Degradation

- **Lignin peroxidase** is an enzyme that catalyzes the removal of a single electron from an atom in the ring structures of lignin. This oxidation step creates a free radical and leads to a series of uncontrolled and unpredictable reactions that end up splitting the polymer into smaller units.

- The uncontrolled oxidation reactions triggered by lignin peroxidase are analogous to the uncontrolled oxidation reactions that occur when wood burns. The uncontrolled nature of the reactions means that the oxidation of lignin cannot be harnessed to drive the production of ATP. However, by oxidizing lignin, the hyphae gain access to huge supplies of energy-rich cellulose within.

Cellulose Digestion

- **Cellulases** are enzymes secreted into the extracellular environment by fungi. Some cleave long strands of cellulose into the disaccharide called cellobiose. Other cellulases are equally specific, and together they eventually convert cellulose into glucose.

- Biologists have purified seven different cellulases from the fungus *Trichoderma reesei*. These enzymes could be used to make paper production more efficient.

Variation in Reproduction

Spores as Key Reproductive Cells

- The **spore** is the most fundamental reproductive cell in fungi. Spores are the dispersal stage of the fungal life cycle and are produced during both asexual and sexual reproduction.

Multiple Mating Types

- Instead of having just two sexes, a single fungal species may have tens of thousands.

How Does Fertilization Occur?

- Only members of the Chytridiomycota produce gametes, and no fungus produces gametes that are different enough to be called sperm and egg.

- The process of sexual reproduction begins when hyphae from two individuals fuse to form a hybrid hypha. When cytoplasm fuses, **plasmogamy** is said to occur. When the two individuals stay independent, the mycelium is **heterokaryotic**. When the nuclei fuse, it is called **karyogamy**.

Asexual Reproduction

- During asexual reproduction, spore-forming structures are produced by a haploid mycelium, and spores are generated by mitosis. The offspring are clones.

Four Major Types of Life Cycles

1. The Chytridiomycota are the only fungi to exhibit alternation of generations. Swimming gametes fuse and form a diploid zygote.

2. The haploid hyphae of Zygomycota come into contact with a different individual, and their nuclei fuse and develop a tough, resistant coat. Meiosis occurs, and the products grow into a structure that produces haploid spores. When spores are released and germinate, they grow into new mycelia.

3. In Basidiomycota, hyphae of mated individuals fuse and then the cytoplasms fuse, but the nuclei remain independent. The diploid nucleus undergoes meiosis, and haploid spores mature from the meiotic products. Sexual reproduction concludes when the spores are expelled and dispersed by the wind.

4. A dikaryotic mycelium, containing one nucleus from each parent, emerges and eventually grows into a complex reproductive structure with a distinctive cell called the **ascus** (sac) at its tip. The nuclei fuse inside the sac, and spores are produced.

✔ CHECK YOUR UNDERSTANDING

(c) How is cellulose degraded by fungi?

_____

_____

_____

_____

31.4 Key Lineages of Fungi

Microsporidia

- All of the estimated 1200 species are single celled and parasitic.

- They are distinguished by their polar tube, which allows them to enter the interior of cells they parasitize.

Chytrids

- Largely aquatic, members of this group are the only fungi that can produce cells that are motile.

- An epidemic of chytrid infections may be responsible for catastrophic declines that recently occurred in frog populations all over the world.

Zygomycetes

- Many members of this group are saprophytic, but some are important parasites of insects and spiders.

- The black bread mold *Rhizus stolonifer* is a common household pest, and some species of *Mucor* are used in the production of steroids.

Glomeromycota

- All AMF are members of this group.

- Because grasslands and tropical forests are among the most productive habitats on Earth, the AMF are enormously important to both human and natural economies.

Basidiomycota (Club Fungi)

- Basidiomycetes are important saprophytes and are the only lineage capable of digesting lignin.

- EMF are enormously important in forest management. Mushrooms are also cultivated for food. Some of the toxins found in poisonous mushrooms are used in medicine.

Ascomycota (Sac Fungi)

Lichen-Formers

- The fungal hyphae form a dense protective layer that shields the photosynthetic species; in return, the cyanobacterium or alga provides carbohydrates to the fungus.

- Lichens dominate the Arctic and Antarctic tundras. They are a major source of food, and they break down minerals from rocks—the first step in soil formation.

Non-Lichen-Formers

- Most are saprophytic, but parasitic forms exist.
- Some species are capable of growing on jet fuel or paint, and they can be used along with bacteria in efforts to clean up contaminated sites.
- *Penicillium* is an important source of antibiotics.

✔ CHECK YOUR UNDERSTANDING

(d) What is the significance of the microsporidians belonging within the fungi?

_____

_____

_____

B. MASTERING CHALLENGING TOPICS

It is usually difficult to understand, initially, the different associations between fungi and other organisms. Ectomycorrhizal fungi (EMF) and arbuscular mycorrhizal fungi (AMF) are often confused in the student's mind, and it is even harder to explain examples once the terms have been confused. If you focus on the word *ecto* as meaning "outside," this should at least help you to get the two terms organized properly. Even though both may appear to penetrate the root tissue, EMF do not penetrate the root cell wall; AMF actually do penetrate the root cell. Once you have learned this concept, pick a detailed example of each type and work through it. By understanding an example of each fungi and being able to explain it to a classmate, lab partner, or friend, you will begin to master the material.

C. ASSESSING WHAT YOU'VE LEARNED

1. The associations between fungi and the roots of land plants are called:
   a. saprophytic associations.
   b. parasitic associations.
   c. mycorrhizal associations.
   d. mycelial associations.

2. Saprophytes are fungi that make their living by:
   a. digesting dead plant material.
   b. digesting live plant material.
   c. living inside the roots of land plants.
   d. living on the leaves of vascular plants.

3. The filaments that make up the multicellular fungi are called:
   a. yeasts.
   b. hyphae.
   c. septa.
   d. gametes.

4. Which of the following is *not* one of the four major groups of fungi?
   a. Zygomycota
   b. Chitinomycota
   c. Basidiomycota
   d. Ascomycota

5. The microsporidians are:
   a. a monophyletic group within the fungi.
   b. a paraphyletic group within the fungi.
   c. the closest living relatives to the fungi.
   d. the closest extinct relatives to the fungi.

6. Which of the following symbiotic relationships offers benefits to *both* the host and the invader?
   a. Parasitism
   b. Commensalism
   c. Mutualism
   d. All of the above

7. Which of these items best describes ectomycorrhizal fungi?
   a. A mutualistic relationship where the fungi grow into the cells of plant roots
   b. A mutualistic relationship where the fungi grow around the plant roots
   c. A commensal relationship where the fungi grow into the cells of plant roots
   d. A commensal relationship where the fungi grow around the plant roots

8. How do fungi digest their food?
   a. They ingest particles and break them down.
   b. They absorb complex molecules directly.
   c. They have a rudimentary gut.
   d. They secrete enzymes extracellularly onto their food and then absorb the broken-down products.

9. Fungi have the ability to break down:
   a. carbohydrates.
   b. cellulose.
   c. lignin.
   d. all of the above

10. When nuclei of two different individual fungi fuse, this is called:
    a. karyogamy.
    b. plasmogamy.
    c. nuclear fusion.
    d. homokaryotism.

CHAPTER 31—ANSWER KEY

Check Your Understanding

(a) Mushrooms are consumed in many cultures. The yeast *S. cerevisiae* is essential in the manufacturing of bread, soy sauce, cheese, beer, wine, whiskey, and other products. Enzymes derived from fungi are being used to improve the characteristics of foods ranging from fruit juice and candy to meat.

(b) Fungi likely evolved from an aquatic ancestor. Fungi are more closely related to animals than to plants. Most animals and fungi produce chitin. Also, the flagella that develop in chytrid spores and gametes are very similar to those observed in animals. Finally, glycogen is used as an energy storage molecule in both animals and fungi.

(c) Cellulases are secreted into the extracellular environment by fungi. Some cleave long strands of cellulose into the disaccharide called cellobiose. Other cellulases are equally specific and, together, they eventually convert cellulose into glucose.

(d) This was an important result because microsporidians cause serious disease in bee colonies, silkworm colonies, and people with AIDS. With this evidence, perhaps fungicides may prove helpful in combating these microsporidian infections.

Assessing What You've Learned

1. c; 2. a; 3. b; 4. b; 5. a; 6. c; 7. b; 8. d; 9. d; 10. a

Looking Forward—Key Concepts in Later Chapters

Emergence—Chapter 35

Like the emerging viruses in **Chapter 35**, fungi that recently switched host organisms show the same outbreak trend.

An Introduction to Animals

Embryonic Tissue Layers—Chapter 22

The basic layers of embryonic tissues and their different paths of development are explained in **Chapter 22** and utilized in Chapter 32 to characterize the evolution of animals.

Homologous Structures—Chapter 24

These structures have been traced to a common ancestor and used to identify points of origin on an evolutionary tree.

Natural Selection—Chapter 24

Natural selection is explained in detail in **Chapter 24** and is applied to the many evolutionary processes of the animal kingdom in Chapter 32. The prevalence of this term in this section should indicate its importance.

KEY CONCEPTS

- Animals are multicellular, heterotrophic eukaryotes that lack cell walls and ingest their prey.

- Fundamental changes in morphology and development occurred as animals diversified.

- The four major groups of animals are non-bilaterian lineages, two protostome groups (Lophotrochozoa and Ecdysozoa), and the deuterostomes.

- Within major groups, diversification was based on innovative ways of sensing the environment, feeding, and moving.

- Methods of sexual reproduction vary widely among animal groups, and many species can reproduce asexually. Metamorphosis is also common in some animal groups.

A. CHAPTER OUTLINE

32.1 Why Do Biologists Study Animals?

- Animals are heterotrophs and are the dominant consumers and predators in both aquatic and terrestrial habitats.

- Animals are a particularly species-rich and morphologically diverse lineage of multicellular organisms on the tree of life.

- Humans depend on wild and domesticated animals for food.

- Efforts to understand human biology depend on advances in animal biology.

32.2 How Do Biologists Study Animals?

Analyzing Comparative Morphology

- The origin and early evolution of animals were based on four aspects: tissues, the nervous system, a body cavity, and early development.

The Origin and Diversification of Tissues

- The number of tissue layers that exists in an embryo helps to identify body plans. A **tissue** is a highly organized and functionally integrated group of cells.

- Although sponges lack many types of tissues, they do have an **epithelium**—a layer of tightly joined cells that covers the surface.

- **Diploblasts** are animals whose embryos have two types of tissues. Only two groups of diploblastic animals are alive today: the cnidarians and ctenophores.

- **Triploblasts** are animals whose embryos have three types of tissues. In triploblasts, the **ectoderm** gives rise to the skin and nervous system, the **endoderm** gives rise to the digestive tract, and the **mesoderm** gives rise to the circulatory system, muscle, and internal structures.

Nervous Systems, Body Symmetry, and Cephalization

- A basic feature of a multicellular body is whether it has a plane of symmetry. All animals except sponges exhibit either **radial** or **bilateral** symmetry. Animals with radial symmetry have at least two planes of symmetry. Bilaterally symmetric organisms face their environment in one direction.

- This structure allowed **cephalization**—the evolution of a head, where structures for feeding, sensing the environment, and processing information are concentrated.

Why Was the Evolution of a Body Cavity Important?

- A body cavity creates a medium for circulation, along with space for internal organs. Fluid-filled chambers are central to the operation of a **hydrostatic skeleton** (**Figure 32.6**). This gave bilaterally symmetrical organisms the ability to move efficiently.

- Diploblasts do not have a body cavity, and neither does the triploblast phylum of flatworms.

- Roundworms and rotifers have a **pseudocoelom** that forms between the endoderm and mesoderm.

- In all other triploblasts, the body cavity forms from within the mesoderm itself and is lined with cells from the mesoderm. Muscle and blood vessels can then form on either side of the body cavity or **coelom**.

What Are the Protostome and Deuterostome Patterns of Development?

- Except for the echinoderms, all true coelomates are bilaterally symmetric and have three embryonic tissue layers. This huge group can be divided into **protostomes** (arthropods, mollusks, and segmented worms) and **deuterostomes** (vertebrates and echinoderms).

- In protostomes, the early cell division pattern is spiral cleavage; deuterostomes show radial cleavage.

- **Gastrulation** is a series of cell movements that results in the formation of the three embryonic layers. This invagination of cells creates a pore that opens to the outside (**Figure 32.7a**). In **protostomes** this pore becomes the mouth, and the anus forms later in development. In **deuterostomes** this initial pore becomes the anus, and the mouth forms later in development.

- Protostomes have two blocks of mesoderm beside the gut; deuterostomes have layers of mesodermal cells located on either side of the gut.

The Tube-within-a-Tube Design

- Over 99 percent of animal species alive are bilaterally symmetric triploblasts that have coeloms.

- The outer tube forms the body wall, and the inner tube is the animal's gut. The mesoderm in between forms the muscles and organs.

CHECK YOUR UNDERSTANDING

(a) Describe the two main patterns of organization of embryonic tissues.

_____

_____

_____

_____

Evaluating Molecular Phylogenies

- The phylogenetic tree is based on genes for ribosomal RNA and several proteins (**Figure 32.9**). Several important observations emerge from the data:

  1. A group of protists called the choanoflagellates are the closest living relatives of the animals.

  2. Sponges are the sister group to all other animals and they are paraphyletic.

  3. Ctenophora and Cnidaria are a monophyletic group.

  4. The endoderm and ectoderm were the first embryonic tissues to evolve, and radial symmetry evolved before bilateral symmetry.

  5. The most ancient group of bilaterally symmetric triploblasts (the Acoelomorpha) lacked a coelom entirely. The mesoderm preceded the evolution of a coelom.

  6. The major event in the evolution of the Bilateria was the split between protostomes and deuterostomes.

  7. A fundamental split occurred within the protostomes: (1) the Ecdysozoa and (2) the Lophotrochozoa.

  8. Flatworms are protostomes and acoelomate, but this trait is derived, meaning they evolved from a coelomate ancestor.

  9. Segmentation evolved several times independently in the bilaterians.

(b) What is the difference between radial and bilateral symmetry?

_____

_____

_____

_____

32.3 What Themes Occur in the Diversification of Animals?

- Innovating methods for sensing the environment, feeding, and moving triggered the diversification of species within many animal lineages.

Sensory Organs

Variation in Sensory Abilities

- Birds and sea turtles can sense the Earth's magnetic field to aid in navigation.
- Some aquatic predators can sense electric fields.
- Many birds can sense barometric pressure in order to avoid storms.

Feeding

How Animals Feed: Four General Tactics

- **Suspension feeders** feed by filtering out food particles suspended in water. Cilia on gills pump water in one siphon and out the other. Food particles are trapped by the gills and swept toward the mouth by cilia.
- **Deposit feeders** eat their way through a substrate. For example, earthworms swallow soil as they move through it and leave mineral material behind as feces.
- Food for deposit feeders consists of soil-dwelling bacteria, protists, fungi, and archaea, along with **detritus**—the dead and partially decomposed remains of organisms.
- **Fluid feeders** suck or mop up liquids such as nectar, plant sap, blood, or fruit juice.
- They often have mouthparts that allow them to pierce a structure so they can withdraw the fluids inside.
- **Mass feeders** take chunks of food into their mouths.

- The structure of the mouthparts correlates with the type of food pieces that are harvested or ingested.

What Animals Eat: Three General Sources

- **Herbivores** are animals that digest algae or plant tissues. **Carnivores** are animals that feed on animals. **Detritivores** are animals that feed on dead organic matter.
- Herbivores and carnivores can be subclassified as predators or parasites.
- **Predators** kill other organisms for food by using an array of mouthparts and hunting strategies.
- **Parasites** are much smaller than their victims and often harvest nutrients without causing death.
- **Endoparasites** live inside their hosts. They are often wormlike in shape and can be extremely simple morphologically. The tapeworms found inside humans have no digestive system; however, most endoparasites ingest their food and have a digestive tract.
- **Ectoparasites** live outside their hosts. They usually have grasping mouthparts that allow them to pierce their host's exterior and suck the nutrient-rich fluids inside. A louse is an example of an insect ectoparasite (**Figure 32.16b**).

Movement

Types of Limbs: Unjointed and Jointed

- Unjointed limbs (such as in velvet worms) are sac-like.
- Jointed limbs move when muscles that are attached to a skeleton contract or relax.
- Ecdysozoans have an exoskeleton, and vertebrates have an endoskeleton.

Are All Animal Appendages Homologous?

- Biologists have concluded that at least a few of the same genes are involved in the development of _all_ appendages observed in animals.
- The current hypothesis is that all animal appendages have some degree of genetic homology, and that they are all derived from appendages that were present in a common ancestor.

Reproduction

- Animal reproduction is extremely diverse.
- Animals can reproduce sexually or asexually.
- Sexual reproduction can occur via internal or external fertilization.
- Eggs (**ovoviviparous**) or embryos (**viviparous**) can be retained in the female's body during development, or eggs may be laid outside the body (**oviparous**).

Life Cycles

- **Metamorphosis** is the change from juvenile to adult body type and one of the most spectacular innovations in life cycles.
- **Hemimetabolous** development is a subtle change from metamorphosis; it refers to the limited morphological difference between juvenile and adult. It is a one-step process of sexual maturation. Grasshoppers undergo hemimetabolism as a series of molts gradually change it from a wingless immature juvenile (**nymph**) to a sexually mature adult capable of flight.
- **Holometabolous** development is a two-step process, from larvae to pupae to adult, involving dramatic changes in morphology and habitat use. This process is found in 10 times as many species as the hemimetabolous process.

CHECK YOUR UNDERSTANDING

(c) What are the various types of metamorphosis in insects, and how do they differ from each other?

_____

_____

_____

_____

_____

32.4 Key Lineages of Animals: Non-bilaterian Groups

Porifera (Sponges)

- All are suspension feeders, and although the adults are sessile, larvae can swim with the aid of flagella.
- Reproduction can be asexual or sexual.
- Sponge cells are totipotent, meaning that an isolated cell has the capacity to develop into a complete adult.

Cnidaria (Jellyfish, Corals, Anemones, Hydroids)

- Cnidarians are radially symmetric diploblasts consisting of ectoderm and endoderm layers that sandwich gelatinous material known as mesoglea.
- The gut has only one opening. A specialized cell, a **cnidocyte**, is used for prey capture.
- Most life cycles have a sessile polyp form and a mobile medusa.

- Polyps can reproduce asexually by budding, fission, or fragmentation. During sexual reproduction, fertilization takes place in open water.

Ctenophora (Comb Jellies)

- Ctenophores are transparent, ciliated, gelatinous diploblasts that live in marine habitats.
- They are predators, and adults move via beating of cilia.
- Most species have male and female organs and routinely self-fertilize, though fertilization is external.

Acoelomorpha

- These animals lack a coelom. They are bilaterally symmetric worms that have distinct anterior and posterior ends and are triploblastic.
- They feed on detritus and prey on small animals.
- They can reproduce asexually by fission or budding. Sexual reproduction is also possible; fertilization is internal, and fertilized eggs are laid outside the body.

CHECK YOUR UNDERSTANDING

(d) Organize the basal lineages of animals into their tissue types and symmetry types.

_____

_____

_____

_____

_____

B. MASTERING CHALLENGING TOPICS

In this chapter, the terminology referring to body tissue layers and body cavity organizations can be confusing, especially when attempting to match these classifications with different animal groups. If these terms are not mastered during the study of this chapter, it will make the next two chapters on animals even more confusing. Try to study the tissue layers first (endoderm, ectoderm, and mesoderm); next, list the animal groups that are diploblastic and triploblastic. Then define coelom, acoelomate, and pseudocoelom and try to list animal groups that match these terms. Finally, try to combine both sets and groups to get an overall summary, perhaps in a chart or other diagram. Explaining the trend to a classmate will help you organize your thoughts and prepare you for incorporating these terms into the next two chapters.

C. ASSESSING WHAT YOU'VE LEARNED

1. Diploblasts and triploblasts are terms that describe the number of:
   a. invaginations during development.
   b. head regions during development.
   c. tissue layers during development.
   d. cell types during development.

2. In triploblasts, the body cavity or coelom forms:
   a. from within the mesoderm.
   b. between the endoderm and mesoderm.
   c. between the ectoderm and mesoderm.
   d. from within the endoderm.

3. During gastrulation, the initial pore eventually forms the mouth; this is a developmental process that defines the lineage as:
   a. the coelomates.
   b. the deuterostomes.
   c. the protostomes.
   d. the gastrosomes.

4. What is the earliest-branching lineage of animals?
   a. The insects
   b. The sponges
   c. The echinoderms
   d. The chordates

5. Animals that eat and digest algae or plant materials are called:
   a. herbivores.
   b. carnivores.
   c. detritivores.
   d. parasites.

6. Which parasites live inside their hosts?
   a. Endoparasites
   b. Ectoparasites
   c. Mesoparasites
   d. None of the above

7. The current hypothesis regarding animal limbs is that
   a. animal limbs are not homologous structures.
   b. all animal limbs have some degree of genetic homology.
   c. animal limbs all perform the same function.
   d. animal limbs all arose at the same time.

8. Which specific term describes the change from larvae to pupa to adult?
   a. Hemimetabolous
   b. Metabolism
   c. Heterometabolous
   d. Homometabolous

9. A cell such as a sponge cell can be isolated and form a complete adult. The term used to describe this cell is:
   a. omnipotent.
   b. totipotent.

c. heteropotent.
d. allipotent.

10. The motile form of the Cnidaria is called the
    a. medusa.
    b. polyp.
    c. cnidocyte.
    d. mesoglea.

CHAPTER 32—ANSWER KEY

Check Your Understanding

(a) Diploblasts are animals whose embryos have two types of tissues. Only two groups of diploblastic animals are alive today: the cnidarians and ctenophores. Triploblasts are animals whose embryos have three types of tissues. In triploblasts, the ectoderm gives rise to the skin and nervous system, the endoderm gives rise to the digestive tract, and the mesoderm gives rise to the circulatory system, muscle, and internal structures.

(b) A basic feature of a multicellular body is whether it has a plane of symmetry. All animals exhibit either radial or bilateral symmetry. Animals with radial symmetry have at least two planes of symmetry. Bilaterally symmetric organisms face their environment in one direction.

(c) Metamorphosis is the change from juvenile to adult body type. This change can either be subtle or spectacular. Hemimetabolous development is a subtle change from metamorphosis; its name refers to the limited morphological difference between juvenile and adult. It is a one-step process of sexual maturation. Grasshoppers undergo this process as a series of molts gradually change it from a wingless, immature juvenile to a sexually mature adult capable of flight. Holometabolous development is a two-step process, from larvae to pupae to adult, involving dramatic changes in morphology and habitat use. This is found in 10 times as many species as the hemimetabolous process.

(d) The Porifera are mostly asymmetrical and have only epithelial tissues; therefore, they lack the tissue organization of other animals. The Cnidaria are radially symmetric diploblasts. The Ctenophora are bilaterally symmetric diploblasts. The Acoelomorpha are bilaterally symmetric triploblasts.

Assessing What You've Learned

1. c; 2. a; 3. c; 4. b; 5. a; 6. a; 7. b; 8. d; 9. b; 10. a

Looking Forward—Key Concepts in Later Chapters

Mechanisms of Movement—Chapter 41

Movement is used to explain the diversity of feeding strategies used in animals. **Chapter 41** focuses on the mechanisms of animal movement in detail.

Reproduction—Chapter 48

Chapter 32 begins to describe many reproductive strategies of animals. **Chapter 48** goes into further detail and focuses on vertebrate reproduction.

Protostome Animals

KEY CONCEPTS

- 🔑 Protostomes are a monophyletic group divided into two major subgroups: the Lophotrochozoa and the Ecdysozoa.

- 🔑 Although many protostomes have wormlike bodies, the Mollusca and Arthropoda have complex and distinctive features.

- 🔑 Mollusks and Arthropods inhabit a wide range of environments.

- 🔑 Key events triggered the diversification of protostomes: several independent water-to-land transitions, a diversification in appendages and mouthparts, and the evolution of metamorphosis.

A. CHAPTER OUTLINE

33.1 An Overview of Protostome Evolution

- Phylogenetic studies have long supported the hypothesis that protostomes are a monophyletic group.

What Is a Lophotrochozoan?

- The 13 phyla of lophotrochozoans include the mollusks, annelids, and flatworms.

- The name came from two structures found in some but not all of the phyla in the lineage.

- A **lophophore** is a specialized structure that rings the mouth and specializes in suspension feeding. It consists of tentacles that have ciliated cells on their surface.

- **Trochophores** are a type of larvae common to marine mollusks, annelids that live in the ocean, and several other oceanic phyla. The larvae have a ring of cilia around their middle that aids in locomotion.

What Is an Ecdysozoan?

- Ecdysozoans have an exoskeleton and grow by molting.

- The most prominent phyla in this group are the roundworms and the arthropods.

✓ CHECK YOUR UNDERSTANDING

(a) Compare and contrast lophotrochozoans with ecdysozoans.

_____

_____

_____

_____

_____

_____

33.2 Themes in the Diversification of Protostomes

How Do Body Plans Vary among Phyla?

- All protostomes are triploblastic and bilaterally symmetric, and all of them undergo embryonic development in a similar way.
- Radical changes occurred in coelom formation as protostomes diversified.

The Arthropod Body Plan

- Arthropods have segmented bodies that are organized into prominent regions called **tagmata**.
- Arthropod locomotion is based on muscles that apply force against an exoskeleton.
- Arthropods have a spacious body cavity called the **hemocoel** that provides space for internal organs and circulation of fluids.

The Molluscan Body Plan

- The **foot** is a large muscle that is located at the base of the animal and is usually used in movement.
- The **visceral mass** is the region that contains the main internal organs and external gill.
- The **mantle** is a tissue layer that covers the visceral mass and that in many species secretes a shell made of calcium carbonate.

Variation among Body Plans of the Wormlike Phyla

- The body plan is extremely similar in these phyla, and they are distinguished by specialized mouthparts.
- **Annelids** include a group called the Echiurans (spoon worms) that burrow into marine mud and feed using an extended structure called a **proboscis**.

- **Priapulids** (penis worms) burrow in substrate but act as sit-and-wait predators. This worm everts its toothed, cuticle-lined throat to grab prey.
- **Nemerteans** (ribbon worms) are active predators; they have a spear-like proboscis that can extend or retract.

The Water-to-Land Transition

- It is clear that water-to-land transitions occurred multiple times as protostomes diversified.
- The following are a few adaptations that allowed protostomes to exchange respiratory gases and avoid drying out:
  1. Roundworms and earthworms live in moist soil and exchange gases across their high-surface-area skin.
  2. Arthropods and many mollusks have gills or other respiratory structures inside their bodies.
  3. Insects have a waxy layer to prevent water loss.

Adaptations for Feeding

- Various feeding strategies are reflected in the wide diversity of mouthparts.
- Some protostomes have jaws and mouth; others have a proboscis. Arthropods show the most diversity—they can pierce, suck, grind, bite, mop, chew, engulf, cut, and mash (**Figure 33.8**).

Adaptations for Moving

- Variation in movement depends on (1) the presence or absence of limbs and (2) the type of skeleton that is present.
- A number of important evolutionary innovations:
  1. The evolution of the jointed **limb** allowed Arthropod movement in these unique ways.
  2. The insect wing is one of the most important adaptations in the history of life.
  3. In mollusks, waves of muscle contractions sweep down the length of the large, muscular foot, allowing individuals to glide.
  4. Cephalopods fill a cavity surrounded by their mantle and expel water through a siphon to propel them through the water.

Adaptations in Reproduction

- Regarding variation in protostome life cycles, the most basic issue is whether a juvenile's body is rearranged during metamorphosis to form an adult.
- Reproduction is highly variable, from sexual to asexual. Many crustacean species can reproduce asexually via **parthenogenesis**.

✔ **CHECK YOUR UNDERSTANDING**

(b) Describe a few unique ways of protostome movement.

_____

_____

_____

_____

_____

33.3 Key Lineages: Lophotrochozoans

Rotifera (Rotifers)

- Rotifers have a cluster of cilia at their anterior called a **corona**, which they use for suspension feeding.
- Females produce unfertilized eggs by mitosis; the eggs then hatch into new, asexually produced individuals (parthenogenesis).

Platyhelminthes (Flatworms)

- This phylum includes (1) the free-living flatworms, (2) the endoparasitic tapeworms, and (3) the endo- or ectoparasitic flukes.
- Flatworms are unsegmented, do not have a coelom, and have no circulatory system. Their high surface-to-volume ratio allows for efficient absorption of nutrients from the host.
- In many cases, the life cycles of tremadotes and cestodes involve two or three distinct host species.

Annelida (Segmented Worms)

- Annelids have a segmented body plan and a coelom that functions as a hydrostatic skeleton.
- They are divided into two major subgroups:
  1. The Polychaeta are named for their numerous, bristle-like extensions called chaetae, which extend from appendages called parapodia.
  2. The Clitellata are comprised of the oligochaetes (which include earthworms) and leeches.
- Annelids can crawl, burrow, or swim using their hydrostatic skeletons.

Mollusca (Mollusks)

- This is by far the most species-rich and morphologically diverse group in the Lophotrochozoa.
- They have a specialized body plan based on a muscular foot, a visceral mass, and a mantle that may or may not secrete a calcium carbonate shell.

Bivalvia (Clams, Mussels, Scallops, Oysters)

- Bivalves have two separate shells, which are hinged.
- They are suspension feeders and use gills for respiration.
- Most are sessile, but some can move using their muscular foot.
- Only sexual reproduction occurs. Externally fertilized eggs develop into trochophore larvae. They metamorphose into a **veliger**, which continues to feed before metamorphosing into the adult form.

Gastropoda (Snails, Slugs, Nudibranchs)

- Gastropods are named for the large, muscular foot on their ventral side.
- In many species the **radula** functions like a rasp to scrape away algae.
- Some species can reproduce via parthenogenesis, but most reproduction is asexual.

Polyplacophora (Chitons)

- These animals are named for the eight calcium carbonate plates along their dorsal side.
- They are marine and feed using a radula. They move using a muscular foot, just as in the gastropods.
- Sexes are separate and fertilization is external.

Cephalopoda (Nautilus, Cuttlefish, Squid, Octopuses)

- Cephalopods have a well-developed head and a foot that is modified to form long, muscular tentacles.
- Except for the nautilus, most have highly reduced shells or none at all.
- They are highly intelligent predators that hunt by sight. They have a radula and a beak for biting.
- They have separate sexes and highly elaborate courtship rituals. Males deposit sperm encased in a spermatophore.

✔ **CHECK YOUR UNDERSTANDING**

(c) Describe one unique structure within the group Mollusca.

_____

_____

_____

_____

_____

33.4 Key Lineages: Ecdysozoans

Nematoda (Roundworms)

- Nematodes are unsegmented worms with a pseudocoelom.
- Several species are common parasites of humans. Pinworms infect about 40 million people in the United States.
- Sexes are separate, and asexual reproduction is rare or unknown. Internal fertilization is used with egg laying and direct development.

Arthropoda (Arthropods)

- Arthropods are easily the most successful lineage of eukaryotes.
- The body is organized into a distinct head and trunk. A **compound eye** contains many light-sensing structures, each with its own lens.
- Most arthropods have a pair of antennae on their head.

Myriapods (Millipedes, Centipedes)

- Myriapods have relatively simple bodies with a series of short segments.
- They have mouthparts that can bite and chew. Millipedes are detritivores, and centipedes are predators.
- Sexes are separate, and fertilization is internal. Males deposit sperm packets that are picked up by the female.

Insecta (Insects)

- Insects are distinguished by having a **head, thorax**, and **abdomen.** In most species, two pairs of wings are mounted on the dorsal side of the thorax.
- Mating usually takes place through direct copulation.

Chelicerata (Spiders, Ticks, Mites, Horseshoe Crabs, Daddy Longlegs, Scorpions)

- This group is named for appendages called **chelicerae** found near the mouth.
- The most prominent group is the arachnids (spiders, scorpions, mites, and ticks).
- Spiders, scorpions, and daddy longlegs capture and sting their prey. Mites and ticks are ectoparasitic.
- Sexual reproduction via internal fertilization and direct development is the most common in this group.

Crustaceans (Shrimp, Lobster, Crabs, Barnacles, Isopods, Copepods)

- The segmented body of the crustacean is divided into the cephalothorax (which combines the head and thorax) and the abdomen.
- Many crustaceans have a **carapace**—a platelike section of their exoskeleton that covers and protects the cephalothorax.

- Crabs and lobsters have **mandibles** that can bite or chew.
- Sexual reproduction via internal fertilization is the norm.
- Many species include a larval stage called a **nauplius**, which is usually planktonic.

✔ CHECK YOUR UNDERSTANDING

(d) Compare and contrast the body plan of insects and crustaceans.

_____

_____

_____

_____

_____

B. MASTERING CHALLENGING TOPICS

The most difficult topic in this chapter is the organization of all the key lineages of protostomes. The sheer number of phyla and their characteristics present a large body of information that can be incredibly confusing to remember and recall on a test. Although strict memorization can get the job done, it can often yield less-than-desirable results on certain types of tests. The preferred way to learn this material is to take the time to outline the major divisions of the lineages first, starting with what makes a protostome a protostome. Then, distinguish the Lophotrochozoans from the Ecdysozoans, even if you can name only a few things that are different. Last, begin to fill in the key phyla under each group, starting with the easiest to remember. Looking at pictures and explaining characteristics to classmates will also help you in learning all of these key lineages.

C. ASSESSING WHAT YOU'VE LEARNED

1. Rotifers have a cluster of cilia at their anterior called a:
   a. pore.
   b. radula.
   c. corona.
   d. lophophore.

2. The group Lophotrochozoans is so named because:
   a. they all have a feeding structure called a lophophore.
   b. they all have a type of larvae called a trochophore.
   c. they are all wormlike.
   d. Answers a and b are both correct.

3. Platyhelminthes include:
   a. free-living flatworms.
   b. endoparasitic tapeworms.
   c. endoparasitic and ectoparasitic flukes.
   d. all of the above

4. The Annelida are divided into two major subgroups called:
   a. Polyplacophora and Monoplacophora.
   b. Polychaeta and Polyplacophora.
   c. Polychaeta and Clitellata.
   d. Polychaeta and Mollusca.

5. The most species-rich and morphologically diverse group of Lophotrochozoa are:
   a. Mollusca.
   b. Annelida.
   c. Nematoda.
   d. Polyplacophora.

6. Which of the following is *not* a Bivalvia?
   a. Clam
   b. Mussel
   c. Snail
   d. Scallop

7. The most intelligent of the mollusks are the:
   a. Bivalvia.
   b. Gastropoda.
   c. Cephalopoda.
   d. Polyplacophora.

8. Which of the following are unsegmented worms with a pseudocoelom?
   a. Nematodes
   b. Annelids
   c. Cephalopods
   d. Platyhelminthes

9. Spiders belong to the major group called:
   a. Myriapods.
   b. Chelicerata.
   c. Insecta.
   d. Crustaceans.

10. The anterior region of the Crustaceans is called the:
    a. thorax.
    b. head.
    c. cephalum.
    d. cephalothorax.

CHAPTER 33—ANSWER KEY

Check Your Understanding

(a) The two monophyletic groups of protostomes are the Lophotrochozoa and the Ecdysozoa. Lophotrochozoa are characterized by two structures found in some but not all of the phyla in the lineage. A **lophophore** is a specialized structure, consisting of tentacles with ciliated cells on their surface, that rings the mouth and specializes in suspension feeding. **Trochophores** are a type of larvae common to marine mollusks, annelids that live in the ocean, and several other oceanic phyla. The larvae have a ring of cilia around their middle that aids in locomotion. In contrast, ecdysozoans have an exoskeleton and grow by molting.

(b) Difference in body movement mechanisms depends on (1) the presence or absence of limbs and (2) the type of skeleton that is present. The evolution of the joined limb allowed movement in these unique ways, along with others:
   - The insect wing is one of the most important adaptations in the history of life.
   - In mollusks, waves of muscle contractions sweep down the length of the large, muscular foot, allowing individuals to glide.
   - Squid fill a cavity surrounded by their mantle and expel water through a siphon to propel themselves through the water.

(c) Bivalves have two separate shells and metamorphose into a veliger stage. Gastropods, polyplacophorans, and cephalopods all have a radula for feeding.

(d) Insects and crustaceans both have an exoskeleton. Insects have a head, thorax, and abdomen while crustaceans have a fused cephalothorax and an abdomen.

Assessing What You've Learned

1. c; 2. d; 3. d; 4. c; 5. a; 6. c; 7. c; 8. a; 9. b; 10. d

Looking Forward—Key Concepts in Later Chapters

Deuterostome Development—Chapter 34

The protostomes develop in a dramatically different way than the other major lineage of bilaterians, the deuterostomes, do.

Mechanisms of Movement—Chapter 41

Movement is used to explain the diversity of feeding strategies used in animals. **Chapter 41** focuses on the mechanisms of animal movement in detail.

Gas Exchange—Chapter 44

In the transition to land, animals had to solve the problem of exchanging gas and preventing drying out. As **Chapter 44** will detail, land animals exchange gases with the atmosphere readily as long as a moist body surface is exposed to air.

Deuterostome Animals

KEY CONCEPTS

- Echinoderms are radially symmetric as adults and have a water vascular system and an endoskeleton.

- Echinoderms are among the most important predators and herbivores in marine environments.

- All vertebrates have a skull and an extensive endoskeleton. Their diversification was driven by the evolution of jaws and limbs.

- Vertebrates are the most important large-bodied predators and herbivores in marine and terrestrial environments.

- Humans and chimpanzees diverged from a common ancestor that lived in Africa 6 to 7 million years ago. Since then, at least 14 humanlike species have existed.

A. CHAPTER OUTLINE

34.1 What Is an Echinoderm?

The Echinoderm Body Plan

- Although all deuterostomes are bilaterally symmetric, echinoderms reverted to a type of radial symmetry called pentaradial symmetry. Recall that echinoderm larvae are bilaterally symmetric.

- Another unique feature of the enchinoderms is their **water vascular system** and associated **tube feet** and **podia.**

- Radial symmetry as adults, an endoskeleton of calcium carbonate, and the water vascular system are all synapomorphies.

How Do Echinoderms Feed?

- Echinoderms suspension feed, deposit feed, or harvest algae or other animals. In most cases, the podia play a key role in obtaining food.

Key Lineages

Asteroidea (Sea Stars)

- They have star-shaped bodies with five or more long arms radiating from a central region that contains the mouth, stomach, and anus (**Figure 34.6**).

- Sea stars are predators or scavengers. Some species can pull apart bivalves.

Echinoidea (Sea Urchins and Sand Dollars)

- Sea urchins have globular bodies and long spines, and they crawl along substrates (**Figure 34.7a**). Most are herbivores of algae.

- Sand dollars are flattened and disk-shaped, have short spines, and burrow in soft sediments (**Figure 34.7b**). They use mucus-covered podia to collect food particles in the sand.

CHECK YOUR UNDERSTANDING

(a) Describe the structure and function of the water vascular system in echinoderms.

_____

_____

_____

_____

34.2 What Is a Chordate?

- This phylum is defined by the presence of these traits:
  1. Pharyngeal gill slits
  2. A bundle of nerve cells that runs the length of the body and forms a **dorsal hollow nerve cord**
  3. A stiff and supportive but flexible rod called a **notochord**
  4. A muscular, **post-anal tail**

Three "Subphyla"

1. **Urochordates** are also called tunicates. The pharyngeal gill slits function in suspension feeding, while the notochord functions as a simple endoskeleton.

2. **Cephalochordates** are also called lancelets or amphioxus. They are small, mobile suspension feeders that look like a little fish. Their notocords stiffen their bodies to aid in swimming.

3. **Vertebrates** include sharks, several lineages of fishes, amphibians, reptiles (including birds), and mammals. The dorsal hollow nerve cord is elaborated into the spinal cord. The notochord helps organize the body plan during embryonic development.

Key Lineages: The Invertebrate Chordates

Cephalochordata (Lancelets)

- The notochord is retained in adult lancelets, where it functions as an endoskeleton.

- They take water in through the mouth and trap food particles in their pharyngeal gill slits.

Urochordata (Tunicates)

- Tunicates have an exoskeleton-like coat of polysaccharide, called a tunic, a U-shaped gut, and two openings called siphons.

- They use their pharyngeal gill slits to suspension feed.

CHECK YOUR UNDERSTANDING

(b) What are the major similarities and differences between chordates and vertebrates?

_____

_____

_____

_____

34.3 What Is a Vertebrate?

- Vertebrates are a monophyletic group distinguished by two traits: **vertebrae** and a **cranium**.

- The vertebrae protect the spinal column, and the cranium protects the brain and sensory organs.

An Overview of Vertebrate Evolution

The Vertebrate Fossil Record

- About 540 million years ago, the earliest vertebrates had streamlined bodies and appeared to have skulls made of **cartilage**.

- In the Ordovician period (480 mya), the first fossils to contain **bone** appear. They looked like plates (exoskeleton) and were presumably used as armor.

- First vertebrates with **jaws** appear about 440 mya. This gave vertebrates the ability to bite and chew.

- The transition to land is dated at about 365 mya. These were the first **tetrapods**—animals with four limbs.
- About 20 mya, the first amniotes were present. An **amniotic egg** is one with a watertight shell or case enclosing a membrane-bound food supply, a water supply, and a waste repository.

Evaluating Molecular Phylogenies

- **Figure 34.12** provides a phylogenetic tree based on morphology and DNA.
- The closest living relatives of the vertebrates are the cephalochordates.
- The most basal groups of chordates lack the skull and vertebral column that define the vertebrates, and the most ancient lineage of vertebrates lacks jaws and bony skeletons.

✓ **CHECK YOUR UNDERSTANDING**

(c) What did bone first look like in the fossil record of vertebrates?

_____

_____

_____

_____

Key Innovations

The Vertebrate Jaw

- The leading hypothesis is that natural selection acted on the gill arches by increasing their size and modifying their orientation.
- Three lines of evidence:
  1. Both gill arches and jaws consist of flattened bars of bony or cartilaginous tissue that hinge and bend forward.
  2. During development, the same population of cells gives rise to the muscles that move the jaws and the muscles that move gill arches.
  3. Both jaws and gill arches derive from specialized cells in the embryo—neural crest cells.

The Tetrapod Limb

- The hypothesis that tetrapod limbs evolved from fish fins is supported by fossil evidence as well as molecular genetic evidence.

- Once the limb evolved, natural selection elaborated it into structures used for running, gliding, crawling, burrowing, or swimming.

Feathers and Flight

- Birds are a part of the monophyletic group called dinosaurs.
- A series of adaptations:
  1. Most dinosaurs have a flat sternum but birds have a keeled sternum on which to attach flight muscles.
  2. Birds are extraordinarily light for their size. They have reduced the number of bones and made them hollow.
  3. Birds are endothermic and can remain active year-round.

The Amniotic Egg

- These eggs have shells that minimize water loss because the embryo develops inside.
- The egg contains a membrane-bound supply of water in a protein-rich solution called **albumen** (**Figure 34.19**). The embryo is enveloped in a membrane called the **amnion**. The **yolk sac** is the membranous pouch that contains the nutrients, and the **allantois** is the pouch that holds waste material. The **chorion** is the membrane that provides the surface for gas exchange.

The Placenta

- The mother retains the egg in her body. The placenta is an organ that is rich in blood vessels, and it facilitates the flow of oxygen and nutrients from mother to offspring (**Figure 34.20**).

Parental Care

- Parental care is any action by a parent that improves the ability of its offspring to survive—feeding, keeping young and/or eggs warm, and protection.
- The most extensive parental care is exhibited by mammals and birds.

✓ **CHECK YOUR UNDERSTANDING**

(d) Describe the main characteristics of the amniotic egg.

_____

_____

_____

_____

Key Lineages

Myxinoidea (Hagfish) and Petromyzontoidea (Lampreys)

- Hagfish and lampreys belong to independent lineages, although both groups lack jaws. Hagfish lack any sort of vertebral column, but lampreys have small pieces of cartilage along their length.
- Hagfish are scavengers and predators (**Figure 34.23a**) that deposit feed on the carcasses of dead fish and whales.
- Lampreys are ectoparasites—they attach to fish and suck blood and other body fluids. They are also anadromous—they spend their adult life in the ocean and swim up streams to breed.

Condrichthyes (Sharks, Rays, Skates)

- These species are distinguished by their cartilaginous skeleton. Most species are marine, and sharks are top predators of marine ecosystems.

Actinopterygii (Ray-Finned Fishes)

- These fish have fins supported by long, bony rods arranged in a ray pattern. They are the most ancient group of vertebrates that have a skeleton made of bone.
- They avoid sinking in the water with the aid of a gas-filled **swim bladder**.
- The most important subgroup of ray-finned fishes is the Teleosti, with 27,000 species.

Actinistia (Coelocanths) and Dipnoi (Lungfish)

- Although these are independent lineages, they are often grouped together and called **lobe-finned fishes**.
- They represent a crucial evolutionary link between the ray-finned fishes and the tetrapods.

Amphibia (Frogs, Salamanders, Caecilians)

- Adults of most species feed on land, but many species have to lay their eggs in water.
- The larvae undergo metamorphosis into the adult form, which is mostly carnivorous.

Mammalia

- All mammals have hair or fur and mammary glands for lactation.

Mammalia > Monotremata (Platypuses, Echidnas)

- The most ancient group of mammals living. They lay eggs and have lower metabolic rates than other mammals.

Mammalia > Marsupiala (Marsupials)

- Although females have a placenta, the young are born poorly developed. They crawl to the female's nipples, where they suck milk. They stay there until they become independent.

Mammalia > Eutheria (Placental Mammals)

- The six (of the 18) most species-rich orders are rodents, bats, insectivores, artiodactyls, carnivores, and primates.
- At birth, the young are much better developed than marsupials, although there is still a prolonged period of parental care.

Reptilia > Lepidosauria (Lizards, Snakes)

- Most lizards have well-developed jointed legs, but snakes are limbless, evolving from limbed ancestors.
- Most lay eggs, although a few give birth to live young that have hatched from the egg and begun development.

Reptilia > Testudinia (Turtles)

- Distinguished by a shell comprised of bony plates that fuse to the vertebrae and ribs.
- Their skulls are more similar to those of amphibians than to those of other reptiles. Although they lack teeth, the jawbone and lower skull form a bony beak.

Reptilia > Crocodilia (Crocodiles, Alligators)

- Only 21 species known. They have eyes and nostrils located on the top of their skulls and can sit underwater for extended periods.
- They are top predators and, although they are oviparous, parental care is extensive.

Reptilia > Aves (Birds)

- They are the only **endotherms** within this group. This means their metabolic rate and insulation rate are so high that they retain a high constant body temperature.
- Most birds are omnivores, and almost all species can fly.
- Birds are oviparous, but parental care is extensive.

✔ **CHECK YOUR UNDERSTANDING**

(e) What distinguishes the three orders of mammals?

_____

_____

_____

_____

34.4 The Primates and Hominins

The Primates

- The **prosimians** are composed of lemurs, tarsiers, pottos, and lorises. Most of these animals live in trees and are active at night.
- The **anthropoids** include the New World monkeys found in Central and South Americas, the Old World monkeys that live in Africa and tropical regions of Asia, and the great apes—orangutans, gorillas, chimpanzees, and humans.
- Primates are distinguished by having their eyes on the front of the face.
- The lineage of the great apes is called the **hominids**.
- Bipedalism is the synapomorphy that defines the **hominins**—a monphyletic group comprising *Homo sapiens* and more than a dozen extinct, bipedal relatives.

Fossil Humans

- The fossil record of humanlike hominids is not nearly complete, but it is rapidly improving. These species can be organized into four general groups (**Figure 34.38**):
    1. Three species of small apes called **gracile australopithecines** lived from 4.12 to 2.4 million years ago and were bipedal.
    2. Three distinct species of **robust australopithecines** lived from 2.7 to 1.0 million years ago and were also bipedal. The robust forms had more massive teeth and jaws, very large cheekbones, and a sagittal crest.
    3. The earliest species in the genus *Homo* are called **humans** and date from 2.4 to 1.5 million years ago. *Homo* species have flatter, narrower faces and smaller jaws, and their braincases are as much as three times larger than those of australopithecines.
    4. More recent species of *Homo* date from 1.2 million years ago to the present. This group includes **Cro-Magnons** and **Neanderthals**.
- Humans did not evolve through a simple, steady progression from a chimpanzee-like ancestor. Instead, a complex radiation of bipedal hominids occurred in Africa during the past 4 to 5 million years.

The Out-of-Africa Hypothesis

- Because the first lineage to branch off leads to descendent populations that live in Africa today, it is logical to infer that the ancestral population also lived in Africa (**Figure 34.39**).
- The lineages branched off to form three monophyletic groups. These lineages are thought to descend from three major waves of migration that occurred as *Homo sapiens* population dispersed from east Africa to:
    1. north Africa, Europe, and central Asia
    2. northeast Asia and the Americas, and
    3. southeast Asia and the South Pacific.
- This hypothesis also contends that *Homo sapiens*, Neanderthals, *H. erectus*, and *H. floriensis* never interbred.

✔ CHECK YOUR UNDERSTANDING

(f)   Explain the out-of-Africa hypothesis.

_____

_____

_____

_____

_____

_____

_____

B. MASTERING CHALLENGING TOPICS

The particulars of hominid evolution are confusing since they consist of hypotheses that are constantly being tested and updated. This is the inherent problem in studying the particulars of a branch of science that is unfolding as textbooks are being printed and classes are being taught. The best place to start is **Figure 34.41** and then move to **Figure 34.42**. Draw the lineages presented in this figure, and try to work out the timeline of presence and extinction of the different hominid species. Next, work through the hypotheses presented in this chapter and the experimental evidence for each.

C. ASSESSING WHAT YOU'VE LEARNED

1. The sections of echinoderm tube feet that project outside the body are called:
   a. podia.
   b. pseudopods.
   c. postiums.
   d. endoskeleton.

2. Which of the following statements regarding the Chordates is *false*?
   a. Chordates have pharyngeal gill slits.
   b. Chordates do not have a notochord.
   c. Chordates have a dorsal hollow nerve cord.
   d. Chordates have a muscular tail.

3. The leading hypothesis regarding the evolution of the vertebrate jaw is that it arose from natural selection acting on the
   a. gills.
   b. gill arches.
   c. opercula.
   d. fins.

4. The hypothesis that tetrapod limbs evolved from fish fins is:
   a. rubbish.
   b. only slightly supported.
   c. supported only by fossil evidence.
   d. supported by fossil and molecular genetic evidence.

5. The organ rich in blood vessels that facilitates the flow of oxygen and nutrients from the mother to offspring is called the:
   a. allantois.
   b. yolk.
   c. placebo.
   d. placenta.

6. The key difference between lampreys and sharks is:
   a. lampreys lack jaws.
   b. lampreys lack cartilage.
   c. lampreys do not have tails.
   d. lampreys cannot swim.

7. The key difference between Condrichthyes and Actinopterygii is:
   a. Actinopterygii have fins.
   b. Actinopterygii have jaws.
   c. Actinopterygii have true bone.
   d. Actinopterygii have no tail.

8. Endothermy is found among the Reptilia only within the group:
   a. Testudinia.
   b. Lepidosauria.
   c. Crocodilia.
   d. Aves.

9. The only mammalian group to lay eggs is the
   a. Monotremata.
   b. Marsupiala.
   c. Eutheria.
   d. Primates.

10. The prosimians include all of the following *except*:
    a. lemurs.
    b. gibbons.
    c. tarsiers.
    d. lorises.

CHAPTER 34—ANSWER KEY

Check Your Understanding

(a) The echinoderm body contains a series of fluid-filled tubes and chambers called the water vascular system. Tube feet are elongated, fluid-filled structures; podia are sections of the tube feet that project outside the body. Echinoderms have an endoskeleton—a hard supportive structure inside the body.

(b) Chordates are a phylum of animals defined by the following characteristics:
   1. pharyngeal gill slits,
   2. a stiff and supportive but flexible rod called a notochord,
   3. a bundle of nerve cells that runs the length of the body and forms a dorsal hollow nerve cord, and
   4. a muscular post-anal tail.

   Vertebrates, however, are a subphylum within the chordates. Vertebrates include sharks, several lineages of fishes, amphibians, reptiles (including birds), and mammals. The dorsal hollow nerve cord is elaborated into the spinal cord. The notochord helps organize the body plan during embryonic development. Vertebrates are a monophyletic group distinguished by two traits: vertebrae and a cranium.

(c) Bone first appeared as an exoskeleton rather than an endoskeleton. These bones looked like plates and presumably looked like armor. Animals in this lineage were the size and shape of large fish.

(d) Amniotic eggs have shells that minimize water loss as the embryo develops inside. The egg contains a membrane-bound supply of water in a protein-rich solution called albumin. The embryo is enveloped in a membrane called the amnion. The yolk sac is the membranous pouch that contains the nutrients, and the allantois is the pouch that holds waste material. The chorion is the membrane that provides the surface for gas exchange.

(e) The three orders of mammals are the monotremes, the marsupials, and the eutherians. All of these orders have fur and mammary glands. The monotremes lay eggs and have lower metabolic rates than the other two orders. The marsupials give birth to a very underdeveloped offspring, which continues to develop in a pouch. The eutherians use a placenta to exchange nutrients and waste between mother and embryo.

(f) Because the first lineage to branch off leads to descendent populations that live in Africa today, it is logical to infer that the ancestral population also lived in Africa. The lineages branched off to form three monophyletic groups. These lineages are thought to descend from three major waves of migration that occurred as *Homo sapiens* population dispersed from east Africa to:

- north Africa, Europe, and central Asia
- northeast Asia and the Americas, and
- southeast Asia and the South Pacific.

This hypothesis also contends that *Homo sapiens*, Neanderthals, *H. erectus*, and *H. floriensis* never interbred.

Assessing What You've Learned

1. a; 2. b; 3. b; 4. d; 5. d; 6. a; 7. c; 8. d; 9. a; 10. b

Looking Forward—Key Concepts in Later Chapters

Mechanisms of Movement—Chapter 41

Movement is used to explain the diversity of feeding strategies used in animals. **Chapter 41** focuses on the mechanisms of animal movement in detail.

Gas Exchange—Chapter 44

In the transition to land, animals had to solve the problem of exchanging gas and preventing drying out. As **Chapter 44** will detail, land animals exchange gases with the atmosphere readily as long as a moist body surface is exposed to air.

Viruses

KEY CONCEPTS

- Viruses are tiny, noncellular parasites that infect virtually every type of cell.

- Viruses cannot perform metabolism on their own, meaning outside a parasitized cell.

- Although viruses are diverse morphologically, they are classified into enveloped and nonenveloped.

- The replication cycle is similar for all viruses.

- In terms of diversity, the key feature of viruses is the nature of their genetic material.

Current Viral Epidemics in Humans: HIV

- Human immunodeficiency virus (HIV) will surpass influenza as the most deadly of diseases. This virus is the cause of **acquired immune deficiency syndrome (AIDS)**.

How Does HIV Cause Disease?

- HIV parasitizes immune cells called T cells. The number produced by the body does not keep pace with the number of T cells destroyed by HIV.

- When the T cell drops, the body becomes less able to fight incoming pathogens.

What Is the Scope of the AIDS Epidemic?

- AIDS has already killed 28 million people, and it is estimated that 33 million people worldwide are infected with HIV.

A. CHAPTER OUTLINE

35.1 Why Do Biologists Study Viruses?

Recent Viral Epidemics in Humans

- Viruses have caused the most devastating **epidemics** in recent human history. The Spanish flu of 1918–1919 was the most devastating epidemic—20 to 50 million people died.

✔ CHECK YOUR UNDERSTANDING

(a) Are viruses living organisms? Justify your answer.

_____

_____

_____

35.2 How Do Biologists Study Viruses?

Analyzing Morphological Traits

- Viruses can either be (1) enclosed by just a shell of protein called a **capsid** or (2) enclosed by both a capsid and a membrane-like **envelope**.

- An important distinction among viruses is therefore whether they are **enveloped** or **nonenveloped**.

- A vaccine is a preparation that primes a host's immune system to respond to a specific type of virus.

Analyzing Variation in Growth Cycles: Replicative and Latent Growth

- Viruses all infect their host cells in one of two ways: via replicative growth or in a dormant form referred to as latency in animal viruses or lysogeny in bacteriophages.

- All viruses undergo replicative growth (the **lytic cycle**), but some can halt this cycle.

- A mature virus particle is called a **virion**; the life cycle is complete when the virions exit the cell, usually killing the host cell in the process.

- **Figure 35.8b** diagrams a **lysogenic replication cycle**, or **lysogeny**, in a bacteriophage.

- It is not possible to treat a latent infection with **antivirals**, because the viruses are quiescent.

Analyzing the Phases of the Replicative Cycle

- Six phases are common to replicative growth in virtually all viruses: (1) entry into a host cell, (2) transcription of the viral genome and production and processing of viral proteins, (3) replication of the viral genome, (4) assembly of a new generation of virions, (5) exit from the infected cell, and (6) transmission to a new host.

How Do Viruses Enter a Cell?

- Many plant viruses are transmitted through mouthparts of sucking insects.

- In bacterial or animal cells, viruses gain entry by binding to specific molecules on cell walls or plasma membranes.

Producing Viral Proteins

- A virus must exploit the host cell's biosynthetic machinery in order to make the ribosomes and tRNAs necessary for translating its own mRNAs into proteins.

- Production of viral proteins begins soon after a virus enters a cell and continues after the viral genome is replicated.

- RNAs that code for a virus's envelope proteins follow a route through the cell identical to the RNAs of the cell's transmembrane proteins.

How Do Viruses Copy Their Genomes?

- Most DNA viruses copy their genomes by using their own DNA polymerase enzyme. This protein synthesizes copies of the viral genome using nucleotides provided by the host.

- Most RNA viruses use a viral enzyme called **RNA replicase**. RNA replicase is an RNA polymerase that synthesizes RNA from an RNA template, using ribonucleotides provided by the host cell.

- In other RNA viruses, the genome is transcribed from RNA to DNA by a viral enzyme called **reverse transcriptase**. Reverse transcriptase is a DNA polymerase that makes a double-stranded DNA from a single-stranded RNA template.

- **Retroviruses** are viruses that reverse-transcribe their genome. Enzymes then insert this copied DNA (cDNA) into a stretch of host-cell chromosome. The viral genes are then transcribed to mRNA, which is translated into proteins by the host cell's ribosomes.

Assembly of New Virions

- Once the viral genome has been replicated and the viral proteins produced, they are transported to locations where a new generation of virions assembles inside the infected host cell.

- During assembly, the capsid forms around the viral genome.

Exiting an Infected Cell

- Viruses leave the host cell in one of two ways: by budding from cellular membranes or by bursting out of the cell.

How Are Viruses Transmitted to New Hosts?

- If the host cell is part of a multicellular organism, the new generation of particles travels through the blood or lymph. Antibodies or macrophages can then attempt to destroy them before they infect another host cell.

- Natural selection favors alleles that allow viruses to replicate within a host and be transmitted to a new host.

- Transmission of HIV has taken place through direct exchange of blood or semen between an uninfected person and an infected person.

- HIV is quickly destroyed if it is heated or dried; therefore, it does not survive long outside the host tissues.

✓ **CHECK YOUR UNDERSTANDING**

(b) What are retroviruses, and what specific viral enzyme do they use?

_____

_____

_____

_____

35.3 What Themes Occur in the Diversification of Viruses?

The Nature of the Viral Genetic Material

- Researchers were able to separate the protein and nucleic acid components of the tobacco mosaic virus (TMV). They showed in this virus that RNA—*not* DNA—functions as the genetic material.

- Subsequent research revealed an amazing diversity of viral genome types.

- In some groups of viruses, like measles and flu, the genome consists of RNA. In others, like herpes and smallpox, the genome is composed of DNA. RNA and DNA genomes of viruses can either be **positive sense** (single stranded) or **negative sense** (double stranded). In **positive-sense** viruses, the genome contains the same sequences as the mRNA required to produce viral proteins. In **negative-sense** viruses, the base sequences in the genome are complementary to those in viral mRNAs. **Ambisense** viruses contain both positive- and negative-sense sections.

- Viruses can have as few as three loci (tymoviruses) or up to hundreds, as in the genome of smallpox.

Where Did Viruses Come From?

- Many biologists suggest that viruses may be derived from plasmids and transposable elements. Viruses are distinguished from plasmids because they encode proteins that form a capsid and allow the genes to exist outside the cell.

- The "escaped-gene sets" hypothesis states that these elements descended from clusters of genes that physically escaped from bacterial or eukaryotic chromosomes long ago.

- The degeneration hypothesis states that organisms gradually degenerated into viruses by gradually losing the genes required to synthesize ATP and other compounds.

- Some researchers have suggested that RNA genomes of some viruses were among the first inhabitants of the RNA world.

Emerging Viruses, Emerging Diseases

- Hantavirus and Ebola are examples of emerging diseases—illnesses that suddenly affect significant numbers of individuals in a host population.

- Hantavirus and Ebola are considered emerging viruses because they had switched from their traditional host species to a new host—humans.

- In an outbreak, physicians need to (1) identify the agent causing the new illness and (2) determine how it is being transmitted.

✓ **CHECK YOUR UNDERSTANDING**

(c) Compare and contrast negative-sense and positive-sense DNA. How do these DNA types relate to the virus types?

_____

_____

_____

_____

35.4 Key Lineages of Viruses

Double-Stranded DNA (dsDNA) Viruses

- This is a large group of 21 families and 65 genera, of which smallpox and herpes are the most familiar.

- These viruses parasitize hosts throughout the tree of life, with the exception of land plants.

- The viral genes have to enter the nucleus to be replicated. Thus, they can infect only cells that are actively dividing, such as those that line the respiratory tract or urogenital tract.

RNA Reverse-Transcribing Viruses (Retroviruses)

- There is only one family called the retroviruses, which includes HIV.

- Species are known to parasitize only vertebrates—specifically birds, fish, or mammals.

- When the virus's RNA genome and reverse transcriptase enter a host cell's cytoplasm, the enzyme catalyzes the synthesis of a single-stranded cDNA from the original RNA. Reverse transcriptase makes this cDNA double stranded. The DNA then integrates with the host chromosome.

Double-Stranded RNA (dsRNA) Viruses

- There are 7 families and a total of 22 genera.
- Species in this group infect a wide variety of hosts, including fungi, land plants, insects, vertebrates, and bacteria.
- For humans, the most important viruses in this group cause disease in major crops.
- Once in the host cell, the double-stranded DNA synthesizes RNA, which is then translated into viral proteins.

Negative-Sense Single-Stranded RNA ([−]ssRNA) Viruses

- There are 7 families and 30 genera in this group.
- The single-stranded virus genome is complementary to the viral mRNA.
- Widely varying plants and animals are parasitized by these viruses. Flu, mumps, measles, Ebola, hantavirus, and rabies all belong to this group.
- Once in a host cell, a viral RNA polymerase uses the negative-sense template to make viral mRNA. The viral mRNAs are then translated to form viral proteins and new negative-sense single-stranded RNA copies.

Positive-Sense Single-Stranded RNA ([+]ssRNA) Viruses

- This is the largest group known, with 81 genera organized into 21 families.
- Since the sequence of bases is the same as in mRNA, it does not need to be transcribed before proteins are produced.
- Most of the commercially important plant viruses belong to this group. This group also includes colds, polio, and hepatitis A, C, and E.
- Single-stranded RNA is immediately translated into viral proteins.

✓ CHECK YOUR UNDERSTANDING

(d) Which of the main viruses discussed in this section is officially extinct, and to which group does it belong?

_____

_____

_____

_____

B. MASTERING CHALLENGING TOPICS

In your first-year biology course, you are being exposed to many new terms in a short time. Often, these terms are introduced with each new branch of biology. In this chapter, the HIV virus is used as an example in describing a viral replication cycle and the process that researchers use in combating a deadly virus. To make matters more complex, HIV is only one virus, and thus only one type of life cycle is explained. To better understand the life cycle of HIV and how it differs from the cell cycle, it would first be beneficial to review the processes of DNA replication and the processes of transcription and translation. Then, by consulting the diagrams in Chapter 35, make a list of differences and similarities between the viral life cycle and the cell cycle. Finally, try to apply this to the HIV virus so you can easily explain to a classmate, lab partner, or friend why we haven't yet found a successful vaccine for HIV.

C. ASSESSING WHAT YOU'VE LEARNED

1. The basic viral replication cycle that usually results in the death of the host cell is called:
   a. lysogeny.
   b. the lysogenic cycle.
   c. the lytic cycle.
   d. the translation cycle.

2. Viruses can enter a cell by:
   a. transmission through mouthparts of sucking insects.
   b. binding to specific proteins on cell membranes.
   c. diffusion.
   d. only a and b

3. The RNA polymerase that synthesizes RNA from an RNA template by using ribonucleotides provided by the host cell is called:
   a. RNA replicase.
   b. reverse transcriptase.
   c. retrovirus transcriptase.
   d. RNA transcriptase.

4. Viruses have to do which of the following in order to make viral proteins?
   a. Ramp up their metabolism
   b. Exploit the host cell's biosynthetic machinery
   c. Enter a dormant phase
   d. Create their own biosynthetic machinery

5. Viruses leave a host cell by:
   a. budding from the cell membrane.
   b. bursting out of the cell.
   c. forming spores.
   d. only a and b

6. The escaped-gene hypothesis states that viral elements descended from clusters of genes that
   a. physically escaped from bacterial or eukaryotic organisms.
   b. degenerated over evolutionary time.
   c. merged with other viral particles to become active.
   d. none of the above

7. The main difference between emerging viruses and emerging diseases is:
   a. emerging diseases involve only one isolated case.
   b. emerging diseases involve a significant number of infected individuals.
   c. emerging viruses involve a significant number of infected individuals.
   d. none of the above

8. The group of viruses to which smallpox belongs is:
   a. double-stranded DNA (dsDNA) viruses.
   b. RNA reverse-transcribing viruses.
   c. double-stranded RNA (dsRNA) viruses.
   d. negative-sense single-stranded RNA ($[-]$ssRNA) viruses.

9. The group of viruses to which HIV belongs is:
   a. double-stranded DNA (dsDNA) viruses.
   b. RNA reverse-transcribing viruses.
   c. double-stranded RNA (dsRNA) viruses.
   d. negative-sense single-stranded RNA ($[-]$ssRNA) viruses.

10. Negative-sense single-stranded RNA must do the following in order to replicate successfully:
    a. use the negative-sense template to make viral proteins.
    b. use the negative-sense template to make viral RNA.
    c. use the negative-sense template to make viral DNA.
    d. use the negative-sense template to make replicase.

CHAPTER 35—ANSWER KEY

Check Your Understanding

(a) Viruses are an obligate intracellular parasite. They are not made up of cells; therefore, they are not organisms. Viruses are referred to as particles or agents.

(b) Retroviruses are viruses that reverse-transcribe their genome. Enzymes then insert this copied DNA (cDNA) into a stretch of host-cell chromosome. The viral genes are then transcribed to mRNA, which is translated into proteins by the host cell's ribosomes. In other RNA viruses, the genome is transcribed from RNA to DNA by a viral enzyme called reverse transcriptase. Reverse transcriptase is a DNA polymerase that makes a double-stranded DNA from a single-stranded RNA template.

(c) RNA and DNA genomes of viruses can either be positive sense (single stranded) or negative sense (double stranded). In positive-sense viruses, the genome contains the same sequences as the mRNA required to produce viral proteins. In negative-sense viruses, the base sequences in the genome are complementary to those in viral mRNAs.

(d) Smallpox is currently extinct in the wild, and the only remaining samples of the virus are stored in research labs. Smallpox is a double-stranded DNA (dsDNA) virus.

Assessing What You've Learned

1. b; 2. d; 3. a; 4. b; 5. d; 6. a; 7. b; 8. a; 9. b; 10. b

Looking Forward—Key Concepts in Later Chapters

The Immune System—Chapter 49

In Chapter 35, we just touch on the role of T cells and HIV. **Chapter 49** explains just how crucial these cells are to the immune system's response to invading bacteria and viruses. Chapter 49 also explores epidemic control through vaccination and explains why researchers haven't been able to design an effective vaccine against HIV.

Plant Form and Function

Photosynthesis—Chapter 10

This chapter details how plants, along with algae, cyanobacteria, and a variety of protists, obtain the energy and carbon they need to grow and reproduce.

The Plant Cell—Chapter 7

Chapter 7 introduced a generalized version of the plant cell. This information should be reviewed before tackling the specialized cells discussed in Chapter 36.

KEY CONCEPTS

- The vascular body consists of a root system that absorbs water and key ions, and a shoot system that absorbs carbon dioxide and sunlight.

- Variation in body size and shape allows different species to harvest water, light, and other resources in unique ways.

- Primary growth occurs when cells located at the tips of each root and shoot divide and enlarge.

- Primary growth lengthens roots and shoots and gives rise to the three primary tissue systems that are specialized for protection, food production and storage, and transport.

- In some species, secondary growth occurs, which widens the root or shoot.

- Secondary growth adds transport tissue and provides structural support.

A. CHAPTER OUTLINE

36.1 Plant Form: Themes with Many Variations

- The root system anchors the plant and takes in water and nutrients from the soil while the shoot system harvests light and CO_2 from the atmosphere to produce sugars.

The Importance of Surface Area/Volume Relationships

- A plant is more efficient as an absorbance-and-synthesis machine when it has a large surface area relative to its volume.

The Root System

- Many root systems have a vertical section called a **taproot**, as well as numerous **lateral roots**.

- The root system functions to anchor the plant in the soil, absorb water and ions from the soil, conduct water and selected ions to the shoot, and store material produced in the shoot for later use.

Morphological Diversity in Root Systems

- Root systems are incredibly diverse.

- As an example, the prairie plants of North America have a vast array of root systems. Some have a **fibrous** root system, while others have a taproot. Others contain very large reservoirs of food stored in the form of starch.

- This diversity allows prairie plants to coexist with less competition for soil resources, and most individuals can therefore survive intense water stress.

- In this case, natural selection has favored structures that minimize competition for water and nutrients.

Phenotypic Plasticity in Root Systems

- Morphological diversity in roots occurs within species as well as among species.

- Roots show a great deal of **phenotypic plasticity**, meaning that they are changeable, depending on environmental conditions.

Modified Roots

- In **Figure 36.5**, you can begin to appreciate the huge diversity of modified roots.

The Shoot System

- The shoot system consists of one or more **stems**, with each stem consisting of **nodes** and **internodes**. A **leaf** is an appendage attached to a node, and the leaf functions as a photosynthetic organism.
- Sometimes, an **axillary bud** may develop a branch. At the tip of each stem and branch is an **apical bud**.

Morphological Diversity in Shoot Systems

- The shoot systems of land plants range in size from mosses and liverworts (a few mm tall) to redwood trees that reach heights of over 100 m.
- Their enormous variation allows plants to harvest light at different locations and thus minimize competition.

Phenotypic Plasticity in Shoot Systems

- Shoots respond to variation in light ability by altering their shape and size to maximize the number of photons absorbed in a given environment.

Modified Shoots (Figure 36.8)

- Cactus stems are often modified into water-storage organs, and the stems contain the plant's photosynthetic tissue.
- **Stolons** are modified stems that run over the soil surface, producing roots and leaves at each node.
- **Rhizomes** are similar to stolons—they are stems that grow horizontally, producing new plants at nodes. However, these stems grow *below* ground.
- **Tubers** are rhizomes that are modified as carbohydrate storage organs.
- **Thorns** are modified stems that help protect the plant from attacks by large herbivores.

The Leaf

- The leaf is composed of the expanded portion called the **blade** and a stalk called the **petiole**.

Morphological Diversity in Leaves

- The arrangement of leaves can vary as much as overall shape.
- Needlelike leaves are interpreted as adaptations that minimize **transpiration** in water-short habitats.

Phenotypic Plasticity in Leaves

- On the same tree, **shade leaves** are relatively large and broad while **sun leaves** are smaller.

Modified Leaves (Figure 36.9)

- It is important to recognize that not all leaves function primarily in photosynthesis.
- Cactus spines are modified leaves that protect the stem.
- The leaves of the flowerpot plant are used as homes by ant colonies.
- The thick leaves of succulents are used to store water.
- The tendrils enable the garden pea and other vines to climb.

✔ CHECK YOUR UNDERSTANDING

(a) Describe four main types of modified shoots.

_____

_____

_____

_____

36.2 Primary Growth Extends the Plant Body

- Plants grow because they have **meristems**. **Apical meristems** are located at the tip of each root and shoot. As the apical meristematic cell divides, one of the daughter cells remains in the meristem, allowing the meristem to persist, while the other cell specializes.

How Do Apical Meristems Produce the Primary Plant Body?

- A tissue is a group of cells functioning as a **tissue**.
- There are three primary meristematic cell types:
  1. **Protoderm** gives rise to the dermal tissue system or epidermis.
  2. **Ground meristem** gives rise to the ground tissue system, which makes up the bulk of the plant body.
  3. **Procambium** gives rise to the vascular tissue system, which provides support and transports water, nutrients, and photosynthetic products.

How Is the Primary Root System Organized?

- The growing region of the root is covered by protective cells—**the root cap**. These cells also secrete **mucigel** for lubrication.

- Three distinct cell populations exist behind the root cap:
  1. The **zone of cellular division** contains the apical meristem.
  2. The **zone of cellular elongation** is where cells increase in length.
  3. The **zone of cellular maturation** is where older cells complete their differentiation. In this region, **root hairs** increase surface area and actively absorb water and nutrients.

How Is the Primary Shoot System Organized?

- On the top of the stem is the shoot apical meristem.
- The primary meristematic cells give rise to the dermal, ground, and vascular tissues.
- Vascular tissues are found in groups called **vascular bundles**.
- The ground tissue inside the vascular bundles is called the **pith**, while the ground tissue outside the vascular bundles is called the **cortex**.

✔ **CHECK YOUR UNDERSTANDING**

(b) Describe the origin of the three main primary tissues.

_____

_____

_____

36.3 A Closer Look at the Cells and Tissues of the Primary Plant Body

- Plant cells are surrounded by a stiff cell wall made of strong cellulose fibers.
- Plant cells have one or more storage **vacuoles** that contain cell sap and help the cell maintain its proper volume.
- Photosynthetic plant cells have **chloroplasts** for photosynthesis.
- Plant cells are connected by gaps in the cell wall and cell membrane called **plasmodesmata**.

The Dermal Tissue System

- Epidermal cells secrete the **cuticle**—a waxy layer that protects the leaves and reduces water loss.
- In addition to water loss, the cuticle forms a barrier to protect the plant from **pathogens**.

- To allow CO_2 to enter photosynthetically active tissues, most land plant leaves have **stomata**—a hole produced by two bean-shaped **guard cells**.
- **Trichomes** are appendages from epidermal cells that function to minimize water loss and attacks by herbivores.

The Ground Tissue System

- Ground tissue is made up of cells below the epidermis and is made up of parenchyma stiffened by collenchyma and sclerenchyma.
- This makes up the bulk of the plant body and is the primary location of photosynthesis and carbohydrate storage.
- **Parenchyma cells** serve as storage cells for starch deposits in roots, stems, and leaves. They are filled with chloroplasts and are the primary site of photosynthesis.
- **Collenchyma cells** function mainly in support. They have thickened primary walls and serve to stiffen leaves and stems.
- **Sclerenchyma cells** also stiffen stems and other structures, but they are distinguished by thickened **secondary cell walls** that are strengthened by tough **lignin** molecules.
- There are two types of sclerenchyma cells:
  1. **Fibers** are usually associated with vascular tissue and are extremely elongated.
  2. **Sclereids** are relatively short, have variable shapes, and often function in protection.

The Vascular Tissue System

- **Vascular tissue** is made up of two complex conducting tissues called xylem and phloem.
- **Xylem** is made up primarily of sclerenchyma cells and conducts water and dissolved ions from the root system to the shoot system.
- Xylem contains two types of conducting cells:
  1. **Tracheids** are dead cells that are long and slender with tapered ends. The sides and ends of the tracheids have pits that reduce resistance to water flow.
  2. **Vessel elements** are shorter and wider and have perforations that produce little resistance to water flow.
- **Phloem** is made up primarily of parenchyma cells and conducts sugar, amino acids, chemical agents, and other substances throughout the plant body.
- The cells that make up phloem are alive at maturity:
  1. **Sieve-tube members** are long, thin cells that have perforated ends called sieve plates. They lack major organelles.

2. **Companion cells** are attached to sieve-tube members by plasmodesmata. They provide materials to the sieve-tube members to maintain the cytoplasm and membrane.

✔ CHECK YOUR UNDERSTANDING

(c) What is the difference between collenchyma cells and sclerenchyma cells?

_____

_____

_____

36.4 Secondary Growth Widens Roots and Shoots

- Whereas primary growth increases the length of roots and shoots, **secondary growth** increases width. It increases the amount of conducting tissue available and provides structural support.

- Without the structural support provided by secondary growth, roots could not anchor and stems would fall over.

What Is a Cambium?

- **Lateral meristems** are also called **secondary meristems** or **cambium**. A cambium differs from apical meristems in two ways:

  1. A cambium forms cylinders that run the length of a root or stem. Apical meristems are localized at root and shoot tips.

  2. In a cambium, the cells divide perpendicularly to the long axis of the root or stem, which increases width. Cells in apical meristems divide in a plane parallel to the long axis of the root or stem.

- **Vascular cambium** is a ring of meristematic cells just inside the cork cambium.

- **Cork cambium** is a ring of meristematic cells located near the perimeter of the stem.

What Does Vascular Cambium Produce?

- The vascular cambium produces both phloem and xylem, which, as new layers form, develop into **secondary phloem** and **secondary xylem**.

- Secondary phloem contributes to bark, and secondary xylem forms the material we call **wood**.

- Much more secondary xylem is produced than secondary phloem. As a result, mature woody stems are dominated by wood.

What Does Cork Cambium Produce?

- The cork cambium produces **cork cells** to the outside and a smaller layer of cells called the **phelloderm**. Together, these cells make up the **periderm**.

- These cells eventually die, forming a protective layer for the mature root or shoot.

- This is an important component for **bark**. Bark actually refers to all cells outside the vascular cambium.

- Gas exchange can still occur between the atmosphere and the living tissues inside the stem through small, spongy segments of the bark called **lenticels**.

The Structure of a Tree Trunk

- Trees are perennial—they live for many years.

- A period of **dormancy** occurs in winter or dry seasons. When growth is rapid during the spring or wet season, large thin-walled cells are produced in the secondary xylem. When dormancy is approached, the secondary xylem cells are smaller and have thick walls.

- When growth is seasonal, **annual growth rings** can be observed in cross sections of wood.

- The inner xylem region is called **heartwood**, while the outer xylem is called **sapwood**.

✔ CHECK YOUR UNDERSTANDING

(d) Explain why a nail in a tree trunk will not move up the trunk as the tree grows.

_____

_____

_____

B. MASTERING CHALLENGING TOPICS

The main aspect of this chapter that creates problems is the overwhelming amount of anatomical terms. Before you can understand the function of anatomy, you need to understand the anatomy itself. Drawing diagrams and labeling them is the best way to learn anatomical terms. Adding the functions on top of the anatomy will help to solidify all the terms and organize them. To completely prove you have mastered the anatomy and function of angiosperms, go out and collect various samples of plants and test yourself. This is a great way to see firsthand the variation of form and function.

C. ASSESSING WHAT YOU'VE LEARNED

1. The term for an individual changing in response to the environment is:
   a. natural selection.
   b. allele frequency modulation.
   c. phenotypic plasticity.
   d. artificial selection.

2. Which of the following is a modified stem that runs over the soil surface?
   a. Stolon
   b. Rhizome
   c. Tuber
   d. Thorn

3. Which of the following is a modified stem used for carbohydrate storage?
   a. Stolon
   b. Rhizome
   c. Tuber
   d. Thorn

4. Plants in desert or northern habitats tend to have leaves that are needlelike because:
   a. they reduce herbivory.
   b. they reduce sun damage.
   c. they increase water absorption.
   d. they reduce water loss.

5. The undifferentiated cells responsible for plant growth are called:
   a. meristematic cells.
   b. parenchyma cells.
   c. mesophyll cells.
   d. apical cells.

6. Which of the following cells serve as storage cells for starch deposits in roots, stems, and leaves?
   a. Parenchyma cells
   b. Meristematic cells
   c. Collenchyma cells
   d. Sclerenchyma cells

7. Which of the following are water-conducting cells that are dead at maturity?
   a. Sclerenchyma cells
   b. Tracheids
   c. Sieve-tube members
   d. Sclereids

8. The waxy layer that is secreted by the epidermis is called the:
   a. dermis.
   b. stoma.
   c. cuticle.
   d. sterol.

9. Which of the following is *not* a group of cells behind the root cap?
   a. The zone of cellular division
   b. The zone of cellular elongation
   c. The zone of cellular communication
   d. The zone of cellular maturation

10. Wood is:
    a. produced by the cork cambium.
    b. secondary xylem.
    c. secondary phloem.
    d. produced by the apical meristem.

CHAPTER 36—ANSWER KEY

Checking Your Understanding

(a) **Stolons** are modified stems that run over the soil surface, producing roots and leaves at each node. **Rhizomes** are similar to stolons—they are stems that grow horizontally, producing new plants at nodes. However, these stems grow *below* ground. **Tubers** are rhizomes that are modified as carbohydrate storage organs. **Thorns** are modified stems that help protect the plant from attacks by large herbivores.

(b) The three main tissue types originate from apical meristem tissue. As the apical meristematic cell divides, one of the daughter cells remains in the meristem, allowing the meristem to persist, while the other cell specializes initially into one of the three main tissue types.

(c) Collenchyma cells have thickened primary walls and serve to stiffen leaves and stems. Sclerenchyma cells also stiffen stems and other structures, but they are distinguished by thickened secondary cell walls that are strengthened by tough lignin molecules.

(d) The tree grows at the root and shoot meristems. Over time, this might give the illusion that the tree is coming out of the ground and moving the nail higher, but this is incorrect. As the tree grows in height at the shoot tips, secondary growth increases the girth of the trunk. The nail stays in the same spot.

Assessing What You've Learned

1. c; 2. a; 3. c; 4. d; 5. a; 6. a; 7. b; 8. c; 9. c; 10. b

Looking Forward—Key Concepts in Later Chapters

Water—Chapter 37

Plants must have water as a source of electrons to run photosynthesis. In **Chapter 37**, you will see that plants also need water to keep their cells in proper working order.

Plant Nutrition—Chapter 38

Plants must obtain nitrogen, phosphorus, potassium, magnesium, and a host of other nutrients to synthesize nucleic acids, enzymes, phospholipids, and the other macromolecules needed to build and run cells. **Chapter 38** explores how plants acquire these key elements.

Sensory Response—Chapter 39

A shoot's growth is controlled by sophisticated sensory and response systems. **Chapter 39** explores how plants sense certain wavelengths of light and respond by bending and growing in that direction.

Reproductive Structures—Chapter 40

Twigs and stems are structures that arrange leaves in space. **Chapter 40** explores the reproductive structures housed in twigs and stems.

Water and Sugar Transport in Plants

Leaf Structure—Chapter 36

Recall that the surfaces of leaves are dotted with structures called stomata. Chapter 37 begins to discuss the functions of the stomata and apply the anatomy learned in **Chapter 36**.

Osmosis—Chapter 8

Water will begin moving into or out of a cell via osmosis—water moves from regions of low solute concentrations to regions of high solute concentrations. Chapter 37 discusses osmosis in relation to the plant and its transport of fluids.

Hydrogen Bonding—Chapter 2

Water molecules are polar and interact with one another through hydrogen bonding. Under an air-water interface, in the body of a solution, all of the water molecules present are surrounded by other water molecules and form hydrogen bonds in all directions. The water molecules on the surface, however, can form hydrogen bonds in only one direction—with the water molecules below them.

CAM and C_4 Photosynthesis—Chapter 10

Chapter 10 introduced two novel biochemical pathways that are found in species native to deserts and other hot, dry habitats. These pathways are called crassulacean acid metabolism (CAM) and C_4 photosynthesis. CAM plants are able to continue photosynthesizing even though their stomata close during the heat of the day.

Proton Pumps—Chapter 9 and 10

As **Chapter 9** and **10** indicated, proteins like these are called proton pumps, or more formally H^+-ATPases. Their activity establishes a large difference in the charge and hydrogen ion concentration on either side of the membrane.

KEY CONCEPTS

- Water moves from areas of high water potential to areas of low water potential.

- Water's potential energy in plants consists of a solute component and a pressure component.

- Plants lose water to transpiration when stomata are open and photosynthesis is occurring.

- Water moves from soil and roots to leaves along a water gradient.

- Evaporation of water from leaves is driven by the Sun and creates a negative pressure that pulls water up.

- In phloem, sugars are transported from tissues that release sugars (sources) to tissues that use or store sugars (sinks).

A. CHAPTER OUTLINE

37.1 Water Potential and Water Movement

- Loss of water via evaporation from the aerial parts of a plant is called **transpiration**.

What Is Water Potential?

- Water potential indicates the potential energy that water has in a particular environment compared to control conditions.

- Water always flows from areas of high water potential to areas of low water potential.

- **Water potential gradient** is the overall movement of water when water potential differences within a series are contrasted. Plants tend to gain water from the soil and lose it to the atmosphere.

What Factors Affect Water Potential?

- A **solution** is a homogeneous liquid mixture containing **solutes**.
- **Osmosis** is the movement of water across membranes in response to differences in water potential.
- The **solute potential** or **osmotic potential** is the difference in solute concentrations between two cells.
- **Turgor pressure** is the force of the cell membrane swelling and pushing against the cell wall. The cell wall exerts an equal and opposite force called **wall pressure**.
- **Pressure potential** is the sum of all the types of pressure on water.

Calculating Water Potential

- The water potential is the pressure potential plus the solute potential, calculated as follows:

$$\psi = \psi_p + \psi_s$$

- The unit of measurement is a pressure unit called the **megapascal** (**MPa**).
- In other words, the potential energy of water in a particular location is the sum of the pressure potential and the solute potential that it experiences.

Water Potentials in Soils, Plants, and the Atmosphere

- The water contained within a leaf or root system or plant has a pressure potential and a solute potential, just as the water inside a cell does. The water potentials of plant tissues, air, and soil are changing all the time in response to heat, cold, rain, and so on.
- If the water potential in the space that surrounds a cell drops, water moves out of a cell in response, causing it to shrink. If the cells do not quickly regain turgor pressure, dehydration and death result.
- Because water moves along the water potential gradient, plants tend to gain water from the soil and lose it to the atmosphere.

✔ CHECK YOUR UNDERSTANDING

(a) Explain the difference between water potential and water potential gradient. What is the significance of this difference?

_____

_____

_____

_____

37.2 How Does Water Move from Roots to Shoots?

Movement of Water and Solutes into the Root

- Look at **Figure 37.6** and note the different tissues. Starting at the outside of the root and working inward, they are the **epidermis** and **root hairs**, the **cortex**, the **endodermis**, the **pericycle**, and the **vascular tissues**.
- When water enters a root along a water potential gradient, it does so through the root hairs. Before it reaches the vascular tissue, it can be transported through three pathways:
  1. The **transmembrane route** is based on flow through aquaporin proteins.
  2. The **apoplast** lies within cell walls, which are porous.
  3. The **symplast** consists of the continuous connection through cells that exist via plasmodesmata.
- **Endodermal** cells are tightly packed and secrete a waxy layer called the **Casparian strip**. It prevents water from creeping through the walls of the endodermal cells (i.e., blocks the apoplastic pathway).
- When water is forced to travel through the plasma membrane of endodermal cells, it can be filtered. The plasma membrane can also prevent water from flowing out of the vascular tissue.

Water Movement via Root Pressure

- The endodermis is also responsible for **root pressure**—active transport of ions brings more water into the root that is lost via transpiration.
- Water droplets can even be forced out of leaves—**guttation**.

Water Movement via Capillary Action

- Capillary movement occurs in response to these forces:
  1. **Surface tension** is a downward pull that exists on water molecules at an air-water interface.
  2. **Adhesion** is the attraction of unlike molecules.
  3. **Cohesion** is the mutual attraction among like molecules, such as the hydrogen bonding among water molecules.
- The water inside a capillary, at the interface with the air, forms a concave boundary layer called a **meniscus**.

The Cohesion-Tension Theory

- The leading hypothesis to explain water movement in vascular plants is called the **cohesion-tension theory** (**Figure 37.10**).

- The leaf area just below the stoma is filled with moist air such that, when the stoma opens, the humid air is exposed to the atmosphere. Water exits through the stomatal pore as a result.

- The water potential of the atmosphere is extremely low compared to the water potential of the space inside the leaf, meaning that there is a steep water potential gradient between the leaf interior and the air.

- If the water potential gradient is especially steep, then water molecules leave the surface rapidly and the menisci in the cell walls become more concave.

The Role of Surface Tension in Water Transport

- The key idea of the cohesion-tension theory is that the negative force or pull (tension) generated at the air-water interface is transmitted through the water present in the leaf cells to the water present in the xylem, on to the water in the vascular tissue of the roots, and finally on to the water in the soil.

- This continuous transmission of pulling force is possible because (1) water is present throughout the plant and because (2) all of the water molecules present hydrogen-bond to one another (cohesion).

What Evidence Do Biologists Have for the Cohesion-Tension Theory?

- Researchers also confirmed that water tension in trees is great enough to actually make tree trunks shrink. The dendrograph is the instrument used to measure trunk diameter.

- If you take a leaf and cut its petiole, the watery fluid in the xylem withdraws from the edge. This confirms that the xylem sap is under tension.

- By placing a leaf or branch in an airtight container and then applying a steadily increasing external pressure, the pressure required to push the **xylem sap** back to the cut surface can be measured. This pressure is equal to the tension exerted on the xylem sap (**Figure 37.11**).

- **Key research**—Chunfang Wei, Melvin Tyree, and Ernst Steudle set out to study the cohesion-tension theory using a root bomb and a xylem pressure probe. This technique allowed the investigators to record changes in xylem pressure instantly and directly (**Figure 37.12**).

- First, they added pressure to the root systems of their experimental plants using a root bomb. Then, they released pressure on the root system and began to alter light levels.

- Pulses of pressure applied through the root bomb produced sharp decreases in the tension exerted on the xylem sap. The tension on the xylem sap increased each time that light intensity was increased.

CHECK YOUR UNDERSTANDING

(b)   Explain what cohesion means and why it is important to plant physiology.

_____

_____

_____

37.3 Water Absorption and Water Loss

- One of the most important features of the cohesion-tension theory is that it does **not require the expenditure of energy** by plants. The Sun furnishes the energy that breaks hydrogen bonds between water molecules at the air-water interface inside leaves and causes transpiration.

- The balance between conserving water and maximizing photosynthesis is termed the photosynthesis-transpiration compromise.

Limiting Water Loss

- To cope with water loss, the oleander has (**Figure 37.13**):
  1. a thick cuticle on the upper surface of the leaves
  2. several layers of epidermis
  3. stomata located in deep pits on the undersides of leaves

Obtaining Carbon Dioxide under Water Stress

- **Crassulacean acid metabolism** (CAM) plants are able to continue photosynthesizing even though their stomata are closed during the day. CAM plants open their stomata at night and store the CO_2 that diffuses into their tissues by adding organic acids.

- When photosynthesis begins during the day, CO_2 is released from organic acids to **rubisco**—the enzyme that initiates the Calvin cycle.

- C_4 plants use CO_2 so efficiently that they are able to keep their stomata closed more than competing plants can. CO_2 is transferred to **bundle-sheath cells**, where rubisco is abundant.

(c) Why must a plant compromise between photo synthesis and transpiration?

_____

_____

_____

37.4 Translocation

- **Translocation** is the movement of sugars through a plant. Sugars move from sources to sinks. In vascular plants, a **source** is defined as a tissue where sugar enters the phloem. A **sink** is a tissue where sugar exits the phloem.

- During the growing season, leaves and stems that are actively photosynthesizing and producing sugar in excess of their own needs act as sources. Early in the growing season, though, the situation is reversed.

- ^{14}C experiments have made two important generalizations possible:

  1. Sugars are translocated very rapidly.

  2. There is a strong correspondence between the physical location of certain sources and certain sinks.

The Anatomy of Phloem

- Phloem is made up of **sieve-tube elements** and **companion cells**. Both cells are alive at maturity. In most plants, sieve-tube elements lack nuclei and many major organelles; they are connected to one another, end to end, by perforated structures called **sieve plates** (**Figure 37.16**).

- Sieve plates create a direct connection between the cytoplasm of adjacent cells. Companion cells, in contrast, have nuclei and a large number of ribosomes, mitochondria, and chloroplasts.

The Pressure-Flow Hypothesis

- **Key research**—Ernst Münch proposed a mechanism for translocation in 1926. The **pressure-flow hypothesis** states that events at source tissues and at sink tissues create a steep pressure potential gradient in phloem.

- The water in phloem moves down this gradient, and sugar molecules are carried along by **bulk flow**—a mass movement of molecules along a pressure gradient.

- The net result of these events in **Figure 37.17** is high turgor pressure near the source and low turgor pressure near the sink. Because the pores of the sieve plate have no membranes, the difference in pressure potential drives phloem sap from source to sink. The pressure contrast is responsible for a one-way flow of sucrose molecules.

Phloem Loading

- At source cells, sugar has to be loaded into sieve-tube elements against a concentration gradient. Therefore, loading requires the use of ATP and a membrane transport mechanism. Unloading also requires ATP and a second membrane transport system.

How Are Sucrose and Other Solutes Transported across Membranes?

- **Passive transport** occurs when ions or molecules move across a plasma membrane by diffusion.

- **Facilitated diffusion** uses protein **channels** or **carriers** that permit the passage of specific ions or molecules across the plasma membrane.

- **Active transport** occurs when ions or molecules move across the plasma membrane against their concentration gradient—this requires the use of energy in the form of ATP.

How Are Sugars Concentrated in Sieve-Tube Members at Sources?

- The observation of strong pH differences between the interior and exterior of phloem cells led researchers to suspect that phloem loading depends on an H^+-ATPase (**proton pumps**).

Where Are H^+-ATPases Located?

- Proton pumps are found in a variety of organisms, including bacteria, fungi, and animals. Researchers found that the *Arabidopsis* (a widely studied plant model from the mustard family) genome actually codes for 10 different pump proteins. One of these genes, called *AHA3*, appeared to be expressed primarily in vascular tissues.

- **Key research**—Natalie DeWitt and Michael Sussman used this information and hypothesized that *AHA3* encoded the proton pump responsible for phloem loading. Their goal was to treat *Arabidopsis* leaves with the AHA3 antibody, examine treated leaves under the electron microscope, and determine exactly where the pump proteins are located.

- The proton pumps responsible for phloem loading are found almost exclusively in the membranes of companion cells.

- This result supported the following model for phloem loading:

  1. Proton pumps in the membranes of companion cells create a strong gradient that favors flow of protons into companion cells.

  2. A cotransporter protein in the membrane of companion cells uses the proton gradient to bring in sucrose.

3. Once inside companion cells, sucrose travels into sieve-tube members via direct cytoplasmic connection.

Phloem Unloading

- Sugar transport is an energy-demanding process, but the membrane proteins involved and mechanism of movement vary among different types of sinks within the same plant, as well as among different species.

✓ CHECK YOUR UNDERSTANDING

(d) Why would an organism require the use of secondary active transport?

_____

_____

_____

B. MASTERING CHALLENGING TOPICS

Now that you have mastered the anatomy of vascular plants from Chapter 36, you begin to learn the physiology of fluid transport in Chapter 37. The terminology can be quite confusing, especially on a test. Try to explain, using plain language, how water and sugars move through a plant. Then apply the appropriate terms in your explanation. List the following terms, and make sure you are able to tell the difference between them:

- solute potential
- turgor pressure
- wall pressure
- pressure potential
- water potential
- water potential gradient

Try using each of these terms in a sentence that explains their meaning, and then try to compare and contrast the terms. If you can master the meaning of these terms, you are well on your way to understanding the physiology of fluid transport in vascular plants.

C. ASSESSING WHAT YOU'VE LEARNED

1. The force of the cell membrane swelling and pushing against the cell wall is called the:
   a. turgor pressure.
   b. wall pressure.
   c. pressure gradient.
   d. solute potential.

2. When the solute potential and the pressure potential of a cell are added together, the resulting quantity is called the cell's
   a. turgor pressure.
   b. wall pressure.
   c. pressure gradient.
   d. water potential.

3. Water potential gradient is the:
   a. overall movement of solutes.
   b. evaporation of water from the leaves.
   c. overall movement of water.
   d. pressure gradient minus the water potential.

4. The barrier inside the root that prevents water from leaking out of the vascular tissue is the:
   a. epidermis.
   b. Casparian strip.
   c. apoplast.
   d. root hair.

5. The pathway for water that lies within the cell walls is the:
   a. apoplastic pathway.
   b. symplastic pathway.
   c. Casparian pathway.
   d. endodermal pathway.

6. Cohesion is the:
   a. mutual attraction among like molecules.
   b. mutual attraction among unlike molecules.
   c. repelling of like molecules.
   d. repelling of unlike molecules.

7. A movement in response to pressure can also be called:
   a. solute flow.
   b. bulk flow.
   c. transpiration.
   d. CAM metabolism.

8. Which type of plants open their stomata at night to store CO_2?
   a. C_4 plants
   b. Rubisco plants
   c. CAM plants
   d. Transpiring plants

9. Translocation is the movement of:
   a. water through a plant.
   b. ions through a plant.
   c. sugars through a plant.
   d. xylem though a plant.

10. Where are the proton pumps responsible for phloem loading located?
    a. On the membranes of companion cells
    b. On the membranes of sieve-tube members
    c. On the membranes of root cells
    d. On the membranes of root hairs

CHAPTER 37—ANSWER KEY

Check Your Understanding

(a) When the solute potential and the pressure potential of a cell are added together, the resulting quantity is called the cell's water potential. It can be thought of as the tendency of water to move from one location to another. Water potential gradient is the overall movement of water when water potential differences within a series are contrasted. Plants tend to gain water from the soil and lose it to the atmosphere.

(b) Cohesion is the mutual attraction among like molecules, such as the hydrogen bonding among water molecules. Negative force or pull (tension) generated at the air-water interface is transmitted through the water present in the leaf cells to the water present in the xylem, on to the water in the vascular tissue of the roots, and finally on to the water in the soil. This continuous transmission of pulling force is possible because water is present throughout the plant and because all of the water molecules present hydrogen-bond to one another.

(c) Photosynthesis requires that the stomata on the leaves be open. The greater the requirement for photosynthesis, the greater the leaf surface area and the more stomata per unit area. More surface area and more stomata mean greater transpiration rates. There is a compromise because transpiration can damage a plant if there is not enough water.

(d) Secondary active transport allows an organism to build up a concentration gradient using active transport (energy), then use that gradient to transport passively (no energy) some other ion or molecule against its concentration gradient, typically using cotransport.

Assessing What You've Learned

1. a; 2. d; 3. c; 4. b; 5. a; 6. a; 7. b; 8. c; 9. c; 10. a

Looking Forward—Key Concepts in Later Chapters

Creating Antibodies—Chapter 49

Researchers raised antibodies to the AHA3 protein using the types of techniques introduced in **Chapter 49**. As that chapter points out, an antibody is a polypeptide that binds to a specific protein.

Plant Nutrition

Looking Back—Key Concepts from Earlier Chapters

Root Anatomy—Chapter 36

The general features of root anatomy were introduced in **Chapter 36**. Chapter 38 investigates further how nutrition is obtained through the roots.

Root Function—Chapter 37

The specifics of root function in water uptake were analyzed in **Chapter 37**. Again, this is broadened in Chapter 38 to include roots and plant nutrition.

Phospholipid Bilayer—Chapter 6

The interior of the phospholipid bilayer is uncharged; therefore, it resists the passage of ions. Some of the proteins found in certain cells span the bilayer and act as channels that allow the transit of specific ions. The ions are discussed in terms of plant nutrition in Chapter 38.

Mycorrhizae—Chapter 31

The fungi that live in close association with roots are called mycorrhizae. **Chapter 31** introduced two major types of mycorrhizae and presented evidence that these fungi transfer nitrogen and phosphorus, respectively, from soil to plant roots. This is further discussed in Chapter 38 in terms of nutrition.

KEY CONCEPTS

- In addition to needing carbon dioxide and water, plants require an array of essential nutrients to support growth.

- Essential nutrients are available as ions and are taken up by roots.

- Nutrient absorption occurs via specialized proteins in membranes of root cells.

- Most plants also obtain nitrogen or phosphorus from fungi associated with their roots.

- Some species of plants have specialized methods of obtaining nutrients, including associations with nitrogen-fixing bacteria, parasitism, and carnivory.

A. CHAPTER OUTLINE

38.1 Nutritional Requirements of Plants

Which Nutrients Are Essential?

- An **essential nutrient** should fulfill the following criteria:

  1. It is required for normal growth and reproduction—the plant cannot complete its life cycle without this nutrient.

  2. It is required for a specific structure or metabolic function.

- **Macronutrients** are the building blocks of nucleic acids, proteins, carbohydrates, phospholipids, and other key molecules required in relatively large quantities.

- Carbon, hydrogen, and oxygen make up about 96 percent of the dry weight of the plant.
- N, P, and K are particularly important because they are often **limiting nutrients**.
- **Micronutrients** (e.g., selenium) are required in relatively small quantities and usually function as cofactors for specific enzymes.

What Happens When Key Nutrients Are in Short Supply?

- Researchers used **hydroponic growth** to explore the effect of copper deficiency on tomatoes. They found that the copper-deprived individuals had stunted shoots and roots, dark foliage, curled leaves, and no flowers.
- Copper has since been revealed as a cofactor or component in several enzymes involved in redox reactions.

✔ **CHECK YOUR UNDERSTANDING**

(a)   What are the criteria for an essential nutrient in plants?

_____

_____

_____

38.2 Soil: A Dynamic Mixture of Living and Nonliving Components

- The process of soil building begins with solid rock. Water, wind, and organisms continually break tiny pieces off large rocks—this process is called **weathering**.
- As organisms occupy the substrate, they add their waste products and carcasses—this organic matter is **humus**.
- **Texture** refers to the proportions of different-sized particles present in soil and is important because:
  1. Texture affects the ability of roots to penetrate.
  2. Texture affects the soil's ability to hold water.
  3. Texture and water content dictate availability of oxygen.

The Importance of Soil Conservation

- **Soil erosion** occurs when soil is carried away from a site by wind or water.

- When plant cover is removed for forestry, farming, or suburbanization, soil erosion is accelerated.
- Techniques that maintain long-term soil quality and productivity are collectively called **sustainable agriculture**.

What Factors Affect Nutrient Availability?

- The elements required for plant growth exist in the soil as **ions**. Ions with negative charges (anions) usually dissolve in water because they interact with water molecules via hydrogen bonding. Ions with positive charges (cations) often interact with the negative charges found in organic matter and on the surfaces of the tiny, sheetlike particles called clay—they are less readily available.
- As solutes, negatively charged ions are available to plants for absorption; however, they are also easily washed out of the soil by rain. The loss of nutrients via washing is called **leaching**.
- **Cation exchange** occurs when protons or other cations bind to negative charges on soil particles and cause bound cations such as magnesium or calcium to be released to nearby roots.

✔ **CHECK YOUR UNDERSTANDING**

(b)   Describe the process of leaching and its effects. How can leaching be prevented?

_____

_____

_____

38.3 Nutrient Uptake

Mechanisms of Nutrient Uptake

- Some of the proteins found in certain cells span the lipid bilayer of a cell membrane and act as channels that allow the transit of specific ions.
- The large surface area of **root hairs** holds large numbers of membrane proteins, which contact the soil and selectively facilitate the passage of ions into the cell.

Establishing a Proton Gradient

- **Proton pumps** use ATP and pump an excess of H^+ ions on the exterior of the plasma membrane of the root-hair cells.

- This separation of charge is measured as a **voltage** across the membrane, and these pumps maintain a **membrane potential**.

- The effect of concentration and electrical charge on an ion is called the **electrochemical gradient**. When an electrochemical gradient favors the movement of an ion, no energy expenditure is required and the movement is described as passive.

Using a Proton Gradient to Transport Ions

- In the membranes of root hairs, the electrical gradient established by the proton pumps is strong enough to favor entry of positive ions.

- Researchers found that K^+ transport was passive; they found that the plant could import K^+ even when the outside concentrations of K^+ were extremely low.

- Cotransporters use the proton gradient to bring in an anion such as NO_3^- along with a proton.

Nutrient Transfer via Mycorrhizal Fungi

- Fungi that live in close association with roots are called **mycorrhizae**, and they are **symbiotic**.

- Many plants living in northern forests receive large quantities of nitrogen from fungi that wrap themselves around the epidermal cells of roots and radiate out into the surrounding soil. Plants that live in grasslands and in tropical forests receive much of the phosphorus they need from species of fungi whose bodies actually penetrate into the plant root interior.

Mechanisms of Ion Exclusion

- Many of the metal ions found in soils are poisonous to plants; even essential nutrients can become toxic if they are present in high levels.

Passive Exclusion

- The Casparian strip prevents some ions from entering the symplast and reaching the xylem.

- Passive exclusion also occurs in root hairs. If root hairs lack a membrane protein required for a certain ion to enter the cell, that ion won't enter.

- Researchers suggest that individuals from salt-tolerant populations have fewer sodium channels in their root hairs than do salt-intolerant individuals.

Active Exclusion

- Plants that grow on waste rock and soils from copper-mining operations experience large concentration gradients that favor an influx of this nutrient. How do plants neutralize excess nutrients?

- **Metallothioneins** are small proteins that bind to metal ions and prevent them from acting as a poison.

- The membrane that surrounds the vacuole—the **tonoplast**—contains transport proteins that move sodium ions from the cytosol into the vacuoles, where they cannot poison the enzymes.

CHECK YOUR UNDERSTANDING

(c)   What is a membrane potential?

_____

_____

_____

_____

38.4 Nitrogen Fixation

- Only selected bacteria are able to take up N_2, convert it to ammonia, and use it to fuel growth. This conversion process is called **nitrogen fixation**. Nitrogen fixation requires a series of specialized enzymes and cofactors, including an enzyme called nitrogenase.

- Bacteria in the genus *Rhizobium* are located in the roots of pea plants in structures called **nodules**. The bacteria provide the plant with ammonia, while the legume provides the bacteria with sugar and protection.

How Do Nitrogen-Fixing Bacteria Colonize Plant Roots?

- Roots do not contain a population of rhizobia from the moment of germination. The rhizobia must colonize the plant.

- Colonization is a complex process involving a series of specific interactions between the rhizobia and the legume.

- Recognition is possible because the root hairs of the pea family plants contain compounds called flavonoids. When rhizobia contact the flavonoids, they produce **Nod factors**, which in turn bind to the proteins on the membrane surface of the root hairs.

- Once Nod factors bind to the root-hair surface, they set off a chain of events that leads to a dramatic morphological change in the legume.

- As **Figure 38.15** shows, rhizobia proliferate around the tip of the root hair and then enter the epidermal cell through an invagination in the root-hair membrane called an **infection thread**.

- The infected cortex cells begin to divide rapidly, forming root nodules.

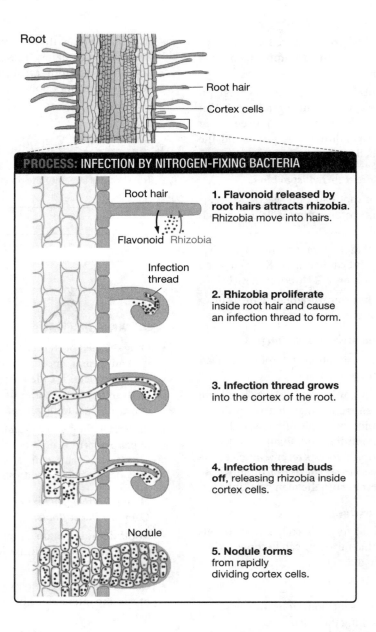

Root
— Root hair
— Cortex cells

PROCESS: INFECTION BY NITROGEN-FIXING BACTERIA

Root hair

1. Flavonoid released by root hairs attracts rhizobia. Rhizobia move into hairs.

Flavonoid Rhizobia

Infection thread

2. Rhizobia proliferate inside root hair and cause an infection thread to form.

3. Infection thread grows into the cortex of the root.

4. Infection thread buds off, releasing rhizobia inside cortex cells.

Nodule

5. Nodule forms from rapidly dividing cortex cells.

- Genetic changes also occur; for example, the production of **leghemoglobin** allows oxygen to bind to it rather than binding to and poisoning nitrogenase.

✔ **CHECK YOUR UNDERSTANDING**

(d) Why would certain plants allow fungi to inhabit their roots? Are there any disadvantages to the plant?

_____

_____

_____

38.5 Nutritional Adaptations of Plants

Epiphytic Plants

- **Epiphytes** are plants that never make contact with the soil and usually grow in the leaves or branches of trees. They absorb most of the nutrients they need from rainwater that collects in their tissues or in the crevices in bark.

Parasitic Plants

- In most cases, parasitic plants use photosynthesis to make their own sugars and tap the root systems of their hosts for water and essential nutrients.

Carnivorous Plants

- **Carnivorous** plants trap insects and other animals, kill them, and absorb the prey's nutrients.

- The Venus flytrap is an angiosperm native to bog habitats in the southeastern United States. It makes its carbohydrates via photosynthesis; but, since bogs have very low nitrogen, it traps organisms for nitrogen.

✔ CHECK YOUR UNDERSTANDING

(e) What is different about nutrient absorption in epiphytes compared to other plants?

_____

_____

_____

_____

B. MASTERING CHALLENGING TOPICS

You will find the concept of gradients to be the most confusing topic in this chapter. Understanding the movement of ions based on concentration and based on charge is actually quite straightforward, if they are separate entities. Once both concentration and charge are incorporated in one group of ions, predicting their movement begins to get quite complicated. Adding a selectively permeable membrane—like the plasma membrane of a cell—and then considering different ion types can confuse even the well-versed physiologist. Luckily, you only need to understand the basics, so make sure that you do. As with any other concept, it is best to start simple and gradually add complexity. Make sure you understand the concepts of concentration gradients and electrical gradients before you attempt to study the electrochemical gradient. Stick with one type of ion, and then try to set up gradients with other ion types and predict their movement. Save the study of the movement of mixed ions for a physiology class.

C. ASSESSING WHAT YOU'VE LEARNED

1. Which of the following is *not* a criterion for an essential nutrient?
   a. It is required for growth and reproduction.
   b. A cofactor is needed for absorption.
   c. No other element can substitute for it.
   d. It is required for a specific structure or metabolic function.

2. Researchers can use which one of the following techniques to explore plant nutrient deficiencies?
   a. Hydroponics
   b. Sun exposure
   c. Crop rotation
   d. Hyperbaric chambers

3. The process of water and wind breaking off tiny pieces of rock is called:
   a. soil erosion.
   b. humus.
   c. blasting.
   d. weathering.

4. Soil texture is important because:
   a. texture affects the ability of roots to penetrate.
   b. texture affects the soil's ability to hold water.
   c. texture and water content dictate oxygen availability.
   d. All of the above answers are correct.

5. Which of the following statements regarding soil erosion is *false*?
   a. Soil erosion is a very slow, natural process.
   b. Removal of trees can accelerate soil erosion.
   c. Suburbanization does not affect soil erosion.
   d. Farming can accelerate soil erosion.

6. The loss of nutrients via washing by rain is called:
   a. erosion.
   b. filtering.
   c. leaching.
   d. binding.

7. Fungi that live in close association with roots are called:
   a. rhizomes.
   b. mycorrhizae.
   c. rhizobium.
   d. rice.

8. Passive exclusion of ions occurs mainly by:
   a. the Casparian strip.
   b. the root hairs.
   c. both a and b
   d. sodium channels.

9. Nitrogen fixation is the ability to:
   a. convert N_2 to ammonia.
   b. convert ammonia to N_2.
   c. absorb N_2.
   d. metabolize nitriles.

10. When rhizobia contact the flavonoids of pea root hairs:
    a. the rhizobia produce Nod factors.
    b. the rhizobia produce flavonoids.
    c. the pea root hairs produce Nod factors.
    d. nothing happens.

CHAPTER 38—ANSWER KEY

Check Your Understanding

(a) An essential nutrient should fulfill the following criteria:

1. It is required for normal growth and reproduction—the plant cannot complete its life cycle without this nutrient.

2. It is required for a specific structure or metabolic function.

(b) As solutes, negatively charged ions are available to plants for absorption; but they are also easily washed out of the soil by rain. The loss of nutrients by means of washing is called leaching. Ions with positive charges, in contrast, often interact with negative charges found in organic matter and on the surfaces of the tiny, sheetlike particles called clay. As a result, the presence of organic matter and clay in soil slows leaching.

(c) Proton pumps use ATP and pump an excess of H^+ ions on the exterior of the plasma membrane of the root-hair cells. This separation of charge is measured as a voltage across the membrane, and these pumps maintain a membrane potential.

(d) Certain plants allow fungi to inhabit their roots so that they have a mutualistic or symbiotic relationship. The plant can gain nitrogen from the fungi and, in return, the fungi receive sugars and protection.

(e) An epiphyte is a plant that never makes contact with the soil. They absorb most of the nutrients they need from rainwater that collects in their tissues or in the crevices in bark.

Assessing What You've Learned

1. b; 2. a; 3. d; 4. d; 5. c; 6. c; 7. b; 8. c; 9. a; 10. a

Looking Forward—Key Concepts in Later Chapters

Mutualism—Chapter 50

A mutually beneficial relationship was shown between mycorrhizae and plants in Chapter 38. This topic is further explored in **Chapter 50**.

Bog Habitats—Chapter 50

Bog habitats are notoriously poor in nutrients—particularly in nitrogen. **Chapter 50** explains why this is so, while Chapter 38 gives an example of a plant adaptation to this nutrient-poor environment.

Plants and Stimuli—Chapter 39

Chapter 39 focuses on how a signal travels from the hairs on a leaf to the rest of the plant, and it examines how plants respond to light.

Ammonia Fertilizers—Chapter 55

The use of ammonia fertilizers has been causing serious pollution problems, as discussed in **Chapter 55**. For this reason, there has been intense interest in the phenoenon of nitrogen fixation.

Plant Sensory Systems, Signals, and Responses

KEY CONCEPTS

○━ Plants are selective about the information they process. They perceive a variety of environmental stimuli that affect their ability to grow and reproduce.

○━ When sensory cells receive a signal, they transduce the signal and respond by producing hormones that carry information to target cells elsewhere in the body.

○━ Target cells respond to hormonal stimulation in ways that increase the ability of the plant to survive and reproduce.

○━ Hormones are also responsible for regulating how plants grow throughout their lives, especially in response to changes in environmental conditions.

A. CHAPTER OUTLINE

39.1 Information Processing in Plants

- Plants have sophisticated systems for collecting information about their environment and responding in ways to maximize their chances of surviving, thriving, and producing offspring.

How Do Cells Receive and Transduce an External Signal?

- When a receptor changes shape in response to a stimulus, the information changes from an external signal to an intracellular signal—**signal transduction**.

- **Phosphorylation cascades** are triggered when the change in the receptor protein's shape leads to the addition of a phosphate group from ATP to the receptor or a nearby protein.

- **Second messengers** are produced when hormone binding results in the release of an intracellular signal from storage areas.

- A **hormone** is an organic compound that is produced in small amounts in one part of the plant and transported to target cells in another region of the plant, where it causes a physiological response.

- Signal transduction in a receptor cell often results in the release of a hormone that carries information to responder cells.

How Are Cell-Cell Signals Transmitted?

- The molecules that function as hormones are wildly diverse in structure.

- Plant hormones have several features in common, but they (1) can elicit a response only if a cell has an appropriate receptor, and (2) are active at extremely small concentrations.

How Do Cells Respond to Cell-Cell Signals?

- The signal in the cell's periphery is rapidly transduced to increased activity inside the cell via a signal transduction pathway.

- This results in the production of many phosphorylated proteins or the release of many second messengers.

- Therefore, the original signal can be amplified many times over.

- When cells respond to a hormone, the change in their activity helps the plant cope with the environmental change sensed by the receptor cell.

✔ **CHECK YOUR UNDERSTANDING**

(a) What is transduction, and why is it important to plants?

_____

_____

_____

39.2 Blue Light: The Phototropic Response

- Charles Darwin and his son (1881) performed the first experiments showing that plants respond to light. They also concluded that the blue part of the spectrum was involved.

- A **coleoptile** is a modified leaf that forms a sheath protecting the stems and leaves of young grasses.

- **Phototropism** is any type of directed movement in response to light. Plants are positively phototrophic.

Phototropins as Blue-Light Receptors

- A **pigment** is a molecule that absorbs certain wavelengths of light.

- Researchers found a membrane protein that is abundant in the tips of emerging shoots and that becomes phosphorylated in response to blue light. But was this protein the membrane receptor itself?

- Researchers began analyzing *A. thaliana* mutants that do not show a phototrophic response to blue light. The modified gene, named *PHOT1*, encodes the sunlight detector in plants. A phototropic response is triggered when the protein expressed by this gene is phosphorylated.

- Most recent research indicates that there are multiple blue-light receptors related to PHOT1—called **phototropins**.

Auxin as the Phototropic Hormone

- Although the Darwins published their hypothesis about chemical signals in 1881, Peter Boysen-Jensen confirmed it in 1913. He cut the tips off young oat shoots and placed either a porous block or nonporous block between the shoot and tip. The stems with the porous block showed normal phototropism, so he concluded that the phototropic signal was indeed chemical.

- Fritz Went later collected the phototrophic hormone in gelatinous blocks of agar. Stems responded by bending away from the source of the hormone, without a source of light present.

- The physical basis of the phototrophic response is cell elongation. Cells on the side of the shoot opposite to the source of light elongate in response to the phototrophic hormone. Because it promotes cell elongation, Went named the hormone **auxin** (from the Greek *auxein*, to increase).

Isolating and Characterizing Auxin

- Researchers succeeded in isolating and characterizing auxin. It is called IAA or indole acetic acid.

The Cholodny-Went Hypothesis

- This hypothesis contends that auxin produced in the tips of coleoptiles is shunted from one side of the tip to the other, in response to light; then it is transported down the shoot.

- Others proposed an alternative hypothesis: that auxin is broken down by blue light, which produces an asymmetric distribution.

- Winslow Briggs grew corn seedlings in the dark, cut off their tips, and placed the tips on agar blocks. He either kept these in the dark or exposed them to light from one side. Later, he put agar blocks on one side of the decapitated shoots. In response, the shoots from each treatment bent the same amount.

- This result is inconsistent with the auxin destruction hypothesis. If light destroys auxin, then the block exposed to light should be much less effective in inducing bending.

- Briggs generated additional evidence in support of the Cholodny-Went hypothesis by dividing tips fully or partially with mica. He showed that auxin had been transported from one side to the other, as in **Figure 39.7**.

The Cell-Elongation Response

- Researchers proposed that once ABP1 is bound to auxin, the signal transduction cascade leads to an increase in the number of membrane H^+-ATPases.

- Because the pH of the cell wall goes down when H^+-ATPases are active, the **acid-growth hypothesis** resulted.

- Two things have to happen for a plant cell to get larger:

  1. Water has to enter the cell. Proton pumps could lead to water entry because potassium and other positively charged ions often enter a cell after protons are pumped out. Water follows K^+ via osmosis.

  2. The cell wall must also expand. Cell wall expansion occurs in an **acidic** environment. Researchers believe that proteins called **expansins** are somehow responsible.

✔ **CHECK YOUR UNDERSTANDING**

(b) Briefly describe how auxin works to cause a plant to bend and grow toward light.

_____

_____

_____

39.3 Red and Far-Red Light: Germination and Stem Elongation

The Red/Far-Red Switch

- Red light drives photosynthesis just as blue light does.

- Far-red wavelengths are not absorbed strongly by photosynthetic pigments.

- Far-red light can also act as an on-off switch for seed germination.

Phytochromes as Red/Far-Red Receptors

- **Photoreversibility** occurs when light causes a photoreceptor pigment to change shape. Each conformation would be responsible for a different response.

- **Phytochromes** from young corn shoots were found to be photoreversible. When they were placed in a solution and exposed to red or far-red light, phytochrome color switched from blue to blue-green and black.

How Were Phytochromes Isolated?

- Young corn and bean plants lengthen their stems in response to light deprivation or exposure to excessive far-red light.

- Purified proteins from corn coleoptiles each hold a small pigment molecule that absorbs light in the red and far-red parts of the spectrum.

✔ **CHECK YOUR UNDERSTANDING**

(c) What is the difference between a pigment and a phytochrome?

_____

_____

_____

39.4 Gravity: The Gravitropic Response

- **Gravitropism** is the ability of plants to move in response to gravity. Charles and Francis Darwin (1881) found that roots stop responding to gravity if their **root caps** are removed.

- Researchers recently demonstrated that by killing tiny blocks of cells in *Arabidopsis* with a laser beam, they were able to determine that the cells directly under the epidermal cells at the tip of the root are the most important for initiating the gravitropic response.

The Statolith Hypothesis

- The **statolith hypothesis** holds that **amyloplasts** (starch-storing organelles) are the primary gravity sensors in plants. The idea is that gravity pulls the amyloplasts to the bottom of cells, where the force of the amyloplasts on the cell membrane activates pressure or stretch receptors that initiate the gravitropic response.

- Although recent experiments strongly support the statolith hypothesis, the receptor itself has yet to be found.

Auxin as the Gravitropic Signal

- Root cap cells that sense changes in the direction of gravitational pull respond by changing the distribution of auxin in the root tip.
- Auxin flows down the middle of the root and then toward the perimeter and away from the root cap.
- The differences in auxin concentrations result in differences in cell elongation (**Figure 39.13**).

CHECK YOUR UNDERSTANDING

(d) Explain the statolith hypothesis.

_____

_____

_____

39.5 How Do Plants Respond to Wind and Touch?

- Some plants can respond by touching or moving. This response, called **thigmotropism**, can be fast.
- A touch receptor cell transduces a mechanical signal into an electrical signal.
- An example of fast plant movement is a Venus flytrap.

CHECK YOUR UNDERSTANDING

(e) Describe in more detail how the Venus flytrap moves its leaves.

_____

_____

_____

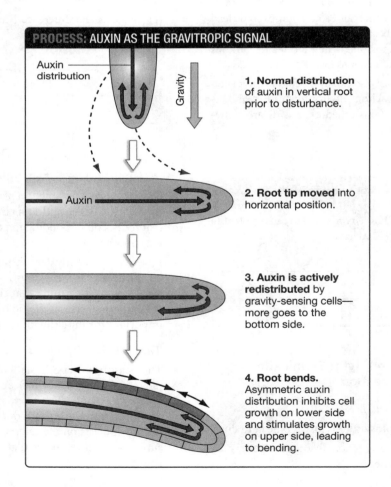

PROCESS: AUXIN AS THE GRAVITROPIC SIGNAL

Auxin distribution

Gravity

1. Normal distribution of auxin in vertical root prior to disturbance.

Auxin

2. Root tip moved into horizontal position.

3. Auxin is actively redistributed by gravity-sensing cells—more goes to the bottom side.

4. Root bends. Asymmetric auxin distribution inhibits cell growth on lower side and stimulates growth on upper side, leading to bending.

39.6 Youth, Maturity, and Aging: The Growth Responses

- Controlling growth in response to changes in age or environmental conditions is one of the most important aspects of information processing in plants.

Auxin and Apical Dominance

- **Apical dominance** is a growth in which the majority of stem elongation occurs at the apical meristem of the main shoot. The presence of this topmost meristem inhibits growth by apical meristems that are present lower down on the plant, in nodes.
- Apical dominance occurs because a continuous flow of auxin from the tips of growing shoots to the tissues below signals the direction of growth. If the signal stops, it means that growth has been interrupted. In response, lateral branches sprout and begin to take over for the main shoot.

Polar Transport of Auxin

- Transport of auxin is **polar**, or unidirectional. It travels in an "inverted umbrella" pattern. At the root, auxin is redistributed before it travels back up the plant.

What Is Auxin's Overall Role?

- Auxin clearly plays a key role in controlling growth via apical dominance, phototropism, and gravitropism. Auxin has other important effects.
- Fruit development is influenced by auxin produced by seeds within the fruit.
- Falling auxin concentrations are involved in the **abscission**, or the shedding of leaves and fruits, associated with **senescence**, or aging. Auxin interacts with ethylene in these processes.
- The presence of auxin in growing roots and shoots is essential for the proper differentiation of xylem and phloem cells in vascular tissue and the development of vascular cambium.
- Auxin stimulates the development of adventitious roots in tissue cultures and cuttings.

Cytokinins and Cell Division

- **Cytokinins** are a group of plant hormones that promote cell division.
- Cytokinins are synthesized in root tips, young fruits, seeds, growing buds, and other developing organs.
- Cytokinins act as growth hormones by activating the genes that keep the cell cycle going. In the absence of cytokinins, cells arrest at the G_1 checkpoint in the cell cycle and cease growth.

Gibberellins and ABA: Growth and Dormancy

- **Dormancy** is a temporary state of reduced or no metabolic activity.
- Growth responses and dormancy are mediated by two main hormones, **abscisic acid** (**ABA**) and the **gibberellic acids** (**GAs**).

Gibberellins Stimulate Shoot Elongation

- In stems, gibberellins appear to promote both cell elongation and rates of cell division. In seeds, GAs activate the transcription of digestive enzymes that support germination and growth.
- In these tissues, GA action must be coordinated with the effects of cytokinins and auxin.
- The strong association between shoot elongation and gibberellin dosage gave researchers an important tool for dissecting how GAs work.
- By using forward genetics, researchers began with mutant phenotypes and attempted to characterize the loci responsible for the defect.
- The locus responsible for the stem-length differences came to be known as *Le*. Early work on *Le* mutants showed that they attain normal height if they are treated with the gibberellin called GA_1.
- Researchers confirmed that the *Le* locus encodes an enzyme involved in GA synthesis by finding a locus in the pea genome that encodes an enzyme called 3β-hydroxylase. This enzyme adds a hydroxyl group to GA_{20} to produce GA_1. Mutant enzymes were unable to convert GA_{20} to GA_1.

Gibberellins and ABA Interact during Seed Dormancy and Germination

- By applying hormones to seeds, researchers learned that in many plants, ABA is the signal that inhibits seed germination and gibberellins are the signal that triggers embryonic development.
- During the germination of a barley seedling, **α-amylase** is released from a tissue called the **aleurone layer**. α-amylase is significant because it acts as a digestive enzyme that breaks the bonds between sugar subunits of starch. The enzyme diffuses into the carbohydrate-rich endosperm tissue and releases sugars that can be transported and used by the growing embryo.
- Adding GA to the aleurone layer increases the production of α-amylase; adding ABA decreases α-amylase levels.
- Summary of hormone action from GA and ABA research:
  1. A cell's response to a hormone often occurs because specific genes are turned on or off.

2. Hormones rarely act on DNA directly. Instead, a receptor on the surface of a cell usually receives the message and responds by initiating a chain of events that leads to gene activation or repression.

3. Different hormones interact at the molecular level because they induce different transcription activators and repressors.

ABA Closes Guard Cells in Stomata

- Early work on stomatal opening and closing suggested that ABA is involved in communicating from roots to leaves about water conditions. Applying ABA to the exterior of guard cells causes them to close.

- Researchers were able to confirm two important predictions:

  1. ABA concentrations in roots on the dry side of the pot were extraordinarily high.

  2. ABA concentrations in the leaves of experimental plants were much higher than in the controls.

- These results suggested that ABA from roots is transported to leaves, and that it actually does serve as an early warning system of drought stress.

- Changes in cell shape are related to changes in the activity of H^+-ATPases in the plasma membrane. See the previous section on stem elongation and the acid-growth hypothesis.

Brassinosteroids and Body Size

- Growth spurts are triggered by surges in steroid hormones called brassinosteroids.

- The hormones were originally discovered in *Brassica napus*, a crop plant that is the source of canola oil.

- They are steroids.

- In contrast to other steroid molecules that bind to intracellular receptors, brassinosteroids bind to receptors on the membrane surface.

- The genes for synthesis of brassinosteroids appear to be homologous with genes required for steroid hormones in animals—they are derived from a similar gene in a common ancestor of plants and animals.

Ethylene and Senescence

- **Senescence** is the regulated process of aging, decline, and eventual death of an entire organism or particular organs, such as plant fruits and leaves.

- **Ethylene** is strongly associated in three aspects of senescence in plants:

  1. Fruit ripening, which eventually leads to rotting
  2. The fading of flowers
  3. Abscission, or the detachment of leaves

- The **abscission zone** is a region of the leaf petiole that is more sensitive to ethylene in the tissue. As a result, it degrades first and the leaf breaks off at this point.

✓ CHECK YOUR UNDERSTANDING

(f) How do GA and ABA interact to promote seed germination, and what cues determine the on/off switch for germination?

_____

_____

_____

39.7 Pathogens and Herbivores: The Defense Responses

How Do Plants Sense and Respond to Pathogens?

- The ability to cause disease is called **virulence**, and disease-causing agents are called **pathogens**.

- Responses to attacks are called **induced defenses**.

- The rapid and localized death of one or a few infected cells is called the **hypersensitive response** (HR).

The Gene-for-Gene Hypothesis

- Plants have disease-resistant genes known as **resistance (R) loci**. The genes associated with virulence or antivirulence in pathogens are known as **avirulence (avr) loci**.

- The idea that *R* and *avr* gene products interact in a specific way is called the **gene-for-gene hypothesis**. *R* genes produce receptors, and *avr* genes produce **ligands**—molecules that bind to receptors.

Why Is the Existence of So Many Resistance Genes and Alleles Significant?

- *R* genes are similar in structure, and within a population of plants there are usually many different alleles at each *R* locus.

- Different alleles allow plants to recognize different proteins from the same pathogen.

The Hypersensitive Response (HR)

- The binding of an *R* gene product to an avr protein triggers the production of NO and **reactive oxygen intermediates** (ROIs).

- ROIs disrupt enzymes in cells at the point of infection and thus kill those cells.

An Alarm Hormone Extends the HR

- Direct application of salicylic acid to tissues triggers **systemic acquired resistance (SAR)**—a slower, widespread set of events.
- **Methyl salicylate (MeSA)** (a molecule derived from salicylic acid) increases dramatically after tissues are infected with a pathogen.

How Do Plants Sense and Respond to Herbivore Attack?

- **Protease inhibitors** found in plants makes herbivores sick. They can then detect this compound by taste and avoid the plant.
- **Systemin** is a peptide that triggers a series of reactions that eventually result in the synthesis of jasmonic acid, which initiates the production of defense compounds.

Pheromones Released from Plant Wounds Recruit Help from Wasps

- An organism that is free living as an adult but parasitic as a larvae is called a **parasitoid**.
- Plants produce **pheromones**—chemical messengers that are synthesized by an individual and released into the environment—to attract wasps in response to attack by caterpillars.

✔ CHECK YOUR UNDERSTANDING

(g) Describe a wound response hormone and how it is activated.

_____

_____

_____

B. MASTERING CHALLENGING TOPICS

The world of plant hormones is a complicated one, and this chapter just begins to scratch the surface of complexity. The key concepts you must understand are that a hormone can have (1) a multitude of actions, depending on the location in the organism; and (2) a multitude of interactions, depending on both the location and the nature of the interactive chemical. You will encounter more actions and interactions of ABA and the GAs, so it is important to learn their present functions as discussed in this chapter. Organize both hormones in a chart and list their actions, locations of those actions, and possible interactions with each other. Continue to describe their actions in a broad sense and then proceed to describe these at the molecular level, citing the research used to determine each hormone. This process will be of great benefit as you move to the chapters dealing with hormones in animals.

C. ASSESSING WHAT YOU'VE LEARNED

1. Any directed movement in response to light is called:
   a. photoperiodism.
   b. tropism.
   c. phototropism.
   d. gravitropism.

2. An organic substance produced in small quantities in one part of the plant and transported to target cells, causing a physiological response, is:
   a. sodium.
   b. a protein.
   c. a hormone.
   d. a nucleotide.

3. Which plants would be likely to bloom in midsummer?
   a. Day-neutral plants
   b. Short-day plants
   c. Long-day plants
   d. None of the above

4. Which molecules hydrolyze ATP and catalyze the addition of a phosphate group?
   a. Pigments
   b. Phototropins
   c. Phosphate genes
   d. Protein kinases

5. The technical name for blue-light receptors is:
   a. phytochromes.
   b. phototropins.
   c. pigments.
   d. transducins.

6. The process that translates a signal from the environment into an energy form that the plant decodes is:
   a. signal transduction.
   b. signal deciding.
   c. signal transmission.
   d. neural transmission.

7. The hypothesis stating that amyloplasts are the primary gravity sensors in plants is the
   a. statolith hypothesis.
   b. gravitational pressure hypothesis.
   c. amyloplast hypothesis.
   d. none of the above

8. What is the best explanation for how cell elongation occurs?
   a. Active pumping of water
   b. Active pumping of ions
   c. Active pumping of ions, passive movement of water
   d. Active pumping of ions, active movement of water

9. Which of the following is *true* about auxin transport?
   a. Auxin transport is random.
   b. Auxin transport is bidirectional.
   c. Nothing is known about the specifics of auxin transport.
   d. Auxin transport is unidirectional.

10. The two main hormones that mediate growth responses and dormancy are:
    a. auxin and cytokinins.
    b. gibberellins and cytokinins.
    c. gibberellins and auxin.
    d. gibberellins and abscisic acid.

CHAPTER 39—ANSWER KEY

Check Your Understanding

(a) Transduction is the changing of energy from one form into another. In a plant, a signal from the environment, such as light, can be transduced into chemical or electrical energy. It is important for plants to be able to respond to various environmental stimuli.

(b) In response to light, auxin moves to the opposite side of the plant and causes cell growth and thus stem elongation. This uneven stem elongation then causes the plant to bend toward the light.

(c) Plants contain a wide variety of pigments that absorb light. These molecules include the chlorophylls and carotenoids involved in photosynthesis. A phytochrome is a specific type of pigment that is known for being photoreversible.

(d) The statolith hypothesis holds that amyloplasts (starch-storing organelles) are the primary gravity sensors in plants. The idea is that gravity pulls the amyloplasts to the bottom of cells, where the force of the amyloplasts on the cell membrane activates pressure or stretch receptors that initiate the gravitropic response.

(e) The Venus flytrap voltage change has a characteristic action potential pattern when an organism lands on the receptor hairs. The action potential is a very rapid change in membrane potential—from negative to positive. In response to touch, membranes of receptor hairs on the trap surface depolarize. Depolarization of receptor cells triggers action potentials in other cells across the leaf. When action potentials reach effector cells on the outer trap surface, the cells swell and the trap shuts.

(f) ABA is the signal that inhibits seed germination and gibberellins are the signal that triggers embryonic development. The appropriate state for the on/off switch is determined by environmental cues such as temperature, moisture, and light.

(g) A wound response hormone is called systemin. It has been confirmed to move from damaged sites to undamaged sites. Systemin initiates a series of reactions that eventually produce jasmonic acid, which in turn initiates the production of many insecticide-type compounds, including protease inhibitors.

Assessing What You've Learned

1. c; 2. c; 3. c; 4. d; 5. b; 6. a; 7. a; 8. c; 9. d; 10. d

Looking Forward—Key Concepts in Later Chapters

Fruit Ripening—Chapter 40

As **Chapter 40** will show, fruit ripening is interpreted as an adaptation that enhances the attractiveness of fruits to birds, mammals, and other animals that disperse seeds to new locations.

The Action Potential—Chapter 45

In **Chapter 45** we will examine the molecular mechanisms responsible for the action potential in much more detail.

Plant Reproduction

Looking Back—Key Concepts from Earlier Chapters

Inbreeding—Chapter 25

Self-fertilization is the most extreme form of inbreeding. Inbreeding is defined as reproduction among relatives. **Chapter 25** explored the consequences of inbreeding in detail and presented data showing that inbred offspring have poor fitness.

Diploid and Haploid Cells—Chapters 11 and 12

Except for certain red, brown, and green algae, plants are the only organisms that have both a multicellular form that is diploid and a multicellular form that is haploid. Recall from **Chapters 11** and **12** that diploid cells have two copies of each chromosome, while haploid cells have one.

Life Cycles of Plants—Chapter 30

In flowering plants, the gametophyte generation is tiny, short lived, and dependent on the sporophyte for nutrition. Chapter 30 explored the diversity of life cycles found in plants in more depth.

Tissue Layers—Chapter 32

The three tissue types formed during embryogenesis were discussed in detail in **Chapter 32**. Chapter 40 shows the relationship between these tissue types and the originating cells.

KEY CONCEPTS

- Plants undergo alternation of generations, in which a diploid sporophyte phase alternates with a haploid gametophyte phase.

- Sporophytes produce spores by meiosis. Gametophytes produce gametes by mitosis.

- In angiosperms, male and female gametophytes are microscopic and are produced inside flowers.

- Male pollen grains are portable while female gametophytes are encased in an ovary and are retained in the flower.

- When pollen grains land on a flower, they deliver sperm cells that fertilize the egg produced by the female gametophyte.

- Seeds contain an embryo and a food supply surrounded by a coat.

- In angiosperms, the walls of the ovary develop into a fruit that encloses the seed(s). In many cases fruits function in seed dispersal.

A. CHAPTER OUTLINE

40.1 An Introduction to Plant Reproduction

Sexual Reproduction

- Most plants reproduce sexually. **Sexual reproduction** is based on the reduction division known as **meiosis** and on **fertilization**, the fusion of haploid cells called **gametes**.

- **Sperm** are small cells from the male that contribute genetic information in the form of DNA but few or no nutrients to the offspring. Female gametes are called **eggs**, which contribute a store of nutrients to the offspring.

- **Outcrossing** occurs when male and female gametes are exchanged between individuals of the same species. When **self-fertilization** occurs, a sperm and an egg from the same individual unite to form a progeny.

- The primary advantage of **selfing** is that successful **pollination** is virtually assured.

The Land Plant Life Cycle

- Plants are the only organisms that have both a multicellular form that is diploid and a multicellular form that is haploid. This life cycle is called **alternation of generations**.

- The diploid phase is called the **sporophyte**, and the haploid phase is called the **gametophyte**.

- A **spore** is a cell that grows directly into an adult individual. A **gamete** is a reproductive cell that must fuse with another gamete before growing into a new individual.

- Meiosis occurs in sporophytes and results in the production of haploid spores, which are produced in structures called **sporangia**.

- Spores divide by mitosis to form multicellular, haploid gametophytes.

- Fertilization occurs when two gametes fuse to form a diploid zygote. The zygote grows by mitosis to form the sporophyte.

- Study the following diagram (**Figure 40.2**) to make sure you understand these life-cycle intricacies.

Asexual Reproduction

- **Asexual reproduction** does not involve meiosis or fertilization. It leads to offspring that are genetically identical to the parent plant. The major advantage of asexual reproduction is its high efficiency.

(a) Liverworts: Gametophytes are large and long lived; the sporophyte is small and short lived.

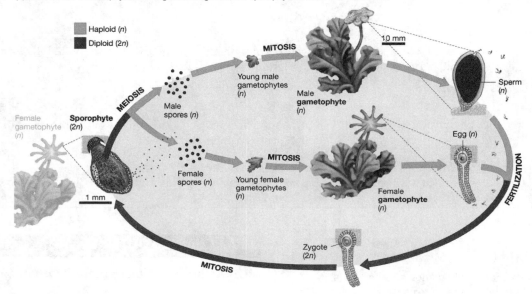

(b) Angiosperms: Sporophyte is large and long lived; gametophytes are small (microscopic) and short lived.

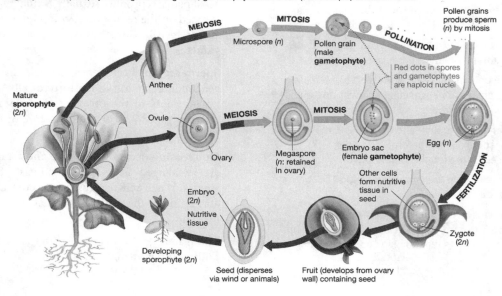

- However, if a fungus or other disease-causing agent infects an individual, it will probably succeed in infecting the plant's asexual offspring as well.

(a) What is the difference between outcrossing and self-fertilization?

_____

_____

_____

40.2 Reproductive Structures

When Does Flowering Occur?

- A **flower** is a modified shoot that develops from a compressed stem and highly modified leaves.
- Flower formation can be stimulated by external or internal cues, or both.
- **Photoperiodism** is any response by an organism that is based on the photoperiod—the relative lengths of day and night.
- Plants fall into three categories:
  1. **Long-day plants** bloom in midsummer when days are at their longest.
  2. **Short-day plants** bloom in spring, late summer, or fall.
  3. **Day-neutral plants** flower without regard to photoperiod.
- Clock proteins rise during the day and trigger expression of a gene called *CONSTANS* (*CO*).
- In long-day plants, high levels of CO stimulate production of the flowering hormone.
- In short-day plants, high levels of CO inhibit production of the flowering hormone.

The General Structure of the Flower

- **Sepals** are leaflike structures that comprise the outermost part of the flower.
- **Petals** are arranged in a stem in a whorl. They are often brightly colored and advertise the flower to visually oriented animals such as bees, wasps, and hummingbirds.
- The **nectary** produces sugar-rich fluid called **nectar**, which is harvested by many of the animals that visit.
- An entire group of petals is called a **corolla**. The male reproductive structure of angiosperms is called a

stamen, while the female reproductive structure is called the **carpel**.

- Flowers that contain both stamens and carpels are referred to as **perfect**. If they have either stamens or carpels, they are called **imperfect**.
- When imperfect flowers have both stamens and carpels on the same plant (but not the same flower), they are called **monoecious**.
- If an individual plant produces either stamens or carpels (but not both), the species is **dioecious**.

How Are Female Gametophytes Produced?

- The carpel consists of three parts: the **stigma**, the **style**, and the **ovary**. The **ovule**, inside the ovary, is where meiosis takes place and where the female gametophyte is produced. Four nuclei result from meiosis, but three of them degenerate. The remaining haploid nucleus is the **megaspore**.
- The megaspore divides by mitosis to produce eight **nuclei** (the number varies among species), which segregate to different positions and form seven cells, comprising the **embryo sac**. The most important elements of the embryo sac are the **egg** and the **polar nuclei**. The egg is located near the base of the structure by an opening called the **micropyle**.

How Are Male Gametophytes Produced?

- The **stamen** consists of the **anther** and a **filament**. Cells inside the anther undergo meiosis. Each of the haploid cells that results is called a **microspore**.
- Each microspore becomes a **pollen grain**. When mature, a pollen grain consists of a small **generative cell** enclosed within a large *vegetative cell*. The vegetative cell develops a *hard coat*, which protects the cell contents and generative cell when pollen grains are shed.
- Pollen grains *do not* contain sperm. Instead, they are tiny **gametophytes**. The haploid generative cell will produce sperm cells via mitosis.

(b) Compare and contrast a spore and a gamete.

_____

_____

_____

40.3 Pollination and Fertilization

Pollination

- **Pollination** is the transfer of pollen from an anther to a stigma.

- In many angiosperms, pollen is carried from the anther of one individual to the stigma of a different individual (**cross-pollination**) on the body of an insect, bird, or bat that moves from flower to flower. Animal pollination is an example of **mutualism**.

Selfing versus Outcrossing: Costs and Benefits

- Selfing ensures pollination while outcrossing increases genetic diversity and thus defense against pathogens.

- Plants may have mechanisms to prevent selfing, including:

  1. *Temporal avoidance*—male and female gametophytes mature at different times.

  2. *Spatial avoidance*—monoecious species only have one type of gametophyte per plant while dioecious species may separate anthers and stigma to reduce the chances of selfing.

  3. *Molecular matching*—proteins on the surface of the pollen and stigma may have to interact for pollination to be successful.

Why Did Pollination Evolve?

- The more recently evolved plant groups do not need water in order for sexual reproduction to occur. As a result, the evolution of pollen made these species less dependent on wet habitats and allowed colonization of dryer habitats.

- Pollination evolved into a much more precise process when animals began to act as pollinators. Insect pollination is an important adaptation because it makes sexual reproduction much more efficient.

Does Pollination by Animals Encourage Speciation?

- Insect pollination appears to be closely associated with the formation of new flower and insect species.

- A biologist documented differences in alpine skypilot flowers. In the lower timberline regions, flowers are small and have a skunk-like odor, while in the tundra regions, flowers are large and smell sweet.

- Large bumblebees pollinate the tundra flowers, while small flies pollinate the timberline flowers. Flower morphology is in tune to attract the type of insect that is abundant in that area. These two types of flowers are on their way to evolving into two distinct species.

Fertilization

- The male gametophyte produces a long projection called a **pollen tube** that grows down the length of the style.

- This growth is directed by chemical attractants released by **synergids**—cells in the female gametophyte that lie close to the egg.

- When the pollen tube reaches the micropyle, the sperm are discharged into the ovule.

- **Fertilization** occurs when a sperm and an egg actually unite to form a diploid zygote.

- In angiosperms, an event called **double fertilization** takes place.

- One sperm unites with the egg nucleus to form the zygote. The other sperm nucleus moves through the embryo sac and fuses with the two polar nuclei to form a single triploid ($3n$) cell.

- The triploid cell resulting from the second fertilization begins a series of mitotic divisions that form tissue called **endosperm**, which stores nutrients.

✔ CHECK YOUR UNDERSTANDING

(c) What is double fertilization, and what makes it different from the typical fertilization process?

_____

_____

_____

40.4 The Seed

- As a seed matures, the embryo and endosperm develop inside the ovule and become surrounded by a covering called the **seed coat**. At the same time, the ovary around the ovule develops into a fruit.

Embryogenesis

- **Embryogenesis** is the process by which a single-celled zygote becomes a multicellular embryo.

- When a zygote divides, it produces two **daughter cells**: the **basal cell** divides to form a row of single cells, and the **terminal cell** is the parent cell of all the cells in the embryo.

- The terminal cell and its progeny divide and sort into three groups:

  1. The exterior layer forms **the protoderm**.

  2. The cells just inside the exterior layer form the **ground meristem**.

  3. A group of cells (**procambium**) in the core of the embryo becomes the vascular tissue.

- **Cotyledons** are the seed leaves that take up the nutrients in the endosperm and store them. The

hypocotyl is the initial stem, and the **radicle** is the first root structure.

- By the time the embryo matures, the three tissue types have differentiated and the structures just mentioned have formed. The seed tissues then dry, and the embryo becomes **quiescent**.

The Role of Drying in Seed Maturation

- Water loss is an adaptation that prevents seeds from germinating on the parent plant.

- Researchers have shown that as water leaves the seed during drying, sugars begin to interact with the hydrophilic parts of cell components. These interactions stabilize the proteins and membranes before damage occurs.

Fruit Development and Seed Dispersal

- After fertilization occurs in angiosperms, the cells that make up the ovary develop a structure called the **pericarp**. This protects the seeds. The mature structure is called a **fruit**.

- **Simple fruits** like the apricot develop from a single flower that contains a single carpel. **Aggregate fruits** like the raspberry develop from a single flower containing many separate carpels. **Multiple fruits** like the pineapple develop from many flowers with many carpels.

- Most dry fruits are either dispersed by wind or simply fall to the ground. Animals are the most common dispersal agent of fleshy fruits.

Seed Dormancy

- Even though water and oxygen are available, seeds may not germinate; this process is called **seed dormancy**.

- Researchers have studied lotus plant seeds from an old dried-up lakebed in China. The oldest seed that germinated was 1300 years old.

- Dormancy is usually a feature of seeds from species that inhabit seasonal environments.

- Dormancy is usually interpreted as an adaptation that allows seeds to remain viable until conditions improve.

What Role Does ABA Play in Dormancy?

- ABA levels have not been universally correlated with dormancy. Researchers have concluded that there is no single universal mechanism for initiating and maintaining seed dormancy.

How Is Dormancy Broken?

- For some seeds, the seed coat must be disrupted, or **scarified**, for germination to occur. For others, exposure to water, light, oxygen, or fire will initiate germination.

- The cue that triggers germination is a reliable signal that conditions for seedling growth are favorable for a particular species in a particular environment.

Seed Germination

- Seeds do not germinate without water, because water uptake is the first event in germination.

- During the *phase 1* of water uptake, oxygen consumption and protein synthesis in the seed increase dramatically, but no new messenger RNAs are transcribed. Based on these observations, biologists have concluded that the earliest events in germination are driven by mRNAs that are stored in the seed.

- During the *phase 2*, when water uptake stops, newly transcribed mRNAs appear and are translated into protein products. Mitochondria also begin to multiply. In effect, seeds take up enough water to hydrate their existing proteins and membranes and then begin to manufacture the proteins and mitochondria needed to support growth.

- During *phase 3*, water uptake resumes as growth begins. This second bout of water uptake enables cells to enlarge and the embryo to burst from the seed coat.

B. MASTERING CHALLENGING TOPICS

The concepts in this chapter are fairly straightforward, but there are many new anatomical terms and physiological processes to learn. The best way to begin studying the anatomy is by drawing pictures and labeling them. Once you are comfortable with the anatomy, move on to the processes of pollination, fertilization, and embryogenesis. Using the figures in your textbook, map out each process while drawing panels and describing each step. Use the proper terms as you have learned them for each structure. Last, try describing each process with only words, no pictures. This will help you to understand the concepts and not just simply memorize facts.

C. ASSESSING WHAT YOU'VE LEARNED

1. Which kind of flower contains both male and female structures?
   a. Perfect flower
   b. Imperfect flower
   c. Haploid flower
   d. Diploid flower

2. Which kind of reproduction increases the genetic variation of the offspring?
   a. Asexual reproduction
   b. Alternation of generations
   c. Sexual reproduction
   d. Sporophyte production

3. What are the leaflike structures that comprise the outermost part of the flower?
   a. The petals
   b. The sepals
   c. The nectary
   d. The leaves

4. The sugar-rich fluid in the flower that is harvested by many animals is called:
   a. juice.
   b. nectar.
   c. pollen.
   d. fruit.

5. Which of these statements about reproductive structures is *false*?
   a. The stamen is the female reproductive structure of angiosperms.
   b. The carpel consists of the stigma, the style, and the ovary.
   c. The stamen consists of the anther and filament.
   d. The pollen grains are tiny gametophytes.

6. When pollen is carried from the anther of one individual to the stigma of a different individual, it is called:
   a. self-pollination.
   b. self-fertilization.
   c. cross-pollination.
   d. cross-fertilization.

7. Animal pollination is an example of:
   a. parasitism.
   b. competition.
   c. commensalism.
   d. mutualism.

8. "One sperm unites with an egg nucleus, and another sperm unites with two polar nuclei." This scenario best describes:
   a. fertilization in gymnosperms.
   b. double fertilization in angiosperms.

   c. pollination in angiosperms.
   d. fertilization in angiosperms.

9. Which of these statements about seed dormancy is *false*?
   a. ABA is the universal hormone initiating seed dormancy.
   b. Seeds can lie dormant for hundreds of years.
   c. Dormancy is usually a feature of seeds from species that inhabit seasonal environments.
   d. Dormancy is usually interpreted as an adaptation that allows seeds to remain viable until conditions improve.

10. Which of the following is absolutely required for seed germination?
   a. Fire
   b. Light
   c. Water
   d. Soil

CHAPTER 40—ANSWER KEY

Check Your Understanding

(a) Outcrossing occurs when male and female gametes are exchanged between individuals of the same species. When self-fertilization occurs, a sperm and an egg from the same individual unite to form a progeny.

(b) A spore is a reproductive cell that grows into a new individual directly. A gamete is a reproductive cell that must fuse with another gamete before growing into a new individual.

(c) Fertilization occurs when a sperm and an egg actually unite to form a diploid zygote. In angiosperms, an event called double fertilization takes place. One sperm unites with the egg nucleus to form the zygote. The other sperm nucleus moves through the embryo sac and fuses with the two polar nuclei to form a single triploid ($3n$) cell.

Assessing What You've Learned

1. a; 2. c; 3. b; 4. b; 5. a; 6. c; 7. d; 8. b; 9. a; 10. c

Looking Forward—Key Concepts in Later Chapters

Disease—Chapter 50

If a fungus or other disease-causing agent infects a big bluestem individual, it will probably succeed in infecting the plant's asexual offspring as well. As **Chapter 50** will show, plants fight disease with a wide variety of molecules.

Animal Form and Function

Natural Selection and Adaptation—Chapter 24

As **Chapter 24** explained, an adaptation is a trait that allows individuals to survive and reproduce better than individuals that lack this trait. This chapter also introduced some of the techniques that biologists use to establish that traits have a genetic basis.

Genetic Drift—Chapter 25

Natural selection is not the only process that leads to changes in allele frequencies over time. Evolution occurs via the random process called genetic drift—the movement of alleles into and out of populations by migration (gene flow) and the constant introduction of new alleles by mutation.

Form and Function—Chapter 7

The shape of proteins often relates to their role as enzymes or structural components of the cell. **Chapter 7** pointed out that strong correlations exist between the structure and function of the rough ER, Golgi apparatus, mitochondrion, chloroplast, and other organelles.

KEY CONCEPTS

- Structure correlates with function at all biological levels.
- Surface-area-to-volume relationships govern many aspects of biology, including metabolism, digestion, the delivery of nutrients, and the removal of wastes.
- Animals maintain homeostasis by employing regulatory systems such as negative feedback inhibition.
- Animals are able to control internal body temperature by using sophisticated thermoregulatory systems.

A. CHAPTER OUTLINE

41.1 Form, Function, and Adaptation

- Adaptive evolution occurs when the frequency of alleles subject to natural selection increases from one generation to the next *and* the resulting changes in the characteristics of the population lead to higher average fitness in a particular environment.

CHECK YOUR UNDERSTANDING

(a) What processes lead to changes in allele frequency in a population?

_____

_____

_____

The Role of Fitness Trade-Offs

- One of the most important constraints on adaptation involves **trade-offs**. This may involve expenditures of time or energy.

- Comparative physiologists have investigated the predicted trade-off in egg size and egg number by manipulating these parameters in side-blotched lizards (**Figure 41.1**). They found that egg size was inversely related to egg number in these manipulated animals. Furthermore, larger eggs gave rise to larger offspring that had higher survival rates than smaller offspring. Females that produced an intermediate number of intermediate sized eggs generated the highest total number of surviving offspring, thus demonstrating that there was a trade-off between egg size and egg number.

Adaptation and Acclimatization

- **Adaptation** occurs when there are genetic changes in a population in response to natural selection. In contrast, **acclimatization** occurs when an individual's phenotype changes in response to a change in environmental conditions.

CHECK YOUR UNDERSTANDING

(b) Using examples, explain the difference between adaptation and acclimatization.

_____

_____

_____

(c) Why aren't all traits adaptive? Give an example.

_____

_____

_____

41.2 Tissues, Organs, and Systems: How Does Structure Correlate with Function?

- Researchers studying medium ground finches on the Galápagos island of Daphne Major found that beak size and shape vary among individuals, and that this trait is heritable. During a major drought, much of the population disappeared. The researchers found that survivors tended to have much deeper beaks than did the birds that died. What was the likely cause of this enduring morphological trait?

- Natural selection is reflected in a strong correlation between structure and function. This relationship is true at multiple levels: structure correlates with function at molecular, cellular, and tissue levels as well as at the gross anatomical level (**Figure 41.2**).

Structure-Function Relationships at the Molecular and Cellular Levels

- Protein structures correlate with their function as enzymes, structural cell components, or membrane channels.

- A cell's function can be derived from its composition and shape.

Tissues Are Groups of Similar Cells That Function as a Unit

- A **tissue** is a group of cells with the same structure and function.

- There are four basic types of tissues (**Figure 41.3**):

  1. **Connective tissue** is made up of cells that are loosely arranged in a liquid, jellylike, or solid extracellular matrix. **Loose connective tissue** serves as a packing material between organs. **Cartilage** and **bone** support the body, and **blood** transports material throughout the body.

  2. Nerve cells or neurons make up **nervous tissue**. Although their shapes vary widely, all neurons have connections to other cells and deliver signals in the form of electrical impulses.

  3. **Muscle tissue** functions in both voluntary and involuntary movement. Muscle cells are packed with specialized proteins that move in response to **phosphorylation**.

  4. The **epithelial tissues** cover the outside of the body and line the surfaces of organs. This tissue consists of layers of tightly packed cells.

Organs and Organ Systems

- An **organ** is a structure that serves a specialized function and consists of several tissues (**Figure 41.7a**).

- An **organ system** consists of tissues and organs that work in conjunction to perform a function (**Figure 41.7b**).

CHECK YOUR UNDERSTANDING

(d) Explain the difference between a tissue and an organ.

_____

_____

_____

41.3 How Does Body Size Affect Animal Physiology?

Surface Area/Volume Relationships: Theory

- As a cell gets larger, its volume increases much faster than its surface area. Examine **Figure 41.9** and convince yourself that this is true.

- Given this relationship, the transport processes that occur across the cell membrane would have to support disproportionately more and more cell volume as the cell size increased.

Surface Area/Volume Relationships: Data

- **Metabolic rate** is defined as the overall rate of energy consumption by an individual. Because it is usually based on aerobic respiration, it is measured in units of ml O_2 consumed per hour.

Comparing Mice and Elephants

- Even though an elephant consumes a great deal more oxygen per hour than a mouse does, if you looked at 1 gram of mouse tissue, it would consume about 12 times more oxygen per hour than 1 gram of elephant tissue would (**Figure 41.10**).

- The leading hypothesis to explain this pattern is based on surface area/volume ratios. Many aspects of metabolism, digestion, the delivery of nutrients, and the removal of wastes depend on the exchange of materials across surfaces.

Changes during Development

- As the Atlantic salmon grows from newly hatched larvae to adult fish, it switches from oxygen diffusion across the body surface to oxygen diffusion across the gills to accommodate its decrease in skin surface area in relation to its volume (**Figure 41.11**).

Adaptations That Increase Surface Area

- Because gills have flattened, sheetlike structures called lamellae, the organ has an extremely high surface area relative to its volume (**Figure 41.12**). Diffusion of gasses from water into blood can take place rapidly enough to keep up with the growth in volume of the developing fish.

- If the function of a cell or tissue depends on diffusion, its structure most likely has a shape that increases its surface area relative to its volume. Can you explain how the structure of the digestive and circulatory systems are consistent with these observations?

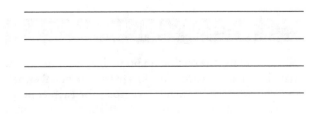

CHECK YOUR UNDERSTANDING

(e) Why does surface area increase more slowly than volume in a growing organism?.

_____

_____

_____

_____

41.4 Homeostasis

- **Homeostasis** is the maintenance of relatively constant chemical and physical conditions in an animal's cells, tissues, and organs.

Homeostasis: General Principles

Two Approaches to Achieving Homeostasis

- **Conformation** is when an individual's internal condition closely matches that of its external environment.

- **Regulatory homeostasis** occurs when an individual maintains its internal state, within limits, despite changes in its external environment.

The Role of Epithelium

- Epithelium plays a vital role in creating an internal environment that is dramatically different from the external environment, and in maintaining physical and chemical conditions inside an animal that are relatively constant. Its most basic function is to control the exchange of materials across surfaces.

Why Is Homeostasis Important?

- Maintaining a constant internal environment is important for optimal function of enzymes and membrane permeability.

The Role of Regulation and Feedback

- Achieving homeostasis requires a regulatory system that has a specific **set point**, or normal value. The temperature that you set on the dial of a room's thermostat is an example of a set point.

- Regulatory systems typically consist of three components: a sensor, an integrator, and an effector (**Figure 41.13**).

- A **sensor** is a structure that senses some aspect of the external or internal environment.

- An **integrator** is a component of the nervous system that evaluates the incoming sensory information and "decides" if a response is necessary to achieve homeostasis.

- An **effector** is any structure that helps to restore the desired internal condition.

- Regulatory systems depend on **negative feedback** to maintain constant conditions. Negative feedback occurs when the regulatory system makes a change in the opposite direction to a change in internal conditions (equivalent to a heater turning on and raising temperature when a room has cooled off).

41.5 How Do Animals Regulate Body Temperature?

Mechanisms of Heat Exchange

- Animals exchange heat with the environment by varying combinations of conduction, convection, radiation, and evaporation (**Figure 41.14**).

- **Conduction** is the direct transfer of heat between two physical bodies that are in contact with each other. The rate at which conduction occurs depends on the surface area, the steepness of temperature difference, and how well each body conducts heat.

- **Convection** occurs when air or water moves over the body surface. As the speed of water or air flow increases, so does the rate of heat transfer.

- **Radiation** is the transfer of heat between two bodies that are not in direct physical contact. The major source of radiant energy is the Sun.

- **Evaporation** is the phase change that occurs when liquid water becomes a gas. It leads to heat loss only. As a result, water is an efficient coolant on a hot day.

Variation in Thermoregulation

- Many animals can control their body temperature through the process of **thermoregulation** (**Figure 41.15**). An **endotherm** produces heat in its own tissue, while an ectotherm relies on heat *gained from the environment*. Birds and mammals are endotherms; most other animals are ectotherms.

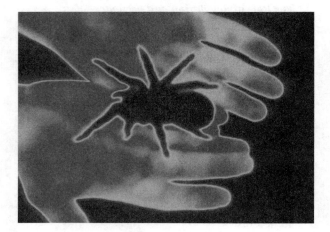

- **Homeotherms** keep their body temperature constant, whereas **heterotherms** experience changes in body temperature. Heat is produced by muscle activity during movement and sometimes by the involuntary muscle contractions known as shivering.

Endothermy and Ectothermy: A Closer Look

- Compared to endotherms, ectotherms have low metabolic rates.

- Endotherms are able to maintain enzymes at optimal temperatures at all times, so mammals and birds can remain active in winter and at night. Due to their high metabolic rates and insulation, endotherms are also able to sustain very high levels of aerobic activities, such as running or flying.

- To fuel their high metabolic rates, endotherms have to obtain large quantities of energy-rich food.

- Ectotherms are able to thrive with much lower intakes of food. They can also use a greater proportion of their energy intake to support reproduction.

- Chemical reactions are temperature dependent, so muscular activity and digestion slow down dramatically as body temperature drops. As a result, ectotherms are more vulnerable to predation in cold weather than endotherms are.

Temperature Homeostasis in Endotherms

- In mammals, this process is coordinated by neurons in the hypothalamus. Neural signals from temperature receptors in the skin (the "sensors") are interpreted by the hypothalamus (the "integrator"), which then causes varying responses, ranging from metabolic to behavioral (the "effectors") (**Figure 41.16**).

- Air conducts heat poorly; thus, it is a good insulator. Endothermic animals have elaborate external structures that trap air, slow the rate of heat exchange, and conserve body heat.

- Water is a good conductor; thus, the body temperature of most aquatic invertebrates and fish is the same as the water they inhabit.

Countercurrent Heat Exchangers

An Example: Gray Whale Tongues

- Some aquatic mammals and birds have **countercurrent heat exchangers**, where arteries and veins lie beside each other and freely exchange heat (**Figure 41.17**). Explain how this anatomical arrangement conserves heat in gray whale tongues.

A Multiplier Effect

- Although the temperature differential at any one point in a countercurrent heat exchanger may be small, the overall temperature differential from one end to the other of the system is large. Thus, sometimes the system is called **countercurrent multipliers**.

B. MASTERING CHALLENGING TOPICS

When studying the section on body size and scaling, you may find that some of the concepts are rather abstract, and you will need concrete examples for each concept described. Surface area/volume relationships cannot simply be written down and studied. You should draw the examples listed in your text to learn what it means for the volume of a cell or organism to outgrow its surface area.

The study of allometry is very complex, and this chapter only touches on the basics. It is intuitive to understand that an elephant consumes more oxygen than a mouse does, but do you really understand why the mouse consumes more oxygen per gram of tissue than an elephant does? As you imagine these comparisons between animals, also place the animals in their natural environment and try to apply what you find about their anatomy and physiology to how they interact with their environment. Maintaining a "big picture" of these concepts in addition to learning everything in little chunks will also help you in understanding this chapter.

C. ASSESSING WHAT YOU'VE LEARNED

1. Which of the following descriptions is an example of *acclimatization*?
   a. Over many generations, the average length of giraffe necks has increased.
   b. The optimal temperature at which goldfish can swim at maximum speed will decrease if the goldfish is maintained in cold water.
   c. The average brain size of the ancestors of modern humans has increased dramatically over time.
   d. The emergence of the seed represents an important event in the evolution of land plants.

2. Consider several objects of different size that are heated and then allowed to cool. Which of the following would cool at the fastest rate?
   a. An object with a surface area of 20 and a volume of 10
   b. An object with a surface area of 40 and a volume of 30
   c. An object with a surface area of 8 and a volume of 2
   d. An object with a surface area of 10 and a volume of 30

3. What happens to the mass-specific metabolic rate of an organism as it gets bigger?
   a. Mass-specific metabolic rate decreases.
   b. Mass-specific metabolic rate remains constant.
   c. Mass-specific metabolic rate increases.

4. Patrick Wells and Alan Pinder found that the percentage of oxygen uptake by the gills increases as an Atlantic salmon grows from a small larva to an adult fish. Which of the following could explain this result?
   a. The gills provide a much lower surface-to-volume ratio, so that gas transfer becomes more efficient as the organism gets larger.
   b. The surface area of skin is much larger than the surface area of the gills.

   c. As an individual grows, the skin surface area decreases in relation to its volume (the surface-to-volume ratio drops). To keep the individual from suffocating, gills must take over most of the gas exchange activity.
   d. As an individual grows, the skin surface area increases in relation to its volume (the surface-to-volume ratio increases). To keep the individual from suffocating, gills must take over most of the gas exchange activity.

5. Which of the following describes an allometric relationship?
   a. A relationship in which two quantities change at the same rate
   b. A relationship in which one quantity increases while one quantity decreases
   c. A relationship in which two quantities both decrease
   d. A relationship in which two quantities change at different rates

6. The relationship between heart mass and body mass differs for dogs and cats. Which of the following statements explains these differences?
   a. Cat hearts are smaller because they have smaller bodies. The heart does not have to work as hard in a smaller animal.
   b. There is no difference in the size of dog and cat hearts when you compare animals of similar size.
   c. Cats are much more active than dogs. This increased activity is reflected in the data when comparing heart mass to body mass for dogs and cats.
   d. The hunting style of dogs involves long-distance chases. Cats use short, quick sprints to ambush their prey. The increased activity in the hunting style of dogs requires a larger heart.

7. Which of the following terms implies the maintenance of relatively constant physical conditions?
   a. Homeostasis
   b. Acclimatization
   c. Adaptation
   d. Isometry

8. Which of the following terms describes an animal that maintains body temperature by producing heat in its own tissues?
   a. Isotherm
   b. Ectotherm
   c. Endotherm

9. Which of the following describes an ectotherm?
   a. A person shivering in the cold
   b. A turtle basking in the sun on a log
   c. A small animal increasing its rate of oxidation of fat in brown adipose tissue to produce heat
   d. A hormonally induced increase in basal metabolism for the purpose of generating heat

10. The direct transfer of heat between two physical bodies that are in contact with each other is:
    a. evaporation.
    b. radiation.
    c. convection.
    d. conduction.

11. Which of the following would explain the observation that marine mammals maintain very thick layers of insulating fat?
    a. The fat helps prevent heat loss through the process of evaporation.
    b. The fat helps prevent heat loss through the process of conduction.
    c. The fat prevents heat loss through the process of convection.

12. When researchers compare the cells in tissues collected from endotherms and ectotherms of similar size, what can they expect to find?
    a. Endotherm cells will have a much greater density of mitochondria.
    b. Ectotherm cells will have a much greater density of mitochondria.
    c. Endotherm cells will have a much greater density of ribosomes.
    d. Ectotherm cells will have a much greater density of ribosomes.

CHAPTER 41—ANSWER KEY

Check Your Understanding

(a) Changes in allele frequency in a population can occur by natural selection, genetic drift, gene flow, and mutation.

(b) Adaptation occurs when a population changes in response to natural selection. For example, the heavy fur of a mammal living at high altitudes has evolved over time to allow the animal to survive winter climates. Acclimatization occurs when an individual changes in response to a change in environmental conditions. For example, a mammal migrating in spring from a low-altitude valley to high mountain slopes will develop more red blood cells to cope with the lower oxygen at higher altitude.

(c) Animals possess a wide variety of structures that were present in ancestral populations but are not currently adaptive. Vestigial traits like this are widespread. Some traits exist as holdovers from structures that appear early in development. Human males have rudimentary mammary glands only because nipples form in the early embryo before sex hormones begin influencing the development of organs.

(d) A tissue is a group of cells with the same structure and function. An organ is a structure that serves a specialized function and consists of several tissues.

(e) In a cell, the surface area grows as a power of 2; the volume increases as a power of 3. As a result, its volume increases much faster than its surface area. Given this, the transport processes that occur across the cell membrane would have to support disproportionately more and more cell volume as the cell size increased.

(f) Allometry occurs when changes in body size are accompanied by disproportionate changes in anatomical structures or physiological processes. In skeletal size and body size, the volume involved is the mass that must be supported by the skeleton. The area involved is the cross-sectional area of the bones supporting the mass. As mass increases, the amount of bone required for support has to increase disproportionately.

Assessing What You've Learned

1. b; 2. c; 3. a; 4. c; 5. d; 6. d; 7. a; 8. c; 9. b; 10. d; 11. b; 12. a

Looking Forward—Key Concepts in Later Chapters

Membrane Transport and Surface Area—Chapter 42

Nutrients such as glucose must diffuse into the cell, and waste products such as urea and carbon dioxide must diffuse out. As **Chapter 42** explains, the rate at which these and other molecules and ions diffuse depends in part on the amount of surface area available. In contrast, the rate at which nutrients are used and waste products are produced depends on the volume of the cell. **Chapter 42** will also explore how the gill functions as an adaptation for increased surface area.

The Vertebrate Eye—Chapter 46

The importance of historical constraint will surface again in the discussion of the vertebrate eye in **Chapter 46**. Although nonbiologists sometimes use the vertebrate eye as an example of a "perfect" adaptation, optometrists can attest that the organ is often defective.

Chapter 42

Water and Electrolyte Balance in Animals

Looking Back—Concepts from Earlier Chapters

Osmosis and Diffusion—Chapter 6

Chapter 6 introduced the processes called diffusion and osmosis. Diffusion describes the movement of substances from regions of higher concentration to regions of lower concentration. The movement of water from regions of higher to lower concentration is called osmosis.

Gas Exchange in Plants—Chapter 36

Water balance in land animals mirrors the situation in land plants. Land plants also lose water as an inevitable by-product of gas exchange. As **Chapter 36** pointed out, land plants lose huge quantities of water in the course of obtaining CO_2 through their stomata.

Surface-to-Volume Ratio—Chapter 41

Small organisms have a high surface-area-to-volume ratio. Insects have a relatively large surface area with which to lose water, and a small volume in which to retain it.

KEY CONCEPTS

- Animals must actively prevent losing or gaining too much water and electrolytes.

- Aquatic animals have chloride cells that are important in the regulation of internal water and salt balance.

- Insects have special structures such as spiracles, chitin of the exoskeleton, and Malphighian tubules to minimize water loss.

- Terrestrial vertebrates have kidneys that act to maintain internal water and electrolyte balance.

A. CHAPTER OUTLINE

42.1 Osmoregulation and Osmotic Stress

- Substances move from regions of higher concentrations to lower concentrations by the process of **diffusion** (**Figure 42.1a**).

- The diffusion of water from areas of higher to lower concentration is called **osmosis** (**Figure 42.1b**). The concentration of dissolved substances in a solution, measured in moles per liter, is referred to as its **osmolarity**.

What Is Osmotic Stress?

- Cells and tissues that are gaining or losing too much water are under **osmotic stress**.

- The ability to maintain homeostasis with respect to water and electrolyte balance in the face of osmotic stress is called **osmoregulation**.

Osmotic Stress in Seawater

- Marine fish are **hypotonic** to seawater, so water tends to flow out of the gill epithelium. Marine fish drink large quantities of seawater and excrete salt so that their cells do not shrivel and die (**Figure 42.2**).

- Marine animals must rid the body of excess ions they gain from drinking seawater (**Figure 42.2**).

(a) How do freshwater fish and saltwater fish differ in their osmotic stresses?

_____

_____

_____

Osmotic Stress in Freshwater

- In freshwater fish, gill epithelium is **hypertonic** in relation to the surrounding water (**Figure 42.3**). Gill epithelial cells can gain water via osmosis.

- Freshwater animals must replace electrolytes that are lost by obtaining them in food or from the surrounding water.

Osmotic Stress on Land

- Terrestrial animals tend to lose water to the environment through evaporation from body surfaces and as water vapor during breathing, and lose electrolytes in urine and feces (**Figure 42.4**).

- Water and electrolyte balance is maintained by drinking, eating, or gaining metabolic water.

How Do Cells Move Electrolytes and Water?

- Solutes move across membranes by passive or active transport. **Passive transport** often occurs via channels or by transmembrane proteins that act as carriers (**Figure 42.5a**).

- **Active transport** requires membrane proteins that act as a pump, requiring energy in the form of **ATP** (**Figure 42.5b**).

- **Secondary active transport** uses the electrochemical gradient established by pumps to drive **cotransporters**, **symporters**, and **antiporters**.

42.2 Water and Electrolyte Balance in Aquatic Environments

How Do Sharks Excrete Salt?

- The shark **rectal gland** secretes a concentrated salt solution. Ions can be concentrated in this way only if they are actively transported against a concentration gradient.

The Role of Na^+/K^+-ATPase

- Epithelial cells of the shark rectal gland contain Na^+/K^+-**ATPase,** which is essential for salt excretion. Disabling Na^+/K^+-ATPase with a plant compound,

ouabain, prevented shark rectal glands from producing a concentrated salt solution.

A Molecular Model for Salt Excretion

- In 1977, Silva and colleagues developed the following molecular *model for salt excretion:*

1. Na^+/K^+-*ATPase* pumps sodium ions out of epithelial cells across the basolateral surface and into the surrounding extracellular fluid (**Figure 42.6**). The pump creates a large electrochemical gradient favoring the diffusion of Na^+ into the cell.

2. A Na^+/Cl^- cotransporter, powered by the gradient favoring Na^+ diffusion, brings these two ions from the extracellular fluid into epithelial cells across their basolateral surfaces.

3. Although sodium ions are pumped back out by Na^+/K^+-ATPase, chloride ion concentrations build up inside the cell as a result of the cotransport process. A chloride channel located in the apical membrane of the epithelial cells allows Cl^- to diffuse out down its concentration gradient.

4. Sodium ions also diffuse into the lumen of the gland, following their charge and concentration gradient. But instead of passing through the epithelial cells as Cl^- does, Na^+ diffuses out through spaces between the cells.

- Using the steps provided here, can you draw the model for salt excretion?

(b) How did the study of the shark rectal gland yield practical applications for human medicine?

_____

_____

_____

A Common Molecular Mechanism Underlies Many Instances of Salt Excretion

- Marine birds and reptiles drink salt water and must excrete NaCl. In their nostrils, they have **salt-excreting glands** that function much like the shark rectal gland.

- Because marine teleost (bony) fish are hypertonic to seawater, salt constantly diffuses in through their **gills**. Their gills contain specialized cells that are configured precisely like the cells lining the shark rectal gland (**Figure 42.2**).

- Cells similar to those of the rectal glands of sharks are responsible for transporting salt in the **kidneys** of mammals.

How Do Freshwater Fish Osmoregulate?

Salmon and Sea Bass as Model Systems

- Some fish such as salmon and sea bass experience both saltwater and freshwater environments during their lifetime. Researchers have capitalized on this to investigate how fish modulate their **chloride cells** to deal with different osmotic stresses.

A Freshwater Chloride Cell?

- Researchers have evidence to support the hypothesis that there is a freshwater version of the classical chloride cell that allows fish to deal with different osmotic stresses:

  1. The location of Na^+/K^+-ATPase on gills of salmon living in freshwater and saltwater is different.
  2. Different forms of Na^+/K^+-ATPase on gills may be activated depending on whether a fish inhabits freshwater or salt water.
  3. $Na^+/Cl^-/K^+$ cotransporters switch from the basolateral side of chloride cells to the apical side when sea bass are moved from saltwater to freshwater environments (**Figure 42.7**).

✓ CHECK YOUR UNDERSTANDING

(c) Why is a migrating salmon such an interesting animal with respect to osmotic stress?

_____

_____

_____

_____

_____

42.3 Water and Electrolyte Balance in Terrestrial Insects

How Do Insects Minimize Water Loss from the Body Surface?

- Water loss is an inevitable by-product of respiration. Insects have a relatively large surface area with which to lose water and a small volume in which to retain it.
- An insect's **trachea** connects with the atmosphere through openings called **spiracles**. Muscles just inside each spiracle close or open the pore. This is an important adaptation for minimizing water loss (**Figure 42.8a**).
- The insect exoskeleton consists of chitin and layers of protein (**Figure 42.8b**). This combination of chitin and protein is known as **cuticle**. A layer of wax covers the surface, which is highly impermeable to water.

✓ CHECK YOUR UNDERSTANDING

(d) What structures and processes are used by terrestrial insects to minimize water loss from the body?

_____

_____

_____

_____

Types of Nitrogenous Wastes: Impact on Water Balance

- **Ammonia** is a by-product of **catabolic** reactions. Those animals that excrete ammonia directly usually lose a lot of water. Fish detoxify ammonia by diluting it to a low concentration in watery urine.

Forms of Nitrogenous Waste Vary among Species

- Freshwater fish dilute ammonia and excrete it in watery urine.
- Freshwater and saltwater fish excrete ammonia by diffusion across gills.
- Humans convert ammonia to nontoxic **urea** and excrete it. Urea requires water for excretion, however, but not as much as excreting ammonia directly.
- Birds, reptiles, and terrestrial arthropods convert ammonia to **uric acid**. Because uric acid is not soluble in water, it can be excreted as a dry paste.

Why Do Nitrogenous Wastes Vary among Species?

- The type of nitrogenous waste produced by an animal correlates with the amount of osmotic stress it endures (**Table 42.1**).

- There is a trade-off between the energetic cost of excreting urea or uric acid and the benefit of conserving water.

(e) Why is uric acid such an efficient waste product?

_____

_____

_____

_____

Maintaining Homeostasis: The Excretory System

- To avoid osmotic stress, insects must carefully regulate the composition of a blood-like fluid called **hemolymph.**

Filtrate Forms in the Malpighian Tubules

- Pre-urine formation occurs in the **Malpighian tubules,** which have a large surface area and are in direct contact with the hemolymph (**Figure 42.9a**).

- The epithelial cells of the tubules transport potassium ions into the tubules, and water follows by osmosis to form the filtrate.

The Hindgut: Selective Reabsorption of Electrolytes and Water

- Reabsorption results in formation of hypertonic urine, conservation of water, and elimination of nitrogenous wastes.

- Because the urine that is finally expelled from the anus is often strongly hyperosmotic to pre-urine, a large amount of water must be reabsorbed from the filtrate. Selective reabsorption of electrolytes and water occurs in the hindgut. Na^+/K^+-ATPase and chloride pumps actively drive the reabsorption process (**Figure 42.9b**).

Regulating Water and Electrolyte Balance: An Overview

- There are no mechanisms for actively pumping water. Water moves between cells or body compartments via osmotic gradients.

(f) What is a Malpighian tubule, and which animals have them?

_____

_____

_____

_____

- The formation of the filtrate is not selective.

- In contrast to filtrate formation, reabsorption is highly selective. Only valuable ions and molecules are reabsorbed.

- Reabsorption is tightly regulated by the membrane pumps and channels involved in reabsorption.

- Reabsorption of water is based on osmotic gradients created by ion pumps.

42.4 Water and Electrolyte Balance in Terrestrial Vertebrates

The Structure of the Kidney

- The paired **kidneys** receive considerable blood flow via the **renal artery.** Urine formed in each kidney is carried to the **bladder** via a long, thin tube called the **ureter.** Urine leaves the bladder and exits the body via the **urethra** (**Figure 42.10**).

Filtration: The Renal Corpuscle

- **Nephrons** are the functional units of the kidney— they perform the work in maintaining water and electrolyte balance (**Figure 42.11**).

- Urine formation begins in the **renal corpuscle** (**Figure 42.12a**). Water and small solutes are pushed out of the capillary's pores, through the slits in the surrounding cells, and into the fluid-filled space inside **Bowman's capsule** (**Figure 42.12b**). Because proteins, cells, and other large components of blood would not fit through the pores, they would not enter the nephron.

- Pressure is much higher inside the capillaries than it is in the surrounding capsule. This pressure differential forces water and solutes out of the blood and into the capsule space.

- Can you predict how glomerular filtration would be affected by high blood pressure?

Reabsorption: The Proximal Tubule

- Fluid leaves the Bowman's capsule and enters a convoluted structure called the **proximal tubule**. The fluid entering this area has the composition of plasma minus the blood cells and large proteins.

- The epithelial cells of the tubule have a prominent series of small projections called **microvilli** facing the lumen (**Figure 42.13a**).

- Na^+/K^+-ATPase in the basolateral membranes removes intracellular Na^+ and creates a gradient for Na^+ entry. In the apical membrane, Na^+-dependent cotransporters simultaneously bind Na^+ and another solute such as glucose, an amino acid, or Cl^-.

- Water follows the movement of these solutes by osmosis. It leaves the proximal tubule via membrane proteins called **aquaporins** (**Figure 42.13b**).

✔ CHECK YOUR UNDERSTANDING

(g) What are aquaporins, and how do they relate to water and ion balance?

_____

_____

_____

Creating an Osmotic Gradient: The Loop of Henle

- In 1942 Werner Kuhn hypothesized that the loop of Henle functions as a countercurrent multiplier, setting up an osmotic gradient. He proposed that an exchange of water and solutes occurs between each of the segments in the loop and cells outside.

Testing Kuhn's Hypothesis

- A strong gradient in osmolarity existed from the cortex to the medulla, as predicted in Kuhn's hypothesis (**Figure 42.14**).

How Is the Osmotic Gradient Established?

- In the ascending loop of Henle, Na^+ and Cl^- constitute at least 60 percent of the solutes; urea is about 10 percent. In contrast, in the collecting duct both ions are rare, and the major solute is urea (**Figure 42.15b**).

- Na^+ and Cl^- are actively pumped out of the ascending limb.

- The descending limb is highly permeable to water but almost completely impermeable to solutes. The thick ascending limb, in contrast, is highly permeable to Na^+ and Cl^-, moderately permeable to urea, and almost completely impermeable to water.

A Comprehensive View of the Loop of Henle

- As water flows down the descending limb, the fluid inside the loop loses water to the tissue surrounding the nephron.

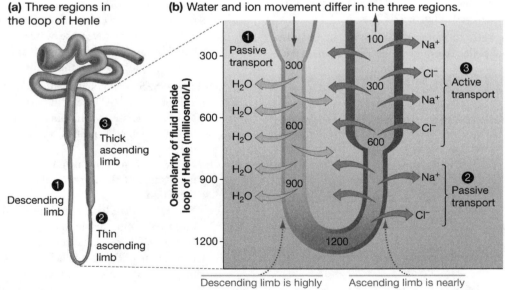

(a) Three regions in the loop of Henle

(b) Water and ion movement differ in the three regions.

Descending limb is highly permeable to water but impermeable to solutes

Ascending limb is nearly impermeable to water but highly permeable to Na^+ and Cl^-

- The fluid inside the nephron begins to lose Na^+ and Cl^-, and as it ascends it encounters surrounding tissues with progressively lower osmolarity.
- In the thick ascending limb, additional Na^+ and Cl^- ions are actively transported out of the nephron.

The Vasa Recta Removes Water and Solutes That Leave the Loop of Henle

- The **vasa recta** is a network of blood vessels that run along the loop of Henle. The vasa recta helps to return water and electrolytes to the body and maintain the osmotic gradient at the descending limb.

The Collecting Duct Leaks Urea

- The innermost section of the collecting duct is permeable to urea, which allows for the steep osmotic gradient in the space surrounding the nephron.

Regulating Water and Electrolyte Balance: The Distal Tubule and the Collecting Duct

- The amount of Na^+, Cl^-, and water reabsorbed in the distal tubule and collecting duct varies with the animal's condition and is under **hormonal control**.
- Aldosterone activates sodium pumps to allow for the reabsorption of Na^+ in the distal tubule.
- If an individual is dehydrated, a molecule called **antidiuretic hormone (ADH)** is released from the brain. When ADH interacts with cells lining the distal tubule and collecting duct, aquaporin channels are inserted into the membrane. The cells thus become highly permeable to water, and large amounts of water are reabsorbed (**Figure 42.17**).
- What happens to water reabsorption in the distal tubule and collecting duct in the absence of ADH?

CHECK YOUR UNDERSTANDING

(h) How does ADH affect water balance in the human kidney?

_____

_____

_____

_____

_____

B. MASTERING CHALLENGING TOPICS

The entire chapter has a central theme of sodium-potassium pumps. They are always involved in some way with water and ion balance. Once you have learned this basic tenet, the difficult part is understanding the most complex process of water and ion balance—which takes place in the vertebrate kidney. By working through this chapter and learning how more primitive vertebrates and invertebrates maintain water and ion homeostasis, you can focus on the vertebrate kidney. Just like the chapter and study guide organize the functions of the kidney, so should you. Study the filtration process first, then reabsorption, and then secretion, as you trace the path of urine formation beginning at the renal corpuscle and ending at the collecting duct. Since the formation of urine is such a dynamic process, make sure to note changes in urine formation under stressful conditions—dehydration, high blood pressure, and response to the various hormones discussed.

C. ASSESSING WHAT YOU'VE LEARNED

1. The major challenges to water and ion balance for most marine vertebrates are:
   a. gaining water from and losing ions to the environment.
   b. gaining water and ions from the environment.
   c. losing water to and gaining ions from the environment.
   d. losing water and ions to the environment.

2. Marine elasmobranchs, like the shark, avoid water loss to the environment by:
   a. drinking seawater and using their rectal gland to excrete excess ions.
   b. maintaining the osmolarity of their body fluids as hypotonic relative to seawater.
   c. gaining water through their gills and rectal gland.
   d. gaining water through their gills.

3. Which of the following statements accurately describes the role of Na^+/K^+-ATPase in the shark rectal gland?
   a. Na^+/K^+-ATPase, localized in the apical membrane of the lumen cells, pumps Na^+ into the lumen of the rectal gland.
   b. Na^+/K^+-ATPase, localized in the basolateral membrane of the lumen cells, pumps Na^+ into the blood.
   c. Na^+/K^+-ATPase, localized in the apical membrane, pumps Na^+ into the blood.
   d. Na^+/K^+-ATPase, localized in the basolateral membrane of the lumen cells, pumps Na^+ out of the blood and into the lumen cells.

4. The major site for osmoregulation in teleosts (bony fishes) is the:
   a. rectal gland.
   b. gill.
   c. kidney.
   d. integument (skin).

5. An animal that lives in a dry environment needs to minimize water loss and will thus most likely secrete nitrogenous waste in the form of:
   a. uric acid.
   b. urea.
   c. creatinine.
   d. ammonia.

6. Which of the following statements correctly summarizes the primary role of the Malpighian tubule system of insects in formation of filtrate?
   a. K^+ ions are actively secreted into the lumen of the Malpighian tubules; other electrolytes, water, and waste products follow passively.
   b. Na^+ and Cl^- ions are actively transported into the lumen of the Malpighian tubules; water and waste products follow.
   c. Cl^- is actively transported into the lumen by the actions of chloride cells; Na^+, waste products, and water follow.
   d. Waste products such as uric acid are actively secreted into the lumen, and water follows.

7. Which of the following statements concerning the composition of the filtrate formed by the renal corpuscle of the human kidney is correct?
   a. Blood and filtrate are identical in composition.
   b. Filtrate has a higher concentration of the waste product urea than blood does.
   c. The filtrate is similar to blood but without proteins and blood cells.
   d. The filtrate has a lower concentration of glucose than blood does.

8. Which of the following represents the primary driving force for the filtration of blood within the renal corpuscle?
   a. Pressure within the glomerular capillaries is lower than pressure in Bowman's capsule.
   b. Pressure within the glomerular capillaries is greater than pressure in Bowman's capsule.
   c. The collecting duct creates a suction that draws fluid into Bowman's capsule.
   d. The filtration slits are present in the glomerular capillaries.

9. Which statement most accurately describes the primary role of the loop of Henle in urine formation?
   a. The loop of Henle deposits Na^+ and Cl^- in the medullary region of the kidney, increasing its osmolarity.
   b. The hormone ADH acts on the loop of Henle to increase water reabsorption.
   c. The ascending limb of the loop of Henle contributes to the high osmolarity of the medullary region by depositing urea.
   d. The loop of Henle is responsible for most of the Na^+ reabsorbed by the nephron tubule.

10. The amount of Na^+ excreted in the urine is primarily determined by:
    a. the amount of Na^+ secreted into the proximal convoluted tubule.
    b. the amount of Na^+ reabsorbed by the proximal convoluted tubule.
    c. the amount of ADH released by the posterior pituitary gland.
    d. hormonal regulation of Na^+ reabsorption in the distal convoluted tubule.

11. Which of the following would result in the production of hypotonic urine?
    a. Dehydration in an individual
    b. An increase in the permeability of the collecting duct to water
    c. A decrease in ADH release by the brain
    d. A decrease in water reabsorption by the loop of Henle

12. From an evolutionary perspective, which of the following adaptations makes the most sense?
    a. Long loops of Henle in the kidneys of a river otter
    b. Long loops of Henle in the kidneys of a desert fox
    c. An extensive Malpighian tubule system in a freshwater beetle
    d. Renal corpuscles that are few in number and small in size in the kidney of a freshwater fish

CHAPTER 42—ANSWER KEY

Check Your Understanding

(a) Freshwater animals must replace electrolytes that are lost by obtaining them in food or from the surrounding water. Marine animals must rid the body of excess ions they gain from drinking seawater.

(b) Investigators realized that the amino acid sequence of CFTR (cystic fibrosis transmembrane regulator) was 80 percent identical to the shark chloride channel. It was their first hint that CFTR involved chloride transport.

(c) Young chum salmon living in salt water have a large number of chloride cells at the base of their gill filaments. In contrast, most of the chloride cells observed in freshwater chum are located in the sheetlike lamellae that extend from the base of gill filaments. This allows researchers to study the distinctive challenges of osmoregulating in freshwater and in salt water.

(d) Water loss is an inevitable by-product of respiration. Insects have a relatively large surface area with which to lose water, and a small volume in which to retain it. The tracheal system connects with the atmosphere at openings called spiracles. Muscles just inside each spiracle close or open the pore. This is an important adaptation for minimizing water loss. The insect exoskeleton consists of chitin and layers of protein. A layer of wax covers the surface, which is highly impermeable to water.

(e) Birds, reptiles, and terrestrial arthropods convert ammonia to uric acid. Because uric acid is not soluble in water, it can be excreted as a dry paste. In this way, the maximum amount of water can be conserved compared to excreting other types of nitrogenous wastes.

(f) Insects must carefully regulate the composition of a blood-like fluid called hemolymph. Malpighian tubules have a large surface area and are in direct contact with the hemolymph. They are the water and ion balance organs in insects. J. A. Ramsay and colleagues found that K^+ accumulated inside the tubules against a concentration gradient. They hypothesized that cells in the membranes of Malpighian tubules contain a pump that transports K^+ into the organ.

(g) Na^+/K^+-ATPase in the basolateral membranes removes intracellular Na^+ and creates a gradient for Na^+ entry. In the apical membrane, Na^+-dependent cotransporters simultaneously bind Na^+ and another solute such as glucose, an amino acid, or Cl^-. Water follows the movement of these solutes via osmosis. It leaves the proximal tubule through membrane proteins called aquaporins.

(h) The amount of Na^+, Cl^-, and water that is reabsorbed in the distal tubule and collecting duct varies with the animal's condition and is under hormonal control. If an individual is dehydrated, a molecule called antidiuretic hormone (ADH) is released from the brain. When ADH interacts with the cells lining the distal tubule and collecting duct, aquaporin channels are inserted into the membrane. As a result, the cells become highly permeable to water and large amounts of water are reabsorbed.

Assessing What You've Learned

1. c; 2. d; 3. b; 4. b; 5. a; 6. a; 7. c; 8. b; 9. a; 10. d; 11. c; 12. b

Looking Forward—Concepts in Later Chapters

Membrane Transport and Surface Area—Chapter 44

Nutrients such as glucose must diffuse into the cell, and waste products such as urea and carbon dioxide must diffuse out. As explained in **Chapter 44**, the rate at which these and other molecules and ions diffuse depends partly on the available surface area. In contrast, the rate at which nutrients are used and waste products are produced depends on the volume of the cell. This chapter will also explore how the gill functions as an adaptation for increased surface area, and how animals exchange gases in detail.

The Sodium-Potassium Pump—Chapter 45

The best-characterized membrane protein pump involving ions is the sodium-potassium pump, also known as Na^+/K^+-ATPase. **Chapter 45** discusses how this enzyme was discovered and explores its function in the transmission of electrical signals in animals. That discussion also introduces a plant defense compound called ouabain. Ouabain is toxic to animals because it binds to Na^+/K^+-ATPase, prevents it from functioning, and poisons transmission of nerve impulses.

Chapter 43

Animal Nutrition

Looking Back—Concepts from Earlier Chapters

Osmosis and Diffusion—Chapter 6

Chapter 6 introduced the processes called diffusion and osmosis. Diffusion describes movement of substances from regions of higher concentration to regions of lower concentration. Movement of water from regions of higher to lower concentration is called osmosis.

Surface-to-Volume Ratio—Chapter 41

Organs such as the small intestine increase surface area in order to increase absorption efficiency.

Glycogen—Chapter 5

Recall that glycogen is a polysaccharide made up of glucose molecules and that glucose is the preferred starting compound for the production of ATP via cellular respiration.

Adaptive Radiation—Chapter 25

Cichlids inhabiting the Rift Lakes of East Africa are a spectacular example of adaptive radiation. Recall from **Chapter 25** that adaptive radiation refers to diversification of a single ancestral population into many species, each of which lives in a different habitat or utilizes a distinct feeding method.

Embryonic Tissue Layers—Chapter 22

The interior tube of the GI tract communicates directly with the environment. Embryologically, it derives from the hollow tube that forms as sheets of cells invaginate during gastrulation.

Aquaporins—Chapter 42

In an attempt to identify the mechanism of water absorption in the large intestine, researchers focused on aquaporins. Recall that aquaporins are water channels and are common in the descending loop of Henle of the kidney.

KEY CONCEPTS

- Nutrients in the form of amino acids, vitamins, elements, and organic compounds are necessary for animals to stay healthy.

- Specialized mouthparts of animals function to capture foods that are a source of nutrients.

- Nutrients are digested and absorbed during complex processes that begin in the mouth and end in the large intestine.

- Dysfunction in nutritional homeostasis can result in disease states such as diabetes.

A. CHAPTER OUTLINE

43.1 Nutritional Requirements

Meeting Basic Needs

- **Nutrients**, substances needed by an organism to remain alive, are contained within **food**. An organism needs chemical energy in the form of reduced carbon compounds and the elements and molecules needed to synthesize body components and sustain cells.

- Recommended Daily Allowances (RDAs) are placed on packaging in the United States to specify amounts of each essential nutrient that must be ingested to meet the needs of practically all healthy people. Specifically:

1. **Essential amino acids** are the 8 (out of a total of 20) amino acids that are required to manufacture proteins, but that cannot be synthesized by humans. We can synthesize the other 12 amino acids.

2. **Vitamins** are carbon-containing molecules that function as coenzymes in critical reactions (**Table 43.1**).

3. **Electrolytes** form ions in solution. They influence osmotic balance and are required for normal membrane function.

4. **Essential elements** cannot be synthesized by the body, but they serve a wide variety of functions (**Table 43.2**).

✓ **CHECK YOUR UNDERSTANDING**

(a) What are the four areas of emphasis of the Recommended Daily Allowances?

_____

_____

_____

_____

_____

Studying Nutrient Requirements

- For non-human animals, researchers determine basic nutritional requirements by systematically varying the type of ions or molecules that the animal consumes. Due to ethical reasons, the study of nutritional requirements of humans is based on observation of the health of people who have different dietary intakes.

43.2 Capturing Food: The Structure and Function of Mouthparts

- Four strategies by which animals obtain food are:
  1. **Suspension feeders** use structures such as cilia to filter food from the surrounding water.
  2. **Deposit feeders** swallow sediments or other types of deposited materials.
  3. **Fluid feeders** suck or lap up fluids such as blood, nectar, or sap.
  4. **Mass feeders** use mouthparts to catch and manipulate food.

Mouthparts as Adaptations

- The mouthparts of animals are adapted to different diets. The diversity of tooth shapes in mammals correlates with the foods they must chew and swallow (**Figure 43.2a**).

- The structure of skull and jaw bones in snakes has allowed them to swallow their food whole (**Figure 43.2b**).

A Case Study: The Cichlid Jaw

- Cichlids have undergone **adaptive radiation** of the pharyngeal jaw in order to exploit the wide variety of food sources in their environment. Cichlids are able to use their pharyngeal jaws to both swallow and bite (**Figure 43.3**). Also, their pharyngeal jaw is equipped with toothlike structures that allow them to further crush and manipulate a variety of foods (**Figure 43.4**).

✓ **CHECK YOUR UNDERSTANDING**

(b) Give an example of an animal with a specialized mouth for feeding.

_____

_____

_____

43.3 How Are Nutrients Digested and Absorbed?

An Introduction to the Digestive Tract

- Digestion takes place in the **alimentary canal**, also known as the **gastrointestinal tract**.

- Food is ingested and wastes are eliminated through the same opening in **incomplete digestive tracts** (**Figure 43.5**). In contrast, in **complete digestive tracts**, food enters through the mouth opening and waste leaves through the anus (**Figure 43.6**).

- Advantages of a tubelike, complete digestive tract include:
  1. Ingestion of large pieces of food.
  2. Compartmentalization of chemical and physical processes of digestion.
  3. Continuous ingestion and digestion of foods.

An Overview of Digestive Processes

- Mechanical breakdown occurs primarily in the mouth and stomach, but chemical breakdown begins with enzymes in the saliva (**Figure 43.7**).

- Chemical breakdown of carbohydrates starts in the mouth, while digestion of proteins begins in the stomach. Carbohydrates, lipids, and proteins are completely processed in the small intestine, where they are absorbed.

- Waste products then move to the large intestine, where more water is absorbed, before they are excreted from the body as **feces**.

The Mouth and Esophagus

Digestion Starts in the Mouth

- **Amylase**, contained in saliva, cleaves the bonds that link glucose monomers in starch, glycogen, and other glucose polymers.

- **Salivary glands** also release water and glycoproteins called **mucins**. The combination of water and mucus makes food soft and slippery enough to be swallowed.

Peristalsis Moves Material Down the Esophagus

- Food then enters the esophagus and is propelled by a wave of contractions called **peristalsis** (**Figure 43.8**). Food then enters the stomach.

A Modified Esophagus: The Bird Crop

- In birds, the **crop** is a widened segment of the esophagus that allows for storage and digestion of foods. The crop is adaptive in that it allows birds to shorten their feeding period to avoid predation, and it allows some birds to benefit from the fatty acids that result from bacterial metabolism.

✔ **CHECK YOUR UNDERSTANDING**

(c) What is peristalsis, and where in the body does it occur?

_____

_____

_____

The Stomach

The Stomach as a Site of Protein Digestion

- When the stomach is filled by a meal, muscular contractions result in churning that mixes the contents (**Figure 43.9**). The predominant acid in the stomach is **hydrochloric acid**. There is also the enzyme **pepsin**, which breaks down proteins. Pepsinogen is converted to active pepsin by contact with the acidic environment of the stomach.

Which Cells Produce Stomach Acid?

- **Parietal cells** are the source of the HCl in gastric juice, whose pH can be as low as 1.5. **Goblet cells** secrete **mucus**, which lines the gastric epithelium and protects the stomach from damage by HCl (**Figure 43.10a**).

How Do Parietal Cells Secrete HCl?

- The enzyme **carbonic anhydrase** is present in high concentration in parietal cells. In solution, the carbonic acid that is formed immediately dissociates to form the bicarbonate ion and a proton.

- Protons formed by dissociation of carbonic acid are actively pumped into the lumen of the stomach. Chloride ions enter parietal cells in exchange for bicarbonate ions and then move into the lumen via a chloride channel (**Figure 43.10b**).

Ulcers as an Infectious Disease

- A bacterium called *Helicobacter pylori* is the cause of **ulcers** in the lining of the stomach or duodenum.

The Ruminant Stomach

- Cattle, sheep, deer, and other ruminants have complex, four-chambered stomachs (**Figure 43.11**). The largest chamber, the **rumen**, serves as a vat for fermenting food before food is regurgitated and rechewed as **cud**. It then passes to the **reticulum** for bacterial breakdown of cellulose, to the **omasum** for water removal, and to the **abomasum** (true stomach) for further chemical digestion.

The Avian Gizzard

- The **gizzard** is a type of modified stomach that aids in digestion of large and hard materials swallowed by birds.

✔ **CHECK YOUR UNDERSTANDING**

(d) Which stomach cells secrete acid?

_____

_____

The Small Intestine

- In the small intestine, partially digested material mixes with secretions from the pancreas and liver and begins a journey of 6 meters.

Folding and Projections Increase Surface Area

- The large surface area of the small intestine is accounted for by its folded structure, **villi**, and **microvilli** that exist on the cells on the surface of the villi (**Figure 43.12**).

- The large surface area allows for efficient nutrient absorption.

Protein Processing by Pancreatic Enzymes

- The activating enzyme in the intestinal juice is **enterokinase,** which activates a pancreatic enzyme called **trypsinogen**, yielding **trypsin**. Trypsin triggers the activation of other protein-digesting enzymes secreted by the pancreas (**Figure 43.13**).

What Regulates the Release of Pancreatic Enzymes?

- **Secretin** is produced in the small intestine in response to the arrival of food to the stomach. This hormone induces a flow of bicarbonate ions (HCO_3^-) from the pancreas to the small intestine, where they neutralize stomach acids. The discovery of secretin was important because it confirmed that digestion is under both neural and hormonal control.

- **Cholecystokinin** is also produced by the small intestine. It stimulates secretion of digestive enzymes from the pancreas, liver, and gallbladder.

CHECK YOUR UNDERSTANDING

(e) How is the hormone secretin involved in the digestion process?

_____

_____

_____

_____

How Are Carbohydrates Digested and Transported?

- Amylase released by the pancreas aids in the digestion of carbohydrates into monosaccharides such as glucose.

- Experiments using purified mRNAs from rabbit intestinal cells resulted in the following model for glucose absorption:

  1. An electrochemical gradient is established by Na^+/K^+-ATPase in the basolateral membrane of epithelial cells in the small intestine.

  2. Glucose and sodium enters the cell via a cotransporter.

  3. The basolateral membrane contains glucose carriers that allow glucose to diffuse into blood vessels.

Digesting Lipids: Bile and Transport

- The pancreatic secretions include digestive enzymes that act on fats and carbohydrates as well as proteins. **Lipase** breaks certain bonds present in complex fats.

- Can you outline the steps in digestion of lipids in the small intestine? (See **Figure 43.15**.)

- **Bile salts** are synthesized in the liver and secreted in a complex solution called **bile** that is stored in the **gallbladder**. When bile enters the small intestine, it raises the pH and emulsifies fats so that they can be digested.

CHECK YOUR UNDERSTANDING

(f) How and where is fat digested?

_____

_____

_____

_____

_____

How Is Water Absorbed?

- Water follows the solutes into the epithelium passively, via osmosis. This movement of water is the mechanistic basis of an extremely important medical strategy called **oral rehydration therapy**.

The Cecum and Appendix

- The **cecum** in rabbits, rodents, and some marsupials is an outpocketing of the digestive tract at the start of the large intestine where cellulose fermentation occurs.

- In humans, the **appendix** plays a role in the body's defense against bacteria and viruses.

The Large Intestine

- The primary function of the large intestine is to compact wastes that remain and absorb enough water to form **feces**.

Aquaporins Play a Key Role in Water Reabsorption

- To date, four distinct transmembrane water channels called **aquaporins** have been found in the large intestine of rats, mice, or humans. AQP3 and AQP4 are located in the basolateral membrane of cells in the epithelium of the large intestine of the rat.
- Mechanisms of water absorption in the large intestine are not well understood, but aquaporins are hypothesized to have a major role.

Variations in Structure and Function

- Plant-eating animals, due to their large consumption of cellulose, have extremely large colons, while insects lack both colons and rectums, and some fish have no large intestine.
- Amphibians, reptiles, and birds have only one orifice for excretion. In these species, urine empties directly into the cloaca, an enlarged portion of the large intestine.

43.4 Nutritional Homeostasis—Glucose as a Case Study

- People with **diabetes mellitus** experience abnormally high glucose levels in their blood.

The Discovery of Insulin

- Banting and Best (1921) removed a pancreas from a dog, purified an extract made from it, and injected the extract into a diabetic dog; its blood glucose stabilized. They had discovered **insulin**.

Insulin's Role in Homeostasis

- **Insulin** is a hormone produced in the pancreas when blood glucose levels are high. It travels through the bloodstream and binds to receptors on cells throughout the body. In response, the cells increase their rate of glucose uptake and respiration.
- If blood glucose levels fall, cells in the pancreas secrete a hormone called **glucagon**. Cells in the liver and skeletal muscle catabolize glycogen, and cells that store lipids catabolize fatty acids. The result is that glucose levels in the blood rise.

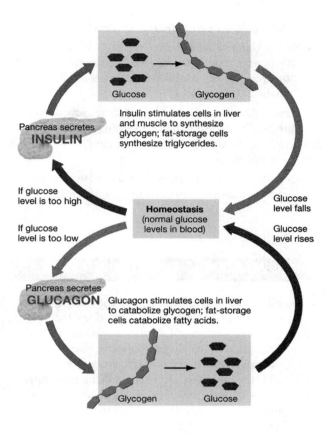

Diabetes Can Take Several Forms

- **Diabetes mellitus** develops in people who do not synthesize insulin or who have defective versions of the insulin receptor. Type I diabetes mellitus is treated with insulin injections; type II diabetes is managed mainly via diet.

✔ **CHECK YOUR UNDERSTANDING**

(g) What is the functional difference between insulin and glucagon?

_____

_____

_____

_____

_____

The Type 2 Diabetes Mellitus Epidemic

- In the United States, about 6.6 percent of people ages 20–74 have type 2 diabetes. Among the Pima, almost 50 percent of the adults over 35 years old have type 2 diabetes (**Figure 43.17**).

- Researchers have concluded that some individuals are genetically predisposed to developing the disease, but evidence is strong that environmental conditions have an impact.

- Nutrition-related diseases such as obesity are being linked to diabetes, especially in young people.

B. MASTERING CHALLENGING TOPICS

Following food through the digestive system is a fairly simple process, but the chemical reactions that occur during this process, coupled with the neural and hormonal signals, can quickly get confusing. Make sure you have mastered the anatomy of the digestive system before you tackle the physiology. It will be difficult to study the functions of the liver if you can't picture where it's located in relation to the digestive system and the body.

Begin by following a piece of food through the digestive system, starting at the mouth. Note the chemical reactions occurring at each stage, and any neural or hormonal signals involved. The most confusing section will be the small intestine because secretions from the liver and pancreas are involved with digesting food entering from the stomach. Draw arrows mapping the hormonal and neural signals so you are clear on which way the signal is traveling.

C. ASSESSING WHAT YOU'VE LEARNED

1. Which of the following represent examples of electrolytes?
   a. K^+ and Cl^-
   b. Valine and glycine
   c. Niacin and folate
   d. Glucose and fructose

2. As you've probably noticed, recommendations for basic nutritional requirements for humans change fairly frequently. Which of the following statements provides the most likely explanation?
   a. As humans evolve, their nutritional requirements change.
   b. The nutritional information provided on packaged food labels contains many errors.

   c. It is difficult to perform controlled experiments on humans in order to make reliable nutritional recommendations.
   d. Food producers have a major impact on the types of food humans consume.

3. Which of the following strategies best describes how humans obtain food?
   a. Suspension feeders
   b. Deposit feeders
   c. Fluid feeders
   d. Mass feeders

4. Which of these adaptations to the pharyngeal jaw system of cichlids does *not* represent an example of how mouthparts are adaptations to allow for efficient feeding?
   a. Because the pharyngeal jaws articulate with the braincase, they can be used for biting as well as for moving food to the back of the throat.
   b. Because the lower pharyngeal jaw exists as two separate pieces, the pharyngeal jaws are better able to grasp prey.
   c. The pharyngeal jaws of cichlids have more muscles attached, improving their use for biting prey.
   d. The toothlike projections of the cichlid pharyngeal jaws vary in shape and size depending on the primary food source of each species.

5. Chemical digestion begins in the mouth. Which of the following best describes the type of digestion that occurs in the mouth?
   a. Lipase begins to break down proteins.
   b. Amylase begins to break down starch.
   c. Pepsin begins to break down proteins.
   d. Amylase begins to break down fats.

6. Which of these statements concerning acid production by the stomach is *not* correct?
   a. The enzyme carbonic anhydrase, located in the parietal cells, catalyzes the formation of carbonic acid from carbon dioxide and water.
   b. Carbonic acid is actively secreted by parietal cells into the stomach lumen.
   c. Protons are actively pumped into the stomach lumen by the parietal cells.
   d. Chloride ions move into the lumen of the stomach through chloride channels located in the parietal cells.

7. Based on what you've learned about the role of the parietal cells in HCl production in the stomach, predict the relative acidities of these three compartments: A = the stomach lumen; B = blood supply arriving to the stomach; C = blood supply leaving the stomach.
   a. pH A > pH C > pH B
   b. pH B > pH C > pH A
   c. pH A > pH B > pH C
   d. pH C > pH B > pH A

8. Which of the following does *not* represent a function of HCl in the stomach?
   a. HCl cleaves peptide bonds.
   b. HCl activates digestive enzymes such as pepsin.
   c. HCl acts as a barrier to pathogens present in food.
   d. HCl disrupts the tertiary and secondary structure of proteins, enhancing their enzymatic degradation.

9. Which of the following represents the correct sequence for protein digestion and absorption?
   a. In the stomach, pepsin degrades large polypeptides to smaller polypeptides. In the small intestine, the smaller polypeptides are further broken down into amino acids by the actions of enzymes such as trypsin, secreted from the pancreas. Amino acids are absorbed in the small intestine by an Na^+-amino acid cotransport mechanism.
   b. In the stomach, enterokinase degrades large polypeptides to smaller polypeptides. In the small intestine, the smaller polypeptides are degraded to amino acids by the actions of pancreatic proteases, such as trypsin, and brush border enzymes. Amino acids are absorbed in the small intestine by an Na^+-amino acid cotransport mechanism.
   c. In the stomach, pepsin degrades large polypeptides to small polypeptides. In the small intestine, the smaller polypeptides are broken down into amino acids by the actions of pancreatic enzymes, such as trypsin, and brush border enzymes. Amino acids are absorbed in the large intestine by an Na^+-amino acid cotransport mechanism.
   d. In the mouth, large polypeptides are degraded to smaller polypeptides by the actions of amylase. In the small intestine, the smaller polypeptides are further broken down into amino acids by the actions of pancreatic enzymes, such as trypsin, and brush border enzymes. Amino acids are absorbed in the small intestine by an Na^+-amino acid cotransport mechanism.

10. Which of the following represents the correct sequence for fat digestion?
    a. The bulk of fat digestion takes place in the small intestine with the release of lipases from the gallbladder. Lipases cleave fats into fatty acids and other small lipids that are absorbed by the small intestine.
    b. The bulk of fat digestion takes place in the small intestine. Bile released from the gallbladder helps to break up large fat globules to smaller ones, while lipases released from the pancreas cleave the fats to fatty acids and other small lipids. Absorption of these small lipids occurs via an Na^+-fatty acid cotransport mechanism in the small intestine.
    c. The bulk of fat digestion takes place in the small intestine. Bile released from the gallbladder helps to break up large fat globules to smaller ones, while amylase released from the pancreas cleaves the fats to fatty acids and other small lipids. Absorption of these small lipids occurs via an Na^+-fatty acid cotransport mechanism in the small intestine.
    d. The bulk of fat digestion takes place in the small intestine. Bile released from the gallbladder helps to break up large fat globules to smaller ones, while lipases released from the pancreas cleave the fats to fatty acids and other small lipids. Absorption of these small lipids occurs using fatty-acid-binding proteins located in the small intestine.

11. Which of these statements concerning pancreatic secretion is *false*?
    a. The introduction of HCl into the upper regions of the small intestine stimulates pancreatic secretion.
    b. Nerves innervating the upper region of the small intestine sense the arrival of food from the stomach and signal the pancreas to secrete enzymes.
    c. The arrival of food into the small intestine from the stomach stimulates release of the hormone secretin, which triggers pancreatic secretion.
    d. Digestive enzymes secreted by the pancreas are activated by an enzyme, called enterokinase, secreted by the small intestine.

12. All of the following occur with the condition known as diabetes *except*:
    a. high blood levels of insulin.
    b. high blood levels of glucose.
    c. reduced rates of glucose uptake by cells.
    d. presence of glucose in the urine.

CHAPTER 43—ANSWER KEY

Check Your Understanding

(a) In 1943, the Recommended Daily Allowances (RDAs) were published to specify the amount of each essential nutrient that must be ingested to meet the needs of practically all healthy people. The following were their focus:

1. Of the 20 amino acids required to manufacture proteins, humans can synthesize 12. The other 8 must be obtained from food and are called essential amino acids.
2. Vitamins are carbon-containing molecules that function as coenzymes in critical reactions.
3. Essential elements serve a wide variety of functions.
4. Electrolytes form ions in solution; they influence osmotic balance and are required for normal membrane function.

(b) Human teeth include sharp canines for tearing meat and flattened molars for crushing seeds, roots, and other sources of carbohydrates.

(c) Peristalsis is a wave of contraction that propels food through the digestive system. It occurs primarily in the esophagus and the intestines.

(d) Parietal cells in the lining of the stomach produce hydrochloric acid.

(e) Secretin is a hormone produced in the small intestine in response to the arrival of food to the stomach. This hormone induces a flow of bicarbonate ions (HCO_3^-) from the pancreas to the small intestine, where they neutralize stomach acids.

(f) The pancreatic secretions include digestive enzymes that act on fats and carbohydrates, as well as proteins. Lipase breaks certain bonds present in complex fats. Bile salts are synthesized in the liver and secreted in a complex solution called bile, which is stored in the gallbladder. When bile enters the small intestine, it raises the pH and emulsifies fats so that they can be digested.

(g) The primary function of the large intestine is to compact wastes that remain and absorb enough water to form feces.

(h) Insulin is a hormone produced in the pancreas when blood glucose levels are high. It travels through the bloodstream and binds to receptors on cells throughout the body. In response to insulin, the cells increase their rate of glucose uptake and respiration. If blood glucose levels fall, cells in the pancreas secrete a hormone called glucagon. In response to glucagon, cells in the liver and skeletal muscle catabolize glycogen and cells that store lipids catabolize fatty acids. The result is that glucose levels in the blood rise.

Assessing What You've Learned

1. a; 2. c; 3. d; 4. b; 5. b; 6. b; 7. d; 8. a; 9. a; 10. d; 11. b; 12. a

---

Looking Forward—Concepts in Later Chapters

Membrane Transport and Surface Area—Chapter 44

Nutrients such as glucose must diffuse into the cell, and waste products such as urea and carbon dioxide must diffuse out. As **Chapter 44** explains, the rate at which these and other molecules and ions diffuse depends partly on the amount of surface area available. In contrast, the rate at which nutrients are used and waste products are produced depends on the volume of the cell. This chapter also explores how the gill functions as an adaptation for increased surface area, and how animals exchange gases in detail.

The Sodium-Potassium Pump—Chapter 45

The best-characterized membrane protein pump involving ions is the sodium-potassium pump, also known as Na^+/K^+-ATPase. **Chapter 45** discusses how this enzyme was discovered and explores its function in the transmission of electrical signals in animals. That discussion also introduces a plant defense compound called ouabain. Ouabain is toxic to animals since it binds to Na^+/K^+-ATPase, prevents it from functioning, and poisons transmission of nerve impulses.

Muscle Types—Chapter 46

The action of peristalsis and churning is unconscious. **Chapter 46** explains the differences between the structure and function of different types of muscle and related nerve actions.

Gas Exchange and Circulation

Chapter 44

Looking Back—Concepts from Earlier Chapters

Osmosis and Diffusion—Chapter 6

Chapter 6 introduced the processes called diffusion and osmosis. Diffusion describes the movement of substances from regions of higher concentration to regions of lower concentration. The movement of water from regions of higher to lower concentration is called osmosis.

Surface-to-Volume Ratio—Chapter 41

Organs such as the small intestine have a large internal surface area to increase absorption efficiency.

Water Loss—Chapter 42

As **Chapter 42** explained, many marine mammals lose water across their respiratory surface via osmosis; animals that live in freshwater lose ions and other solutes by diffusion. Chapter 42 also introduced the trachea of insects. Biologists found that when spiracles were kept open experimentally, insects were likely to die of dehydration.

Carbonic Anhydrase—Chapter 43

Recall from **Chapter 43** that carbonic anhydrase catalyzes the formation of carbonic acid from carbon dioxide and water. As a result, CO_2 that diffuses into red blood cells is quickly converted to bicarbonate ions and protons.

KEY CONCEPTS

- In animals, the respiratory and circulatory systems are involved in the transport of oxygen to cells and carbon dioxide from cells.

- Oxygen and carbon dioxide behave differently in air and water.

- Gas-exchange organs are designed to maximize the rate of diffusion of oxygen and carbon dioxide across the respiratory surface.

- The heart pumps blood, which transports gases, nutrients, hormones, and wastes throughout the body.

A. CHAPTER OUTLINE

44.1 The Respiratory and Circulatory Systems

- The transport of O_2 from environment to cells, and of CO_2 from cells to environment, requires both the respiratory and circulatory systems (**Figure 44.1**). Four stages are involved:

  1. **Ventilation**, where air or water moves through a specialized gas-exchange organ (e.g., lungs or gills)
  2. **Gas exchange**, where O_2 and CO_2 diffuse between air or water and blood at the respiratory surface
  3. **Circulation**, where dissolved O_2 and CO_2 are transported throughout the body
  4. **Cellular respiration**, where O_2 and CO_2 diffuse between blood and cells

44.2 Air and Water as Respiratory Media

How Do Oxygen and Carbon Dioxide Behave in Air?

- **Partial pressure** is the pressure of a particular gas in a mixture. To calculate this, multiply the percent composition of that gas by the total pressure exerted by the entire mixture.

(a) The composition of air at 760 mm Hg is 20.95 percent oxygen, 78.89 percent nitrogen, and 0.03 percent carbon dioxide. What are the partial pressures for each of these gases? As you climb Mt. Everest, what happens to these gas percentages and these gas partial pressures?

_____

_____

_____

_____

How Do Oxygen and Carbon Dioxide Behave in Water?

- A liter of air can contain up to 209 ml of O_2 while a liter of water may contain up to 7 ml of O_2. To extract the same amount of oxygen, water breathers must breathe 30 times more water than air.

- Water is also about 1000 times denser than air. As a result, water breathers expend considerably more energy to ventilate their respiratory surfaces.

What Affects the Amount of Gas in a Solution?

- The amount of gas that dissolves in a liquid depends on several factors:
  1. Solubility of the gas in that liquid
  2. Temperature of the liquid
  3. Presence of other solutes
  4. Partial pressure of the gas in contact with the liquid

- Oxygen has very low solubility in water; thus blood contains a carrier molecule that preferentially binds to oxygen.

- As the temperature of the water increases, the amount of gas that dissolves decreases. Cold-water habitats have much more oxygen available than warm-water habitats do.

- Because seawater has a much higher concentration of solutes than freshwater, it can hold much less dissolved gas.

- Gas will leave a liquid if the partial pressure in a liquid is higher than that in the adjacent gas.

What Affects the Amount of Oxygen Available in an Aquatic Habitat?

- Photosynthetic organisms can increase oxygen levels, while organisms undergoing cellular respiration can deplete oxygen in aquatic habitats.

- Shallow bodies of water are better oxygenated due to a higher surface-area-to-volume ratio.

- In the absence of mixing of water, water near the surface has higher levels of oxygen than water near the bottom.

- Rapids and waterfalls tend to be the most highly oxygenated of all aquatic environments because there is mixing of air and water, and a larger surface area is exposed to air.

(b) What are the three factors that determine the amount of gas that will dissolve in water? Which factor contributes the most in the difference between O_2 and CO_2 dissolving in water?

_____

_____

_____

_____

44.3 Organs of Gas Exchange

Design Parameters: The Law of Diffusion

- Gas diffusion rate, defined by Fick's law of diffusion (**Figure 44.3**), depends on five factors:
  1. Solubility of the gas
  2. Temperature
  3. Surface area available for diffusion
  4. Difference in partial pressures of the gas across the respiratory medium
  5. Thickness of the barrier to diffusion

(c) Which of the factors affecting rate of gas diffusion has the potential to be quickly altered by an animal? Explain your answer.

_____

_____

_____

_____

How Do Gills Work?

- **Gills** are outgrowths of the body surface that are used for gas exchange in aquatic animals.
- Among invertebrates, the structure of gills is extremely diverse, whereas the gills of bony fish are similar in structure (**Figure 44.4**).

How Do Fish Ventilate Their Gills?

- To move water through the gill structures, fish open and close their mouths and the stiff flap of tissue that covers the gills. Fish can also force water through their gills by swimming with their mouths open; this is called **ram ventilation**.
- Fish gills are composed of **gill filaments** that extend from gill arches. Gill filaments are composed of sheetlike structures called **gill lamellae** (**Figure 44.5**).

The Fish Gill Is a Countercurrent System

- Movement of water over the gills is **unidirectional**, which sets up a **countercurrent exchange system** (**Figure 44.6**). This ensures that the difference in the amount of O_2 and CO_2 in water versus blood is large over the entire respiratory surface.

How Do Insect Tracheae Work?

- **The tracheal system** transports air close enough to cells for gas exchange to take place directly. As a result, insects do not require a circulatory system to transport gases from a respiratory structure to the tissues.
- Tracheae connect to the exterior via an opening called a **spiracle** that can be closed. Spiracles are interpreted as adaptations to minimize water loss.

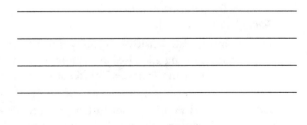
CHECK YOUR UNDERSTANDING

(d) Describe the tracheal system of insects. How does the tracheal system of insects differ from the mammalian respiratory system?

_____

_____

_____

_____

How Do Vertebrate Lungs Work?

Lung Structure and Ventilation Vary among Species

- Mammalian lungs are divided into tiny sacs called **alveoli** (**Figure 44.9**). Each human lung contains about 150 million of these structures. It has about 100 m^2 of surface area, and the barrier of diffusion is only 0.2 μm thick.
- Frogs and other amphibians have lungs that are sacs lined with blood vessels.
- In the simple lungs of snails and spiders, air movement takes place by diffusion only.
- Vertebrates actively ventilate their lungs by pumping air via muscular contractions.

Ventilation of the Human Lung

- During inhalation, the downward motion of the diaphragm causes the pressure within the chest cavity to decrease relative to the atmosphere. The drop in pressure in the chest cavity allows air to flow into the lungs.
- During exhalation, the diaphragm and rib muscles relax, causing the chest cavity volume to decrease, and the elastic nature of the lung causes it to contract. Both events force air out of the lungs.

Ventilation of the Bird Lung

- Airflow through the avian lung is unidirectional. Air flows into the trachea and enters two large air sacs posterior to the lungs during inhalation. As the bird exhales, the air from the air sacs enters the parabronchi. During the next inhalation it exits the parabronchi and enters air sacs anterior to the lung. Finally, air leaves the anterior air sacs through the trachea during the final exhalation (**Figure 44.11**).
- The avian respiratory system is highly efficient because dead space is reduced, gas exchange occurs during both inhalation and exhalation, and there is a crosscurrent pattern of circulation in the lung.

Homeostatic Control of Ventilation

- The rate of breathing is established by the **medullary respiratory center**, which stimulates the rib and diaphragm muscles to contract about 12–14 times per minute in humans.
- The medullary respiratory center receives neural input from sensory receptors sensitive to O_2 levels in arterial blood. The cells of the respiratory center are also sensitive to pH changes that result from changes in CO_2.
- Signals received by the medullary center stimulate breathing when O_2 levels fall and CO_2 levels rise. Thus, through a negative-feedback system, O_2 and CO_2 delivery are maintained and blood pH is stabilized.

44.4 How Are Oxygen and Carbon Dioxide Transported in Blood?

- Blood is a **connective tissue** that consists of cells in a fluid extracellular matrix called **plasma**. **Formed elements** in the blood include red blood cells, platelets, and several types of white blood cells.

- The functions of blood include:
    1. Carrying O_2 and CO_2 between cells and lungs
    2. Transporting nutrients from the digestive tract to other tissues in the body
    3. Moving waste products to the kidney and liver for processing
    4. Conveying hormones from glands to target tissues
    5. Delivering immune system cells to sites of infection
    6. Distributing heat from deeper organs to the surface
- **White blood cells** are part of the immune system and fight infections. **Platelets** are involved in clot formation.
- **Red blood cells** contain an oxygen-carrying molecule called **hemoglobin**.

CHECK YOUR UNDERSTANDING

(e) List the functions of the blood.

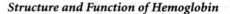

_____

_____

_____

_____

Structure and Function of Hemoglobin

- Hemoglobin is a tetramer consisting of four polypeptide chains, each bound to an iron-containing nonprotein group called **heme**. Each heme molecule contains an iron ion that can bind an oxygen molecule (**Figure 44.12**).

What Is Cooperative Binding?

- Blood leaving the lungs has a P_{O_2} of 100 mm Hg, while muscle and other tissues have P_{O_2} levels of about 40 mm Hg at rest. This partial pressure difference drives the unloading of O_2 from the hemoglobin to the tissues via diffusion.
- Cooperative binding is when the binding of each successive oxygen molecule to a subunit of a hemoglobin molecule causes a conformational change in the protein that increases the likelihood

that oxygen molecules will bind to the remaining subunits. This phenomenon is represented by the sigmoidal **oxygen-hemoglobin equilibrium curve** (**Figure 44.13**).

Why Is Cooperative Binding Important?

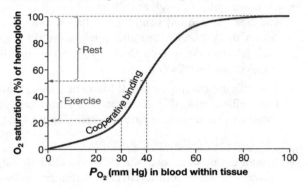

(a) **With cooperative binding**, large amounts of O_2 are delivered to resting and exercising tissues.

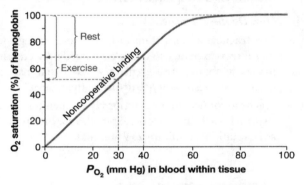

(b) **Without cooperative binding**, smaller amounts of O_2 would be delivered to resting and exercising tissues.

- Because of the sigmoidal relationship, hemoglobin responds quickly and effectively to small changes in oxygen demand.

How Do Temperature and pH Affect Hemoglobin?

- Hemoglobin is also sensitive to changes in pH and temperature. During exercise, the temperature and the PCO_2 increase in the tissue. This results in a drop in pH.
- Decreases in pH and increases in temperature alter hemoglobin's conformation such that it is more likely to release O_2 at all values of PO_2. This is known as the **Bohr shift** (**Figure 44.15**).

(f)   What is the Bohr shift?

_____

_____

_____

_____

Oxygen Delivery Is Extremely Efficient

- Biologists wanted to determine how the oxygen transport system in fish responded to sustained exercise. Arterial O_2 levels remained fairly constant as swimming speed increased, but O_2 levels in the venous blood dropped steadily as swimming speed increased. In hard-working tissues, the combination of increased temperature, lower pH, and lower P_{O_2} caused hemoglobin to become almost completely deoxygenated.

Comparing Hemoglobins

- Fetal hemoglobin, structurally and functionally, is different from adult hemoglobin. Fetal hemoglobin has a higher affinity for oxygen, which ensures that the fetus has an adequate supply of oxygen during development.

CO_2 Transport and the Buffering of Blood pH

The Role of Carbonic Anhydrase and Hemoglobin

- **Carbonic anhydrase** in red blood cells catalyzes the formation of carbonic acid from carbon dioxide and water. As a result, CO_2 that diffuses into red blood cells is quickly converted into bicarbonate ions and protons (**Figure 44.17**). The same reaction occurs much more slowly in the plasma surrounding the red blood cells.
- When Hb is not carrying O_2, it has a high affinity for protons. As a result, Hb takes up much of the H^+ that is produced by the dissociation of carbonic acid. In this way, Hb acts as a buffer—a compound that minimizes changes in pH.

What Happens in the Lungs?

- At the lungs, CO_2 diffuses from the blood into the alveoli and the PCO_2 in the blood declines. Bicarbonate is converted back to CO_2, which then diffuses into the alveoli and is exhaled from the lungs. Hb picks up O_2 during inhalation and the cycle begins again.

(g)   Describe the significance of carbonic anhydrase in carbon dioxide elimination.

_____

_____

_____

_____

44.5 The Circulatory System

What Is an Open Circulatory System?

- One or more hearts pump hemolymph into an artery that empties into many small, open, fluid-filled spaces. Hemolymph is returned to the heart when it relaxes and creates suction.
- Hemolymph is not tightly constrained inside vessels, so the overall pressure in the system is low.

What Is a Closed Circulatory System?

- Blood is completely contained within blood vessels and flows in a continuous circuit through the body under pressure generated by the heart.

Which Lineages Have Closed Circulatory Systems?

- Vertebrates and other lineages such as annelids and cephalopods possess closed circulatory systems to support intense muscular activity.

Types of Blood Vessels

- In a closed circulatory system, the heart ejects blood into a large artery, usually called the **aorta**, which carries blood into the distribution vessels, or **arteries**.
- Small arteries lead into arterioles, which are surrounded by sphincters that help regulate blood flow.
- **Capillaries** are the smallest blood vessels. With walls only a single cell layer thick, the capillaries allow gases, nutrients, and wastes to pass between blood and cells.
- **Veins** and **venules** return blood to the heart. Larger veins contain one-way valves that prevent backflow of blood.

What Is Interstitial Fluid?

- **Interstitial fluid** is constantly augmented by water and proteins from blood that leak out between the cells forming the walls of the capillaries.

- Two forces that drive interstitial fluid across capillary walls are hydrostatic force and osmotic force (**Figure 44.20**).

The Role of the Lymphatic System

- The lymphatic system is a collection of vessels that branches throughout the body. The fluid inside the lymph vessels is called lymph and is eventually returned to the body.

How Does the Heart Work?

- The number of distinct chambers in the heart has increased as vertebrates diversified. The circulatory system in fish forms a single loop; in all other vertebrates, there are two separate circuits—to the lungs and the body (**Figure 44.21**).

- In humans, the atrium receives blood returning from circulation and the ventricle generates force to propel the blood through the system.

- The right ventricle powers the blood to the lungs and back to the heart in a section called the **pulmonary circulation**. Blood flow from the atrium to the ventricle moves only one way due to tissue flaps called **valves**. The left ventricle pumps blood to the entire body and back, or to the **systemic circulation**.

- A cardiac cycle consists of one complete contraction phase, called **systole**, and one complete relaxation phase, called **diastole**.

- **Blood pressure** is the force that blood exerts on the walls of arteries, capillaries, and veins.

The Cardiac Cycle

- Peak pressure is due to the contraction of the ventricle and is called **systolic pressure**. The lowest pressure occurs at the end of ventricular relaxation and is called **diastolic pressure**.

- A typical blood pressure might be 120/80 (systolic/diastolic). People with pressures of 150/90 and higher are considered to have high blood pressure or **hypertension**.

✔ **CHECK YOUR UNDERSTANDING**

(h) What is the difference between systole and diastole, and how does this correlate with the pulsatile pressure in arterial blood?

_____

_____

_____

_____

Electrical Activation of the Heart

- The "pacemaker cells" that initiate contraction are located in a region called the **SA (sinoatrial) node**.

- The pace at which the SA node initiates electrical signals determines heart rate. The pace, in turn, is regulated by electrical signals from the brain.

- Heart rate may be altered by electrical signals from emotional centers in the brain or by signals from the chemical messengers called epinephrine and norepinephrine.

✔ **CHECK YOUR UNDERSTANDING**

(i) What is the electrical event that regulates heart rate?

_____

_____

_____

_____

Patterns in Blood Pressure and Blood Flow

- As the same volume of blood travels through a greater cross-sectional area, the pressure in the fluid drops. Overall, blood pressure is highest in the artery that leads away from the left ventricle (100 mm Hg) and lowest in the veins that return blood from the body to the right atrium (10 mm Hg; **Figure 44.26**).

- Sensory cells are located next to an artery in the neck, an artery near the heart, and the wall of the heart. These cells act as **baroreceptors**, or pressure receptors. When they detect a change in blood pressure, they trigger electrical signals that change the heart's output and vessel diameter.

B. MASTERING CHALLENGING TOPICS

The two most difficult topics in this chapter have to do with the respiratory properties of blood. The first topic deals with the partial pressures of respiratory gases. This concept is central to respiratory physiology, so you must feel comfortable using partial pressures to describe elements of respiration and know the difference between partial pressures and percent composition in respiratory gases. In air, think of percent composition as constant but partial pressures as changing with altitude. Run through a few calculations, and follow the partial pressures of O_2 and CO_2 in the blood going from the lungs to the tissues and back. Have classmates ask you questions and try to stump each other. Mastering this concept will make future lessons in respiratory physiology more enjoyable.

The second topic that most students have difficulty with is the sigmoidal curve of hemoglobin and O_2 loading/unloading. Always think of the loading and unloading of O_2 as a dynamic process. Think of the binding of one oxygen molecule like a person getting into a car. Once this person is inside, he or she can open the doors of the car to allow three other people easier access to the car. In this way, when one oxygen binds to Hb, it changes the conformation such that it is very easy for the next three O_2 molecules to bind. This is the primary reason for the sigmoidal curve in Hb-O_2 loading.

C. ASSESSING WHAT YOU'VE LEARNED

1. Which of the following most accurately describes the gas composition of the atmosphere?
   a. Nitrogen > oxygen > carbon dioxide > inert gases (such as argon)
   b. Oxygen > nitrogen > carbon dioxide > inert gases (such as argon)
   c. Nitrogen > oxygen > inert gases (such as argon) > carbon dioxide
   d. Oxygen > carbon dioxide > nitrogen > inert gases (such as argon)

2. Of these gases found in the atmosphere—nitrogen, oxygen, argon, and carbon dioxide—which are considered to be physiologically active?
   a. Carbon dioxide and oxygen
   b. Nitrogen, carbon dioxide, and oxygen
   c. Nitrogen and oxygen
   d. Oxygen only

3. All of the following represent examples of how the design of respiratory organs conforms to the dictates of Fick's law of diffusion except:
   a. the respiratory surface of the human lung is 70 square meters.
   b. the lung epithelium and blood vessels are extremely thin.
   c. the epithelium of both internal and external gills is in direct contact with oxygen-bearing water.
   d. the flow of blood through the gills is in the same direction as the flow of water over the gill surface.

4. Which of the following represents a disadvantage of breathing air (at lower elevations) versus breathing water?
   a. A liter of water holds less O_2 than a liter of air.
   b. Water is denser than air.
   c. Carbon dioxide is quite soluble in liquids such as water.
   d. Air-breathing animals are better able to retain heat generated by their bodies.

5. The insect tracheal system differs from the mammalian respiratory system in all of the following ways except:
   a. O_2 is delivered by hemolymph, not blood, to cells.
   b. the insect tracheal system uses spiracles to reduce evaporative water loss.
   c. there is no respiratory pigment for gas transport.
   d. muscles of locomotion help to ventilate the respiratory system.

6. Which of the following statements most accurately describes the structure of hemoglobin?
   a. Hemoglobin consists of a single polypeptide chain with a heme center containing a copper ion to which a molecule of O_2 binds.
   b. Hemoglobin consists of a single polypeptide chain with a heme center containing an iron ion to which a molecule of O_2 binds.
   c. Hemoglobin consists of four polypeptide chains, each of which has a heme group containing a copper ion. One hemoglobin molecule can bind a total of four O_2 molecules.
   d. Hemoglobin consists of four polypeptide chains, each of which has a heme group containing an iron ion. One hemoglobin molecule can bind a total of four O_2 molecules.

7. Which of the following best describes the significance of the sigmoidal relationship between P_{O_2} levels and hemoglobin saturation?
   a. As tissue P_{O_2} levels decline, the amount of O_2 released increases proportionately.
   b. As it circulates, hemoglobin releases the same amount of oxygen to all tissues.
   c. As tissue P_{O_2} levels decline with an increase in activity, the oxygen reserve increases.
   d. Once tissue P_{O_2} levels drop below 40 mm Hg, the amount of oxygen released by hemoglobin increases greatly.

8. Which of the following best describes the phenomenon known as the Bohr shift?
   a. As tissue CO_2 levels increase, pH decreases, resulting in an increase in O_2 delivery to the tissues.
   b. As tissue CO_2 levels decrease, pH decreases, resulting in an increase in O_2 delivery to the tissues.
   c. As tissue CO_2 levels increase, pH increases, resulting in a decrease in O_2 delivery to the tissues.
   d. As tissue CO_2 levels decrease, pH increases, resulting in an increase in O_2 delivery to the tissues.

9. Which of the following situations would result in the greatest degree of O_2 saturation for hemoglobin, assuming P_{O_2} remains constant?
   a. Increased CO_2 levels, decreased temperature
   b. Increased CO_2 levels, increased temperature
   c. Decreased CO_2 levels, decreased temperature
   d. Decreased CO_2 levels, increased temperature

10. The transport of CO_2 by the blood is primarily dependent on the:
    a. solubility of CO_2 in blood.
    b. presence of carbonic anhydrase in red blood cells.
    c. ability of hemoglobin to bind and transport CO_2.
    d. ability of other blood proteins to bind CO_2.

11. Hemoglobin acts as a buffer by:
    a. binding O_2 and CO_2.
    b. binding CO_2.
    c. binding H^+.
    d. binding bicarbonate ions.

12. One of the main advantages of a closed over an open circulatory system is the:
    a. ability to direct the flow of blood to certain tissues.
    b. use of a heart to create pressure gradients necessary for blood flow.
    c. ability to return blood to the heart.
    d. use of a respiratory pigment to transport O_2.

13. What prevents the atria and the ventricles from contracting at the same time?
    a. Pacemaker cells located in the atria fire before the pacemaker cells in the ventricles.
    b. It takes time for epinephrine to diffuse from the atria to the ventricles to trigger contraction.
    c. The electrical signal generated in the right atrium is delayed at the AV node before passing to the ventricles.
    d. The Na^+ channels responsible for initiating ventricular contraction are inactivated and need to return to activated configuration to be electrically stimulated.

14. Your blood pressure is reported to you as two values (e.g., 120/80). These represent:
    a. the pressure at the beginning of atrial contraction and the beginning of ventricular contraction.
    b. peak pressure during ventricular contraction and lowest pressure at the end of the ventricular relaxation phase.
    c. peak pressure at the end of atrial contraction and lowest pressure at the end of atrial relaxation.
    d. the pressure at the end of atrial contraction and at the end of ventricular contraction.

CHAPTER 44—ANSWER KEY

Check Your Understanding

(a) $760 \times 0.2095 = 159.2$ mm Hg
$760 \times 0.7889 = 599.6$ mm Hg
$760 \times 0.0003 = 0.228$ mm Hg

The percentages of these gases would remain the same as you climbed up in altitude; however, the partial pressures of these gases would all decrease because the total atmospheric pressure decreases as you increase altitude.

(b) The amount of gas that dissolves in a liquid depends on several factors:
1. the solubility of the gas in that liquid,
2. the temperature of the liquid, and
3. the presence of other solutes.
   CO_2 is almost 30 times more soluble in water than O_2 is. As a result, fish rid themselves of CO_2 more easily than they can obtain oxygen.

(c) The rate of diffusion depends on (1) solubility of the gas, (2) temperature, (3) surface area available for diffusion, (4) difference in partial pressures of the gas across the respiratory medium, and (5) thickness of the barrier to diffusion. Of these, (1) is a physical-chemical factor and is unalterable. Factors (3) and (5) represent characteristics of the gas-exchange organs, which cannot be altered in the short term. An animal could alter factor (2) if it were a cold-blooded animal, by moving to a colder or warmer environment; it could alter factor (4) by increasing ventilation of the gas exchange organs.

(d) In insects, the tracheal system transports air close enough to cells for gas exchange to take place directly. As a result, insects do not require a circulatory system to transport gases from a respiratory structure to the tissues. Tracheae connect to the exterior via an opening called a spiracle that can be closed. Spiracles are interpreted as adaptations to minimize water loss.

(e) The functions of blood include:
1. Carries O_2 and CO_2 between mitochondria and lungs
2. Transports nutrients from the digestive tract to other tissues in the body
3. Moves waste products to the kidney and liver for processing
4. Conveys hormones from glands to target tissues
5. Delivers immune system cells to sites of infection
6. Distributes heat from deeper organs to the surface

(f) Decreases in pH and increases in temperature alter hemoglobin's conformation such that it is more likely to release O_2 at all values of P_{O_2}. This is known as the Bohr shift.

(g) **Carbonic anhydrase** in red blood cells catalyzes the formation of carbonic acid from carbon dioxide and water. As a result, CO_2 that diffuses into red blood cells is quickly converted into bicarbonate ions and protons. The same reaction occurs much more slowly in the plasma surrounding the red blood cells.

(h) A cardiac cycle consists of one complete contraction phase, called systole, and one complete relaxation phase, called diastole. Peak pressure is due to the contraction of the ventricle and is called systolic pressure. The lowest pressure occurs at the end of ventricular relaxation and is called diastolic pressure.

(i) The "pacemaker cells" that initiate contraction are located in a region called the SA (sinoatrial) node. The pace at which the SA node initiates electrical signals determines heart rate. The pace, in turn, is regulated by electrical signals from the brain.

Assessing What You've Learned

1. c; 2. a; 3. d; 4. c; 5. a; 6. d; 7. d; 8. a; 9. c; 10. b; 11. c; 12. a; 13. c; 14. b

---

Looking Forward—Concepts in Later Chapters

Electrical Signals—Chapter 45

Because intercalated discs of cardiac muscle contain numerous gaps, electrical signals pass directly from one cell to the next. **Chapter 45** explores how electrical signals in animals are initiated and propagated.

Fight-or-Flight—Chapter 47

As part of the fight-or-flight response analyzed in **Chapter 47**, blood is directed away from the skin and digestive system and toward the heart, brain, and muscles.

The Lymphatic System—Chapter 49

The lymphatic system acts as a type of circulatory system. Because its primary function is in defense against disease-causing agents, however, the lymphatic system is analyzed in more detail in **Chapter 49**.

Chapter 45

Electrical Signals in Animals

Osmosis and Diffusion—Chapter 6

Chapter 6 introduced the processes called diffusion and osmosis. Diffusion describes the movement of substances from regions of higher concentration to regions of lower concentration. The movement of water from regions of higher to lower concentration is called osmosis. This chapter also discussed the electrochemical gradient and ion channels.

KEY CONCEPTS

🔑 Neurons are specialized cells that transmit information via electrical signals.

🔑 Electrical signals are generated by a change in membrane potential that is propagated along the axon.

🔑 Electrical signals are transmitted from neuron to neuron in the form of neurotransmitters that are released at synapses between neurons.

🔑 The peripheral nervous system (PNS) relays information, while the central nervous system (CNS) processes information.

A. CHAPTER OUTLINE

45.1 Principles of Electrical Signaling

Types of Neurons in the Nervous System

- **Receptor cells** transmit the information they receive via a **sensory neuron**, and this travels to the brain or spinal cord. Together, the brain and spinal cord form the **central nervous system** (**CNS**).

- The CNS integrates information from many sensory neurons and sends signals to **effector cells** via **motor neurons**. All of the components outside the CNS are part of the **peripheral nervous system** (**PNS; Figure 45.1**).

The Anatomy of a Neuron

- Most neurons have a **cell body**, a highly branched group of short projections called **dendrites**, and one or more long projections called **axons** (**Figure 45.2a**).

- A dendrite receives electrical signals from the axons and dendrites of adjacent cells. The axon then sends the signal to the dendrites of other neurons.

✔ CHECK YOUR UNDERSTANDING

(a) Describe the anatomy of a neuron.

_____

_____

_____

_____

_____

An Introduction to Membrane Potentials

- Because of the separation of ions across a membrane, and because ions carry charge, there is a separation of charge across the membrane called a **membrane potential**.

- If the difference in charges on either side of the membrane is large, so is the membrane voltage.

How Is the Resting Potential Maintained?

- When a neuron is not transmitting an electrical signal but is merely sitting in extracellular fluid at rest, its membrane has a voltage called the **resting potential** (**Figure 45.3**).

- Most proteins inside the cell are negatively charged. The major positively charged ion inside neurons is potassium. In the extracellular fluid, however, sodium and chloride predominate.

- In neurons, only K^+ can cross the membrane easily along its concentration gradient. It does so via potassium channels.

- As K^+ moves from the interior of the cell to the exterior via K^+ channels, the inside of the cell becomes more negatively charged relative to the outside. Eventually, the buildup of negative charge inside the cell begins to attract K^+ and counteract the concentration gradient that favors movement of K^+ out. As a result, the membrane reaches a voltage where there is equilibrium between the concentration gradient that moves K^+ out and the electrical gradient that moves K^+ in.

- The voltage where there is no net movement of K^+ is called the **equilibrium potential**.

- The **Nernst equation** converts energy stored in a concentration gradient into the energy stored as an electrical potential.

✔ CHECK YOUR UNDERSTANDING

(b) How is K^+ involved in the maintenance of the resting membrane potential?

_____

_____

_____

- The equilibrium potential given by the Nernst equation is calculated independently for each ion.

Using Microelectrodes to Measure Membrane Potentials

- Hodgkin and Huxley pioneered the study of electrical signaling in animals. To record membrane voltages in the giant squid axon, they cut off an intact section of the axon, bathed it in seawater or other known solution, and inserted a microelectrode. Their recordings suggested that the giant squid axon had a resting potential of −45 mV, but later work confirmed that most neurons have a resting potential of around −70 mV.

What Is an Action Potential?

A Three-Phase Signal

- The initial event is a rapid **depolarization** of the membrane. Current flow causes the inside of the membrane to become less negative and then positive with respect to the outside.

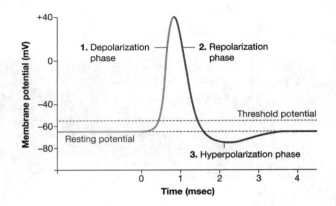

- The membrane then experiences a rapid **repolarization**. The membrane actually becomes more negative than the resting potential. This is called **hyperpolarization**.

An "All-Or-None" Signal That Propagates

- An action potential is all or none. There is no such thing as a partial action potential.

- Action potentials can be triggered by artificially depolarizing the membrane with an injection of electrical current through a microelectrode.

- Action potentials are propagated down the length of the axon.

45.2 Dissecting the Action Potential

Distinct Ion Currents Are Responsible for Depolarization and Repolarization

- The action potential consists of a strong inward flow of Na^+ followed by a strong outward flow of K^+.

How Do Voltage-Gated Channels Work?

- **Voltage-gated channels** are proteins that change conformation in response to charges present at the membrane surface. These changes will open or close the channel.

- Cole and colleagues came up with a technique—**voltage clamping**—that holds the membrane potential of a neuron constant and measures the flows of ions.

- They found that action potentials result from staggered activity of voltage-gated Na^+ and K^+ channels.

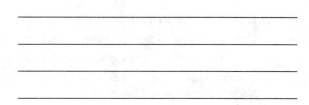

CHECK YOUR UNDERSTANDING

(c)   What are voltage-gated channels?

_____

_____

_____

_____

Patch Clamping and Studies of Single Channels

- Neher and Sakmann perfected a technique called patch clamping whereby they suctioned a microelectrode to the membrane (**Figure 45.7**). They could then document the currents that flowed through individual transmembrane channels.

Positive Feedback Occurs during Depolarization

- The opening of sodium channels exhibits **positive feedback**. This takes place when the occurrence of an event makes the same event more likely to occur.

Using Neurotoxins to Identify Channels and Dissect Currents

- In giant axons from lobsters treated with tetrodotoxin (found in puffer fish), the outward-directed K^+ current was normal but the inward-directed flow of Na^+ was blocked.

How Is the Action Potential Propagated?

- Positive charges inside the cell are repulsed by the influx of Na^+ and negative charges are attracted. As a result, sections of the membrane close to the site of an action potential are depolarized. Nearby voltage-gated channels pop open in response, positive feedback occurs, and an action potential results (**Figure 45.8**).

- Na^+ channels are **refractory**. Once they have opened and closed, they are less likely to open again for a short period of time.

Axon Diameter Affects Speed

- Large axons have fewer sodium channels per unit of membrane surface, which reduces the number of Na^+ molecules that leak out of the axon. This allows for large-diameter axons to propagate action potentials faster than small-diameter axons.

Myelination Affects Speed

- **Myelination** acts as insulation. It prevents charge from leaking out as it spreads down an axon. An unmyelinated area of the neuron called a **node of Ranvier** has a dense concentration of voltage-gated channels. Electrical signals jump from node to node much faster than if the entire neuron were unmyelinated.

45.3 The Synapse

- Otto Loewi (1920) showed that the mechanism that transmits electrical signals from cell to cell involves **neurotransmitters**. He stimulated the vagus nerve and slowed the frog heart. He then took some of the solution that this heart was bathed in and applied it to another heart, and it slowed as well. This result supplied conclusive evidence for the chemical transmission of electrical signals.

CHECK YOUR UNDERSTANDING

(d)   Why does an action potential not go backward?

_____

_____

_____

_____

(e)   How does myelination aid in the propagation of an action potential?

_____

_____

_____

Synapse Structure and Neurotransmitter Release

- The interface between two neurons is called a **synapse**. Just inside the synapse, the axon contains numerous sac-like structures called synaptic vesicles (**Figure 45.11**). At this site, the following steps occur:

  1. The action potential arrives and triggers entry of Ca^{2+}.

  2. In response to Ca^{2+}, synaptic vesicles fuse with membranes and release neurotransmitter.

  3. Ion channels open when neurotransmitter binds the postsynaptic cell potential. This potentially triggers an action potential.

  4. The neurotransmitter is degraded and taken back up by the presynaptic cell.

What Do Neurotransmitters Do?

- Many receptors function as **ligand-gated ion channels**, meaning that a molecule binds to a specific site on the receptor to activate it.

- Some other receptors activate enzymes that lead to the production of a **second messenger**, which may induce changes in enzyme activity, gene transcription, or membrane potential.

✓ CHECK YOUR UNDERSTANDING

(f)   What are ligand-gated ion channels?

_____

_____

_____

Postsynaptic Potentials

- If the receptors at the synapse admit an influx of sodium ions in response to the arrival of the neurotransmitter (NT), then the postsynaptic membrane depolarizes. These are called **excitatory postsynaptic potentials (EPSPs)**.

- In other receptors, binding of the NT leads to the influx of K^+ or Cl^-. These events hyperpolarize the membrane and make action potentials less likely to occur. These are called **inhibitory postsynaptic potentials (IPSPs)**.

✓ CHECK YOUR UNDERSTANDING

(g)   Compare and contrast the EPSP and the IPSP.

_____

_____

_____

- EPSPs and IPSPs are not all-or-none events. They are graded and short lived. The size depends on the amount of NT released at the synapse.

- The sodium channels that trigger action potentials in the postsynaptic cell are located near the start of the axon at a site called the **axon hillock**.

45.4 The Vertebrate Nervous System

- The vertebrate nervous system is functional and anatomically divided into the central nervous system (CNS) and the peripheral nervous system (PNS).

What Does the Peripheral Nervous System Do?

- The PNS consists of two distinct systems (**Figure 45.15**):

  1. A somatic system controls the skeletal muscles.

  2. An autonomic system controls internal processes such as digestion and heart rate.

- The autonomic system is composed of the parasympathetic nerves, which stimulate digestion and other processes, and the sympathetic nerves, which prepare organs for stressful situations.

Functional Anatomy of the CNS

- Anatomically, the brain has four major lobes (temporal, occipital, parietal, and frontal) and two hemispheres (left and right; **Figure 45.18**).

- Functional areas of the brain have been mapped by lesion studies.

- Broca (1861) studied an individual who could understand language but could not speak. After this person's death, Broca found a damaged area in the left front lobe of the cerebrum. Broca's claim that functions are localized to specific brain areas has been verified through extensive efforts to map the cerebrum.

- Wilder Penfield worked with several epileptics scheduled to have seizure-prone areas of their brains surgically removed. While patients were awake, Penfield electrically stimulated portions of the cerebrum and recorded the patients' sensations.

✓ CHECK YOUR UNDERSTANDING

(h)   How were the functional areas of the brain mapped out?

_____

_____

_____

_____

_____

How Does Memory Work?

- Researchers recorded individual neurons in the temporal lobes of humans.

- Work on the molecular basis of memory is founded on two main ideas:
  1. Learning and memory must involve some type of short- or long-term change in neurons responsible for these processes.
  2. It will be much easier to understand what these changes are if an extremely simple system of neurons can be studied.
- Eric Kandel and colleagues have focused on the sea slug *Aplysia californica*. The sea slugs have a simple reflex that can be modified by learning. Follow-up studies have shown that the neurons involved in learning release the NT **serotonin**, and that serotonin causes an EPSP in the motor neuron in the gill.
- More serotonin is released after learning takes place. Therefore, EPSPs are higher and the motor neuron is more likely to generate action potentials.

B. MASTERING CHALLENGING TOPICS

The resting and action potentials are some of the first concepts you will encounter in any physiology or neurology course, and are therefore the foundation of this chapter on electrical signals. The movement of ions generating a resting potential across a membrane is sometimes difficult to visualize as occurring in the human body. Nevertheless, visualizing these ion movements is a great place to start. First, visualize the movement of ions across the membrane in a resting potential and then try to draw this process from memory while explaining it aloud. Once you are comfortable with how the resting potential is maintained, try your skills on the action potential. Concentrate on where it starts, how it is propagated, and where it ends. Visualize the ions moving through channels as the action potential is created, and graph the voltage across the membrane as the action potential proceeds. Last, as you consider how this process works in the entire body, integrate the concepts of EPSP and IPSP so that you are sure not to confuse them in a broader sense.

C. ASSESSING WHAT YOU'VE LEARNED

1. Which of the following describes the role of a dendrite on a nerve cell?
   a. Receives electrical signals
   b. Transmits electrical signals from the neuron
   c. Maintains the DNA of the neuron in a nucleus
   d. Contains the dendritic hillock

2. Which of the following is true about the distribution of ions when a cell is at rest?
   a. Potassium and sodium ion concentrations are higher inside the cell than outside.
   b. Potassium and chloride ion concentrations are higher inside the cell than outside.
   c. Sodium and chloride ion concentrations are higher inside the cell than outside.
   d. Potassium ion concentration is higher inside the cell, and sodium ion concentration is higher outside the cell.

3. What would happen if voltage-gated sodium channels on a resting neuron were to open?
   a. Sodium ions would move into the cell, and the membrane potential would become more negative.
   b. Sodium ions would move into the cell, and the membrane potential would become more positive.
   c. Potassium ions would move out of the cell, and the membrane potential would become more negative.
   d. Sodium ions would move out of the cell, and the membrane potential would become more negative.

4. What would happen if voltage-gated potassium channels on a resting neuron were to open?
   a. Potassium ions would move into the cell, and the membrane potential would become more negative.
   b. Sodium ions would move into the cell, and the membrane potential would become more positive.
   c. Potassium ions would move out of the cell, and the membrane potential would become more positive.
   d. Potassium ions would move out of the cell, and the membrane potential would become more negative.

5. What phase of the action potential is associated with the influx of sodium ions that occurs when voltage-gated sodium ion channels open?
   a. Repolarization
   b. Depolarization
   c. Resting potential
   d. Undershoot

6. The action potential is produced by a combination of an influx of sodium ions and an efflux of potassium ions. Since these ions carry similar charge and are moving in opposite directions, why don't their movements simply cancel each other out so that there is no change in the membrane potential?
   a. Voltage-gated potassium channels open at the same time as voltage-gated sodium channels.
   b. Voltage-gated potassium channels open at the same voltage as voltage-gated sodium channels.
   c. The opening of voltage-gated potassium channels is delayed.
   d. The opening of voltage-gated sodium channels is delayed.

7. What would happen to the resting membrane potential of a neuron if the extracellular fluid were flooded with a high concentration of potassium ions?
   a. The resting membrane potential would become more negative.
   b. The resting membrane potential would become more positive.
   c. There would be no change in the resting membrane potential.

8. In patients who suffer from multiple sclerosis (MS), the myelin surrounding the axons degenerates. Which of the following describes the effect of this loss of myelination?
   a. Action potentials will spread down the axon more quickly if the myelin is removed.
   b. Action potentials will spread down the axon more slowly if the myelin is removed.
   c. Action potentials will spread down the axon more slowly because voltage-gated ion channels will open more slowly.
   d. Action potentials will spread down the axon more quickly because voltage-gated ion channels will open more quickly.

9. Otto Loewi determined that the vagus nerve of the frog releases a neurotransmitter that causes the heart to slow down. Which of the following explains the effect that this neurotransmitter has?
   a. It is causing voltage-gated sodium channels in the heart muscle to open more quickly.
   b. It is causing voltage-gated sodium channels in the heart muscle to become more "sensitive" to changes in voltage.
   c. It is causing an increase in the rate at which action potentials can be generated in the heart muscle.
   d. It is causing a decrease in the rate at which action potentials can be generated in the heart muscle.

10. Eventually, it was discovered that the neurotransmitter being released from the vagus nerve was acetylcholine (ACh). Heart muscle contracts following the depolarization during an action potential. In these cells, the negative resting potential must become less negative and eventually reach what is called a threshold. Once reaching this threshold potential, the action potential will spontaneously generate and the muscle can contract. Which of the following descriptions could explain the effect of acetylcholine on the heart muscle?
   a. The cells become leaky to potassium ions. Potassium ions leak out of the cell, which causes the resting potential to become more negative.
   b. The cells become leaky to sodium ions. Sodium ions leak out of the cell, which causes the resting potential to become more negative.

   c. The cells become leaky to potassium ions. Potassium ions leak into the cell, which causes the resting potential to become more negative.
   d. The cells become leaky to sodium ions. Sodium ions leak into the cell, which causes the resting potential to become more positive.

11. Under which of the following conditions would a ligand-gated ion channel open?
    a. When the membrane potential becomes more positive
    b. When a molecule binds to a specific site on the channel
    c. When the membrane potential becomes more negative

12. The ability of a postsynaptic neuron to generate an action potential (a signal) depends on information that is being received from surrounding presynaptic neurons. These "connections" can either be inhibitory or excitatory. What action would be associated with an inhibitory presynaptic neuron?
    a. The signal from the presynaptic neuron causes an influx of sodium into the postsynaptic neuron.
    b. The signal from the presynaptic neuron causes an efflux of sodium ions from the postsynaptic neuron.
    c. The signal from the presynaptic neuron causes an influx of potassium ions from the postsynaptic neuron.
    d. The signal from the presynaptic neuron causes an efflux of potassium ions from the postsynaptic neuron.

13. Neurotransmitters released from the vagus nerve cause a decrease in heart rate. What part of the vertebrate nervous system would this nerve be a part of?
    a. Parasympathetic nervous system
    b. Sympathetic nervous system
    c. Somatic system

14. Which of the following two structures make up the central nervous system (CNS)?
    a. The spinal cord and parasympathetic system
    b. The brain and sympathetic nervous system
    c. The brain and spinal cord
    d. The spinal cord and sympathetic nervous system

15. A 42-year-old male patient comes into the emergency room with paralysis on the right side of his body. The attending physician also notices that the patient is having trouble with language. Assuming that there is a head injury, which part of the head did the injury occur on?
    a. The back
    b. The left side
    c. The right side

CHAPTER 45—ANSWER KEY

Check Your Understanding

(a) Most neurons have a cell body, a highly branched group of short projections called dendrites, and one or more long projections called axons. A dendrite receives signals from the axons of adjacent cells. The axon then sends the signal to the dendrites of other neurons.

(b) In neurons, only K^+ can cross the membrane easily along its concentration gradient. It does so via potassium channels. As K^+ moves from the interior of the cell to the exterior via K^+ channels, the inside of the cell becomes more negatively charged relative to the outside. Eventually, the buildup of negative charge inside the cell begins to attract K^+ and counteract the concentration gradient that favors movement of K^+ out. As a result, the membrane reaches a voltage where there is equilibrium between the concentration gradient that moves K^+ out and the electrical gradient that moves K^+ in. The voltage where there is no net movement of K^+ is called the equilibrium potential.

(c) Voltage-gated channels are proteins that change conformation in response to charges present at the membrane surface. These changes will open or close the channel.

(d) Na^+ channels are refractory. Once they have opened and closed, they are less likely to open again for a short period of time after.

(e) Myelination acts as insulation. It prevents charge from leaking out as it spreads down an axon. An unmyelinated area of the neuron called a node of Ranvier has a dense concentration of voltage-gated channels. Electrical signals jump from node to node much faster than if the entire neuron were unmyelinated.

(f) Many receptors function as ligand-gated ion channels, meaning that a molecule binds to a specific site on the receptor to activate it.

(g) If the receptors at the synapse admit an influx of sodium ions in response to the arrival of the neurotransmitter (NT), then the postsynaptic membrane depolarizes. These are called excitatory postsynaptic potentials (EPSPs). In other receptors, binding of the NT leads to the influx of K^+ or Cl^-. These events hyperpolarize the membrane and make action potentials less likely to occur. These are called inhibitory postsynaptic potentials (IPSPs). EPSPs and IPSPs are not all-or-none events. They are graded and short lived. The size depends on the amount of NT released at the synapse.

(h) Wilder Penfield worked with several epileptics scheduled to have seizure-prone areas of their brains surgically removed. While patients were awake, Penfield electrically stimulated portions of the cerebrum and recorded the patients' sensations. This technique was used to map the general areas of brain function.

Assessing What You've Learned

1. a; 2. d; 3. b; 4. d; 5. b; 6. c; 7. a; 8. b; 9. d; 10. a; 11. b; 12. d; 13. a; 14. c; 15. b

Looking Forward—Key Concepts in Later Chapters

Senses—Chapter 46

Before we discuss how hormones work, the electrical signals involved in vision, hearing, taste, and movement must be discussed. This is the subject of **Chapter 46**.

Second Messengers—Chapter 47

The second messenger may induce changes in enzyme activity, gene transcription, or membrane potential. The role of second messengers and hormones is explored in detail in **Chapter 47**.

Animal Sensory Systems and Movement

Looking Back—Key Concepts from Earlier Chapters

Osmosis and Diffusion—Chapter 6

Chapter 6 introduced the processes called diffusion and osmosis. Diffusion describes the movement of substances from regions of higher concentration to regions of lower concentration. The movement of water from regions of higher to lower concentration is called osmosis. This chapter also discussed the electrochemical gradient and ion channels.

Gene Hunting—Chapter 19

To find the gene responsible for tasting PTC, biologists compared the distribution of genetic markers observed in "tasters" and "nontasters." The effort narrowed its location down to several candidates. **Chapter 19** introduced how this type of gene hunt is done.

Amoebae and Slime Molds—Chapter 29

Actin and myosin are also responsible for the amoeboid movement observed in amoebae and slime molds and the streaming of cytoplasm observed in algae and land plants.

Pollination—Chapter 40

Certain flowers have patterns only apparent in ultraviolet wavelengths that serve as signals for pollinating insects.

Electrical Signals—Chapter 45

Because intercalated discs of cardiac muscle contain many gaps, electrical signals pass directly from one cell to the next. **Chapter 45** explored how electrical signals in animals are initiated and propagated.

KEY CONCEPTS

- Sensory receptors change sensory input such as light, sound, touch, and odors into electrical signals that are then relayed to the brain for processing.

- Hair cells transduce sound waves into electrical signals to allow for hearing.

- Rods and cones are photoreceptors that transduce light waves into electrical signals to allow for vision.

- Taste cells and olfactory receptor neurons transduce molecules into electrical signals to allow for taste and smell.

A. CHAPTER OUTLINE

46.1 How Do Sensory Organs Convey Information to the Brain?

Sensory Transduction

- Although animals have sensory receptors that detect a remarkable variety of stimuli, they all **transduce** sensory input (light, sounds, touch, and odors) to a change in membrane potential. In this way, different types of information are transduced into a common type of signal.

- The amount of depolarization that occurs in a sound receptor cell is proportional to the loudness of the sound. Louder sounds induce a higher frequency of action potentials than do softer sounds (**Figure 46.2**).

Transmitting Information to the Brain

- Receptor cells tend to be highly specific. For example, each receptor in a human ear responds best to certain frequencies of sound. Therefore, the train of action potentials from a cell contains information about the frequency of the sound, its intensity, and how long the stimulus lasts.

- Each type of sensory neuron sends its signal to a specific portion of the brain. Axons from sensory neurons in the human ear project to a particular area in the side of the brain.

46.2 Hearing

- **Hearing** is the ability to sense the changes in pressure called **sound**. A sound consists of **waves** of air or water pressure. The number of pressure waves that occur in 1 second is called the **frequency**.

How Do Sensory Cells Respond to Sound Waves and Other Forms of Pressure?

- In **pressure receptors**, direct physical pressure on a cell membrane or distortion by bending causes ion channels on a cell membrane to open or close. In response to change in ion flows, the membrane depolarizes or hyperpolarizes; the result is a new pattern of action potentials from the sensory neuron.

- In many pressure-sensing organs, **hair cells** are named for a set of stiff outgrowths, called **stereocilia**, that occur at one end of the cell (**Figure 46.3**). If they bend in one direction in response to pressure, channels open and depolarize the cell. If they bend in the other direction, channels close and the cell hyperpolarizes.

✔ CHECK YOUR UNDERSTANDING

(a) Define a pressure receptor.

_____

_____

_____

_____

The Mammalian Ear

- When sound waves reach the head, they pass through a canal and strike the **tympanic membrane**, which separates the outer ear from the middle ear (**Figure 46.4**). The membrane vibrations are passed to three tiny bones called **ear ossicles**. The last ossicle, called the **stapes**, vibrates against a membrane called the **oval window** that separates the middle ear from the inner ear.

- The oval window vibrates and generates waves in the fluid inside the **cochlea**. These pressure waves are sensed by **hair cells** in the cochlea.

The Middle Ear Amplifies Sounds

- The **size difference** between the tympanic membrane and oval window is important. The amount of vibration induced by sound waves is increased by a factor of 15. The ossicles act as **levers** and further amplify the sound such that the overall amplification factor is 22.

The Cochlea Detects the Frequency of Sounds

- Georg von Békésy (1920s) performed experiments on cochleas painstakingly dissected from human cadavers. He was able to vibrate the oval window and record how the cochlea's internal membranes moved in response. His key finding was that sounds of different frequencies caused the **basilar membrane** to vibrate in specific spots along its length (**Figure 46.6**).

- Certain portions of the basilar membrane vibrate in response to specific pitches and result in the bending of hair-cell stereocilia. In this way, hair cells in a particular place on the membrane respond to sounds of certain frequency.

✔ CHECK YOUR UNDERSTANDING

(b) How does the ear amplify sound?

_____

_____

_____

_____

Sensory Worlds: What Do Other Animals Hear?

- Elephants use **infrasounds** (sound frequencies too low for humans to hear) to communicate.

- Bats use ultrasonic sound to **echolocate** (navigate by sound). Bats with cotton in their ears or their mouths taped shut are unable to fly without running into objects. Blindfolded bats, in contrast, are fully able to navigate.

- Fishes and larval amphibians use hair cells on their **lateral line system** to sense changes in water pressure.

46.3 Vision

The Insect Eye

- Insects possess **compound eyes** composed of many **ommatidia**, which allows insects to have high resolution and good movement detection (**Figure 46.7**). Each ommatidium has about four receptor cells that respond to light.

The Vertebrate Eye

The Structure of the Camera Eye

- The outermost layer of the eye is a tough rind of white tissue called the **sclera** (**Figure 46.8**). The front of the sclera is transparent and forms the **cornea**. Just inside the cornea is a colored, round muscle called the **iris**. The hole in the center of the iris is called the **pupil**.
- Together, the cornea and **lens** focus the light onto the retina in the back of the eye. The **retina** contains a layer of **photoreceptors** and several layers of neurons.

A Flawed Structure?

- The vertebrate eye has a **blind spot** where the optic nerve leaves the retina. No photoreceptor cells exist at the blind spot, thus if light hits this spot, no signal is sent to the brain.

What Do Rod and Cone Cells Do?

- **Photoreceptors** in the vertebrate eye are small rod- or cone-shaped cells called rods and cones. **Rods** are very sensitive to dim light but not to color. **Cones** are much less sensitive to faint light but are stimulated by different colors.

How Do Rods and Cones Detect Light?

- Rods and cones are packed with membrane-rich disks containing large quantities of a transmembrane protein called **opsin**. Each opsin molecule is associated with a much smaller molecule called **retinal**. The two-molecule complex is called **rhodopsin** (**Figure 46.9b**).
- Retinal changes shape when it absorbs a photon of light, leading to a change in opsin's conformation. This in turn leads to a series of events that culminates in a change in the cell's membrane potential.

Color Vision: The Puzzle of Dalton's Eye

- John Dalton and his brother (1794) could not differentiate between the colors red and green (a condition called red-green color blindness; **Figure 46.11**). Dalton hypothesized that the color of the eye fluid is responsible for color vision. His theory was later rejected.

Color Vision: Multiple Opsins

- Biologists analyzed cones and found that there were three different cone types, each having a different type of opsin. These proteins are called the blue, green, and red (or S, M, and L for short, medium, and long wavelength) opsins.
- In 1990, it was discovered that color-blind people lack either functional M or L cones, or both. Hunt analyzed Dalton's DNA and found that he had a perfectly normal *L* allele but lacked a functional *M* allele. As a result, Dalton did not have green-sensitive cones.

CHECK YOUR UNDERSTANDING

(c)   Compare and contrast the rods and cones of the eye.

_____

_____

_____

_____

Sensory Worlds: Do Other Animals See Color?

- A marine fish called a coelacanth lives in water 200 meters deep and has two opsins that respond to the blue region of the spectrum. It is likely that these fish perceive several hues of blue that we would consider one color.
- Humans and other primates have two of three opsins that are sensitive to wavelengths around 550 nm, which allows them to distinguish between the colors of ripe and unripe fruits.
- Birds and insects can see ultraviolet (UV) light, which has shorter wavelengths than the human eye can see. Some flowers have UV patterns that signal to insect pollinators and many birds have UV patterns in their plumage that are used in mate selection.

CHECK YOUR UNDERSTANDING

(d)   Why would the ability of birds and insects to detect UV light be adaptive?

_____

_____

_____

_____

46.4 Taste and Smell

Taste: Detecting Molecules in the Mouth

- A **taste bud** contains about 100 spindle-shaped **taste cells** that synapse to taste neurons (**Figure 46.13**).
- The four basic tastes are salty, sweet, sour, and bitter.

Salty and Sour

- The sensation of **saltiness** is primarily due to sodium ions dissolved in food. These ions flow into certain taste cells through open channels and depolarize their membranes.
- **Sourness** is due to the presence of protons that flow directly into certain taste cells through channels. In general, the lower the pH of food, the more it depolarizes a taste-cell membrane and the more sour the food tastes.

Why Do Many Different Foods Taste Bitter?

- Arthur Fox (1931) blew some phenylthiocarbamide (PTC) into the air and found that a colleague could taste bitterness in the air while he could not. Follow-up research confirmed that the ability to taste PTC is inherited and polymorphic.
- Independently, two teams identified a family of 40–80 genes that encode transmembrane receptor proteins. Each protein in this family binds to just one particular type of bitter molecule. A taste cell can have many different receptor proteins from this family.

What Is the Molecular Basis of Sweetness and Other Tastes?

- Approaches similar to those used to investigate the molecular basis of perceiving bitter tastes have been used to discover the receptors responsible for detecting sweetness as well as glutamate and other amino acids.
- The glutamate receptor is responsible for tasting monosodium glutamate, while a single receptor responds to a variety of sugars.

✔ CHECK YOUR UNDERSTANDING

(e)   Describe and characterize the four basic tastes.

_____

_____

_____

_____

Olfaction: Detecting Molecules in the Air

- **Smell** allows animals to monitor airborne molecules that convey information. Wolves and domestic dogs can distinguish millions of different odors at extremely small concentrations.
- When odor molecules reach the nose, they diffuse into a **mucus layer** in the roof of the structure. There they activate **olfactory receptor neurons** via membrane-bound receptor proteins. Axons from these neurons project up to the **olfactory bulb** of the brain (**Figure 46.14**).
- A gene family containing hundreds of distinct coding regions encodes receptor proteins on the surface of olfactory receptor neurons.
- Each olfactory neuron has only one type of receptor, and neurons with the same type of receptor project to different regions in the olfactory area of the brain.

46.5 Movement

Skeletons

- Skeletons provide attachment sites for muscles and a support system for the body's soft tissues. **Exoskeletons** are hard, hollow structures that envelop the body. **Hydrostatic skeletons** use the pressure of internal body fluids to support the body. **Endoskeletons** are hard structures buried inside the body.

Endoskeleton Structure

- Endoskeletons are composed of cartilage and bone. **Cartilage** is made up of cells scattered in a gelatinous matrix of polysaccharides and scattered protein fibers. **Bone** is made up of cells in a hard extracellular matrix of calcium phosphate with small amounts of calcium carbonate and protein fibers.
- **Joints** or **articulations** are locations where bones meet and interact (**Figure 46.15**).
- The ends of skeletal muscle are often attached to two different bones by **tendons**, which are bands of tough, fibrous connective tissue.

Endoskeleton Function

- Paired flexor-extensor muscles attached to bones contract to allow for the movement of the skeletal system. The **flexor** brings two long bones in an arc toward each other, while the **extensor** straightens the two bones out (**Figure 46.16**).

Muscle Types

- The three types of muscle tissue are:
  1. Skeletal muscle consists of multinucleate cells and appears striped.
  2. Cardiac muscle contains branched cells whose ends are connected by intercalated discs.

3. Smooth muscle is unbranched and lacks myofibrils. This type of muscle is important to the function of the lungs, blood vessels, the digestive system, the urinary bladder, and the reproductive system.

How Do Muscles Contract?

- A **muscle fiber** is a long, thin muscle cell. Within each cell there are many small strands called **myofibrils**.

Myofibril Structure

- The **sarcomere**, which is the functional unit of skeletal muscle, appears as light and dark bands on the myofibril.

The Sliding-Filament Model

- Andrew Huxley and Jean Hanson (1954) proposed that the banding patterns in the sarcomere are actually caused by two types of long filaments—thick and thin—and that the filaments slide past one another during contraction. This explanation became known as the **sliding-filament model**.

- Follow-up research showed that the thin filaments are composed of two coiled chains of a globular protein called **actin**. One end of each chain is bound to the sarcomere; the other interacts with the thick filament.

- Thick filaments are composed of multiple strands of a long protein called **myosin**, which is anchored to the middle of the sarcomere.

How Do Actin and Myosin Interact?

- The myosin head can bind actin and catalyze the hydrolysis of ATP into ADP and a phosphate ion. Myosin and actin are locked together when an animal dies, and its muscles enter the stiff state known as **rigor mortis**. This evidence suggested that ATP must be present for myosin to release from actin.

- Ivan Rayment and colleagues proposed the following model (**Figure 46.20**):

  1. When a molecule of ATP binds to the myosin head, the myosin releases from actin.

  2. When ATP is subsequently hydrolyzed, the conformation of the protein changes. Specifically, the myosin neck straightens out, the head pivots, and the myosin head then binds to actin in a new location.

  3. When inorganic phosphate is released, the myosin neck bends back into its original position. This bending is called the **power stroke** because it moves the entire actin filament.

  4. A new ATP molecule then binds to myosin, and the cycle starts again.

CHECK YOUR UNDERSTANDING

(f)   Why does an animal that dies undergo rigor mortis?

_____

_____

_____

_____

How Does Relaxation Occur?

- Sarcomeres contain proteins called **tropomyosin** and **troponin**. They work together to block the myosin binding sites on actin. As a result, actin and myosin cannot slide past each other (**Figure 46.21**).

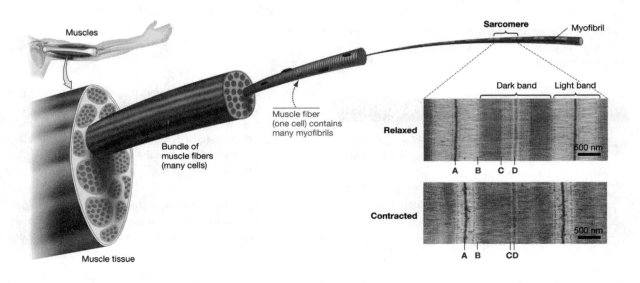

An Overview of Events at the Neuromuscular Junction

1. Action potentials trigger the release of **acetylcholine (ACh)** from the **motor neuron** onto the muscle cell (**Figure 46.22**).

2. Membrane depolarization occurs on the muscle fiber in response to ACh. Action potentials propagate throughout muscle cells via axon-like structures called **T tubules**.

3. T tubules intersect extensive sheets of smooth endoplasmic reticulum called the **sarcoplasmic reticulum (SR)**. As an action potential passes down a T tubule, a protein in the T tubule membrane changes conformation and opens a **calcium channel** in the SR.

4. Calcium ions are released from the SR, causing a conformational change in troponin, which then moves tropomyosin away from the myosin binding sites on actin.

B. MASTERING CHALLENGING TOPICS

You may have had difficulty with the action potential in the last chapter; the difficulty you may have in this chapter relates to the action potential in muscle cells. The sliding-filament model and the electrical and mechanical events that occur in muscle contraction and relaxation are quite difficult to absorb in one sitting. As you prepare for the test, break muscle contraction down into a few discrete sections. First, go over how the action potential gets to the motor neuron and depolarizes the muscle surface. Once you have mastered that part, focus on how the depolarization travels to the actin and myosin filaments and how the release of calcium is involved in initiating muscle contraction. Last, go over the molecular action of muscle contraction itself and how the myosin heads ratchet along the actin filaments. Once you have mastered the process in smaller pieces, try to put it all together and write out all the steps as if you were answering an essay question. This should prepare you for an essay question you may see on your test.

C. ASSESSING WHAT YOU'VE LEARNED

1. When sound strikes a sound-receptor cell, the cell depolarizes. Which of the following could explain why this depolarization occurs?
   a. An influx of chloride ions
   b. An influx of positively charged ions
   c. An efflux of potassium ions
   d. An efflux of sodium ions

2. A light-receptor cell becomes hyperpolarized in response to a flash of light. Which of the following could explain why this hyperpolarization occurs?
   a. The light promotes an influx of sodium ions into the receptor cell.
   b. The light promotes an efflux of chloride ions.
   c. The light promotes the opening of sodium ion channels.
   d. The light promotes the closing of sodium ion channels.

3. When sound waves enter the ear, they cause a membrane called the tympanic membrane to vibrate. These vibrations are then transmitted across the middle ear to another membrane called the oval window. Why is it beneficial to have the middle ear involved in the delivery of these vibrations to the oval window?
   a. The lever action of the sound waves and the smaller surface area of the oval window help to boost sound energy in the ear so that the ear is more sensitive to sound.
   b. The ear ossicles produce electrical signals that help to amplify the sound waves as they are transmitted to the oval window.
   c. The middle ear is not important, and our hearing would be the same if the ear canal led directly to the oval window.
   d. The middle ear helps to dampen the sound energy to prevent the sound energy from damaging the inner ear.

4. In what part of the human ear are sound waves transmitted in a fluid?
   a. The outer ear
   b. The middle ear
   c. The inner ear

5. Which of the following lists represents the pathway of light through the vertebrate eye?
   a. Cornea, iris, pupil, retina, lens
   b. Cornea, iris, pupil, lens, retina
   c. Cornea, iris, lens, retina, pupil
   d. Cornea, lens, pupil, iris, retina

6. What role does the cornea play in the vertebrate eye?
   a. Controls the amount of light entering the eye
   b. Focuses light on the retina
   c. Contains light-sensitive cells that send signals to the brain to process the visual image
   d. The opening through which light passes into the eye

7. John Dalton, an eighteenth-century chemist, spent time working on the cause of the color blindness that he and his brother both experienced. To them,

red and green appeared to be the same color. Dalton incorrectly hypothesized that the problem was in the color of the fluid that filled his eye. We now know that he was missing cones that are important for telling the difference between red and green light. How do we know that he had a normal amount of S opsin cones?

a. S opsin cones reflect red light, and Dalton could see red objects.

b. M and L opsin cones absorb blue light, and Dalton could see blue objects.

c. S opsin cones absorb blue light, and his problem was with the colors red and green.

d. S opsin cones are not involved in sensing color.

8. Taste research has focused on four basic tastes (salty, sour, sweet, and bitter). Which of these tastes is caused by elevating the sodium ion concentration and sensed by a depolarization of taste cells due to an influx of sodium ions?

a. Salty

b. Sour

c. Sweet

d. Bitter

9. Which of the following muscle types is associated with the intestinal tract and arteries?

a. Smooth muscle

b. Skeletal muscle

c. Cardiac muscle

10. Which of the following muscle types is considered voluntary?

a. Smooth muscle

b. Skeletal muscle

c. Cardiac muscle

11. Which of the following muscle types contains multinucleated fibers?

a. Smooth muscle

b. Skeletal muscle

c. Cardiac muscle

12. What component of the sarcomere contains a binding site for ATP?

a. Myosin heads

b. Actin

c. Troponin

d. Tropomyosin

13. Which of the following events produces the "pull" that is associated with myosin and actin sliding past one another?

a. The binding of ATP to the myosin head

b. The hydrolysis of ATP

c. The release of a phosphate from the myosin head after ATP is hydrolyzed

d. The myosin head with an empty ATP binding site

14. Research has shown that a contraction occurs in response to a release of calcium ions from the sarcoplasmic reticulum found in muscle fibers. What is the role of these calcium ions in the contraction?

a. The calcium ions are incorporated into ATP, which fuels the contraction process.

b. The calcium ions cause the thick filaments to shorten. This shortening of the thick filaments is what produces the contraction.

c. The calcium ions bind to the thin filaments and pull on them to produce the contraction.

d. The binding of calcium to troponin causes the tropomyosin to rotate out of the way of the binding sites for myosin on the actin molecules. This allows the myosin to bind to the actin to initiate a contraction.

15. In animal joints, at least two muscles are required to flex and extend at the joint. Which of the following explains why a single muscle cannot act as both a flexor and an extensor?

a. A single muscle would not be strong enough to produce both actions.

b. The muscle would have to move to a different bone in order to produce both actions.

c. A muscle can only "pull" at a joint through a contraction. Since muscles cannot push, a different muscle would be required to move the joint in an opposite direction.

d. The contraction of a muscle that is acting as a flexor is different from the contraction that produces extension.

CHAPTER 46—ANSWER KEY

Check Your Understanding

(a) In pressure receptors, direct physical pressure on a cell membrane or distortion by bending causes ion channels on a cell membrane to open or close. In response to change in ion flows, the membrane depolarizes or hyperpolarizes, and the result is a new pattern of action potentials from the sensory neuron.

(b) The size difference between the tympanic membrane and oval window is important. The amount of vibration induced by sound waves is increased by a factor of 17. The ossicles act as levers and further amplify the sound such that the overall amplification factor is 22.

(c) Photoreceptors in the vertebrate eye are small rod- or cone-shaped cells called rods and cones. Rods are very sensitive to dim light but not to color. Cones are much less sensitive to faint light but are stimulated by different colors.

(d) Birds and insects can see ultraviolet (UV) light, which has shorter wavelengths than the human eye can see. Some flowers have UV patterns that attract insect pollinators, and many birds have UV patterns in their plumage that are used in mate selection.

(e) The four basic tastes are salty, sweet, sour, and bitter. The sensation of saltiness is primarily due to sodium ions dissolved in food. Sourness is due to the presence of protons that flow directly into certain taste cells through channels. For bitterness, researchers identified a family of 40–80 genes that encode transmembrane receptor proteins. Each protein in this family binds to just one particular type of bitter molecule. A taste cell can have many different receptor proteins from this family. Membrane receptors that respond to sugars have yet to be identified by the same techniques employed in the search for bitterness receptors.

(f) The myosin head can bind actin and catalyze the hydrolysis of ATP into ADP and a phosphate ion. Myosin and actin are locked together when an animal dies, and its muscles enter the stiff state known as rigor mortis. This evidence suggests that ATP must be present for myosin to release from actin.

Assessing Your Knowledge

1. b; 2. d; 3. a; 4. c; 5. b; 6. b; 7. c; 8. a; 9. a; 10. b; 11. b; 12. a; 13. c; 14. d; 15. c

---

Looking Forward—Key Concepts in Later Chapters

Second Messengers—Chapter 47

The second messenger may induce changes in enzyme activity, gene transcription, or membrane potential. The role of second messengers and hormones is explored in detail in **Chapter 47**.

Chemical Signals in Animals

KEY CONCEPTS

- ☞ The endocrine system produces and secretes hormones that relay information throughout the body.

- ☞ Hormones carry information that tells cells what to do during development and environmental change.

- ☞ Hormone production and release are tightly regulated by neural input and by other hormones.

- ☞ Hormones act by inducing changes in gene expression or protein activation.

A. CHAPTER OUTLINE

- The **endocrine system** is responsible for the production and secretion of hormones. A **hormone** is a chemical signal that circulates through the blood or other bodily fluids to affect **target cells**.

47.1 Cell-to-Cell Signaling: An Overview

Major Categories of Chemical Signals

- Chemical messengers can function in six ways (**Table 47.1**).
    1. **Autocrine** signals act on the same cell that secreted them.

2. **Paracrine** signals diffuse locally, acting on neighboring cells.

3. **Endocrine** signals are produced by hormones, molecules carried to distant cells by blood or other body fluids.

4. **Neural** signals diffuse extremely short distances between adjacent neurons at synapses.

5. **Neuroendocrine** signals are hormones that are released from neurons but act remotely on tissues other than synapses.

6. **Pheromone** signals are released into the environment and act on a different individual.

Hormone Signaling Pathways

- Hormonal signaling pathways take one of three forms (**Figure 47.1**):

  1. Direct hormonal signaling from an endocrine cell

  2. Direct hormonal signaling from the central nervous system

  3. Central nervous system–to–endocrine system signaling

- Signaling pathways are regulated by **negative feedback** (feedback inhibition), where the product being produced (e.g., hormone) inhibits its own production if high enough levels are present. When levels

then fall, production begins again, thus regulating product levels.

What Makes Up the Endocrine System?

- Hormone-secreting organs are called **endocrine glands**. They are diverse in size, shape, and location.

Chemical Characteristics of Hormones

- There are three types of chemical messengers in animals (**Figure 47.3**):

  1. Polypeptides

  2. Amino acid derivatives

  3. Steroids

Hormone Concentrations Are Low, But Their Effects Are Large

- The key **similarity** in these hormone types is that all are present in extremely small concentrations, yet have large effects. For example, 1000 grams of cow pituitary tissue yield just 0.04 grams of growth hormone.

Only Some Hormones Can Cross Cell Membranes

- The key **difference** in these hormone types is that steroids are lipid soluble while polypeptides and amino acid derivatives are not. As a result, steroids cross cell membranes much more readily than do other types of hormones.

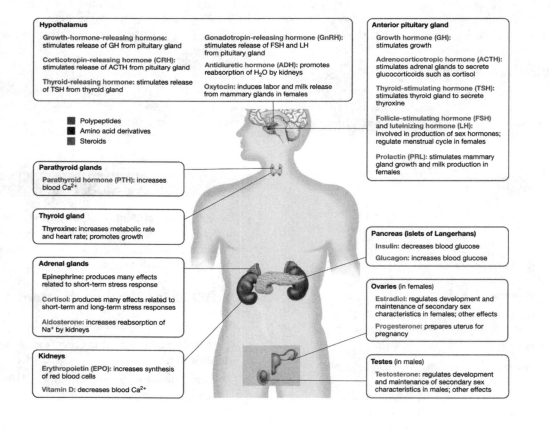

Hypothalamus

Growth-hormone-releasing hormone: stimulates release of GH from pituitary gland

Corticotropin-releasing hormone (CRH): stimulates release of ACTH from pituitary gland

Thyroid-releasing hormone: stimulates release of TSH from thyroid gland

Gonadotropin-releasing hormone (GnRH): stimulates release of FSH and LH from pituitary gland

Antidiuretic hormone (ADH): promotes reabsorption of H_2O by kidneys

Oxytocin: induces labor and milk release from mammary glands in females

■ Polypeptides
■ Amino acid derivatives
■ Steroids

Anterior pituitary gland

Growth hormone (GH): stimulates growth

Adrenocorticotropic hormone (ACTH): stimulates adrenal glands to secrete glucocorticoids such as cortisol

Thyroid-stimulating hormone (TSH): stimulates thyroid gland to secrete thyroxine

Follicle-stimulating hormone (FSH) and luteinizing hormone (LH): involved in production of sex hormones; regulate menstrual cycle in females

Prolactin (PRL): stimulates mammary gland growth and milk production in females

Parathyroid glands

Parathyroid hormone (PTH): increases blood Ca^{2+}

Thyroid gland

Thyroxine: increases metabolic rate and heart rate; promotes growth

Adrenal glands

Epinephrine: produces many effects related to short-term stress response

Cortisol: produces many effects related to short-term and long-term stress responses

Aldosterone: increases reabsorption of Na^+ by kidneys

Kidneys

Erythropoietin (EPO): increases synthesis of red blood cells

Vitamin D: decreases blood Ca^{2+}

Pancreas (islets of Langerhans)

Insulin: decreases blood glucose

Glucagon: increases blood glucose

Ovaries (in females)

Estradiol: regulates development and maintenance of secondary sex characteristics in females; other effects

Progesterone: prepares uterus for pregnancy

Testes (in males)

Testosterone: regulates development and maintenance of secondary sex characteristics in males; other effects

(a) What are the similarities and differences between the three major groups of hormones?

_____

_____

_____

_____

_____

_____

How Do Researchers Identify a Hormone?

- The following steps are usually taken in identifying and purifying hormones:
  1. Remove suspected tissue and study it.
  2. Extract a solution from the tissue in intact animals.
  3. Purify the active ingredient in the raw tissue extract, inject the hormone into animals that cannot produce the molecule, and determine if their symptoms are cured.

(b) What steps are typically taken in the identification and purification of a hormone?

_____

_____

_____

_____

(c) What do you think could go wrong with the hormone identification process?

_____

_____

_____

_____

_____

47.2 What Do Hormones Do?

- Hormones coordinate the activities of cells in response to three general situations:
  1. Environmental challenges
  2. Growth, development, and reproduction
  3. Homeostasis

How Do Hormones Direct Developmental Processes?

T_3's Role in Amphibian Metamorphosis

- Thyroid-stimulating hormone in the amphibian brain's pituitary stimulates the production of the thyroid hormone **triiodothyronine**, or T_3, which is responsible for most of the changes observed in amphibian metamorphosis.

Hormone Interactions Regulate Insect Metamorphosis

- Insect metamorphosis depends upon interactions between **juvenile hormone** (**JH**) and **20-hydroxyecdysone** (**Figure 47.6**), which stimulates developmental changes from larva to pupa to adult in holometabolous insects (insects that pass through larval stages before reaching adulthood).

Sexual Development and Activity in Vertebrates

- **Primary sex determination** in early embryonic development dictates whether the sex organs of an embryo become male or female. Once testes or ovaries develop, they begin producing male- or female-specific hormones.

- **Testosterone** leads to early development of the male reproductive tract and genitalia and inhibits the formation of breast primordia.

- **Estrogens** induce the initial formation of the female reproductive tract, genitalia, and breast tissue.

- Juvenile-to-adult transition—In boys, surges of sex hormones lead to changes that include enlargement of the penis and testes and growth of facial and body hair.

- In girls, increased concentrations of estrogen called **estradiol** lead to the enlargement of breasts, the onset of menstruation, and other changes.

- Seasonal or cyclical sexual activity—Most long-lived animals reproduce seasonally. In many species, environmental cues trigger the release of sex hormones.

Growth in Mammals

- In humans and other mammals, **growth hormone** (**GH**) interacts with sex hormones to stimulate the growth of long bones and vertebrae of the spinal column during puberty.

(d) What are the specific effects of hormone surges in males and females entering puberty?

_____

_____

_____

How Do Hormones Coordinate Responses to Environmental Change?

Short-Term Responses to Stress

- When a person is put into a dangerous situation, hormones regulate both the short-term and long-term response. The short-term reaction is called the **fight-or-flight response** and occurs in conjunction with the activation of the sympathetic nervous system.
- If you were being chased by a grizzly bear, action potentials from sympathetic nerves would stimulate the adrenal medulla and lead to the release of **epinephrine**.

Long-Term Responses to Stress

- **Cortisol** is a hormone produced by the adrenal cortex in response to extended periods of stress. Those who were subjected to long periods of stress had heightened levels of cortisol in their system.

What Does Cortisol Do?

- Cortisol maintains glucose production by:
    1. Stimulating the synthesis of liver enzymes that make glucose
    2. Rendering adipose tissue and muscles resistant to insulin so that glucose is not removed from the bloodstream by these tissues
    3. Promoting the release of fatty acids for use by the heart and muscles
- Glucocorticoids prepare an individual for long-term stress by conserving glucose for use by the brain at the expense of other tissues and organs. They also conserve glucose by actively suppressing wound healing and other aspects of immune system function.

(e) How does cortisol allow an individual to withstand long-term stress?

_____

_____

_____

Long-Term Stress in Non-Human Organisms

- Glucocorticoids induce the proliferation of chloride cells in the gills of salmon that migrate from freshwater to salt water.

How Are Hormones Involved in Homeostasis?

- Messages from sensory receptors travel to the brain for integration, and the messages often travel from integrators to effectors in the form of hormones.

Leptin and Energy Reserves

- Animals store energy in the form of **triglycerides** during periods of decreased food availability.
- In the 1970s, biologists discovered that mice that have mutations in the *obese* and *diabetic* genes were less active, consumed more foods, and were obese compared to their wild-type siblings (**Figure 47.8**). When researchers performed **parabiosis** between homozygous *obese* (*ob/ob*), homozygous *diabetic* (*db/db*), and wild-type animals, and allowed the two to become **parabiotic** partners that shared a circulatory system, and thus hormones and other blood components, they made the following observations:
    1. Joining a *db/db* mouse to either a lean or *ob/ob* mouse caused the lean and *ob/ob* mouse to stop eating and lose weight.
    2. Joining an *ob/ob* mouse to a lean mouse caused the *ob/ob* mouse to eat less food and gain weight less rapidly than one that was joined to another *ob/ob* mouse.
- From the parabiosis experiments, biologists postulated that although *ob/ob* animals and *db/db* animals shared the same phenotype, two different parts of the same hormone-signaling system was inactivated in the two genotypes.
- Biologists confirmed that the *obese* gene encodes for **leptin**, a hormone that is produced by adipocytes, and the *diabetic* gene encodes for leptin receptors. Leptin interacts with leptin receptors in the brain to control feeding behavior. Mutations in either gene result in the inability to control food intake and thus energy reserves.

(f) Explain why a mutation in the *obese* gene can cause the same phenotype as a mutation in the *diabetic* gene.

_____

_____

_____

ADH, Aldosterone, and Water and Electrolyte Balance

- **Antidiuretic hormone (ADH)** causes the retention of water in the body by increasing the permeability of the kidney's collecting ducts to water. Alcoholic beverages inhibit the release of ADH, and thus cause dehydration.

- **Aldosterone** is released in response to low ion concentrations in body fluids. It increases reabsorption of sodium ions in the distal tubules of the kidney in order to maintain electrolyte and water balance in the body.

EPO and Oxygen Availability

- **Erythropoietin (EPO)** is released from the kidneys and other tissues when blood oxygen levels are low. EPO acts to restore homeostasis by stimulating the production of red blood cells, which increase the oxygen-carrying capacity of the blood.

47.3 How Is the Production of Hormones Regulated?

The Hypothalamus and Pituitary Gland

- Physiologists showed that rats suffered from a variety of debilitating symptoms when their pituitary glands were removed. In addition, their genitals, thyroid glands, and adrenal cortexes atrophied. Biologists proposed the existence of a molecule that affects the adrenal gland.

Controlling the Release of Glucocorticoids

- In 1943, **adrenocorticotropic hormone (ACTH)** was purified and characterized, confirming that glucocorticoids are secreted from the adrenal cortex when ACTH is released from the pituitary. ACTH is therefore called a **regulatory hormone**.

- **Corticotropin-releasing hormone** (**CRF**) was later purified and characterized as stimulating the secretion of ACTH from cells of the anterior pituitary.

- The presence of cortisone inhibits the release of ACTH. When the presence of a molecule inhibits its production, **feedback inhibition** is occurring.

- All of these hormones act as regulators, and they are all involved in feedback inhibition. **CRF** triggers ACTH production and ACTH triggers glucocorticoid release, but ACTH also inhibits CRF function through a feedback loop, and the glucocorticoids provide feedback by inhibiting ACTH release (**Figure 47.10**).

Patterns in Glucocorticoid Release

- Normally, CRH level is at the highest in the early morning, with ACTH and cortisol levels following suit. In stressful situations, CRH, ACTH, and cortisol are at a constant high level throughout the day (**Figure 47.11**).

The Hypothalamic-Pituitary Axis—an Overview

- The posterior and anterior sections of the pituitary gland are each influenced by different populations of neurons in the **hypothalamus**. The populations of hypothalamic neurons are called **neurosecretory cells** because they synthesize the release of hormones (**Figure 47.12**).

- The release of hormones by the cells in the hypothalamus is under the control of brain regions responsible for integrating information about the external or internal environment.

The Posterior Pituitary

- The posterior pituitary is actually an extension of the hypothalamus itself. Neurosecretory cells that project from the hypothalamus produce the hormones **ADH** and **oxytocin**, which are then stored in the posterior pituitary. From there, these hormones are released into the bloodstream.

The Anterior Pituitary

- The hypothalamus and anterior pituitary (also known as the master gland) are connected indirectly by blood vessels. In response to the arrival of releasing hormones from the hypothalamus, the anterior pituitary secretes hormones that enter the bloodstream and act on target tissues and glands.

✔ **CHECK YOUR UNDERSTANDING**

(g) Describe the functional relationship between the anterior pituitary, the posterior pituitary, and the hypothalamus.

_____

_____

_____

_____

_____

Control of Epinephrine by Sympathetic Nerves

- Epinephrine and **norepinephrine** belong to a family of molecules called **catecholamines**. Catecholamines function as a neurotransmitter as well as a hormone, and in some cases, the mode of action of hormones and neurotransmitters is similar.

- By altering synapses during memory functions in the brain, neurotransmitters play a central role in learning and memory. Similarly, many hormones exert their effects by activating particular genes in target cells.

47.4 How Do Hormones Act on Target Cells?

Steroid Hormones and Intracellular Receptors

Identifying the Estrogen Receptor

- In 1964 biologists succeeded in isolating the estradiol receptor in laboratory rats (**Figure 47.13**).

- Follow-up experiments confirmed that the estradiol receptor is located in the nucleus, but not associated with the nuclear envelope. Further, it is found in all target tissues including the uterus, hypothalamus, and mammary glands.

- Subsequently it was found that the gene for the estradiol receptor has a sequence and structure that are similar to the receptors for the glucocorticoids, testosterone, and other steroids.

Documenting Changes in Gene Expression

- A distinctive DNA-binding region occurs in the steroid hormone receptor. The region codes for the DNA-binding domain called a **zinc finger**. Zinc fingers are found in all of the proteins in the steroid hormone-receptor family.

- Investigators also confirmed that steroid hormone-receptor complexes bind to specific sites in DNA (**Figure 47.14**). These sites are called **hormone-response elements**. They function as the type of regulatory DNA sequence called an enhancer.

Hormones That Bind to Cell-Surface Receptors

- Epinephrine and the peptide hormones are not lipid soluble, so they have to bind to receptors on the cell surface. The messenger never enters the target cell, so it must activate a receptor on the cell—**signal transduction**.

Identifying the Epinephrine Receptor

- Epinephrine and its agonists produce two distinct patterns of responses because two types of receptors exist—**alpha receptors** and **beta receptors**—each of which has two subtypes.

What Acts as the Second Messenger?

- To understand how epinephrine triggers the release of glucose from target cells, biologists focused on an enzyme called **phosphorylase**. It catalyzes the cleavage of glucose molecules off of glycogen. Phosphorylase exists in active and inactive forms, and as the enzyme (**Figure 47.15**).

- Researchers found that when they added epinephrine to their cell-free system, large amounts of phosphorylase were activated. By purifying components of the cell-free system, they found **cyclic adenosine monophosphate (cAMP)** as the molecule that activated phosphorylase.

- A **second messenger** is a signaling molecule that increases in concentration inside a cell in response to a molecule that binds at the surface.

A Phosphorylation Cascade

- The synthesis of cAMP initiates a chain of events called a **signal transduction cascade**. cAMP is produced from ATP in a reaction that is catalyzed by the enzyme **adenylyl cyclase**. cAMP binds to a protein called **cAMP-dependent protein kinase**. This enzyme responds by phosphorylating an enzyme that then phosphorylates phosphorylase kinase.

- The second messenger cAMP plays a primary role in transmitting the signal from the cell surface to the signaling cascade.

- The primary function of the other events in the sequence is to amplify the signal.

Why Do Different Target Cells Respond in Different Ways?

- The same chemical messenger can trigger different responses in cells from different organs or in cells at different developmental stages because the cells contain different receptors, second messengers, amplification steps, protein kinases, or transcriptionally active genes.

B. MASTERING CHALLENGING TOPICS

The most complicated system to learn in this chapter is the last one you studied—the second messenger system. As with other physiological cascades, break the system down bit by bit, learn each piece, and then integrate the puzzle. For chemical reactions in the cAMP second messenger system, drawing the reactions in the cascade is of great benefit. This is the best way to answer an essay question. Rather than memorize a lot of key points, practice drawing a diagram and you will find great amounts of information to extract.

C. ASSESSING WHAT YOU'VE LEARNED

1. All of the following accurately describe the functions of hormones *except*:
   a. hormones are released into the bloodstream to act on distant targets.
   b. hormones coordinate the activities of groups of cells.
   c. hormones respond to changes in the internal, but not the external, environment.
   d. hormones act only on cells that possess specific receptors for that hormone.

2. All of the following would help in the identification of a suspected hormone *except*:
   a. removal of hormone target tissues.
   b. removal of suspected source (endocrine) gland.
   c. injection of source gland extract to animal that lacks source gland.
   d. purification of source gland extract.

3. All of the following statements concerning the chemical classes of hormones are correct *except*:
   a. steroid hormones are lipid soluble, while polypeptide hormones are water soluble.
   b. steroid hormones readily cross cell membranes, while polypeptide hormones do not.
   c. some amino acid derivative hormones are lipid soluble, while most are water soluble.
   d. steroid hormones are lipid soluble, while polysaccharide hormones are water soluble.

4. Which of the following represents a consequence of the observation that hormones fall into different chemical classes?
   a. Polypeptide hormones bind to receptors located inside target cells.
   b. Steroid hormones bind to receptors located inside target cells.
   c. Steroid hormones bind to receptors located on the surface of target cells.
   d. Most amino acid derivative hormones bind to receptors located inside target cells.

5. Hormones coordinate the activities of cells in response to which of the following?
   a. Maintaining homeostasis
   b. Growth and development
   c. Environmental challenges
   d. All of the above

6. Which of the following situations do you predict might lead to epinephrine release by the adrenal glands, but *not* to cortisol release?
   a. Witnessing a car accident
   b. Preparing for final exams
   c. Skipping meals for two days
   d. Training for a marathon

7. Predict the effects of the administration of ACTH on a normal rat.
   a. Decrease in cortisol release and an increase in CRH release
   b. Increase in cortisol release and a decrease in CRH release
   c. Increase in cortisol release and an increase in CRH release
   d. Increase in growth hormone release and a decrease in CRH release

8. Predict the effects of the administration of ACTH to a rat that has had its pituitary gland removed.
   a. The adrenal glands will grow in size.
   b. The adrenal glands will shrink (or atrophy).
   c. There will be an increase in CRH levels.
   d. There will be a decrease in cortisol production.

9. All of the following effects would be expected to occur if a normal rat was given cortisol injections *except*:
   a. a decrease in size of the adrenal glands.
   b. a decrease in ACTH release.
   c. a decrease in CRH release.
   d. an increase in cortisol release from the adrenal glands.

10. Which of the following provides the best definition of a hormone-response element?
    a. A hormone-response element is a receptor with a hormone bound to it.
    b. A hormone-response element is a DNA-binding region located on the hormone receptor.
    c. A hormone-response element is a DNA sequence that binds a specific hormone-receptor complex.
    d. A hormone-response element is an RNA transcript.

11. What is the significance of the discovery of different hormone receptor types for the hormone epinephrine?
    a. Two types of epinephrine hormone-response elements were identified.
    b. Different tissues responded differently to the presence of epinephrine.
    c. It was discovered that slightly different forms of epinephrine (agonists) were secreted by the adrenal glands.
    d. All target tissues for epinephrine possess all four types of epinephrine receptors.

12. Which of the following molecules serves as a second messenger for epinephrine actions on glucose production?
    a. Phosphorylase
    b. cAMP
    c. Protein kinase
    d. Adenylyl cyclase

13. Which of the following represents the correct sequence of events following the binding of epinephrine to a liver cell?
    a. Increase in cAMP levels; phosphorylation of protein kinase; activation of phosphorylase; decrease in glucose levels
    b. Increase in cAMP levels; activation of protein kinase; inhibition of phosphorylase; decrease in glucose levels

c. Increase in cAMP levels; activation of protein kinase; phosphorylation of phosphorylase; increase in glucose levels

d. Increase in cAMP levels; phosphorylation of protein kinase; inhibition of phosphorylase; increase in glucose levels

14. The primary functions of a hormonally regulated second messenger system are:
    a. to relay a signal from the cell surface to the interior and to stimulate expression of certain genes.
    b. to relay a signal from the cell surface to the interior and to initiate long-term metabolic effects in that cell.
    c. to relay and amplify the signal from the cell surface to the interior.
    d. to change levels of cAMP in target cells.

15. Which of the following best explains the observation that certain hormones, such as epinephrine, can have different effects on different cells?
    a. Second messenger molecules, such as cAMP, can bind to different proteins within cells.
    b. Second messenger molecules, such as cAMP, can result in the transcription of different genes.
    c. Different receptor types can activate the same signal transduction pathway.
    d. Different receptor types can activate different signal transduction pathways.

CHAPTER 47—ANSWER KEY

Check Your Understanding

(a) The key similarity in these hormone types is that all are present in extremely small concentrations, yet have large effects. For example, one kilogram of cow pituitary tissue yields 0.04 grams of growth hormone. The key difference in these hormone types is that steroids are lipid soluble while polypeptides and amino acid derivatives are not. As a result, steroids cross cell membranes much more readily than do other types of hormones.

(b) The following steps are usually taken in identifying and purifying hormones:
    1. Remove suspected tissue and study it.
    2. Extract a solution from the tissue in intact animals.
    3. Purify the active ingredient in the raw tissue extract, inject the hormone into animals that cannot produce the molecule, and determine if their symptoms are cured.

(c) Interference with the process of hormone identification could result from tissue contamination (multiple tissue types in a tissue sample) or from the use of inappropriate solvents for extraction. Test animals may also give ambiguous results unless their current physiological state is fully known.

(d) In boys entering puberty, surges of sex hormones lead to changes that include enlargement of the penis and testes and growth of facial and body hair. In girls, increased concentrations of estrogen (called estradiol) lead to the enlargement of breasts, onset of menstruation, and other changes.

(e) Glucocorticoids prepare an individual for long-term stress by conserving glucose for use by the brain at the expense of other tissues and organs. They also conserve glucose by actively suppressing wound healing and other aspects of immune system function. Cortisol maintains glucose production by: (1) stimulating the synthesis of liver enzymes that make glucose, (2) rendering adipose tissue and muscles resistant to insulin so that glucose is not removed from the bloodstream by these tissues, and (3) promoting the release of fatty acids for use by the heart and muscles.

(f) Although *ob/ob* animals and *db/db* animals share the same phenotype, two different parts of the same hormone-signaling system was inactivated in the two genotypes. The *obese* gene encodes for leptin, a hormone that is produced by adipocytes, and the *diabetic* gene encodes for leptin receptors. Leptin interacts with leptin receptors in the brain to control feeding behavior. Mutations in either genes results in the inability to control food intake and thus energy reserves.

(g) The posterior pituitary is actually an extension of the hypothalamus itself. Neurosecretory cells that project from the hypothalamus produce the hormones ADH and oxytocin, which are then stored in the posterior pituitary. From there, these hormones are released into the bloodstream. The hypothalamus and anterior pituitary are connected indirectly by blood vessels. In response to the arrival of releasing hormones from the hypothalamus, the anterior pituitary secretes hormones that enter the bloodstream and act on target tissues and glands.

Assessing What You've Learned

1. c; 2. a; 3. d; 4. b; 5. d; 6. a; 7. b; 8. a; 9. d; 10. c; 11. b; 12. b; 13. c; 14. c; 15. d

Looking Forward—Key Concepts in Later Chapters

Reproduction—Chapter 48

After discovering what is at the forefront of research in chemical signaling, the way has been paved for exploring how hormones regulate reproduction by humans and other animals.

Animal Mating—Chapter 48

Chapter 48 details how during development, a flush of testosterone or estrogen induces the development of seasonal traits such as singing in male birds and sexual receptivity in female lizards.

Animal Reproduction

Fertilization—Chapter 22

Chapter 22 introduced this process by describing how sperm makes contact with an egg and penetrates its membrane. The discussion in that chapter focused on the molecular mechanisms of fertilization using the sea urchin as a model organism. Recall that contact between sperm and egg triggers the acrosomal reaction, in which enzymes released from the head of the sperm chew a path through the material surrounding the egg membrane.

Countercurrent Exchangers—Chapter 44

Maternal and fetal arteries in the placenta are arranged in a countercurrent fashion. As **Chapter 44** explained, countercurrent flows maintain a concentration gradient that makes diffusion efficient.

Hormones—Chapter 47

Chapter 47 introduced the female sex hormone estradiol and the male sex hormone testosterone. Also recall that the human sex hormones play a key role in three events—the development of the male and female reproductive tract in embryos, the maturation of the reproductive tract during the transition from childhood to adulthood, and the regulation of spermatogenesis and oogenesis in adults.

KEY CONCEPTS

- Asexual and sexual reproduction are two ways in which animals produce offspring.
- Sexual reproduction involves the fertilization of an egg by a sperm.
- Mammalian reproduction is tightly regulated by sex hormones such as testosterone and estradiol.

A. CHAPTER OUTLINE

48.1 Asexual and Sexual Reproduction

How Does Asexual Reproduction Occur?

- Three main mechanisms of asexual reproduction are as follows (**Figure 48.1**):
  1. **Budding** is completed when a miniature version of the parent breaks free and begins to grow on its own.
  2. **Fission** occurs when an individual splits into two or more descendants.

3. **Parthenogenesis** occurs when female offspring develop from unfertilized eggs. Some groups, such as the guppies, rotifers, crustaceans, and lizards, reproduce by parthenogenesis.

✔ CHECK YOUR UNDERSTANDING

(a) What is parthenogenesis, and which animals use it?

_____

_____

_____

Switching Reproductive Modes: A Case History

- *Daphnia* reproduce asexually throughout the spring and summer, with the females producing diploid eggs without sex via parthenogenesis. In late summer or early fall, many females begin producing male offspring by parthenogenesis. Subsequently, sperm from these males fertilize haploid eggs that females produce via meiosis.

What Environmental Cues Trigger the Switch?

- The shortening of days in late summer or fall affects brain sensors that in turn induce the production of males and haploid eggs.

- High population densities, which affect food availability and water quality, are also factors (in addition to day length) in causing this switch (**Figure 48.3**).

Why Do Daphnia Switch between Asexual and Sexual Reproduction?

- Biologists hypothesize that the switch from asexual to sexual reproduction in *Daphnia* is beneficial during stressful situations because sexual reproduction increases genetic diversity among offspring produced. In stressful environments, genetically variable offspring are more likely to survive and reproduce than offspring that are identical to their parents.

Mechanisms of Sexual Reproduction: Gametogenesis

- The mitotic cell divisions, meiotic cell divisions, and developmental events that result in the production of male and female gametes are collectively called **gametogenesis**. This process usually occurs in a sex organ, or **gonad**; the male gonads are called **testes** and the female gonads are called **ovaries**.

- In the male and female gonad, diploid cells called spermatogonia (**Figure 48.4a**) and oogonia (**Figure 48.4b**) divide by mitosis to generate cells that undergo meiosis.

48.2 Fertilization and Egg Development

- **Fertilization** is the joining of a sperm and egg to form a diploid zygote. The most basic aspect of diversity in fertilization is where the union of sperm and egg takes place.

External Fertilization

- Most animals that rely on external fertilization live in aquatic environments and tend to produce exceptionally large numbers of gametes.

- Gametogenesis occurs in response to environmental cues such as lengthening days and warmer water temperatures.

- **Pheromones,** chemical messengers, may play a role in synchronizing gamete release.

Internal Fertilization

- **Internal fertilization** occurs in the vast majority of terrestrial animals as well as in a significant number of aquatic animals.

- Males deposit sperm directly into the female reproductive tract with an organ called the **penis**, or males may package their sperm into a structure called a **spermatophore**, which is then placed into the female's reproductive tract by the male or female.

What Is Sperm Competition?

- Geoff Parker (1970) published results of an experiment that confirmed the existence of **sperm competition**. Whichever male was last to copulate fathered an average of 85 percent of the offspring produced.

✔ **CHECK YOUR UNDERSTANDING**

(b) Compare and contrast external and internal fertilization.

_____

_____

_____

_____

Sperm Competition and Second-Male Advantage

- By introducing a gene into male fruit flies that produced sperm with green tails, researchers were able to establish that the sperm and the fluid that accompanies it of the second male physically displaced from the female storage area the sperm of the first male. This allows for the second-male advantage.

Why Is Testes Size Variable among Species?

- In species where multiple mating is common, males have extraordinarily large testes for their size and produce proportionately larger numbers of sperm. These characteristics increase the chance that a male fertilizes the female's eggs.

- Females are able to change the odds of a male's fertilizing her eggs by choosing which male performs the last copulation before fertilization takes place, and by ejecting sperm from undesirable males.

Unusual Aspects of Mating

- **Femme fatales** are female Australian redback spiders that cannibalize their mating partners to increase copulation time and fertilization rates.

- *Drosophila bifurca* have a body length of about 1.5 mm long, but their sperm are each 6 cm long. The **giant sperm** fill the female's sperm storage area and prevent another male's sperm from entering.

- Male red-billed buffalo weavers have a **false penis** that stimulates the female's reproductive tract.

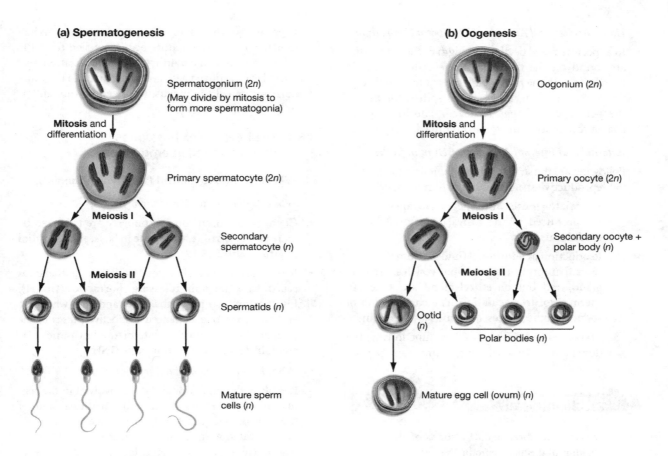

(a) Spermatogenesis

Spermatogonium (2*n*)
(May divide by mitosis to
form more spermatogonia)

Mitosis and
differentiation

Primary spermatocyte (2*n*)

Meiosis I

Secondary
spermatocyte (*n*)

Meiosis II

Spermatids (*n*)

Mature sperm
cells (*n*)

(b) Oogenesis

Oogonium (2*n*)

Mitosis and
differentiation

Primary oocyte (2*n*)

Meiosis I

Secondary oocyte +
polar body (*n*)

Meiosis II

Ootid
(*n*)

Polar bodies (*n*)

Mature egg cell (ovum) (*n*)

- Some bird species appearing to be monogamous practice **infidelity**.

- Individuals of **hermaphroditic** species such as snails and slugs simultaneously receive and deposit sperm during mating. In some species, mucus-covered "love darts" increase the chance of fertilization.

- Some bedbugs use their hypodermic penises to directly inject sperm into the female's body cavity.

Why Do Some Females Lay Eggs while Others Give Birth?

- In **oviparous** animals, the embryo develops in the external environment. In some animals there is no parental care, but a substantial number of species continue to care for their young after the eggs have emerged from the mother's body.

- In **viviparous** species, development takes place within the mother's body. The embryo attaches to the female's reproductive tract and receives nutrition directly from the mother's circulatory system.

- Egg laying probably represents the ancestral condition in the lizard *Sceloporus*, and viviparity evolved from this, probably due to cold or high-altitude habitats.

✔ CHECK YOUR UNDERSTANDING

(c) What are the similarities and differences between oviparity and viviparity?

_____

_____

_____

_____

_____

48.3 Reproductive Structures and Their Functions

The Male Reproductive System

- The external anatomy of the male reproductive system consists of a sac-like **scrotum** and the **penis**. The scrotum holds the testes; the penis functions as the organ of intromission prior to fertilization. These structures are diverse and complex.

How Does External Anatomy Affect Sperm Competition?

- In a species of seed beetle, the length of the spines that are found on the tips of male genitalia is directly correlated with the proportion of eggs fertilized (**Figure 48.8**). The longer the genital spines, the higher the percentage of eggs fertilized. The long spines enable copulation time to be prolonged.

External and Internal Anatomy of Human Males

- Following are the structural components of the internal reproductive structures in the human male:
  1. **Spermatogenesis and sperm storage**—Sperm are produced in the **testes** and stored in the **epididymis**.
  2. **Production of additional fluids**—Complex solutions that form in the **seminal vesicles, prostate gland**, and **bulbourethral gland** are added to sperm prior to **ejaculation**. The combination of sperm and these **accessory fluids** is called **semen**.
  3. **Transport and delivery**—These functions are the domain of the **vas deferens, urethra**, and **penis**.

✔ CHECK YOUR UNDERSTANDING

(d) Describe the three main functions of the internal anatomy of the male reproductive system.

_____

_____

_____

The Female Reproductive System

The Reproductive Tract of Female Birds

- Birds produce eggs in a single ovary through a series of steps beginning with meiosis and maturation of the follicle, fertilization, addition of egg white and membranes, and finally the formation of a calcium carbonate eggshell prior to laying (**Figure 48.10**).

- Dual oviducts exist, but only one is functional.

External and Internal Anatomy of Human Females

- The external anatomy of the female reproductive system features two folds of skin that cover the **clitoris**, the opening of the **urethra**, and the opening of the **vagina** (**Figure 48.11**). The **clitoris** develops from the same population of embryonic cells that gives rise to the penis in males. The **vagina** is where semen is deposited during sexual intercourse and where the fetus is delivered during birth.

- Eggs are produced in the paired **ovaries**. When **ovulation** occurs, a mature egg is expelled from the ovary and enters the **oviduct**, where fertilization may take place. Fertilized eggs are transported to the muscular sac called the **uterus**, where embryonic development continues.

48.4 The Role of Sex Hormones in Mammalian Reproduction

Which Hormones Control Puberty in Mammals?

What Regulates the Gonadal Hormones?

- Changes that occur during puberty are triggered by increased levels of **testosterone** in boys, and **estradiol** in girls (**Figure 48.12**).

- Researchers isolated a hormone from the hypothalamus called **gonadotropin-releasing hormone (GnRH)**. Investigators also noted that boys and girls who were entering puberty experienced distinctive pulses in the pituitary hormones called **luteinizing hormone (LH)** and **follicle-stimulating hormone (FSH)**.

What Regulates the Regulatory Hormones?

- It is likely that nutritional state is involved in the regulation of GnRH. As the general nutritional state of the United States improved, the average age for onset of menstruation in females has decreased to 12 years. Also, currently, girls with large fat stores enter puberty earlier than do thin girls.

✔ CHECK YOUR UNDERSTANDING

(e) What hormonal changes occur during male and female puberty, and what physiological changes do they induce?

_____

_____

_____

_____

Which Hormones Control the Menstrual Cycle in Mammals?

- Although the length of the **menstrual cycle** varies among women, 28 days is about average. In conjunction with changes in the ovary, the lining of the uterus undergoes a dramatic thickening and regression (**Figure 48.13**). **Menstruation**, by definition, is the expulsion of the uterine lining, which designates day 0 in the menstrual cycle.

- **Follicular phase** (14 days)—a **follicle** matures, and **ovulation** occurs when the follicle is mature; it releases its **oocyte** into the oviduct.
- **Luteal phase** (14 days)—the **corpus luteum** forms from the ruptured follicle and subsequently degenerates.

✔ **CHECK YOUR UNDERSTANDING**

(f)   Describe the two phases of the menstrual cycle.

_____

_____

_____

_____

_____

How Do Pituitary and Ovarian Hormones Change during a Menstrual Cycle?

- Dramatic changes in hormone concentrations occur during the cycle (**Figure 48.14**), including estradiol, LH, FSH, and progesterone.
- LH stays constant except during a spike just prior to ovulation, whereas FSH levels are high during the follicular phase and low during the luteal phase.
- Progesterone is low during the follicular phase and high during the luteal phase.

How Do Pituitary and Ovarian Hormones Interact?

- Changes in the concentrations of estradiol and progesterone affect the release of the pituitary hormones LH and FSH.
- The following steps summarize the interaction of these hormones:
  1. As the uterus is shedding its lining, a follicle is being stimulated to develop under the influence of FSH.
  2. As the follicle grows, its production of estradiol begins to increase. While estradiol is low, it suppresses LH secretion via negative feedback inhibition.
  3. When the follicle produces large quantities of estradiol, it begins to exert positive feedback on LH, producing a spike in both hormones.
  4. The LH surge triggers ovulation and ends the follicular phase.
  5. As the corpus luteum develops from the ruptured follicle, it begins to secrete progesterone in response to LH.

6. The rise in progesterone converts the thickened uterine lining to an actively secreting tissue with a well-developed blood supply.
7. If fertilization does *not* occur, the corpus luteum degenerates and progesterone levels fall, causing the uterine lining to degenerate. With progesterone levels down, LH and FSH are released from inhibitory control, their levels rise, and a new cycle starts again.

Manipulating Hormone Levels to Prevent Pregnancy

- Hormone-based birth control methods deliver synthetic versions of progesterone, or progesterone and estradiol, that act to suppress GnRH and FSH release. This prevents LH from spiking during the follicular phase of the cycle, thus preventing ovulation from occurring.

48.5 Pregnancy and Birth in Mammals

Gestation and Early Development in Marsupials

- **Marsupials** differ from **eutherians** in that the corpus luteum is not maintained in the same manner, and no placenta forms.
- In marsupials, the offspring is ejected from the mother's body at the end of the estrous cycle and gestation is completed outside of the mother.

Major Events during Pregnancy

- Once an egg is released from the ovary, it is viable for less than 24 hours. Sperm can survive for 5 days, so sexual intercourse must occur less than 5 days prior to ovulation for pregnancy to result.
- Although an ejaculate may contain hundreds of millions of sperm, only about 100–300 actually reach the oviduct.

Implantation

- As smooth-muscle contractions in the oviduct gradually move the zygote toward the uterus, the cell begins to divide by mitosis.
- Once the embryo has become implanted in the uterine lining, its cells begin synthesizing and secreting a hormone called **human chorionic gonadotropin (hCG)**, which is a chemical messenger that prevents the corpus luteum from degenerating.

The First Trimester

- The mass movement of cells results in the formation of the embryonic tissues called ectoderm, endoderm, and mesoderm. By eight weeks, these tissues have differentiated into the various organs and systems of the body.

- Another key event in the first trimester is the formation of the **placenta**. This structure starts to form on the uterine wall a few weeks after implantation.

The Second and Third Trimesters

- During this time, the remainder of development focuses mainly on growth. The brain and lungs undergo dramatic growth and development.

How Does the Mother Nourish the Fetus?

Oxygen Exchange between Mother and Fetus

- The developing embryo is completely dependent on the mother for oxygen; chemical energy in the form of sugars, amino acids, and other raw materials for growth; and waste removal.

- A mother's blood volume expands by as much as 50 percent during pregnancy, and dramatic increases occur in the stroke volume of the heart.

- The mother's breathing rate and volume also increase to accommodate the fetus's demand for oxygen and its production of carbon dioxide.

- In some species, there is a countercurrent flow in the placenta to increase efficiency of gas exchange between mother and fetus (**Figure 48.18a**).

- The fetus's blood has a higher affinity for oxygen than the mother's blood does (**Figure 48.18b**).

Toxic Chemicals Can Be Transferred from Mother to Fetus

- During the 1950s, thalidomide diffused from the mother into the fetal bloodstream and caused birth defects such as shortening of the arms.

- The consumption of alcohol by the mother during pregnancy results in **fetal alcohol syndrome (FAS),** which is characterized by hyperactivity, severe learning disabilities, and depression. FAS results from the destruction of neurons by ethanol.

Birth

- The rapid growth of the fetus's brain during the last trimester makes birth challenging. The pituitary hormone **oxytocin** is clearly important in stimulating smooth-muscle cells in the uterine wall to begin contractions (**Figure 48.20**).

- During the last stage of birth, the uterus contracts at a low frequency; then, as labor progresses, the cervix at the base of the uterus begins to dilate. Once the cervix is fully dilated, uterine contractions become more forceful, longer, and more frequent. Eventually, the fetus is expelled through the cervix and into the vagina.

- After the baby is delivered, the medical staff clamps and then cuts the umbilical cord that connects the child and the placenta. By gently tugging the cord, they remove the placenta.

- Due to modern-day advances in health care, the mortality rate during childbirth has greatly declined (**Figure 48.21**).

B. MASTERING CHALLENGING TOPICS

You have studied mitosis and meiosis in previous chapters, so make sure you have reviewed these processes extensively. They were presented as difficult topics in the chapters in which they first appeared. As far as new material is concerned, the menstrual cycle is by far the most complex portion of this chapter. As with everything else you've studied, begin by learning how each major hormone changes during this cycle. Then, study the physical and physiological changes that occur because of the hormonal changes. Finally, try to put everything together by learning how all of the hormones interact by positive and negative feedback. Drawing diagrams and explaining them to classmates will help to reinforce these concepts.

C. ASSESSING WHAT YOU'VE LEARNED

1. The production of offspring via unfertilized eggs is called:
   a. gametogenesis.
   b. parthenogenesis.
   c. oogenesis.
   d. spermatogenesis.

2. All of the following conditions led to a switch from asexual to sexual reproduction in the aquatic crustacean *Daphnia except*:
   a. longer day lengths.
   b. low food availability.
   c. high population densities.
   d. season.

3. Many animals, including *Daphnia,* are able to reproduce both sexually and asexually. Which of the following represents the leading hypothesis explaining the advantage associated with the ability to switch to sexual reproduction?
   a. Sexual reproduction is advantageous when energy is readily available.
   b. Sexual reproduction does not require the process of meiosis.
   c. Sexual reproduction produces more genetically diverse offspring capable of surviving stressful environmental conditions.
   d. Sexual reproduction allows for more rapid production of offspring.

4. All of the following represent differences between the processes of spermatogenesis and oogenesis in humans *except*:
   a. the production of oogonia stops before birth, while the production of spermatogonia occurs throughout adult life.
   b. sperm are haploid, while eggs are diploid.
   c. oogenesis produces polar bodies; spermatogenesis does not.
   d. sperm cells are smaller in size than oocytes.

5. All of the following are general characteristics of external fertilization *except*:
   a. production of large numbers of gametes.
   b. use of terrestrial environments.
   c. synchronization of release of gametes.
   d. production of diploid offspring.

6. Which of the following experimental results supports the hypothesis that, for some species, pheromones may coordinate spawning?
   a. Fish maintained in large groups, but not small groups, exhibited coordinated spawning activity.
   b. Only 10 percent of sea cucumbers raised in isolation under natural conditions released their gametes.
   c. Observations of sea stars in the wild have demonstrated the coordination of spawning activity.
   d. Male fruit flies mate with a female only when her eggs have achieved the appropriate level of maturity.

7. Which of the following mechanisms likely play a role in conferring "the second male advantage" during internal fertilization in fruit flies? *Note:* There may be more than one correct choice.
   a. The fluid surrounding the sperm displaces the sperm of other males.
   b. The second male's sperm physically dislodge the first male's sperm from storage areas within the female reproductive tract.
   c. The female reproductive tract produces toxins that eventually harm stored sperm.
   d. The most recently deposited sperm are more vigorous in their swimming movements toward the egg.

8. Which of the following is true of animals for which multiple matings are common?
   a. They tend to live in aquatic habitats.
   b. This characteristic is observed only in insects (such as the fruit fly).
   c. The females tend to produce large numbers of offspring.
   d. The males tend to have large testes.

9. Which of the following best explains the difference between oviparity and viviparity?
   a. Oviparous animals utilize external fertilization, while viviparous animals use internal fertilization.
   b. Oviparous animals do not care for their young following fertilization, while viviparous animals do.
   c. Oviparous animals are egg bearing, and embryonic development takes place outside the mother's body; viviparous animals retain their embryos for development within the mother's body.
   d. Oviparous animals receive nutrition directly from the mother's body during development, while the embryos of viviparous animals receive nourishment from yolk sacs.

10. How have studies of the lizard species *Sceloporus* helped scientists understand how natural selection has favored changes between oviparity and viviparity?
    a. Because oviparity evolved independently in two different groups of lizards, it is thought to represent the ancestral condition.
    b. Viviparous lizard species appear to have evolved from oviparous species.
    c. There appears to be a correlation between cold weather and the evolution of oviparity.
    d. Oviparity is favored in high-altitude environments.

11. The seminal vesicles, bulbourethral glands, and prostate gland contribute all of the following components to the semen *except*:
    a. an energy source in the form of sucrose.
    b. prostaglandins to stimulate smooth-muscle contractions in the female reproductive tract.
    c. alkaline compounds to counteract the acidity of the female reproductive tract.
    d. antibiotic substances.

12. Puberty in males and females is triggered by:
    a. testosterone in males and estrogen in females.
    b. LH release from the hypothalamus in both males and females.
    c. GnRH release from the hypothalamus in both males and females.
    d. GnRH release from the anterior pituitary gland in both males and females.

13. How does a new menstrual cycle become initiated when fertilization fails to occur?
    a. Progesterone levels drop due to regression of the corpus luteum, releasing the hypothalamus and pituitary from feedback inhibition. FSH levels rise.
    b. hCG produced by the uterine lining signals the corpus luteum to regress, releasing the hypothalamus and pituitary from feedback inhibition. FSH levels rise.

c. The nervous system detects the absence of implantation, signaling the release of GnRH from the hypothalamus, which triggers the release of FSH from the pituitary.

d. hCG, produced by the embryo following implantation in the uterus, signals the corpus luteum to regress. Progesterone levels drop and FSH levels rise.

14. Which statement is correct concerning the length of viability of the human egg and sperm?
    a. Both the egg and the sperm are viable for 1 day.
    b. Both the egg and the sperm are viable for 5 days.
    c. Sperm are viable for 1 day, while eggs are viable for 5 days.
    d. Sperm are viable for 5 days, while eggs are viable for 1 day.

15. Which of the following best describes the function of the placenta?
    a. The placenta produces the hormone FSH, needed for embryonic development.
    b. The placenta provides the embryo with a protective cushion.
    c. The placenta allows for gas and nutrient exchange between the maternal and fetal circulations.
    d. The placenta is one of the three main types of embryonic tissues. Eventually, it develops into the fetal heart.

CHAPTER 48—ANSWER KEY

Check Your Understanding

(a) *Daphnia* reproduce asexually throughout the spring and summer; the females produce diploid eggs without sex. The production of offspring via unfertilized eggs is called parthenogenesis. In late summer or early fall, many females begin producing male offspring parthenogenetically. Sperm from these males fertilize haploid eggs that females produce via meiosis.

(b) Most animals that rely on external fertilization live in aquatic environments and tend to produce exceptionally large numbers of gametes. Gametogenesis occurs in response to environmental cues such as lengthening days and warmer water temperatures. Internal fertilization occurs in the vast majority of terrestrial animals as well as in a significant number of aquatic animals. Males deposit sperm directly into the female reproductive tract with an organ called the penis; or males may package their sperm into a structure called a spermatophore, which is then placed into the female's reproductive tract by the male or female.

(c) In oviparous animals, the embryo develops in the external environment. In some animals there is no parental care, but a substantial number of species continue to care for their young after the eggs have emerged from the mother's body. In viviparous species, development takes place within the mother's body. The embryo attaches to the female's reproductive tract and receives nutrition directly from the mother's circulatory system.

(d) The structural components of the internal reproductive structures in the human male:

1. *Spermatogenesis and sperm storage.* Sperm are produced in the testes and stored in the epididymis.

2. *Production of additional fluids.* Complex solutions that form in the seminal vesicles, prostate gland, and bulbourethral gland are added to sperm prior to ejaculation. The combination of sperm and these accessory fluids is called semen.

3. *Transport and delivery.* These functions are the domain of the vas deferens, urethra, and penis.

(e) Changes that occur during puberty are triggered by increased levels of testosterone in boys, and estradiol in girls. Researchers isolated a hormone from the hypothalamus called gonadotropin-releasing hormone (GnRH). Investigators also noted that boys and girls who were entering puberty experienced distinctive pulses in the pituitary hormone, called luteinizing hormone (LH).

(f) (1) Follicular phase (14 days)—a follicle matures, and ovulation occurs when the follicle is mature; it releases its oocyte into the oviduct. (2) Luteal phase (14 days)—the corpus luteum forms from the ruptured follicle and subsequently degenerates.

Assessing What You've Learned

1. b; 2. a; 3. c; 4. b; 5. b; 6. b; 7. a and b; 8. d; 9. c; 10. b; 11. a; 12. c; 13. a; 14. d; 15. c

The Immune System in Animals

KEY CONCEPTS

- In animals, the immune system consists of two types of responses, the innate immune response and the acquired immune response.

- The innate immune response is nonspecific in its destruction of pathogens.

- The acquired immune response is activated when lymphocytes recognize specific foreign antigens.

- Lymphocytes can either directly destroy pathogens or produce antibodies that tag pathogens for destruction.

A. CHAPTER OUTLINE

49.1 Innate Immunity

- **Innate immunity** refers to immune system cells that are ready to respond to foreign invaders at all times.

- An **antigen** is any foreign molecule. Most antigens are proteins or glycoproteins from bacteria or viruses or other invaders, but foreign carbohydrates and lipids also function as antigens.

- The innate immune system responds the same way to all antigens.

Barriers to Entry

- In humans and other animals, the most important barrier to pathogen entry is the **skin**. In addition to a physical barrier, skin cells present a chemical barrier because they secrete **fatty acids** that lower the pH of the surface and prevent bacterial growth.

How Are Openings in the Body Protected?

- **Mucus**, a proteoglycan-rich solution secreted by cells within the epithelium, protects gaps in surfaces where certain organs make contact with the environment. For example, pathogens that are breathed in by animals become stuck in the mucus that lines the airways and cannot damage epithelial cells.

- Lysozymes contained in tears protect the eyes from bacterial infection.

How Do Pathogens Gain Entry?

- Flu viruses infect an individual by penetrating the mucous lining of the respiratory tract. Also, open wounds and other types of trauma allow viruses, bacteria, and fungi entry into your body.

The Innate Immune Response

- The cells involved in the innate response are alerted to the presence of foreign invaders by the presence of certain molecules (**Figure 49.2**). For example, all bacteria have proteins that begin with an N-formylmethionyl molecule instead of methionine. When **pattern-recognition receptors** in the innate immune system detect specific shapes or patterns present in foreign molecules, **leukocytes** (white blood cells) respond.

The Inflammatory Response in Humans

- The main steps in the **inflammatory response** (**Figure 49.3**) are as follows:

  1. Bacteria and other pathogens enter the wound.

  2. At the wound, **platelets** release proteins that form clots to reduce bleeding.

  3. Wounded tissue and macrophages release **chemokines** to guide wound healing factors to the site of injury.

  4. Mast cells secrete factors, such as **histamine**, that mediate **vasodilation** and **vascular constriction**, which bring blood, plasma, and cells to the injured area.

  5. **Neutrophils** secrete factors that kill and degrade pathogens. Neutrophils and **macrophages** remove pathogens by **phagocytosis**.

  6. Macrophages secrete hormones called **cytokinins** that attract immune system cells to the site and activate cells involved in tissue repair.

  7. The inflammatory response continues until the foreign material is eliminated and the wound is repaired.

Innate Immunity in Invertebrates

- Similar to vertebrate innate immune responses, invertebrates such as insects are able to synthesize and secrete antibacterial and antifungal peptides. Also, sea stars have cells that function like neutrophils and macrophages.

49.2 The Acquired Immune Response: Recognition

- An **antibody** is a protein that binds to a specific antigen. The binding of an antibody to an antigen leads to the destruction of the foreign molecule or cell.

- Four key characteristics of the acquired immune response are:

  1. Specificity

  2. Diversity

  3. Memory

  4. Self-nonself recognition

An Introduction to Lymphocytes

- The cells involved in the acquired immune response are called **lymphocytes**.

Immune System Structures

- Lymphocytes form and mature in the primary organs of the immune system: the **bone marrow** and **thymus**.

- The **lymph nodes** and **spleen** are important sites where lymphocytes encounter antigens. The fluid and lymphocytes present in the lymph nodes and ducts are called **lymph**.

Lymphocytes Must Be Activated

- Lymphocytes that do not encounter an antigen during their lifetime migrate throughout the body, but eventually die. However, once lymphocytes encounter their corresponding antigens, they are activated and become known as B cells (**Figure 49.5**).

The Discovery of B Cells and T Cells

- Investigations of the immune system's response to *Salmonella typhimurium* show that antibodies are important in neutralizing the antigen and that the **bursa** is critical for antibody production.

- Mice lacking a thymus developed pronounced defects in their immune system. Follow-up experiments showed that lymphocytes from the bursa and thymus have two different functions.

- The bursa- and thymus-dependent lymphocytes became known as **B cells**, which produce antibodies, and **T cells**, which are involved in an array of functions.

✔ **CHECK YOUR UNDERSTANDING**

(a) What are the bursae, and how do they relate to the immune system?

_____

_____

_____

_____

The Clonal-Selection Theory

- The clonal-selection theory consists of three central claims (**Figure 49.6**):

  1. Each lymphocyte formed in the bone marrow or thymus has a unique receptor on its surface that recognizes one antigen.

  2. When the receptor on a lymphocyte binds to an antigen, the lymphocyte is activated.

3. An activated lymphocyte divides and, in essence, makes clones of itself.

4. Some of these cloned cells persist after the pathogen is eliminated. As a result, they are able to respond quickly and effectively if the infection recurs in the future.

CHECK YOUR UNDERSTANDING

(b) Outline the main steps in the inflammatory response.

_____

_____

_____

_____

The Discovery of B-Cell Receptors

- The **B-cell receptor (BCR)** and antibodies belong to a family of proteins called the gamma globulins and are more specifically called **immunoglobulins** (**Table 49.2**).
- The BCR has three distinct components (**Figure 49.7a**):
  1. The **light chain** of about 25,000 daltons
  2. The **heavy chain** of about 50,000 daltons
  3. A **transmembrane domain** that anchors the protein in the B-cell membrane; antibodies lack this domain and are secreted from the cell

The Discovery of T-Cell Receptors

- The **T-cell receptor (TCR)** binds to antigens only after they have been processed by other immune system cells and presented on their cell membranes. It is composed of a single **alpha chain** and a single **beta chain** (**Figure 49.7b**). The shape is similar to the arm of the BCR molecule.

Antibodies and Receptors Bind to Epitopes

- Antibodies and lymphocyte receptors do not bind to the entire antibody molecule, but to a selected region called an **epitope**. An antigen may have many different epitopes where binding by antibodies and lymphocytes actually takes place.

What Is the Molecular Basis of Antigen Specificity and Diversity?

- Researchers have shown that certain amino acids in the light chain are extremely variable among B cells.

- The presence of specific amino acids at certain positions in the light chain is responsible for the ability of the BCR and antibodies to bind to unique epitopes.
- The binding occurs at the tips of the antibody and BCR molecule's arms.

The Discovery of Gene Recombination

- Dryer and Bennett (1965) suggested that a single gene codes for the C region of the light chain and that a separate gene codes for the V region. They hypothesized that early in the life of a lymphocyte, sequences from the *V* gene are inserted into the *C* gene.
- Hozumi and Tonegawa (1976) reasoned that if light-chain diversity was produced by DNA recombination, then the variable and constant regions of mature B cells should be shorter than in the same region in immature cells.
- Two large DNA fragments from the immature cell hybridized with the entire L chain, and the larger of these two fragments also hybridized with the constant-region gene.
- Biologists proposed that the smaller fragment contained the variable region and that the larger fragment included the constant region. This was strong evidence that DNA recombination brings *V* and *C* genes closer together as B cells mature (**Figure 49.12**).
- Gene recombination allows for BCR and TCR to have a unique amino acid sequence that recognizes a unique epitope on an antigen, allowing for specificity.

How Does the Immune System Distinguish Self from Nonself?

- Researchers introduced B cells and T cells with anti-self receptors into mice. These experimental lymphocytes were eliminated.
- The mechanism for recognition of self is being intensively studied, since it holds the key to **autoimmunity**, where immune system cells turn on the normal cells of the body and destroy them as if they were nonself.

49.3 The Acquired Immune Response: Activation

T-Cell Activation

Antigen Presentation of MHC Proteins

- If bacteria migrate rapidly at a wound site, leukocytes called **dendritic cells** are recruited to the area. The following steps then take place (**Figure 49.13**):
  1. Dendritic cells take up some of the antigens present at the wound.
  2. The ingested antigens enter the endoplasmic reticulum or endosome.

3. The antigens are broken down and become bound to a **major histocompatibility (MHC) protein**.

4. The MHC-antigen complex moves to the cell surface, where it is displayed.

✔ CHECK YOUR UNDERSTANDING

(c) What are the steps involved in antigen presentation of MHC proteins?

_____

_____

_____

_____

How Do T Cells Respond to Antigen-Presenting Cells?

- As CD4$^+$ T cells move through the lymph node in search of an antigen, those with complementary receptors interact with their corresponding epitope on dendritic cells.

- Once a T cell receives signal 1 and 2, it begins to divide and produce a series of daughter cells. This process, called **clonal expansion**, is a crucial step in the acquired immune response because it leads to a large population of lymphocytes capable of responding specifically to the antigen that has entered the body.

Cytotoxic T Cells and Helper T Cells

- If the original T cell is CD4$^+$, the daughters differentiate into one of two types of **helper T cells**: T_H1 and T_H2.

- If the original T cell is CD8$^+$, its daughter cells develop into **cytotoxic T lymphocytes (CTLs)**.

- Helper T cells and CTLs are often referred to as **effector T cells**.

✔ CHECK YOUR UNDERSTANDING

(d) What is clonal expansion?

_____

_____

_____

_____

B-Cell Activation and Antibody Secretion

- Receptors on B cells interact directly with the bacterial or viral antigens that are floating free in lymph or blood. Once a free antigen is bound, B cells **internalize** the molecule and process it via the same Class II pathway used by dendritic cells.

- The second part of the activation process occurs when an activated CD4$^+$ T_H2 cell with a **complementary receptor** arrives. Helper T cell binds to the **receptor–MHC complex** on the B cell.

- The interaction between the B cell and helper T cell supplies an initial activation signal, which is followed by a co-stimulatory signal 2 analogous to the one that occurs during T-cell activation.

49.4 The Acquired Immune Response: Culmination

How Are Bacteria and Other Foreign Cells Killed?

- During the innate response to bacteria that enter a wound, macrophages at the site phagocytize some of the invaders.

- Macrophages also process and present antigens via the MHC Class I and Class II proteins. As a result, macrophages at the site of infection display epitopes on their surfaces that can be recognized by helper T cells.

- If an activated T_H1 cell binds to these antigen-laden macrophages, two things happen:

   1. The phagocytic activity of the macrophages is enhanced.

   2. The T_H1 cells secrete cytokines that kill bacteria and viruses, recruit additional phagocytic cells to the site, and increase the inflammatory response.

- Also, agglutination of foreign cells marks it for destruction by macrophages.

How Are Viruses Destroyed?

- The immune system has two major ways to eliminate viruses, the **cell-mediated response** and the **humoral response**.

Cell-Mediated Response Leads to Cell Suicide

- Once T cells and B cells have been activated, most of the major elements of the acquired immune response are in place.

- Infected cells respond to the arrival of a virus by processing antigens from the invader. Viral antigens are attached to **MHC Class I proteins** and presented on the surface of the infected cell. This sends a message to the immune system to kill the cell.

- In the cell-mediated response, cytotoxic T lymphocytes (CTLs) contain granules that punch holes in cells. CTL makes contact with virus-infected cell and releases granules, causing the cell to lyse.

The Humoral Response Coats Pathogens with Antibodies

- In the humoral response, antibodies coat free virus particles. The virus envelope cannot fuse with the host-cell membrane. The antibody-coated virus is recognized and phagocytized by the macrophage.

✓ CHECK YOUR UNDERSTANDING

(e) Compare and contrast the cell-mediated response and the humoral response.

_____

_____

_____

_____

Why Does the Immune System Reject Foreign Tissues and Organs?

- Organ transplants must be matched to the donor, because the MHC cells of the transplanted tissue will otherwise be recognized as foreign antigens.

- Transplanted human organs are carefully chosen to match the immune system characteristics of the recipients, who are also given immune-suppressing drugs to ensure the success of the transplant operation.

Responding to Future Infections: Immunological Memory

- **Memory cells** do not participate in the initial, or **primary, immune response**. Instead, they provide a surveillance service after the original infection has been cleared.

The Secondary Response Is Strong and Fast

- If the same antigen enters a body a second time, a **secondary acquired immune response** takes place. This response is faster and more efficient than the primary response (**Figure 49.18**).

- The process of **somatic hypermutation** leads to a fine-tuning of the immune response. The B cells that result from somatic hypermutation produce antibodies that bind to the antigen more tightly than their ancestor cells did during the primary response.

Why Does Vaccination Work?

- A **vaccine** contains epitopes from a pathogen, but not the pathogen itself. After vaccination occurs, the body mounts a primary immune response that results in the production of memory cells.

- The three types of vaccines are **subunit vaccines**, **inactivated viruses**, and **attenuated viruses**.

(a) B-cell receptor

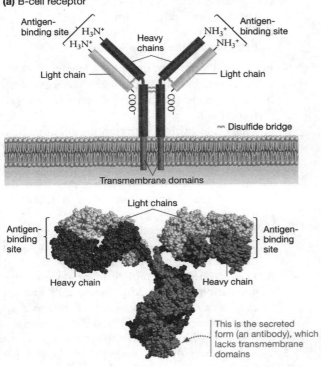

(b) T-cell receptor

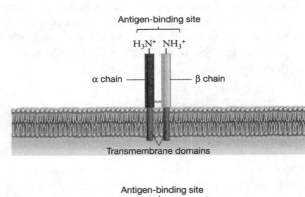

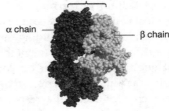

(f) How does a vaccination impart immunity to an individual?

_____

_____

_____

49.5 What Happens When the Immune System *Doesn't* Work Correctly?

Immunodeficiency Diseases

- Children afflicted with severe combined immunodeficiency (SCID) generate abnormal, nonfunctional T-cell and B-cell receptors.

- The **human immunodeficiency virus** (**HIV**) impairs the immune system by attacking and killing CD4$^+$ T cells and macrophages. Eventually HIV-infected people develop **acquired immune deficiency syndrome** (**AIDS**), where they often die from illnesses that usually do not affect people with healthy immune systems.

Allergies

- An allergy occurs when the immune system overreacts to an antigen. The molecules that trigger the release of the IgE class of antibodies during allergic reactions are called **allergens**.

- IgE antibodies trigger a **hypersensitive reaction** where upon first exposure to the allergen, receptors on mast cells and basophils bind to and sensitize IgE antibodies. When the allergen is present later, the cells produce large amounts of histamine, cytokines, and chemokines.

B. MASTERING CHALLENGING TOPICS

The topics covered in this chapter are very visual and involve multistep processes; therefore a lot of diagram drawing will be helpful in learning the immune responses. The most confusing aspect of the immune system is the different responses, including the differences in the B-cell and T-cell activation. Inevitably, compare-and-contrast questions will appear on your exam, so study similarities as well as differences in these systems. As with every chapter, learn the responses and cell types separately until you are comfortable with them. Next, attempt to draw charts outlining similarities and then differences.

C. ASSESSING WHAT YOU'VE LEARNED

1. The barriers faced by invading pathogens are considered part of the innate division of the immune system. Which of the following statements regarding innate immunity is true?
   a. Components of the innate immune system must first be activated to respond to an antigen.
   b. Components of the innate immune system respond the same way to all antigens.
   c. Components of the innate immune system respond in a specific way to each antigen.
   d. The response of the innate immune system depends on the type of proteins or glycoproteins found on the bacteria, viruses, or other invaders.

2. Which of the following would be associated with innate immunity?
   a. Lymphocytes
   b. Antibodies
   c. Inflammation
   d. B cells

3. A very important step in our understanding of acquired immunity came when scientists discovered the receptors on B cells and T cells that bind to antigens. What part of these protein receptor molecules shows the highest variation in amino acid sequence from cell to cell?
   a. The antigen-binding sites
   b. The transmembrane domains
   c. The heavy chain of the B-cell receptor and alpha chain of the T-cell receptor
   d. The light chain of the B-cell receptor and the beta chain of the T-cell receptor

4. In acquired immunity, the epitope is:
   a. the type of bacteria or pathogen that is invading.
   b. the molecule produced by B cells for antigen recognition.
   c. a molecule involved in promoting inflammation.
   d. a selected region of an antigen where antigen recognition occurs.

5. When T cells encounter an antigen, they begin the process of clonal expansion. What is meant by clonal expansion?
   a. The T cell divides rapidly to produce a large population of genetically identical cells that are all capable of recognizing the same antigen.
   b. The T cell divides rapidly to produce a large population of genetically diverse cells that are capable of recognizing a variety of antigens.
   c. The T cell secretes chemicals to destroy the antigen.
   d. The T cells bind to B cells to activate the B cells and promote the production of antibodies that are capable of recognizing the antigen.

6. What process leads to the activation of T cells and triggers the clonal expansion of these cells?
   a. T-cell clonal expansion is triggered when T cells encounter a free-floating antigen in blood or lymph.
   b. T cells are always active and producing clones.
   c. T-cell clonal expansion is triggered when T cells encounter an antigen that is presented by another cell.

7. During HIV infection, the number of cells drops as the virus infects these cells and destroys many of them. What makes HIV selective for this specific type of cell?
   a. HIV contains surface proteins that bind to the $CD4^+$ receptor.
   b. HIV contains surface proteins that bind to the antigen-binding portion of the heavy chain of an antibody receptor.
   c. HIV contains surface proteins that bind to the CD8 receptor.

8. Which of the following is associated with the humoral response to a viral infection?
   a. Activated cells
   b. Cytotoxic T cells
   c. MHC I proteins bound to an antigen found at the surface of an antigen-presenting cell
   d. Plasma cell production of antibodies

9. Which of the following can explain why the secondary immune response is both stronger and faster than the primary immune response?
   a. The response to an antigen is always stronger the second time it is encountered.
   b. Cells remain activated after they are exposed to the antigen in the primary response.
   c. Memory cells were produced during the first exposure to the antigen.
   d. Antibodies produced during the first exposure remain in the blood for decades and can take part in the secondary response.

10. During the secondary immune response, some of the memory B cells that respond migrate to a specialized area in the lymph node, where they undergo rapid mutation. This process is referred to as:
    a. clonal expansion.
    b. clonal selection.
    c. somatic hypermutation.
    d. inflammation.

11. What role does somatic hypermutation play in the immune response?
    a. This helps the body to prevent other infections while the immune system responds to a specific antigen.
    b. This helps the immune system become more efficient as antibodies are produced that can

bind even more tightly to the antigen during the second exposure.
    c. Somatic hypermutation is a defect in the immune system, which produces a secondary immune response that is inefficient.

12. HIV and influenza are viruses that have been difficult to fight with vaccines. Explain why this has been such a problem.
    a. The epitopes on these viruses change frequently as the viruses replicate.
    b. The vaccines would cause serious infections such as HIV infection and the flu.
    c. These viruses do not contain any epitopes on their surface.
    d. These viruses cause infections that have only recently been discovered, and an effective vaccine has not been found yet.

13. During an allergic response, the body produces a large number of IgE antibodies. What do these antibodies bind to en route to producing the well-known symptoms of an allergic response?
    a. T_H1 cells
    b. Plasma cells and memory B cells
    c. T_H2 cells
    d. Mast cells and basophils

14. During the innate response to bacteria that enter a wound, macrophages at the site phagocytize some of the invaders. What are the macrophages doing in this process?
    a. They are engulfing or "eating" the bacteria.
    b. They are producing chemicals that kill the bacteria.
    c. They are producing antibodies that bind to the bacteria.
    d. They are releasing histamine, which causes an inflammatory response around the bacteria.

15. The production and maturation of T cells occurs in two different places in the body. Where do T cells become mature?
    a. Bone marrow
    b. Thyroid
    c. Thymus
    d. Pancreas

CHAPTER 49—ANSWER KEY

Check Your Understanding

(a) Antibodies in the immune system's response to *Salmonella typhimurium* are important in neutralizing the antigen. The bursa is critical for antibody production. The bursa- and thymus-dependent lymphocytes became known as B cells, which produce antibodies, and T cells, which are involved in an array of functions.

(b) The main steps in the inflammatory response (**Figure 49.3**) are as follows:

1. Bacteria and other pathogens enter the wound.

2. At the wound, platelets release proteins that form clots to reduce bleeding.

3. Wounded tissue and macrophages release chemokines to guide wound-healing factors to the site of injury.

4. Mast cells secrete factors, such as histamine, that mediate vasodilation and vascular constriction, which bring blood, plasma, and cells to the injured area.

5. Neutrophils secrete factors that kill and degrade pathogens. Neutrophils and macrophages remove pathogens by phagocytosis.

6. Macrophages secrete hormones called cytokinins that attract immune system cells to the site and activate cells involved in tissue repair.

7. The inflammatory response continues until the foreign material is eliminated and the wound is repaired.

(c) If bacteria migrate rapidly at a wound site, leukocytes called dendritic cells are recruited to the area. The following steps then take place (**Figure 49.13**):

1. Dendritic cells take up some of the antigens present at the wound.

2. The ingested antigens enter the endoplasmic reticulum or endosome.

3. The antigens are broken down and become bound to a major histocompatibility (MHC) protein.

4. The MHC-antigen complex moves to the cell surface, where it is displayed.

(d) Once a T cell receives signal 1 and 2, it begins to divide and produce a series of daughter cells. This is called clonal expansion and is a crucial step in the acquired immune response because it leads to a large population of lymphocytes capable of responding specifically to the antigen that has entered the body.

(e) The immune system has two major ways to eliminate viruses—the cell-mediated response and the humoral response. In the cell-mediated response, cytotoxic T lymphocytes (CTLs) contain granules that punch holes in cells. CTL makes contact with a virus-infected cell and releases granules, causing the cell to lyse. In the humoral response, antibodies coat free virus particles. The virus envelope cannot fuse with the host-cell membrane. The antibody-coated virus is recognized and phagocytized by macrophage.

(f) A vaccine contains epitopes from a pathogen, but not the pathogen itself. After vaccination occurs, the body mounts a primary immune response that results in the production of memory cells.

Assessing Your Knowledge

1. b; 2. c; 3. a; 4. d; 5. a; 6. c; 7. a; 8. d; 9. c; 10. c; 11. b; 12. a; 13. d; 14. a; 15. c

An Introduction to Ecology

KEY CONCEPTS

- Ecology's goal is to explain the abundance and distribution of organisms.

- Ecology is also the branch of biology that provides a scientific foundation for conservation efforts.

- Water depth is the primary factor that limits the distribution and abundance of aquatic species.

- Climate (temperature and moisture) is the primary factor that limits the distribution and abundance of terrestrial species.

- Climate varies with latitude, elevation, and other factors such as proximity to oceans and mountains.

- A species' distribution is constrained by historical and biotic factors, as well as abiotic factors.

A. CHAPTER OUTLINE

50.1 Areas of Ecological Study

- Ecologists study how interactions between organisms and their environment result in a particular species being found in a particular area at a particular population size.

Organismal Ecology

- This ecology focuses on how individual organisms interact with their environment.

- Organismal ecologists explore the morphological, physiological, and behavioral adaptations that allow individuals to live successfully in a particular area.

Population Ecology

- A **population** is a group of individuals of the same species that live in the same area at the same time.

- Population ecologists focus on how the numbers of individuals in a population change over time.

Community Ecology

- A biological **community** consists of the species that interact with each other within a particular area.
- Community ecologists ask questions about the nature of the interactions between species and the consequences of those interactions.

Ecosystem Ecology

- An **ecosystem** consists of all of the organisms in a particular region along with nonliving, or **abiotic**, components.
- Biologists analyze how chemical elements that act as nutrients cycle through ecosystems and how energy flows through the organisms in a region.

How Do Ecology and Conservation Efforts Interact?

- **Conservation biology** is the effort to study, preserve, and restore threatened populations, communities, and ecosystems.

✓ CHECK YOUR UNDERSTANDING

(a) What is the difference between population ecology and community ecology?

_____

_____

_____

_____

50.2 Types of Aquatic Ecosystems

Nutrient Availability

Nutrient levels limit growth rates in the photosynthetic organisms that provide food for other species.

- *Ocean upwelling*—as surface water moves away from the coast, it is steadily replaced by nutrient-rich water moving up from the ocean bottom.
- *Lake turnover*—because 4°C water is the most dense, it sinks during the fall freeze or the spring thaw and brings nutrients from the bottom up.

Water Flow

- Organisms that live in fast-flowing streams must contend with water current while organisms living in stagnant water may have to contend with limited oxygen.

Water Depth

- Because light has a major influence on **productivity**, as water depth increases, light penetration decreases and so does productivity.

Freshwater Environments— Lakes and Ponds

- Lakes and ponds are distinguished by size, and lakes are larger.

Water Depth

- The **littoral zone** consists of the shallow water along the shore.
- The **limnetic zone** is offshore and comprises water that receives enough light to support photosynthesis.
- The **benthic zone** is made up of the substrate.

Water Flow and Nutrient Availability

- Water movement in lakes and ponds is driven by wind and temperature.
- Mixing driven by wind and temperature changes allows well-oxygenated water from the surface to reach the benthic zone and nutrient-rich water from the benthic zone to reach the surface.

Organisms

- Cyanobacteria, algae, and other microscopic organisms are collectively called **plankton** and live in the photic zone.
- Animals that consume **detritus** are common in the benthic zone.

Freshwater Environments—Wetlands

Water Depth

- Wetlands are shallow-water habitats where the soil is saturated with water for at least part of the year and can be distinguished from lakes and ponds by these traits:
  1. They have shallow water.
  2. They have **emergent vegetation** (plants grow above water surface).

Water Flow and Nutrient Availability

- Marshes and swamps have slow but steady flow of water.
- **Bogs** develop in depressions where there is no water flow.
- Bog water is low in oxygen and very acidic.

Organisms

- **Marshes** lack trees and typically feature grasses.
- **Swamps** are dominated by trees and shrubs.
- Because their physical environments are so different, there is little overlap in the types of species found in bogs, marshes, and swamps.

Freshwater Environments—Streams

- **Streams** are bodies of water that move constantly in one direction. Creeks are small streams, and rivers are large.

Water Depth

- Most streams are shallow enough that light reaches the bottom.

Water Flow and Nutrient Availability

- Where it originates at a mountain glacier, lake, or spring, a stream tends to be cold, narrow, and fast. These areas tend to be well oxygenated.
- As it descends, a stream tends to become larger, warmer, and slower. These areas tend to be oxygen poor.

Organisms

- It is rare to find photosynthetic organisms in small, fast-moving streams.
- As streams widen and slow down, conditions become more favorable for growth of algae and plants.

Freshwater/Marine Environments—Estuaries

- An **estuary** is the environment that forms where rivers meet the ocean.
- Salinity has dramatic effects on osmosis and water balance; therefore, the species living in estuaries have adaptations that allow them to cope with variations in salinity.

Marine Environments—The Ocean

Water Depth

- Biologists describe the structure of an ocean by naming six regions:
  1. The **intertidal zone** (littoral zone) is along the shore.
  2. The **neritic zone** extends from the intertidal zone to an ocean depth of about 200 m (the edge of the **continental shelf**).
  3. The **oceanic zone** encompasses the remainder and largest of the ocean environments.
  4. The **benthic zone** is the area at the bottom.
  5. The **photic zone** receives sun to support photosynthesis.
  6. The **aphotic zone** is the area too dark to support photosynthesis.

Water Flow and Nutrient Availability

- Tides and wave action are major influences in the intertidal zone.
- Large-scale currents circulate water in the oceanic zone in response to prevailing winds and the Earth's rotation.

Organisms

- Each zone in the ocean is populated by distinct species.
- In the tropics, shallow portions of the neritic zone may support **coral reefs**—among the most productive environments in the world.

✔ CHECK YOUR UNDERSTANDING

(b) What are the three main factors that influence aquatic ecosystems?

_____

_____

_____

_____

50.3 Types of Terrestrial Ecosystems

- A **biome** is a type of terrestrial ecosystem that is unique to a given region and characterized by a distinct type of vegetation.
- **Climate** is the prevailing long-term weather conditions in a particular region. **Weather** consists of the specific short-term atmospheric conditions of temperature, moisture, sunlight, and wind.
- **Net primary productivity** is the total amount of carbon fixed per year minus the amount of carbon oxidized during cellular respiration. **Aboveground biomass** is the total mass of living plants excluding roots.

Terrestrial Biomes—Tropical Wet Forests

- **Tropical wet forests**, or rain forests, are found in equatorial regions where temperatures and rainfall are high and variation is low. The plants grow all year round.
- Equatorial forests are also renowned for their **species diversity**. A plot of 100 m x 100 m may contain over 200 tree species, and some researchers contend that the rain forests house over 30 million arthropod species.

Terrestrial Biomes—Subtropical Deserts

- Subtropical desert bands exist along 30 degrees of latitude north and south of the equator.
- Most of the world's great deserts, including the Sahara, Gobi, Kalahari, and Australian Outback, lie at this latitude.
- The striking feature of this biome is the low precipitation.

- Desert species generally cope with the extreme temperatures and aridity in one of two ways:
  1. Growing at a low rate year round
  2. Growing rapidly in response to any rainfall

Terrestrial Biomes—Temperate Grasslands

- Precipitation is quite low, and the temperature profiles in temperate grasslands are quite seasonal. Temperature variation is important because it dictates a well-defined growing season. These regions usually exist because conditions are too dry to support tree growth but too cold and seasonal for drought-adapted desert species.

- Although the productivity of temperate grasslands is generally lower than that of forest communities, grassland soils are often highly fertile.

Terrestrial Biomes—Temperate Forests

- In temperate areas with high precipitation, grasslands give way to forests (e.g., the region around Chicago). The abundance of moisture allows trees to dominate the landscape.

- Trees experience a dormant period in the winter, and therefore temperate forests are dominated by deciduous species whose leaves fall in the autumn and regrow in the spring.

- Productivity and diversity of temperate forests are moderate.

Terrestrial Biomes—Boreal Forests

- The **boreal forest**, or **taiga**, stretches across most of Canada, Alaska, Russia, and northern Europe. This climate is characterized by very cold winters, cool, short summers, and extraordinarily high annual variation in temperature. Precipitation is virtually identical to that in the temperate grasslands.

- The subarctic landscape is dominated by highly cold-tolerant conifers, including spruce, pine, fir, and larch trees. Except for larch, these species are evergreen.

- Although aboveground biomass is high, productivity and diversity in boreal forests are low.

Terrestrial Biomes—Arctic Tundra

- The growing season is six to eight weeks at the most; otherwise, temperature is below freezing. The annual precipitation is also very low.

- The arctic is treeless; however, the soils are saturated year round due to the low evaporation rates.

- Tundra has low species diversity, productivity, and aboveground biomass. Most tundra soils are in the perennially frozen state known as **permafrost**.

CHECK YOUR UNDERSTANDING

(c) What would you predict about the productivity and aboveground biomass in a tropical wet forest that has year-round growing conditions?

_____

_____

_____

_____

50.4 The Role of Climate and the Consequences of Climate Change

Global Patterns in Climate

Why Are the Tropics Wet?

- The **Hadley cell** is responsible for making the Amazon River basin wet and the Sahara dry.

- Air that is heated by the strong sunlight along the equator expands and rises. Warm air can also hold a great deal of moisture. When the air cools as it rises, it can no longer hold moisture; thus, it rains.

- As the old, cold air is pushed out of the way, it falls and warms over the desert areas. Thus, little rain occurs in these areas.

Why Are the Tropics Warm and the Poles Cold?

- Areas of the world are warm if they receive a large amount of sunlight per unit area; they are cold if they receive a small amount of sunlight per unit area.

- The Earth's shape dictates that the regions at or near the equator receive a great deal of sunlight per unit area relative to regions that are closer to the poles.

What Causes Seasonality in Weather?

- **Seasons** are regular, annual fluctuations in temperature, precipitation, or both.

- Seasons are the result of Earth's tilt on its axis. Tilting areas toward the Sun causes summer, and tilting areas away from the Sun causes winter.

Mountains and Oceans: Physical Features Have Regional Effects on Climate

- The broad patterns of climate that are dictated by global heating patterns, Hadley cells, and seasonality are overlain by regional effects.

- The most important regional effects are mountains and oceans.
- The presence of mountain ranges tends to produce extremes in precipitation, while the presence of oceans has a moderating influence on temperature.

How Will Global Climate Change Affect Ecosystems?

- Recent increases in average global temperatures may represent the second-most rapid period of climate change ever.
- Scientists use three tools to predict how global warming will affect ecosystems:
    1. *Simulation studies* are based on computer models of weather patterns.
    2. *Observational studies* are based on long-term monitoring at fixed sites around the globe.
    3. *Experiments* are designed to simulate changed climate conditions.

Experiments That Manipulate Temperature

- Many of these experiments have been done in the tundra biome because these regions are projected to experience dramatic changes in temperature and precipitation.
- Results support the conclusion that species diversity decreases and arctic tundra environments are giving way to boreal forest.

Experiments That Manipulate Precipitation

- Simulation and observational studies have supported the conclusion that increases in average global temperature increase the variability in temperature and precipitation.
- Most experiments are performed in the fashion outlined in the figure on the right:

✓ CHECK YOUR UNDERSTANDING

(d) What is the difference between climate and weather?

_____

_____

_____

_____

EXPERIMENT

QUESTION: How does increasing variation in rainfall—without changing total amount—affect grassland biomes?

HYPOTHESIS: Increasing variability in rainfall affects soil moisture, net primary productivity (NPP), and species diversity.

NULL HYPOTHESIS: Increasing variability in rainfall has no effect on grassland biomes.

EXPERIMENTAL SETUP:

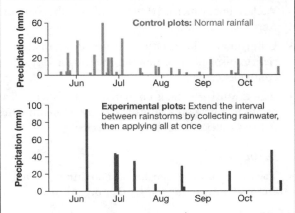

Record soil moisture, aboveground biomass (as an index of NPP), and species diversity, over a 4-year period.

PREDICTION: Soil moisture, NPP, and/or species diversity will differ between experimental plots and control plots.

PREDICTION OF NULL HYPOTHESIS: Soil moisture, NPP, and species diversity will not differ between experimental plots and control plots.

RESULTS:

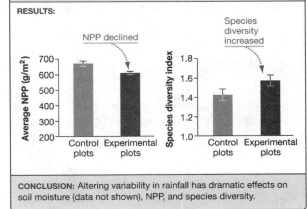

CONCLUSION: Altering variability in rainfall has dramatic effects on soil moisture (data not shown), NPP, and species diversity.

50.5 Biogeography: Why Are Organisms Found Where They Are?

- **Biogeography** is the study of how organisms are distributed in space.
- No one species can survive the full range of environmental conditions present on Earth.

Abiotic Factors

- The distribution of species is also limited by **abiotic factors**—particularly temperature and moisture.

- To understand a species' distribution thoroughly, biologists examine historical and biotic factors in addition to the physical conditions present.

The Role of History

- **Dispersal** refers to the movement of an individual from the place of birth to the location where it lives and breeds as an adult.
- There are many barriers to dispersal, such as the Wallace line, which separates species with Asian and Australian affinities.
- Unfortunately, humans have broken down many barriers to dispersion, sometimes resulting in disastrous consequences.
- If an **exotic** (one that is not native) organism is introduced to a new area, spreads rapidly, and eliminates native species, it is an **invasive species**.

Biotic Factors

- The distribution of species is often limited by **biotic factors**—interactions with other organisms.

Biotic and Abiotic Factors Interact

- It is often difficult to separate the effects of abiotic and biotic factors on a species' range.
- For example, changes in abiotic and biotic factors have allowed cheatgrass to invade native biomes in North America and thrive as an invasive species.

✔ CHECK YOUR UNDERSTANDING

(e) List one additional biotic and abiotic factor not mentioned above that could affect species distribution.

_____

_____

_____

B. MASTERING CHALLENGING TOPICS

The most challenging concept in this chapter, apart from memorizing all the new terms, is the dynamics of the Hadley cell. Remember that the Hadley cell is responsible for maintaining the wet conditions in the Amazon River basin and dry conditions at 30 degrees north and south latitude. The easiest way to learn the dynamics of the Hadley cell is to go over **Figure 50.23** and read the accompanying text. The figure allows you to really

understand the movements of cold air and warm air. Drawing the movements without the aid of the figure and explaining the concept to a fellow student is a sure way to ensure that you have a grasp of this process.

C. ASSESSING WHAT YOU'VE LEARNED

1. The ecology that focuses on how individuals in a group change over time is called:
   a. organismal ecology.
   b. population ecology.
   c. community ecology.
   d. ecosystem ecology.

2. The effort to preserve, study, and restore threatened populations, communities, and ecosystems is called:
   a. community ecology.
   b. biogeography.
   c. conservation biology.
   d. climate biology.

3. The overall factor dictating whether an area will be generally warm or cold is:
   a. the amount and intensity of sunlight.
   b. the amount of wind.
   c. the amount of precipitation.
   d. volcanic activity.

4. Seasonality in climate is caused directly by:
   a. geomagnetic differences year round.
   b. Earth's tilt on its axis.
   c. ocean currents.
   d. Sun position relative to other planets.

5. The most productive biome is:
   a. subtropical deserts.
   b. temperate forests.
   c. boreal forests.
   d. tropical wet forests.

6. The biome dominated by deciduous trees is:
   a. subtropical deserts.
   b. temperate forests.
   c. boreal forests.
   d. tropical wet forests.

7. The biome with the lowest species diversity is:
   a. subtropical deserts.
   b. temperate forests.
   c. boreal forests.
   d. tundra.

8. The wetlands with the slowest water flow and the most acidic water are typically:
   a. marshes.
   b. bogs.
   c. swamps.
   d. streams.

9. The ocean zone over the continental shelf is called the:
   a. oceanic zone.
   b. benthic zone.
   c. intertidal zone.
   d. neritic zone.

10. If an exotic organism is introduced to a new area and becomes successful, it is called a(n):
    a. invasive species.
    b. foreign species.
    c. foreign successful species.
    d. alien species.

CHAPTER 50—ANSWER KEY

Check Your Understanding

(a) A population is a group of individuals of the same species that live in the same area at the same time. Population ecologists focus on how the numbers of individuals in a population change over time. A biological community consists of the species that interact with each other within a particular area. Community ecologists ask questions about the nature of the interactions between species and the consequences of those interactions.

(b) Light has a major influence on productivity: the total amount of carbon fixed by photosynthesis per unit area per year. Water flow and depth also have a major influence on aquatic environments.

(c) The favorable year-round growing conditions in wet tropical forests produce extensive growth, leading to extremely high productivity and above-ground biomass.

(d) Climate refers to the prevailing long-term weather conditions in a particular region. Weather consists of the specific short-term atmospheric conditions of temperature, moisture, sunlight, and wind.

(e) A biotic factor involves living organisms. A bacterium could cause disease, and that could affect species dispersal. Besides temperature and moisture, altitude is another important abiotic factor.

Assessing What You've Learned

1. b; 2. c; 3. a; 4. b; 5. d; 6. b; 7. d; 8. b; 9. d; 10. a

Looking Forward—Key Concepts in Later Chapters

Animal Interactions—Chapters 51–53

Chapters 51, 52, and **53** investigate how different species interact with one another and with their physical environment. Here in Chapter 50, we have introduced the topics.

Chapter 51

Behavioral Ecology

Looking Back—Key Concepts from Earlier Chapters

Behavioral Traits—Chapter 13

To use the terms introduced in **Chapter 13**, most behavioral traits are polygenic—exhibiting a quantitative variation, with individuals differing by degree and many different phenotypes being observed when the population is considered as a whole.

The *for* Gene—Chapters 19 and 20

Recall from **Chapter 20** that the entire fruit-fly genome has been sequenced. **Chapter 19** introduced the techniques that Sokolowski used to map the *for* gene.

Flexible Behavior—Chapter 45

Chapter 45 introduced how learning works at the proximate level and concluded that learning occurs because individual neurons and connections between neurons are modified in response to experience.

KEY CONCEPTS

- Biologists analyze behavior at the proximate and ultimate levels—the genetic and physiological mechanisms and how they affect fitness.

- Individuals can behave in a wide range of ways, and behavior depends on the current conditions.

- Foraging decisions maximize energy gain and minimize costs.

- Females choose mates that provide good alleles and needed resources.

- Animals navigate by using an array of cues.

- Animals communicate with movements, odors, or other stimuli; they may be honest or deceitful.

- Individuals that behave altruistically are usually helping relatives of individuals that help them in return.

A. CHAPTER OUTLINE

51.1 An Introduction to Behavioral Ecology

- **Behavior** is action—a response to a stimulus.

Proximate and Ultimate Causation

- Most behavioral studies start by carefully observing what animals do in response to specific problems or situations.

- **Proximate causes** determine how actions occur.

- **Ultimate causes** determine why actions occur.

Conditional Strategies and Decision Making

- Highly inflexible behavior patterns are called **fixed action patterns** or **FAPs**.

- FAPs are examples of **innate behavior**—types of behavior that are inherited and show little variation based on learning.

- However, it is common for an individual's behavior to change in response to learning and to show flexibility.

- Most animals have a range of actions that they can perform in response to a situation. Animals take in information from the environment and make decisions about what to do.

- **Cost-benefit analysis**—animals appear to weigh the costs and benefits of responding in various ways. The responses are measured in terms of their impact on fitness.

✔ CHECK YOUR UNDERSTANDING

(a) What is the difference between innate and learned behavior?

_____

_____

_____

51.2 What Should I Eat?

Foraging Alleles in Drosophilia melanogaster

- When animals seek food, they are **foraging**.

- Marla Sokolowski discovered an important behavioral trait controlled by a single gene. She noticed that some fruit-fly larvae tended to move away from food after eating, while others remained in place. By breeding "rovers" and "sitters" that possessed other distinct genetic markers, Sokolowski and colleagues were able to map the gene responsible for the behaviors.

Optimal Foraging in White-Fronted Bee-Eaters

- When biologists set out to study why animals forage in a particular way, they usually assume that individuals make decisions that maximize the amount of usable energy they take in, given the costs of finding and ingesting their food, and the risk of being eaten while they're at it.

- **Optimal foraging** defines animals maximizing their feeding strategy.

- Some individuals forage a few meters away from their colony, while others forage hundreds of meters away.

- Individuals foraging far from the colony actually bring back a much larger mass of insects on each trip.

✔ CHECK YOUR UNDERSTANDING

(b) In your own words, describe what optimal foraging means.

_____

_____

_____

51.3 Who Should I Mate With?

Sexual Activity in Anolis Lizards

- *Anolis carolinensis* live in the woodlands of the southeastern United States. Males emerge in January and establish breeding territories, and females become active a month later. By May, females are laying an egg every 10–14 days and by the end of the breeding season will have produced an amount of eggs equaling twice their body mass.

- Testosterone and estradiol produce these behaviors in males and females, respectively. If production of these hormones is the proximate cause, what environmental cues trigger these breeding events?

- Researchers brought a large group of inactive lizards into the laboratory and divided them into five groups:

  1. Single females isolated
  2. Females placed in female-only groups
  3. Females paired with males
  4. Females placed with groups of castrated males
  5. Females placed with groups of uncastrated males

- Females exposed to breeding males began producing eggs much earlier than the females placed in the other treatment groups. Females need to experience spring-like light and temperatures and exposure to breeding males.

- To test this idea, researchers repeated the previous experiment, but added a twist. They placed females with males that had intact dewlaps or with males whose dewlaps had been surgically removed. The result? Females grouped with dewlap-less males were slow to produce eggs—just as slow, in fact, as the females in the first experiment that had been grouped with castrated males. These females had not been courted at all. The result suggests that the dewlap is a key visual signal.

How Do Female Barn Swallows Choose Mates?

- Male barn swallows with particularly long tails are more efficient in flight and thus more successful in finding food.

- Long-tailed males mated earliest—meaning that females preferred them. Early-nesting pairs have time to complete a second nest later in the season.

✓ **CHECK YOUR UNDERSTANDING**

(c) How did researchers investigate the importance of the dewlap, and what was the result?

_____

_____

_____

51.4 Where Should I Live?

- Habitat selection involves a species defining a **territory**.

- **Migration** is defined as the long-distance movement of a population associated with a change of seasons.

How Do Animals Find Their Way on Migration?

Piloting

- **Piloting**, or using familiar landmarks, is used by many species to find their way. Young offspring follow their parents during migrations and appear to memorize the route.

- Homing pigeons can always find their way back home. But if their eyes are covered, they fail to navigate only the final stage, suggesting they may use piloting for the final part of a journey.

Compass Orientation

- To determine where north is, animals appear to use the Sun during the day and stars at night.

- During cloudy conditions, birds appear to use Earth's magnetic field to orient.

Why Do Animals Move with a Change of Seasons?

- Arctic terns nest along the Atlantic coast of North America, fly south along the coast of Africa to wintering grounds off Antarctica, then fly back along the eastern coast of South America.

- Many of the monarch butterflies native to North America spend the winter in the mountains of central Mexico or southwest California.

- Salmon that hatch in rivers along the Pacific Coast of North America and northern Asia migrate to the ocean when they are a few months to several years old. They return to the stream where they hatched to mate and then die.

- Increased access to food is a benefit of migration. There is a high cost, however, in time, energy, and predation risk.

✓ **CHECK YOUR UNDERSTANDING**

(d) What is migration? Give two examples taken from animal behavior.

_____

_____

_____

_____

51.5 How Should I Communicate?

- **Communication** is defined as any process in which a signal from one individual modifies the behavior of a recipient individual.

- A **signal** is any information-containing behavior.

Honeybee Language

- Karl von Frisch suspected that successful food finders communicate the location of food to other individuals.

- He observed bees displaying a "round dance" that contained information about the location of food. Workers got the information about the location of food through the dancer's movements.

- He also observed a "waggle run" that tells the length of distance the workers must fly, as well as the direction.

Modes of Communication

- Communication can be acoustic, visual, olfactory, or tactile.

- The type of communication correlates with a species' habitat.

When Is Communication Honest or Deceitful?

- At the ultimate level, one of the questions that biologists ask about communication concerns the quality of the information. Is the signal reliable?

Deceiving Individuals of Another Species

- The anglerfish has an appendage that looks remarkably like a minnow and dangles near its mouth.
- Predatory *Photinus* fireflies can mimic the pattern of flashes by females of several other species.
- Individuals increase their fitness by providing inaccurate or misleading information to members of a different species.

Deceiving Individuals of the Same Species

- In some cases, natural selection has also favored the evolution of lying to the same species.
- A male bluegill will imitate a female and get close to another male while he is courting an actual female. The other male bluegill then thinks he is courting two females; and while he does this, the female mimic secretly fertilizes the real female's eggs but does not have to care for them.

When Does Deception Work?

- Lying works only when it is relatively rare.
- As deceit becomes extremely common, then natural selection will strongly favor individuals that can detect and avoid or punish liars.

CHECK YOUR UNDERSTANDING

   (e)   Describe what Karl von Frisch found about honeybee dances.

_____

_____

_____

_____

51.6 When Should I Cooperate?

- **Altruism** is an act that has cost to the actor, in terms of his or her ability to survive and reproduce, and a benefit to the recipient of the altruistic act.

Kin Selection

- The **coefficient of relatedness** is a measure of how closely the actor and beneficiary are related. **Hamilton's rule** states that if the benefits of altruistic behavior are high, if the benefits are dispersed to close relatives, and if the costs are low, then alleles associated with altruistic behavior will be favored by

natural selection and will be spread throughout the population.

- **Kin selection** is natural selection that acts through benefits to relatives.
- Individuals can pass their alleles on to the next generation not only by having their own offspring but also by helping close relatives produce more offspring.

Reciprocal Altruism

- **Reciprocal altruism** is an exchange of fitness benefits that are separated in time.
- For example, vervet monkeys are most likely to groom unrelated individuals that have groomed or helped them in the past.
- Vampire bats are most likely to donate blood meals to non-kin that have previously shared food with them.

An Extreme Case: Abuse of Non-Kin in Humans

- Kids who are less than 2 years old are 70 times more likely to be killed in a household with a stepparent than in a household with only biological parents.

CHECK YOUR UNDERSTANDING

   (f)   Compare and contrast altruism and kin selection.

_____

_____

_____

B. MASTERING CHALLENGING TOPICS

The study of animal behavior seems to be a more tangible subject in biology because we can directly observe data with our own eyes as it is occurring. However, there are still a few topics that require extra time and thought. The concepts of **altruism** and **kin selection** are relatively straightforward to define, but even professors have a difficult time applying these terms to the observations they record in nature. It would be wise to locate a current ecology text and learn how to apply altruism or kin selection to examples given in the text. An ecology text will provide you with much more detail and reinforce the knowledge that you have already acquired.

C. ASSESSING WHAT YOU'VE LEARNED

1. A change in behavior that results in a specific experience in the life of an individual is termed:
   a. behavior.
   b. proximate cause.
   c. ultimate cause.
   d. learning.

2. An innate behavior is one that
   a. is not inherited.
   b. shows some variation.
   c. is inherited and shows little variation.
   d. is inherited and shows wide variation.

3. Individuals are trained by experience to give the same response to more than one stimulus. This is termed:
   a. classical conditioning.
   b. imprinting.
   c. proximate conditioning.
   d. sensitization.

4. The recognition and manipulation of facts about the world is termed:
   a. imprinting.
   b. learning.
   c. cognition.
   d. memory.

5. Inflexible behavior is adaptive when
   a. mistakes would be costly.
   b. there are many safe opportunities to learn.
   c. mistakes can be made without dying.
   d. parents can teach their young.

6. The key signal prompting a hormonal response and breeding in female *Anolis* lizards is:
   a. light.
   b. male dewlap display.
   c. female dewlap display.
   d. male hormones.

7. Any process in which the signal from one individual modifies the behavior of a recipient individual is called:
   a. a connecting signal.
   b. communication.
   c. imprinting.
   d. learning.

8. Natural selection favoring the deceit of the same species is best described by which of the following examples?
   a. Anglerfish
   b. Fireflies
   c. Monkeys
   d. Bluegills

9. A movement that results in a change in position is called:
   a. orientation.
   b. taxis.
   c. phototaxis.
   d. instant taxis.

10. An exchange of fitness benefits that are separated in time is termed:
    a. kin selection.
    b. altruism.
    c. reciprocal altruism.
    d. coefficient of relatedness.

CHAPTER 51—ANSWER KEY

Check Your Understanding

(a) Innate behavior is a type of behavior that is inherited and shows little variation with learning. It is also present in an animal without having to learn it. A learned behavior is one that is acquired through a process (perhaps observation). These behaviors can show significant flexibility.

(b) The answer should be in your own words but should include the following concepts. Individuals make decisions that maximize the amount of usable energy they take in, given the costs of finding and ingesting their food, and the risk of being eaten while they're at it. Optimal foraging defines animals maximizing their feeding strategy.

(c) Researchers placed female *Anolis* lizards with males that had intact dewlaps or with those whose dewlaps had been surgically removed. The result? Females grouped with dewlap-less males were slow to produce eggs—just as slow, in fact, as females that had been grouped with castrated males. These females had not been courted at all. The result suggests that the dewlap is a key visual signal.

(d) Migration is defined as the long-distance movement of a population associated with a change of seasons. Arctic terns nest along the Atlantic coast of North America, fly south along the coast of Africa to wintering grounds off Antarctica, then fly back along the eastern coast of South America. Many of the monarch butterflies native to North America spend the winter in the mountains of central Mexico or southwestern California. Salmon that hatch in rivers along the Pacific Coast of North America and northern Asia migrate to the ocean when they are a few months to several years old.

They return to the stream where they hatched to mate and then die.

(e) Von Frisch observed bees displaying a "round dance" that contained information about the location of food. Workers got the information about the location of food through the dancer's movements. He also observed a "waggle run" that tells the length of distance the workers must fly, as well as the direction.

(f) Altruism is an act that has cost to the actor, in terms of his or her ability to survive and reproduce, and a benefit to the recipient. Kin selection is natural selection that acts through benefits to relatives.

Assessing What You've Learned

1. d; 2. c; 3. a; 4. c; 5. a; 6. b; 7. b; 8. d; 9. a; 10. c

Chapter 52

Population Ecology

Looking Back—Key Concepts from Earlier Chapters

Animal Interaction—Chapter 51

Chapter 51 explored how organisms respond to environmental stimuli and how they interact with the environment. Here in Chapter 52, the focus is on population size.

KEY CONCEPTS

- Life tables summarize how likely it is that individuals of each age class in a population will survive and reproduce.

- The growth rate of a population can be calculated from life-table data or from the direct observation of changes in population size over time.

- Researchers observe a variety of patterns when they track changes in populations over time.

- Data from population ecology studies help biologists evaluate prospects for endangered species and design effective management strategies.

A. CHAPTER OUTLINE

52.1 Demography

- A **population** is a group of individuals from the same species that live in the same area at the same time.
- **Population ecology** is the study of how and why the number of individuals in a population changes over time.
- **Demography** is the study of factors that determine the size and structure of populations through time. Analyzing birthrates, death rates, **immigration** rates, and **emigration** rates is fundamental to demography.

Life Tables

- Formal demographic analyses of populations are based on a type of data set called a **life table**, which

summarizes the probability that an individual will survive and reproduce in any given year over the course of its lifetime.

Survivorship

- **Survivorship** is the proportion of offspring produced that survive, on average, to a particular age (l_x, where x represents the age class being considered). It is calculated by dividing the number of individuals in that age class by the number of individuals in the first age class.

$$I_x = N_x/N_0 \text{ (Box 52.1, Eq. 52.1)}$$

- Type I curve—high survivorship throughout life.
- Type II curve—constant mortality throughout life.
- Type III curve—high death rate early in life.

Fecundity

- **Fecundity** is the number of female offspring produced by each female in the population.

$$R_0 = \sum l_x m_x \text{ (Box 52.1, Eq. 52.2)}$$

Lifetime reproduction is a function of fecundity at each age (m_x) and survivorship to each age class (l_x). If R_0 is greater than 1, the population is increasing. If R_0 is less than 1, the population is declining.

The Role of Life History

- Fitness trade-offs occur because every individual has a restricted amount of time and energy at its disposal. For example, this is why it is not possible for females to have high fecundity and high survival.
- An organism's **life history** consists of how the organism allocates resources to growth, reproduction, and activities related to survival.

(a) What is fecundity, and how does it relate to population growth? Include the equation in your answer.

_____

_____

_____

_____

_____

52.2 Population Growth

Quantifying Growth Rate

- A population's growth rate is the change in number of individuals in the population (ΔN) per unit time (Δt).

Exponential Growth

- **Exponential growth** occurs when r does not change over time. It does *not* depend on the number of individuals in the population. Therefore, it is **density independent**.

- In reality, it is not possible for growth to continue indefinitely. Eventually, the habitat would not have enough resources for the number of individuals present. When a population stabilizes at the maximum number that can be supported by the resources available, that population has reached the habitat's **carrying capacity**.

Logistic Growth

- Biologists recognize two types of logistic growth: *discrete* growth and *continuous* growth.

- In calculating discrete growth, biologists use N to symbolize population size. N_0 is the population size at time zero, and N_1 is the population size one breeding interval later.

$$N_1/N_0 = \lambda \text{ (Box 52.2, Eq. 52.3)}$$

where λ is the **finite rate of increase**.

$$N_t = N_0\lambda^t \text{ (Box 52.2, Eq. 52.5)}$$

This equation summarizes how populations grow when breeding takes place seasonally.

- In calculating continuous growth, a population's **per capita increase** is symbolized r and is defined as the per capita birthrate minus the per capita death rate.

$$\lambda = e^r \text{ (Box 52.2, Eq. 52.6)}$$

where e is the natural logarithm, or about 2.72.

$$N_t = N_0 e^{rt} \text{ (Box 52.2, Eq. 52.7)}$$

This equation summarizes how populations grow when they breed continuously. Because r represents the growth rate at any given time and because r and λ are so closely related, biologists routinely calculate r, even for species that breed seasonally.

What Limits Growth Rates and Population Sizes?

- **Density-independent factors** change birthrates and death rates irrespective of the number of individuals in a population.

- **Density-dependent factors** are usually biotic in nature and change in intensity as a function of population size.

- Researchers studied the bridled goby, a coral-reef fish. They stocked artificial reefs with varying densities of adult gobies; then, after 2.5 months, they captured all the gobies and computed the growth rate of individuals, the survival rate, and the immigration rate.

- Adult gobies survive better when population density is *low*. More juvenile gobies immigrate successfully when population density is low. Higher rates of predation and disease might occur in dense populations.

(b) Describe the characteristics of exponential growth.

_____

_____

_____

_____

52.3 Population Dynamics

- Research on population dynamics (changes in populations through time) has uncovered a wide array of patterns in addition to exponential and logistic growth.

How Do Metapopulations Change through Time?

- If individuals from a species occupy many small patches of habitat, they may form many independent populations, or **metapopulations**.

- Ilkka Hanski and colleagues have shown that metapopulations exist in nature. They performed a mark-and-recapture study on the Glanville fritillary butterfly. They found a migration rate of 9 percent, which was high enough to recolonize extinct populations. They also confirmed that some populations had gone extinct while others had been created.

- Therefore, the dynamics of the metapopulation depend on the births and deaths of the smaller populations.

Why Do Some Populations Cycle?

- **Population cycles** are regular fluctuations in size that some populations exhibit.

- Researchers have concluded that hare populations are limited by availability of food as well as by predation, and that both of these factors interact. (See **Figure 52.11.**)

How Does Age Structure Affect Population Growth?

- A population's **age structure**—meaning the proportion of individuals that are at each possible age—dramatically influences the population's growth over time.

Age Structure in a Woodland Herb

- The common primrose has a complex population dynamic:
  1. Populations dominated by juveniles experience rapid growth and then are limited by the shading of larger trees.
  2. The long-term trajectory of the primrose population seems dependent on the frequency and severity of windstorms.
  3. Primroses have metapopulation structure.

Age Structure in Human Populations

- In nations where industrial and technological development is advanced and average incomes are relatively high, the **age pyramid** of the population tends to be even.

- The predicted population structure in 2050 highlights a major policy concern in developed countries: how to care for an increasingly aged population.

- In contrast, the age distribution is "bottom-heavy" in the less-developed countries.

- The projected age distribution in 2050 illustrates the major public policy concerns in these countries: providing education and jobs for an enormous influx of young people who will want to be starting families.

Analyzing Change in the Growth Rate of Human Populations

- The growth rate for humans has increased over time since about 1750, leading to a very steeply rising curve over the past few centuries. The highest values occurred between 1965 and 1970, when population growth averaged 2.04 percent per year.

Will Human Population Size Peak in Your Lifetime?

- Humans are ending a period of rapid growth that lasted well over 500 years. How quickly growth rates decline and the maximum population size ultimately reached will be decided by changes in fertility rates and the course of the AIDS epidemic.

- Extrapolating the world's population to the year 2050 is based on three different fertility rates:
  1. High fertility = nearly 11 billion
  2. Medium fertility = about 9 billion
  3. Low fertility = 7.4 billion

✔ **CHECK YOUR UNDERSTANDING**

(c) What is a metapopulation, and what are some of its characteristics?

_____

_____

_____

_____

_____

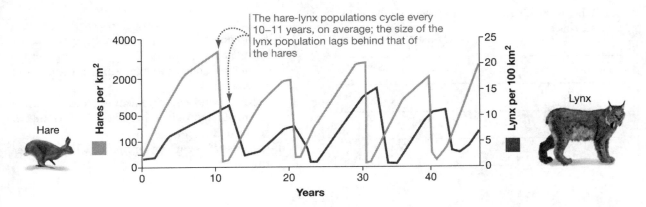

The hare-lynx populations cycle every 10–11 years, on average; the size of the lynx population lags behind that of the hares

Hare

Lynx

52.4 How Can Population Ecology Help Endangered Species?

- When designing programs to save species threatened with extinction, conservationists draw heavily on concepts and techniques from population ecology.

Using Life-Table Data to Make Population Projections

- Population projections allow biologists to alter values for survivorship and fecundity at particular ages and then assess the consequences.
- Studies performed on this aspect support some general conclusions:
  1. Whooping cranes, sea turtles, spotted owls, and many other endangered species have high juvenile mortality, low adult mortality, and low fecundity.
  2. In humans and species with high survivorship in most age classes, rates of population growth are extremely sensitive to changes in age-specific fecundity.

Preserving Metapopulations

- Over time, each population within the larger metapopulation is likely to be wiped out. Migration from nearby populations can reestablish populations in these empty habitats, thereby stabilizing the overall population.
- Areas being protected need to be substantial enough in area to maintain large populations.
- If that is not possible, smaller tracts of land should be connected by corridors of habitat so that migration is possible.
- Land where populations have emigrated should be preserved so that immigration is possible.

CHECK YOUR UNDERSTANDING

(d) What is the difference between demography and a life table?

_____

_____

_____

_____

_____

B. MASTERING CHALLENGING TOPICS

The difficult aspects of this chapter are undoubtedly the mathematical equations and quantitative nature of these topics. For those of us who have trouble with math and quantitative analysis, this chapter will present challenges. The amount of time you will spend on working through equations and solving problems will ultimately depend on how much your instructor focuses on these issues. If you are required to solve equations and word problems associated with population size, mortality, and fecundity, then you will need to invest a significant amount of time in learning these concepts and solving problems. As with any quantitative subject, you need to learn the concepts and the origins of the equations before you can solve word problems off the top of your head. Study qualitatively, learning all the definitions of terms and meanings of the equations, and then start plugging in numbers and solving equations. Do as many problems as you can; then start making up your own problems for your classmates and ask them to do the same. In problem solving, practice makes perfect!

C. ASSESSING WHAT YOU'VE LEARNED

1. The study of how and why the number of individuals in a group of animals changes over time is called:
   a. demography.
   b. population ecology.
   c. population.
   d. fecundity.
2. The proportion of offspring produced that survive, on average, to a particular age is called:
   a. survivorship.
   b. a life table.
   c. demography.
   d. fecundity.
3. An organism's life history
   a. is the number of female offspring produced by each female.
   b. is the probability an organism will survive and reproduce.
   c. consists of how an organism allocates resources to growth, reproduction, and activities related to survival.
   d. consists of how an organism interacts with other organisms.

4. Factors that are usually biotic in nature and change in intensity as a function of population size are termed:
   a. density-dependent factors.
   b. density-independent factors.
   c. density-neutral factors.
   d. exponential growth factors.

5. Some populations exhibit regular fluctuations in size called:
   a. age structure.
   b. population crashes.
   c. population tides.
   d. population cycles.

6. In nations where industrial and technological development are advanced, age distribution is:
   a. top-heavy.
   b. bottom-heavy.
   c. even.
   d. bimodal.

7. Adult gobies survive better when population density is:
   a. low.
   b. high.
   c. intermediate.
   d. dynamic.

8. A huge population of animals consisting of smaller isolated populations is termed a:
   a. megapopulation.
   b. metapopulation.
   c. metropolis.
   d. cohort.

9. The model that estimates the likelihood that a population will avoid extinction for a given time period is termed:
   a. mortality.
   b. age structure.
   c. population viability analysis.
   d. population fecundity.

10. A small, isolated population
    a. will do extremely well if protected.
    b. is unlikely to survive, even within a reserve.
    c. will grow, but extremely slowly.
    d. will explode given the right conditions.

CHAPTER 52—ANSWER KEY

Check Your Understanding

(a) Fecundity is the number of female offspring produced by each female in the population: $R_0 = S\, l_x m_x$. Lifetime reproduction is a function of fecundity at each age (m_x) and survivorship to each age class (l_x). If R_0 is greater than 1, the population is increasing. If R_0 is less than 1, the population is declining.

(b) Exponential growth occurs when r does not change over time. It does *not* depend on the number of individuals in the population. Therefore, it is density independent.

(c) A metapopulation is made up of many small populations isolated in fragments of habitat. Over time, each population within the larger metapopulation is likely to be wiped out. Migration from nearby populations can reestablish populations in these empty habitats, thereby stabilizing the overall population.

(d) Demography is the study of factors that determine the size and structure of populations through time. Formal demographic analyses of populations are based on a type of data set called a life table, which summarizes the probability that an individual will survive and reproduce in any given year over the course of its lifetime.

Assessing What You've Learned

1. b; 2. a; 3. c; 4. a; 5. d; 6. c; 7. a; 8. b; 9. c; 10. b

Looking Forward—Key Concepts in Later Chapters

Animal Interaction—Chapters 53–55

Chapters 53–55 investigate how different species interact with one another and with their physical environment. Here in Chapter 52, our focus is on population size.

Chapter 53

Community Ecology

Looking Back—Key Concepts from Earlier Chapters

Growth Rings—Chapter 36

Researchers estimate the frequency and impact of storms by finding wind-killed trees and determining the date of their death. This can be done by comparing patterns in the growth rings of the dead trees with those of living individuals nearby.

Animal Interaction—Chapter 51

Chapter 51 explored how organisms respond to environmental stimuli and how they interact with the environment. Here in Chapter 53, however, the focus is on large numbers of species within a community.

Population Dynamics—Chapter 52

Chapter 52 explored the dynamics of populations—how and why they grow or decline, and how they change over time and space. Although that chapter considered populations in isolation, the reality is that individuals of different species constantly interact within a community. As a result, the fate of a community may be tightly linked to the other species that share its habitat.

KEY CONCEPTS

- Interactions among species, such as competition, consumption, and mutualism, have two main outcomes: (1) they affect the distribution and abundance of the interacting species, and (2) they affect the evolution of the interacting species.

- The nature of interactions between species frequently changes over time.

- The assemblage of species in a biological community changes over time and is primarily a function of climate and chance historical events.

- Species richness is higher in large islands near continents than in small, isolated islands, due to differences in immigration and extinction.

- Species richness is higher in the tropics and lower toward the poles.

A. CHAPTER OUTLINE

53.1 Species Interactions

- A biological **community** consists of interacting species, usually within a defined area.

- Consumption occurs when one organism eats another.

- Mutualism occurs when two species interact in a way that contributes fitness benefits to both.

Three Themes

- In analyzing species interactions, notice these themes:

  1. Species interactions may affect the distribution and abundance of a particular species.

  2. Species act as agents of natural selection when they interact.

  3. The outcome of interactions among species is dynamic and conditional.

Competition

- **Competition** is a − / − interaction that occurs when individuals use the same resources and when those resources are limited.

- Competition occurring between the same species is called **intraspecific competition**, while **interspecific competition** occurs between individuals from different species.

Using the Niche Concept to Analyze Competition

- A **niche** is the range of resources that a species is able to use or the range of conditions it can tolerate.

What Happens When One Species Is a Better Competitor?

- The principle of **competitive exclusion** states that it is not possible for species within the same niche to coexist. However, if the niches do not overlap completely, then the weaker species should be able to retreat into an area of non-overlap—a mechanism called **niche differentiation**.

- Under **asymmetric competition**, one species suffers a greater fitness decline, whereas under **symmetric competition**, both species have an equal fitness decline.

- The **fundamental niche** is the combination of conditions that the species will occupy in the absence of competitors, and its **realized niche** is that portion of resources used when competition occurs.

Experimental Studies of Competition

- Joseph Connell noticed two species of barnacles with interesting distributions on an intertidal rocky shore of Scotland.

- Connell's study is one of the first experimental examples of how competition can modify a distribution of a species. In this case, the poorer competitor (*Chthamalus*) is restricted to the upper intertidal zone, which is a more severe habitat.

Fitness Trade-offs in Competition

- The key here is that the ability to compete for a particular resource, such as space on rocks or edible seeds of a certain size, is only one aspect of an organism's niche. If individuals are extremely good at competing for a particular resource, then they are probably less good at enduring drought conditions, warding off disease, or preventing predation.

Mechanisms of Coexistence: Niche Differentiation

- Change in resource use is called **niche differentiation**.

- The change in species' traits is called **character displacement**.

Consumption

- Consumption occurs when one organism eats another:
  1. **Herbivory**—When **herbivores** consume plant tissues.
  2. **Parasitism**—When a **parasite** consumes small amounts of tissue from another organism, or **host**.
  3. **Predation**—When a predator kills and consumes most or all of another individual, or **prey**.

Constitutive Defenses

- Some species run or hide in response to predators; others sequester or spray toxins or employ weaponry for defense (**standing** or **constitutive defenses**).

- **Mimicry** occurs when one species closely resemble another species.

- **Müllerian mimicry** occurs when harmful species resemble each other. For example, brightly colored wasps resemble each other and by this, predators tend to avoid these colors.

- **Batesian mimicry** occurs when a harmless species resembles a harmful one.

Inducible Defenses

- Researchers hypothesized that if blue mussels possess **inducible defenses** against crabs, they should occur in the low-flow area where predation pressure is higher. They found that mussels in the high predation area had thicker shells.

Are Animal Predators Efficient Enough to Reduce Prey Populations?

- Species interactions have a strong impact on the evolution of predator and prey populations.

- A wolf control program in Alaska during the 1970s decreased predator abundance to 55% to 80% below pre-control density. Concurrently, the population of moose tripled. This observation suggests that predators reduced this moose population far below the maximum number that the habitat can support.

- In the arctic, tests have shown that hare population is limited by the availability of food as well as by predation, and that food availability and predation intensity interact—the combined effect of food and predation is much larger than their impact in isolation.

Why Don't Herbivores Eat Everything— Why Is the World Green?

- Biologists routinely consider two possible answers to the above question:
  1. ***Top-down control hypothesis***: Herbivore populations are limited by predation and disease.
  2. ***Bottom-up limitation hypothesis***: Plant tissues are well-defended or offer poor nutrition.

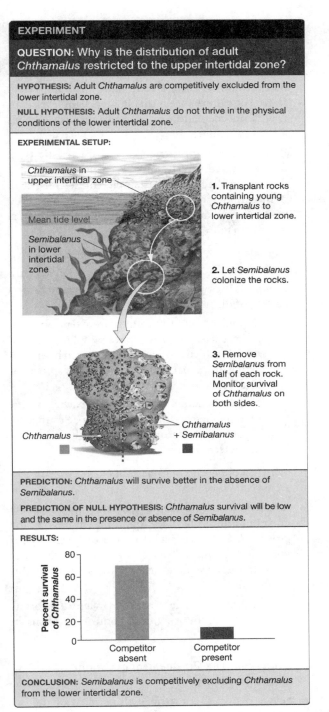

EXPERIMENT

QUESTION: Why is the distribution of adult *Chthamalus* restricted to the upper intertidal zone?

HYPOTHESIS: Adult *Chthamalus* are competitively excluded from the lower intertidal zone.

NULL HYPOTHESIS: Adult *Chthamalus* do not thrive in the physical conditions of the lower intertidal zone.

EXPERIMENTAL SETUP:

Chthamalus in upper intertidal zone

Mean tide level

Semibalanus in lower intertidal zone

1. Transplant rocks containing young *Chthamalus* to lower intertidal zone.

2. Let *Semibalanus* colonize the rocks.

3. Remove *Semibalanus* from half of each rock. Monitor survival of *Chthamalus* on both sides.

Chthamalus

Chthamalus + *Semibalanus*

PREDICTION: *Chthamalus* will survive better in the absence of *Semibalanus*.

PREDICTION OF NULL HYPOTHESIS: *Chthamalus* survival will be low and the same in the presence or absence of *Semibalanus*.

RESULTS:

[Bar graph: y-axis "Percent survival of Chthamalus" from 0 to 80. "Competitor absent" bar ≈ 70, "Competitor present" bar ≈ 14.]

CONCLUSION: *Semibalanus* is competitively excluding *Chthamalus* from the lower intertidal zone.

Adaptation and Arms Races

- A **coevolutionary arms race** between parasites and hosts begins when a parasitic species develops a trait that allows it to survive and reproduce in a host. In response, natural selection favors host individuals that are able to defend themselves against the parasite.

- *Plasmodium* are unicellular protists that cause malaria. They have a complex life cycle that includes the mosquito as the main vector.

- Researchers have found a strong association between a certain allele and protection against malaria. HLA-B53 proteins bind to a particular protein found in *Plasmodium*.

- Researchers have also shown that *Plasmodium* populations in West Africa now have a variety of alleles for the protein recognized by HLA-B53. This suggests that natural selection has favored the evolution of *Plasmodium* strains with weapons to counter HLA-B53.

Can Parasites Manipulate Their Hosts?

- Snails infected with flukes become attracted to light and are therefore more easily eaten by birds, thus continuing the life cycle of the fluke.

Using Consumers as Biocontrol Agents

- The use of predators and parasites as biocontrol agents is a key part of integrated pest management.

- These strategies maximize crop and forest productivity and minimize pesticide use.

Mutualism

- **Mutualisms** are +/+ interactions that involve a wide variety of organisms and rewards. For example, bees visit flowers to harvest nectar and pollen, and the flowers benefit by being fertilized.

- Examples are the association between fungi and roots, bacteria and certain species of plants, and rancher ants that protect aphids in exchange for sugar.

CHECK YOUR UNDERSTANDING

(a) Compare and contrast competition and mutualism.

_____

_____

_____

_____

53.2 Community Structure

How Predictable Are Communities?

The Clements-Gleason Dichotomy

- Clements promoted a view that biological communities are stable, integrated, and orderly entities with a highly predictable composition. He argued that communities develop by passing through a series of predictable stages, culminating in a **climax community**.

- Gleason contended that the community found in a particular area is neither stable nor predictable. He claimed that plant and animal communities are associations of species that just *happen to share climatic requirements*.

An Experimental Test

- Researchers constructed 12 identical ponds, sterilized them, and then inoculated them with various species of zooplankton. They found a total of 61 species in all the ponds but only 31 to 39 species in individual ponds.

- Researchers contended that some species are particularly good at dispersing and are likely to colonize all or most of the available habitats. Other species disperse more slowly and tend to reach only one or a few of the available habitats. As a result, the specific details of community assembly and composition are contingent and unpredictable. As Gleason maintained, communities are a product of chance and history.

Mapping Current and Past Species' Distributions

- Historical data on plant communities showed that groups of species did not change their distributions in close association. Instead, species tended to change their ranges independently of one another.

- Although both biotic interactions and climate are important in determining which species exist at a certain site, chance and history also play a large role.

How Do Keystone Species Structure Communities?

- A **keystone species** has a much greater impact on the surrounding species than its abundance would suggest.

- Wolves, sea stars, and sea otters are examples of species that vary in abundance over space and time, which significantly affects their communities.

> ✓ **CHECK YOUR UNDERSTANDING**
>
> (b) Briefly state Clements's view on community dynamics.
>
> _____
>
> _____
>
> _____

53.3 Community Dynamics

Disturbance and Change in Ecological Communities

- **Disturbance** is an event that removes some individuals or biomass from a community. The important feature of disturbance is that it alters light levels, nutrients, unoccupied space, or some other aspect of resource availability.

- Each community experiences a characteristic type, frequency, and severity of disturbance, known as a **disturbance regime**. For example, forest fires kill all or most of the trees in a boreal forest every 100 to 300 years, whereas disturbances in temperate and tropical forests are more frequent.

How Do Researchers Determine a Community's Disturbance Regime?

- Ecologists use two strategies to determine the natural pattern of disturbance:

  1. They infer long-term patterns from data obtained in short-term analysis. Unless sampling is extensive, it is difficult to avoid errors caused by extrapolating from particularly disturbance-prone or disturbance-free years or areas.

  2. By reconstructing the history of a particular site, flood frequency (for example) can be estimated. Forest fires are most often studied by using historical techniques.

Why Is It Important to Understand Disturbance Regimes?

- Researchers studied the fire history of giant sequoia groves in California. They found that in most of the groves, 20 to 40 fires had occurred each century for the past 2000 years. They also established that fires are extremely frequent and of low severity in this community.

- Biologists are now able to better manage these forests by allowing, monitoring, and controlling burns in this area.

Succession: The Development of Communities after Disturbance

- Severe disturbances remove all or most of the organisms from an area. The recovery that follows is **succession**. **Primary succession** occurs when the disturbance removes the soil and its organisms as well as the organisms above the surface. **Secondary succession** occurs when a disturbance removes some or all of the organisms from an area but leaves the soil intact.

- The specific sequence of species that appears over time is called a **successional pathway**.

- Biologists focus on three factors to predict the outcome of succession in a community:
  1. Particular traits of the species involved
  2. How species interact
  3. Historical and environmental circumstances such as the size of the area involved and short-term weather conditions

The Role of Species Interactions

- **Facilitation** takes place when the presence of an earlier-arriving species makes conditions more favorable for the arrival of certain later species. **Inhibition** is the opposite of facilitation.

- **Tolerance** means existing species do not affect the probability that subsequent species will become established.

A Case History: Glacier Bay, Alaska

- An extraordinarily rapid and extensive glacial recession is occurring at Glacier Bay. In just 200 years, glaciers have retreated about 100 km, thus uncovering an important site for studying succession.

CHECK YOUR UNDERSTANDING

(c) Hypothesize what would happen if California prevented small wildfires from occurring.

_____

_____

_____

_____

_____

53.4 Species Richness in Ecological Communities

Predicting Species Richness:
The Theory of Island Biogeography

- **Species richness** is a simple count of how many species are present in a given community, whereas **species diversity** is a weighted measure that incorporates abundance and presence.

- Islands in the ocean have smaller numbers of species than do areas of the same size on continents.

- Immigration rates should be higher on large islands closer to the mainland, and extinction rates should be highest on small islands that are far from shore.

Global Patterns in Species Richness

- The communities of the tropics contain more species than temperate or subarctic environments do. As latitude increases, species diversity decreases.

- *High productivity in the tropics promotes high diversity.* Increased biomass supports more herbivores, and thus more predators.

- In addition, speciation rates should increase when niche differentiation occurs within populations of herbivores, predators, parasites, and scavengers. This hypothesis is still in the experimental stage.

CHECK YOUR UNDERSTANDING

(d) Compare and contrast species richness and species diversity.

_____

_____

_____

_____

_____

B. MASTERING CHALLENGING TOPICS

The development of communities after disturbance is probably the most complex event described in this chapter. Because communities respond differently after disturbances, you cannot simply memorize one type of succession. Instead, learning various examples will help you understand how succession proceeds in different communities and after different types of disturbances. Also, by understanding the theoretical considerations (traits of species, species interaction, and environment), you will gain an appreciation for the difficulty in determining the exact events of succession in various areas.

C. ASSESSING WHAT YOU'VE LEARNED

1. It is not possible for species within the same niche to coexist. This statement describes:
   a. competition.
   b. niche theory.
   c. competitive exclusion.
   d. mimicry.

2. The combination of conditions that a species will occupy in the absence of competitors is:
   a. the fundamental niche.
   b. the realized niche.
   c. competitive exclusion.
   d. mimicry.

3. Batesian mimicry can be best described as:
   a. when one species closely resembles another.
   b. when a harmless species resembles a harmful one.
   c. when harmful species resemble each other.
   d. when harmless species resemble each other.

4. When predation or disease limits herbivore populations, this is:
   a. top-down control.
   b. the poor nutrition hypothesis.
   c. the plant defense hypothesis.
   d. the disease hypothesis.

5. The + / + interactions that involve a wide variety of organisms and rewards are termed:
   a. predation.
   b. competition.
   c. parasitisms.
   d. mutualisms.

6. An event that removes some individuals or biomass from a community is called:
   a. competition.
   b. predation.
   c. disturbance.
   d. resistance.

7. The recovery that follows a severe disturbance is:
   a. succession.
   b. growth.
   c. colonization.
   d. seeding.

8. The primary difference between primary and secondary succession is:
   a. the complete removal of species during the disturbance.
   b. the complete removal of soil during the disturbance.
   c. time.
   d. nothing—there is no difference.

9. Which one of the following statements is *not* representative of a global pattern in species diversity?
   a. High productivity in the tropics promotes high diversity.
   b. Habitats with simple physical structures have more niches.
   c. Tropical regions have had more time for speciation to occur.
   d. Tropical areas experience fewer disturbance events.

10. Net primary productivity is the:
    a. total amount of photosynthesis per unit area.
    b. total amount of plant mass per unit area.
    c. amount of primary productivity that ends up as biomass.
    d. amount of primary productivity above the ground.

CHAPTER 53—ANSWER KEY

Check Your Understanding

(a) Competition and mutualism are both types of species interactions. Competition is a − / − interaction that occurs when individuals use the same resources and when those resources are limited. Mutualisms are + / + interactions that involve a wide variety of organisms and rewards. For example, bees visit flowers to harvest nectar and pollen, and the flowers benefit by being fertilized.

(b) Clements promoted a view that biological communities are stable, integrated, and orderly entities with a highly predictable composition. He argued that communities develop by passing through a series of predictable stages, culminating in a climax community.

(c) One could hypothesize that by preventing small fires from occurring in giant sequoia groves, the underbrush would continue to grow instead of remaining low. This would continue and not be a physical threat (although biologically, the species diversity could be affected); if a fire were to occur, it would burn longer, higher, and with more intensity such that the sequoia groves would likely be harmed. In smaller, more regular wildfires, the sequoias remain unharmed.

(d) Species richness and species diversity are two different measures of the number of species in a community. Species richness is a simple count of how many species are present in a given community, whereas species diversity is a weighted measure that incorporates abundance and presence.

Assessing What You've Learned

1. c; 2. a; 3. b; 4. a; 5. d; 6. c; 7. a; 8. b; 9. b; 10. c

Chapter 54

Ecosystems

Looking Back—Key Concepts from Earlier Chapters

Chemosynthesis—Chapter 2

In some ecosystems, energy flow begins not with photosynthesis but with chemosynthesis. For example, bacteria that live near the deep-sea vents introduced in **Chapter 2** derive energy from hydrogen sulfide rather than sunlight.

Producers—Chapter 10

Producers form the basis of ecosystems by transforming the energy in sunlight or reduced inorganic compounds into the chemical energy stored in sugars. As **Chapter 10** pointed out, producers are called autotrophs or self-feeders.

Nitrogen Fixation—Chapter 38

Nitrogen fixation results from lightning-driven reactions in the atmosphere or enzyme-catalyzed reactions in bacteria that live in the soil and oceans. **Chapter 38** explains how bacteria fix nitrogen.

Animal Interaction—Chapter 51

Chapter 51 explored how organisms respond to environmental stimuli and how they interact with the environment. Here in Chapter 54, the focus is on large numbers of species within an ecosystem.

Population Dynamics—Chapter 52

Chapter 52 explored the dynamics of populations—how and why they grow or decline, and how they change over time and space. Although that chapter considered populations in isolation, the reality is that individuals of different species constantly interact within a community, and communities within an ecosystem. As a result, the fate of an ecosystem may be tightly linked to the other species that share its habitat.

KEY CONCEPTS

- An ecosystem has four components: (1) the abiotic environment, (2) primary producers, (3) consumers, and (4) decomposers.

- The components of the ecosystem are linked by the movement of energy and nutrients.

- As energy flows from producers to consumers and decomposers, much of it is lost.

- The productivity of terrestrial ecosystems is limited by warmth and moisture, whereas in aquatic ecosystems, nutrient availability is the key constraint.

- To analyze nutrient cycles, biologists focus on the nature of the reservoirs where elements reside and the processes that move elements between reservoirs.

- The burning of fossil fuels has led to rapid global warming; rapid ecological and evolutionary changes are being observed in response.

A. CHAPTER OUTLINE

54.1 How Does Energy Flow through Ecosystems?

- An **ecosystem** consists of the organisms that live in an area along with certain nonbiological components. The relevant abiotic components include the nutrients used by growing organisms and the energy supplied by the Sun.

- **Primary producers** use solar energy or chemical energy contained in reduced inorganic compounds to manufacture their own food (**autotrophs**).
- Primary producers use chemical energy for growth and maintenance.
- Energy that is invested in new tissue is called **net primary productivity** (NPP).

Why Is NPP so Important?

- NPP results in biomass—organic material that non-photosynthetic organisms can eat.
- **Consumers** eat other organisms. **Herbivores** are consumers that eat plants; **carnivores** are consumers that eat animals.
- **Decomposers** obtain energy by feeding on the dead remains of other organisms or waste products.
- Components of an ecosystem are linked by the movement of energy and nutrients.

Solar Power: Transforming Incoming Energy to Biomass

- **Gross primary productivity** is the total amount of photosynthesis in a given area and time period.
- **Gross photosynthetic efficiency** is the ratio of gross photosynthesis to solar radiation in kcal/m^2.

Trophic Structure

- Organisms that obtain their energy from the same type of source occupy the same **trophic level**. Productivity declines from one trophic level to the next.
- A **food chain** describes the species occupying each trophic level in a particular ecosystem.
- A **grazing food chain** is made up of the network of organisms that eat plants, along with the organisms that eat the herbivores.
- Consumers that eat herbivores are called **secondary consumers**.
- The **decomposer food chain** is comprised of species that eat the dead remains of organisms. Plant litter and dead animals, or detritus, are consumed by bacteria, archaea, fungi, protozoa, and millipedes.
- Food chains are often embedded in more complex **food webs** because multiple species are present at several trophic levels.

Energy Flow to Grazers versus Decomposers

- At Hubbard Brook, only about 24 percent of NPP is eaten by consumers; 76 percent is uneaten until the primary producer dies.
- The percentage of dead versus alive is highly variable across space and time.

Energy Transfer between Trophic Levels

- All ecosystems share a characteristic pattern: The total biomass produced each year declines from one trophic level up to the next.
- In chipmunks, 80.7 percent of energy intake goes to maintenance, 17.7 percent is excreted, and only 1.6 percent goes into the production of new tissue (secondary production).
- The amount of biomass produced at the second trophic level must be less than the productivity at the first trophic level. This pattern holds true for the entire food chain and produces a pyramid of productivity.
- Although biomass production varies among communities, the data are representative and show that the biomass production at the upper trophic levels is often only 10 percent of the production at the next lowest level.

Trophic Cascades and Top-Down Control

- When a consumer limits a prey population, this is top-down control.
- When changes in top-down control cause effects two or three links away in a food web, a trophic cascade has occurred.
- Although the Yellowstone wolf population has grown to only 150 individuals, their presence has led to far-reaching changes in the food web.

Biomagnification

- Persistent organic pollutants (POPs) tend to increase in concentration at higher levels in the food chain.
- Toxaphene, an insecticide, was widely used until it was banned in 1986. Toxaphene levels increase dramatically at higher levels in the Arctic food chain.
- There is also new evidence that estrogen-mimicking POPs are impacting fish populations in the Arctic and elsewhere.

Global Patterns in Productivity

What Geographical Areas Are Most Productive?

- **Tropical rain forests** have the highest productivity, whereas desert and arctic regions have the lowest. Marine productivity is the highest along coasts.

What Limits Productivity?

- The overall productivity of terrestrial ecosystems is limited by a combination of temperature and the availability of water and sunlight.
- The overall productivity of marine ecosystems is limited by the availability of nutrients.

- Biologists show that large increases occur in the concentration of chlorophyll *a* in surface waters over a two-week interval after fertilization with iron.

✓ CHECK YOUR UNDERSTANDING

(a) Define an ecosystem.

_____

_____

_____

(b) Compare and contrast a food chain and a food web.

_____

_____

_____

54.2 How Do Nutrients Cycle through Ecosystems?

- The path that an element takes as it moves from abiotic systems through living organisms and back again is referred to as its **biogeochemical cycle**.

Nutrient Cycling within Ecosystems

- The combination of breakdown products and microscopic decomposers forms the soil organic matter. **Soil organic matter** is a complex mixture of partially decomposed detritus rich in a family of carbon-containing molecules called humic acids. (Soil organic matter is also called **humus**.)

- A key feature of nutrient cycling is that nutrients are reused. Not all of the nutrients are reused, however. Nutrients leave ecosystems when animals leave or when flowing wind or water removes nutrients.

What Factors Control the Rate of Nutrient Cycling?

- Decomposition of detritus most often limits the rate at which nutrients move through an ecosystem. Decomposition rate is influenced by:
  1. Abiotic conditions—temperature, precipitation, and so on.
  2. The quality of detritus as a nutrient source for fungi, bacteria, and archaea that accomplish the decomposition.

Sources of Nutrient Loss and Gain

- Nutrient loss is an important characteristic of an ecosystem. Farming, logging, and soil erosion accelerate nutrient loss.

- Vegetation removal has a huge impact on nutrient export. In the absence of plants to recycle nutrients, they are quickly washed out of the soil and lost to the ecosystem.

An Experimental Study

- Researchers compared nutrient loss in a vegetated and devegetated site over the course of four years.

- Nutrient losses were typically 10 times higher in the devegetated site versus the control site.

Global Biogeochemical Cycles

- There are three aspects studied in biogeochemical cycling:
  1. The nature and size of the **reservoirs** where elements are stored for a period of time
  2. The rate of movement between reservoirs and the factors that influence these rates
  3. How different biogeochemical cycles interact

The Global Water Cycle

- Water typically evaporates from the ocean and then precipitates back into the ocean. The key here is that evaporation exceeds precipitation—a net gain of water into the atmosphere.

- This vapor moves over land and precipitates. The cycle is completed by water that moves from land to the oceans via streams and **groundwater**.

- The **water table** is the upper limit below the surface that the ground is saturated with stored water.

The Global Nitrogen Cycle

- The key aspect of the **global nitrogen cycle** is that plants are able to use nitrogen only in the form of ammonium or nitrate ions (NH_4^+ or NO_3^-).

- Nitrogen fixation (converted from N_2 to NH_4^+ or NO_3^-) results from lightning-driven reactions in the atmosphere or enzyme-catalyzed reactions in bacteria that live in the soil and oceans.

- Three sources of human-fixed nitrogen are as follows:
  1. Industrially produced fertilizers.
  2. The cultivation of crops such as soybeans and peas that harbor nitrogen-fixing bacteria.
  3. Release of nitric oxide during the combustion of fossil fuels.

The Global Carbon Cycle

- The **global carbon cycle** involves a relatively rapid movement of carbon between terrestrial ecosystems, the oceans, and the atmosphere.

- Photosynthesis is responsible for taking carbon from the atmosphere and incorporating it into tissue. Respiration releases carbon that has been incorporated into living organisms to the atmosphere.

- Burning fossil fuels moves carbon from an inactive geological reservoir to an active pool—the atmosphere.

- Carbon dioxide is called a **greenhouse gas** because it traps heat that has been radiated from Earth and keeps it from being lost to space—in the same way that the glass of a greenhouse traps heat underneath it.

✓ **CHECK YOUR UNDERSTANDING**

(c) How are photosynthesis and respiration involved in the global carbon cycle?

_____

_____

_____

_____

54.3 Global Warming

Understanding the Problem

- The chain of causation begins with the direct link between the CO_2 released by human fossil fuel use and increases in atmospheric CO_2. It continues with a link between high CO_2 and the trapping of heat. It concludes with the recent and dramatic increases of average global temperatures.

- In 1998, researchers summarized the studies and data available on **global warming** and concluded that current evidence suggests a "discernible human influence on climate."

- In 2001 the new report stated that "there is new and stronger evidence that most of the warming observed over the last 50 years is attributable to human activities."

- In its most recent report, issued in 2007, the Intergovernmental Panel on Climate Change (IPCC) declared that the evidence for global warming is unequivocal and that it is "very likely" due to human-induced changes in greenhouse gases.

- The models currently being used suggest that the average global temperature _will_ undergo additional increases of 1.1–6.48°C by the year 2100.

Positive and Negative Feedback

- Positive feedback occurs when changes due to global warming result in a release of additional greenhouse gases.

- Negative feedback occurs when changes due to global warming result in increased uptake and sequestration of CO_2 and other greenhouse gases.

Impact on Organisms

- Increases in water temperature cause reef-building coral to expel their photosynthetic algae. Corals begin to die of starvation.

- Malarial parasite and copepod distributions are changing.

- Timing of events is changing in many seasonal environments, including early flowering by _Baptisia leucantha_.

- Changing temperatures are causing allele frequencies to change, such as increased fitness to hot habitats in _Drosophila_.

- Polar bears use pack ice to hunt seals. As pack ice retreats throughout the Arctic, polar bears are dying of starvation. If trends continue, polar bears may be extinct by 2100.

Productivity Changes

- If grassland habitats are fertilized with nitrogen, productivity increases but species richness declines over time.

- Nitrate ions derived from fertilizers are washing into the Gulf of Mexico. This has stimulated the growth of planktonic organisms, whose subsequent decomposition has used up available oxygen and triggered the formation of anaerobic "dead zones."

- Increased nitrate runoff from fertilizers has also cause toxic algal blooms.

- Massive phosphate runoff in the 1960s and 1970s resulted in explosive growth in cyanobacteria, large die-off, and eventual lake eutrophication—the conversion of a lake to a highly productive ecosystem with rapid decomposition, low oxygen levels, and rapid filling with decomposing organic matter.

✓ **CHECK YOUR UNDERSTANDING**

(d) Provide an example of how global warming is directly affecting a species.

_____

_____

_____

_____

B. MASTERING CHALLENGING TOPICS

The most complex information you will process in this chapter relates to the biogeochemical cycles. It is not difficult to memorize the facts and specifics about each cycle; however, you may find it difficult to understand concepts when these types of questions are asked on a test. It is not sufficient to simply memorize the carbon or nitrogen cycle; you must learn the concepts and trends as well. For example, what happens to the carbon cycle when fossil fuels are burned? Not only do you need to know that atmospheric levels of CO_2 would increase, but you also need to understand how this will affect the rest of the cycle.

C. ASSESSING WHAT YOU'VE LEARNED

1. Which of the following use solar energy or chemical energy contained in reduced inorganic compounds to manufacture their own food?
   a. Consumers
   b. Primary producers
   c. Secondary producers
   d. Decomposers

2. The areas that typically have the highest productivity are:
   a. tropical rain forests.
   b. grasslands.
   c. arctic regions.
   d. marine ecosystems.

3. The overall productivity in terrestrial ecosystems is limited by:
   a. temperature.
   b. water.
   c. sunlight.
   d. a combination of the above.

4. Productivity at the second trophic level is always:
   a. greater than the productivity at the first trophic level.
   b. less than the productivity at the first trophic level.
   c. equal to the productivity at the first trophic level.
   d. extremely variable compared to the productivity at the first trophic level.

5. Which hypothesis regarding what limits food chain length has the most support?
   a. Energy transfer hypothesis
   b. Stability hypothesis
   c. Environmental complexity hypothesis
   d. All of the above have equal support

6. The path that an element takes as it moves from abiotic systems through living organisms and back again is referred to as its:
   a. biogeochemical cycle.
   b. biological cycle.
   c. nutrient cycle.
   d. geochemical cycle.

7. Which of the following is *not* one of the three aspects studied in biogeochemical cycling?
   a. The nature and size of element reservoirs
   b. The rate of movement between reservoirs
   c. How different biogeochemical cycles interact
   d. How new species create their own biogeochemical cycles

8. Which of the following *most often* limits the rate at which nutrients move through an ecosystem?
   a. Species composition
   b. Decomposition rate
   c. Primary production
   d. None of the above

9. Perturbation of which of the following cycles contributes most to global warming?
   a. The global carbon cycle
   b. The global water cycle
   c. The global nitrogen cycle
   d. All of the above cycles contribute equally

10. The most recent evidence on global warming concludes that:
    a. human activity is not causing global warming.
    b. human activity is possibly causing global warming.
    c. there is strong evidence of human activity causing global warming.
    d. humans are responsible for every catastrophe for the past 1000 years.

CHAPTER 54—ANSWER KEY

Check Your Understanding

(a) An ecosystem consists of the organisms that live in an area along with certain nonbiological components. The relevant abiotic components include the nutrients used by growing organisms and the energy supplied by the Sun.

(b) Food chains and food webs both describe the relationships between species at different trophic levels. A food chain describes the species occupying each trophic level in a particular ecosystem. Food chains are often embedded in more complex food webs because multiple species are present at several trophic levels.

(c) Photosynthesis is responsible for taking carbon from the atmosphere and incorporating it into tissue. Respiration releases carbon that has been incorporated into living organisms to the atmosphere.

(d) Any one of the following examples were in the text:

Increases in water temperature cause reef-building coral to expel their photosynthetic algae. Corals begin to die of starvation.

Malarial parasite and copepod distributions are changing.

Timing of events is changing in many seasonal environments, including early flowering by *Baptisia leucantha*.

Changing temperatures are causing allele frequencies to change, such as increased fitness to hot habitats in *Drosophila*.

As pack ice retreats throughout the Arctic, polar bears are dying of starvation. If trends continue, polar bears may be extinct by 2100.

Assessing What You've Learned

1. b; 2. a; 3. d; 4. b; 5. c; 6. a; 7. d; 8. b; 9. a; 10. c

Biodiversity and Conservation Biology

Inbreeding—Chapter 25

Genetic factors play a role in the fate of populations. You may recall from **Chapter 25** that small populations tend to become inbred, and that inbreeding is usually detrimental to the fitness of organisms.

Extinction Rates—Chapter 27

Chapter 27 analyzed evidence that extinction rates are now accelerating to between 10 to 100 times the normal, or "background," extinction rates, indicating that a mass extinction event is under way.

Animal Interaction—Chapter 51

Chapter 51 explored how organisms respond to environmental stimuli and how they interact with the environment; but here in Chapter 55, the focus is on large numbers of species, their diversity, and their conservation.

Population Dynamics—Chapter 52

Chapter 52 explored the dynamics of populations and metapopulations—how and why they grow or decline, and how they change over time and space. Although that chapter considered populations in isolation, the reality is that individuals of different species constantly interact within a community, and communities within an ecosystem. As a result, the fate of an ecosystem may be tightly linked to the other species that share its habitat.

KEY CONCEPTS

- Biodiversity is quantified at the level of allelic diversity, species diversity, and ecosystem diversity.

- According to recent analyses, the sixth mass extinction in the history of life is occurring due to habitat loss, overexploitation, and global climate change.

- Humans depend on biodiversity for the products that wild species provide and for ecosystem services that protect the quality of the abiotic environment.

- Solutions to the biodiversity crisis include mitigating climate change, protecting key habitats, supporting sustainable development, lowering human population growth and resource use, preserving endangered species, and restoring ecosystems.

A. CHAPTER OUTLINE

55.1 What Is Biodiversity?

Biodiversity Can Be Measured and Analyzed at Several Levels

- Biodiversity defines all the distinctive populations and species living today. There are three levels of biodiversity:

  1. **Genetic diversity**—the total genetic information contained within all individuals of a species.

  2. **Species diversity**—the variety of life-forms on Earth.

  3. **Ecosystem diversity**—the variety of biotic communities in a region along with abiotic components.

- Biodiversity changes through time due to mutations, speciations, and/or extinctions.

How Many Species Are Living Today?

Taxon-Specific Surveys

- Terry Erwin and J. C. Scott used insecticidal fog to knock down species from the top of a rain-forest tree called *Luehea seemannii*. They identified over 900 different species of beetles among the individuals that fell.

- If each of the 50,000 tropical tree species harbors the same number of arthropod specialists, then the world total of arthropod species would exceed 30 million. Based on these types of studies, biologists estimate that there are over 100 million species.

All-Taxa Surveys

- To obtain a more direct estimate of total species numbers, the first effort to find and catalog *all* forms of life present at a single site is now under way. The location is the Great Smoky Mountains National Park in the southeastern United States. A consortium of biologists and research organizations initiated this all-taxon survey in 1998. When it is complete, in 2015, biologists will have a much better database to use in estimating the extent of biodiversity.

CHECK YOUR UNDERSTANDING

(a) What are the three levels of biodiversity?

_____

_____

_____

_____

_____

_____

55.2 Where Is Biodiversity Highest?

Hotspots of Biodiversity and Endemism

- Biologists use the term **hotspot** to denote areas in the tropics with exceptionally high species richness.

- In terms of bird species richness, the Andes Mountains, the Amazon River basin, portions of East Africa, and southwest China are important hotspots.

- Researchers are also interested in understanding which regions have a high proportion of **endemic species**—meaning species that are found in an area and nowhere else.

- In 1999, a team set out to identify regions of the world meeting two criteria:

  1. They contain at least 1500 endemic plant species.

  2. At least 70 percent of their traditional or primary vegetation has been lost.

- Although these areas cover only 2.3 percent of Earth's land area, they contain over 50 percent of all known plant species and 42 percent of all known vertebrate species.

CHECK YOUR UNDERSTANDING

(b) Identify some regions considered hotspots because of their species richness.

_____

_____

_____

_____

_____

_____

55.3 Threats to Biodiversity

- Today, species are vanishing faster than at virtually any other time in Earth's history.

Changes in the Nature of the Problem

- Most extinctions that have occurred over the past 1000 years took place on islands and were caused by overhunting or the introduction of **exotic species**—nonnative competitors, diseases, or predators.

- **Endangered species**, which are almost certain to go extinct unless effective conservation programs are put in place, are now more likely on continents than on islands.

- A recent analysis of 488 endangered species in Canada shows:

  - Virtually all the endangered species are affected by more than one factor.

  - Habitat loss is the single most important factor in the decline of these species.

  - Overharvesting is the dominant problem for marine species, while pollution plays a large role for freshwater species.

  - Factors beyond human control can be important—background extinctions will continue to occur.

Habitat Destruction

- We cause **habitat destruction** by logging and burning forests, damming rivers, dredging and trawling the oceans, plowing prairies, grazing livestock, filling in wetlands, and excavating and by extracting minerals to build housing developments, golf courses, shopping centers, office complexes, airports, and roads.

Habitat Fragmentation

- This activity turns large, contiguous areas of natural habitats into small, isolated fragments. **Habitat fragmentation** concerns biologists for these reasons:

    1. Habitats can be too small to support some species.

    2. It reduces the ability for some animals to disperse.

    3. It creates a large amount of vulnerable edge habitat.

- When habitats are fragmented, the quality *and* quantity of habitat decline drastically.

Stochastic and Genetic Problems in Small Populations

- When populations are fragmented into a metapopulation structure, gene flow between isolated groups is reduced or eliminated.

- Genetic drift is much more pronounced in small populations.

- Small populations become inbred, and inbreeding often leads to lowered fitness—the phenomenon known as inbreeding depression.

How Can Biologists Predict Future Extinction Rates?

Estimates Based on Direct Counts

- The best information comes from fossils of birds.

- Background extinctions occur at a rate of about one extinction per million species alive per year.

- Over the past 1000 years, data suggest that birds have been going extinct at about 100 times the background rate.

- This trend appears to be similar for mammals, reptiles, amphibians, and conifers.

Species–Area Relationships

- **Species–area relationships** measure rates of habitat destruction to projected rates of species loss.

- Researchers compared satellite images of the Amazon Basin from 1978 and 1988. Their analysis showed that 5000 km² was deforested in the Amazon each year during this decade.

- Researchers have used species–area curves to analyze the number of bird species found on islands in the Bismarck Archipelago near New Guinea (**Figure 55.13; BioSkills 7**).

- Biologists want to understand how many species are currently threatened so they can estimate how many will go extinct in the near future.

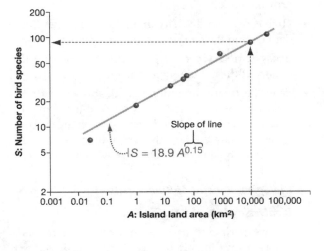

(c) Define habitat fragmentation, and explain why biologists are concerned about it.

_____

_____

_____

_____

_____

55.4 Why Is Biodiversity Important?

Economic Benefits of Biodiversity

- Agricultural scientists are preserving diverse strains of crop plants in seed banks and continue to use wild relatives of domesticated species in breeding programs aimed at improving crop traits.

- The production of almonds, apples, cherries, chocolate, alfalfa, and an array of other crops depends on the presence of wild pollinators.

- Research programs collectively known as bioprospecting focus on assessing bacteria, archaea, plants, and fungi as novel sources of drugs or ingredients in consumer products.

- Strategies for cleaning up oil spills, abandoned mines, and contaminated industrial sites are incorporating bioremediation—the use of bacteria, archaea, and plants to metabolize pollutants and render them harmless.

- Recreation based on visiting wild places, or eco-tourism, is a major industry internationally and is growing rapidly.

- High ecosystem diversity dramatically reduces flood damage and the danger posed by mudslides.

- **Ecosystem services** are processes that increase the quality of the abiotic environment.

- Economists are attempting to quantify the dollar value of ecosystem services as a way of justifying the cost of preserving natural areas, and biologists continue to document how the loss of biodiversity is affecting the quality of the abiotic environment.

Biological Benefits of Biodiversity

Biodiversity Increases Productivity

- Many experiments have shown that species richness has a positive impact on **net primary productivity (NPP)**.

- Several mechanisms may be at work:

  - **Resource use efficiency.** Diverse assemblages of plant species make more efficient use of the sunlight.

  - **Facilitation.** Certain species facilitate the growth of other species by providing them with nutrients, partial shade, or other benefits.

  - **Sampling effects.** In many habitats, it is common to observe that one or two species are extremely productive. Therefore, if the number of species in a study is low, the likelihood of missing the big producer is high.

Does Biodiversity Lead to Stability?

- Stability of a community is the ability to:

  1. Withstand a disturbance without changing.

  2. Recover to former levels of productivity or species richness after a disturbance.

  3. Maintain productivity and other aspects of ecosystem function as conditions change over time.

- **Resistance** is a measure of how much a community is affected by a disturbance. **Resilience** is a measure of how quickly a community recovers following a disturbance.

- Species richness has a strongly positive effect on how ecosystems function.

(d) List at least two economic benefits of preserving biodiversity.

_____

_____

_____

_____

_____

55.5 Preserving Biodiversity

Designing Effective Protected Areas

- A Gap Analysis Program (GAP) analysis tries to identify gaps between geographic areas that are particularly rich in biodiversity and areas that are actually managed for conservation.

- The analysis revealed that many species' ranges occur completely outside any protected areas.

- It also revealed that the 11.5 percent of Earth's surface area that is now being managed for biodiversity will not be enough to conserve many species.

- The leading hypothesis in reserve design is to make sure that strips of undeveloped habitat, called **wildlife corridors**, connect populations that would otherwise be isolated.

- The goals of the corridors are as follows:

  1. Allow areas to be recolonized if a species was lost.

  2. Introduce new alleles that would counteract the deleterious effects of genetic drift and inbreeding.

Beyond Protected Areas: A Comprehensive Approach

- *Sustainable development*—create ways for humans to live off of resources that are constantly being produced rather than mining finite resources.

- It is impossible to preserve diversity if the human population continues to grow and nations continue to use fossil fuels.

- **Ex-situ conservation** is the preserving of species in zoos, aquaria, wildlife ranches, seed banks, or other artificial settings.

  1. The California condor and other species have been rescued from extinction by the use of these settings.

✔ CHECK YOUR UNDERSTANDING

(e) Why would wildlife corridors help in protecting species?

_____

_____

_____

_____

_____

- In 2004, Wangari Maathai won the Nobel Peace Prize with her work founding the Green Belt Movement—a reforestation effort to plant over 1 billion trees beginning in Kenya.
- Since the 1980s, ecosystem restoration efforts in northwestern Costa Rica have produced a popular ecotourist destination.
- Biologists have been using fire to restore grassland ecosystems in the Malpai Borderlands and to restore forest ecosystems in the Guanacaste Reserve.

B. MASTERING CHALLENGING TOPICS

The most challenging aspect of this chapter is the concept of species–area relationships. Curves that are drawn from these relationships indicate how many species may be lost by the destruction of a particular size of habitat. Theoretical and conservation biologists derive these curves from complex data sets so that amateur biologists can calculate the effect of habitat loss on a particular species. The tricky part of these curves is that they are based on a logarithmic scale. If you are not familiar with logarithms, it may be a good idea to review them. If you are familiar with logarithms, this should help you in drawing and analyzing the species–area curves.

C. ASSESSING WHAT YOU'VE LEARNED

1. Research programs focusing on assessing bacteria, archaea, plants, and fungi as novel sources of drugs or ingredients in consumer products are termed:
   a. bioremediation.
   b. bioprospecting.
   c. bioprocessing.
   d. ecosystem services.

2. Why are economists trying to quantify the dollar value of ecosystem services?
   a. Because this allows them to sell abiotic components
   b. Because this allows them to manage the area
   c. Because this allows them to justify the cost of preservation
   d. Because this allows them to prevent habitat loss

3. What is genetic diversity?
   a. The total genetic information contained within all individuals of a species
   b. The total phenotypic information contained within all individuals of a species
   c. The variety of life-forms on Earth
   d. The variety of biotic communities in a region along with abiotic components

4. An area with an exceptionally high species richness is called a(n):
   a. endemic area.
   b. protected area.
   c. threatened area.
   d. hotspot.

5. What defines how much a community is affected by a disturbance?
   a. Hardiness
   b. Toughness
   c. Resistance
   d. Resilience

6. An exotic species that is introduced to a new area, spreads rapidly, and eliminates native species is called a(n):
   a. an exotic species.
   b. an immigrant species.
   c. an invasive species.
   d. a destructive species.

7. Which of the following does *not* constitute habitat destruction?
   a. Fishing
   b. Logging
   c. Burning forests
   d. Filling in wetlands

8. Which of the following is *not* a concern regarding habitat fragmentation?
   a. Habitats can change type.
   b. Habitats can be too small to support some species.
   c. It reduces dispersal ability of some animals.
   d. It creates a large amount of vulnerable edge habitat.

9. Impact on other species from the loss of a different species is called a(n):
   a. linked effect.
   b. dependent effect.
   c. domino effect.
   d. edge effect.

10. Which species descriptor implies that the species will almost certainly go extinct unless swift and decisive action is taken to protect it?
    a. Threatened species
    b. Red-flag species
    c. Extinct species
    d. Endangered species

CHAPTER 55—ANSWER KEY

Check Your Understanding

(a) Genetic diversity—the total genetic information contained within all individuals of a species. Species diversity—the variety of life-forms on Earth. Ecosystem diversity—the variety of biotic communities in a region along with abiotic components.

(b) In terms of bird species richness, the Andes Mountains, the Amazon River basin, portions of East Africa, and southwest China are important hotspots.

(c) Habitat fragmentation is the transformation of large, continuous blocks of habitat into isolated parcels surrounded by pasture, cropland, or urban development. Habitat fragmentation concerns biologists because:

   1. Habitats can be too small to support some species.

   2. It reduces the ability for some animals to disperse.

   3. It creates a large amount of vulnerable edge habitat.

(d) Research programs collectively known as bioprospecting focus on assessing bacteria, archaea, plants, and fungi as novel sources of drugs or ingredients in consumer products. Agricultural scientists are preserving diverse strains of crop plants in seed banks and continue to use wild relatives of domesticated species in breeding programs aimed at improving crop traits. The production of almonds, apples, cherries, chocolate, alfalfa, and an array of other crops depends on the presence of wild pollinators. Strategies for cleaning up oil spills, abandoned mines, and contaminated industrial sites are incorporating bioremediation—the use of bacteria, archaea, and plants to metabolize pollutants and render them harmless. Recreation based on visiting wild places, or ecotourism, is a major industry internationally and is growing rapidly. High ecosystem diversity dramatically reduces flood damage and the danger posed by mudslides.

(e) Wildlife corridors would help recolonize an area if a species was lost. Wildlife corridors would also help to introduce new alleles that could counteract the deleterious effects of genetic drift and inbreeding.

Assessing What You've Learned

1. b; 2. c; 3. a; 4. d; 5. c; 6. c; 7. a; 8. a; 9. c; 10. d